KU-796-520

Author Disclaimer

"While the author and the publisher believe that the information and guidance given in this work are correct, all parties must rely upon their own skill and judgement when making use of it. Neither the author nor the publisher assume any liability to anyone for any loss or damage caused by any error or omission in the work, whether such error or omission is the result of negligence or any other cause. Any and all such liability is disclaimed."

Copyright and Copying

"All Rights Reserved. No part of this publication may be produced, stored in a retrieval system, or transmitted in any form or by any means - electronic, mechanical, photocopying, recording or otherwise - without the prior written permission of the publisher."

Conditions of Acceptance of Advertisements

The Institution reserves the right to refuse to insert any advertisements (even if ordered and paid for) and/or to make alterations necessary to maintain its standards.

It is not guaranteed that any advertisement will be placed in any specified position or on any specified page unless agreement has been entered into and the agreed surcharge paid.

Every effort will be made to avoid errors but no responsibility will be accepted for any mistakes that may arise in the course of publication of any advertisement. These mistakes may include non-insertion, insertions other than those ordered and errors and omissions within the advertisement.

Notice to cancel any advertisement must be received in writing ten days prior to its next scheduled appearance, otherwise a charge will be made.

No responsibility will be accepted for repetitive errors unless the advertiser's correction has been accepted in respect of that error.

No responsibility will be accepted for loss or damage alleged to arise from errors within advertisement copy, non-appearance of an advertisement or delay in forwarding box number replies.

Advertisers are required to ensure that the content of advertisements conforms with all legislation currently in force affecting such matters. They shall further indemnify the publisher in respect of any claims, costs and expenses that may arise from anything contained within the advertisement and published on their behalf by the Institution of Electrical Engineers.

The placing of an order or contract for insertion of an advertisement in any newspaper or journal published by the Institution of Electrical Engineers whether in writing or by verbal or telephone instructions will be deemed an acceptance of each and all the above conditions.

Published by the Institution of Electrical Engineers, London.

Eleventh International Symposium on High Voltage Engineering. Volume 5 of a 5 part set

ISBN 0 85296 719 5   ISSN 0537-9989.

This publication is copyright under the Berne Convention and the International Copyright Convention.
All rights reserved. Apart from any copying under the U.K. Copyright, Designs and Patents Act 1988, Part 1, Section 38, whereby a single copy of an article may be supplied, under certain conditions, for the purposes of research or private study, by a library of a class prescribed by The Copyright (Librarians and Archivists) (Copying of Copyright Material) Regulations 1989: SI 1989/1212, no part of this publication may be reproduced, stored in a retrieval system or transmitted in any form or by any means without the prior permission of the copyright owners. Permission is, however, not required to copy abstracts of papers or articles on condition that a full reference to the source is shown.

Multiple copying of the contents of the publication without permission is always illegal.

© 1999 The Institution of Electrical Engineers

Printed in Great Britain by Short Run Press Ltd. Exeter.

CONTENTS

The Institution of Electrical Engineers is not, as a body, responsible for the opinions expressed by individual authors or speakers

UNIVERSITY OF STRATHCLYDE

30125 00653214 6

ELEVENTH INTERNATIONAL SYMPOSIUM ON
HIGH VOLTAGE ENGINEERING

23 - 27 August 1999

ANDERSONIAN LIBRARY
★
WITHDRAWN
FROM
LIBRARY
STOCK
★
UNIVERSITY OF STRATHCLYDE

VOLUME 5

Topic G   Dielectric diagnostics, expert systems

Topic H   Industrial Applications

**Books are to be returned on or before
the last date below.**

1 8 JUL 2001   3 0 AUG 2002   DUE

1 - 3 MAY 2002 ◇   2 0 MAR 2006

1 7 DEC 2002 2 6 SEP 2002 ◇

6 NOV 2002

2 9 MAY 2003   1 6 FEB 2004

2 4 JUN 2004

1 7 SEP 2003 ◇

LIBREX —

Venue

Stakis Metropole Hotel, L

ORGANISED BY

The Symposium was planned by the University of Wales, with the Universities of Glasgow, Caledonian, UMIST, Southampton, Strathclyde and Sunderland. The organisation was managed by the Science, Education and Technology Division of the Institution of Electrical Engineers.

Support from the following organisations is gratefully acknowledged:

Diagnostic Monitoring Systems
London Electricity
The National Grid Company

UK Organising and Technical Programme Committee

Professor Denis J Allan, Alstom T&D Transformers
Dr Norman L Allen, UMIST
Professor I D Chalmers, University of Strathclyde
Professor Anthony J Davies, University of Wales, Swansea
Professor Anthony E Davies, University of Southampton
Mr Robert Dean, ERA Technology
Mr Rod Doone, Bowthorpe EMP Ltd
Dr Anthony Foreacre, University of Strathclyde
Professor Owen Farish, University of Strathclyde
Dr Abderrahmane Haddad, University of Wales, Cardiff
Dr Brian Hampton, Diagnostic Monitoring Systems
M D Judd, University of Strathclyde
Dr Ian Kemp, Glasgow Caledonian University
Dr Ross Mackinlay, EA Technology
Dr Chris Melbourne, Meggitt Electronic Components
Professor R Miller, University of Wales College Newport
Professor Hugh Ryan, University of Sunderland
Dr Lorenzo Thione, CESI, Italy (representing CIGRE)
Dr Yu Kwong Tong, The National Grid Company plc
Professor Ron T Waters, University of Wales, Cardiff (Chairman)
Professor Jeremy Wheeler, ALSTOM Res Tech Centre

International Steering Committee

F A M Rizk (Canada), Chairman
M Ieda (Japan)
H C Kärner (Germany)
W Mosch (Germany)
R T Waters (UK)
G Zingales (Italy)

M Chamia, CIGRÉ (Sweden)
P Jacob (USA)
T Kawamura (Japan
M Muhr (Austria)
W Zaengl (Switzerland)

International Advisory Committee

M Awad (Egypt)
N N Tikhodeev (Russia)
S B Byoun (Korea)
F Chagas (Brazil)
M Darveniza (Australia)
G R Nagabhushana (India)
J Sletback (Norway)
T Sirait (Indonesia)
L Thione (Italy)
R Velasquez (Mexico)
O Farish (UK)

N Ma (People's Republic of China)
I W McAllister (Denmark)
G Praxl (Austria)
J Reynders (South Africa)
A Sabot (France)
S Grzybowski (USA)
T Horvath (Hungary)
A Kelen (Sweden)
D Kind (Germany)
F H Kreuger (Netherlands)

D
621.31
INT

UNIVERSITY OF STRATHCLYDE
14 SEP 2000
UNIVERSITY LIBRARY

Topic H   Industrial Applications

Addendum

Please note that the copyright of the following papers belongs to :

1.21.S1   **Digital impulse measurements meeting standards while pushing the limits**
T R McComb, J.Dunn, J.Kuffel, Canada

© Government of Canada

2.276.S15   **Computation and experimental results of the grounding model of Three Gorges power plant**
W Xishan, Z Yuanfang, Y Jianhui, C Cixuan, Q Liming, X Jun, S Lianfu, P R China

© Government of People's Republic of China

5.418.S28   **Stableness of electrical operation characteristics of electrostatic lentoid precipitator**
C Shixiu, S Youlin, C Xuegou, P R China

© Government of People's Republic of China

List of Authors

List of Authors

List of Authors

List of Authors

List of Authors

xvii

List of Authors

List of Authors

List of Authors

TESTING AND MONITORING AS BASIS OF THE DIELECTRIC DIAGNOSTIC

Ernst Gockenbach

Schering Institute of High Voltage Technique and Engineering
University of Hannover, Germany

ABSTRACT

The behaviour of the electrical insulating is the main parameter of the maintenance and life time estimation. The diagnostic is based on the knowledge of the insulating material, the measuring technique, the apparatus and the system and asset management. Tests have a defined purpose under given requirements with a clear decision and may be used as reference for the later diagnostic. Monitoring is the recording of the relevant parameter under service conditions with requirements given by the character of the recorded signal. Diagnostic is the evaluation of the signals, continuously recorded by a monitoring system, based on the knowledge of the reference value and the influence of the recorded parameters on the electrical insulating material. The weighting and judging of the parameters with mathematical algorithms allow to create a self-learning system which can be used for maintenance and life time estimation.

I INTRODUCTION

The demand on electrical energy is increasing world-wide, with small rates in the highly industrialised countries and with high rates in the developing countries. In both cases the requirements on the availability and reliability of the apparatus within the electrical energy supply and transmission systems are increasing and one of the key factor for these apparatus is the performance of the electrical insulating material. The electrical insulation determines the electrical strength of the apparatus and in many cases its maintenance requirements and life time. The insulating material may change its behaviour due to some kind of overstress, due to a combination of different stresses like thermal and electrical stress or due to normal ageing behaviour. The influence of some of these parameters is well known and extensively investigated, but there are some parameters whose influence on the performance of the insulating material is unknown and therefore monitoring and diagnostic are required. An unexpected outage of an important component of the energy supply is very costly, the use of an equipment longer as the expected life time is very cost saving. It is therefore obvious that the electrical insulation performance influences the investment and maintenance costs.

The dielectric diagnostic can be a very important and powerful tool to increase the reliability and availability of the components of the electrical energy system and to increase the safety and decrease the costs of the energy supply.

II MAIN ASPECTS OF DIAGNOSTIC

The diagnostic is influenced by four main aspects:

- knowledge of the electrical insulating material
- knowledge of measuring technique including noise suppression
- knowledge of apparatus to be diagnosed
- knowledge of the system and asset management

Normally each aspect will be dealt with in its own research field, but an interdisciplinary consideration is absolutely necessary. The International Conference on Large High Voltage Electrical Systems (CIGRE) has established a number of Working Groups dealing with the main aspects. The apparatus committees have their own Working groups for monitoring of the relevant equipment, the Working Groups of Study Committee 15 „Material for Electrotechnology" has a Joint Working Group with Study Committee 33 „Power System Insulation Coordination" dealing with the insulation monitoring and life time estimation. The Working Group 33-03 „High Voltage Test and Measuring Technique" is responsible for the measuring technique of all the relevant parameters and finally the Study Committees of system aspects have their Working Groups handling the system aspects, the asset management and the integration of the monitoring and diagnostic systems.

III TESTING

The tests of the electrical insulating material are one part of the diagnosis of the electrical insulation. During the manufacturing of an electrical apparatus numerous tests on the behaviour of the electrical insulation will by carried out, like routine tests with AC voltage, lightning and/or switching impulse voltage, partial discharge (PD) measurements, loss factor tan δ measurements, etc. These tests should show that the design of the apparatus is in accordance with the requirements and the pass of the tests confirms the expectation of the proper function during the entire service time. The test results can be used as so-called fingerprints, which give information on the actual situation at that time. These fingerprints can be used as basis for the evaluation of the reference values of the equipment, which is very helpful for the later diagnosis.

A few examples should stress this point. During the lightning and switching impulse tests of a transformer a transfer function can be calculated by the well known procedure. During service an impulse stress of the transformer by lightning overvoltage or by switching operation of a disconnector can be used to calculate again the transfer function and then to compare with the original one in order to check the transfer behaviour of the transformer [1]. The partial discharge tests of high voltage cables or high voltage gas-insulated substations in the factory give an information, that the tested part of the equipment has no PD above a given level at this time and confirm the quality assurance. The PD test value can be used later during on-site tests as reference value, in particular if components are included in the on-site tests which were not part of the manufacturing tests[2].

The tests have a clear defined purpose, e.g. type test, routine test, acceptance test, performance test, etc. and they have to follow given requirements according to international or national standards or to mutual agreement. A lightning impulse tests shall be carry out with the required waveshape and amplitude and the required measuring uncertainty unless the test object or the test equipment do not allow to reach these requirements. After performing the tests only a simple evaluation is necessary in order to state the equipment has passed the test or not, that means a „yes" or „no" decision.

IV MONITORING

Monitoring is the continuously recording of a parameter. Depending on the variation of the parameter the time between the single records can be vary from minutes up to weeks. On-line monitoring is the continuos or repetitive recording during normal operation of the equipment under observation. It is furthermore necessary to monitor all the important parameters at the same time in order to take all the mutual influences. The monitoring contains three main components, a sensor, a transmission device and a recording device. All the components should have at least the same reliability as the equipment to be monitored. If the monitoring should be carry our under on-site conditions and during service the sensors should be installed in the equipment. The transmission and recording system can be installed only for measurements if the installation does not disturb the normal service. In particular cases it may possible to remove an

equipment from the service and to carry out some tests which are part of the monitoring.

Monitoring under on-site condition requires sensors, which can withstand all the service stress and are not harmful for the operation of the equipment in case of a sensor fault. The transmission device must not change the measured signal and pick up of noise in order to keep the required high signal to noise ratio from the sensor. The recording device should be able to record the complete information in amplitude and time resolution and it should also operate properly under service conditions. An example of a monitoring system is the three-phase water monitoring system for XLPE high voltage cables. This system is able to detect water under the polymerised sheath and also to locate the fault point [3, 4].

There are some typical problems for running a monitoring system. First of all the equipment needs very often sensors which are built-in in the apparatus and which can not be installed subsequently. Secondly the sensors should be very often placed on the position which are highly stressed and where the sensor may increase the stress. Furthermore the monitoring system gives normally no benefit in the first years of operation of an equipment, but for a benefit later on during e.g. the ageing time of the electrical insulating material, the information of the monitoring from the beginning of the service life system is necessary in order to judge correctly and reliably the actual situation, which is then part of a diagnosis. Therefore the basis for a good diagnostic is a reliable and sensitive monitoring system which is running the whole service time of the equipment to be judged.

The purpose of the monitoring is to gain knowledge from the monitored apparatus or at least a part of it. In the high voltage technique this part is very often the electrical insulation. The collection of information is combined in the most cases with an evaluation and judgement of the recorded data and may be than called diagnosis. The requirements on a monitoring system are only safety and reliability and the measuring uncertainty depends actually on the required sensitivity to record the desired parameter.

V DIAGNOSIS

The diagnosis of an apparatus needs normally a lot of continuously and simultaneously recorded parameters in order to take care of the mutual influences of the different parameters. A diagnosis of the insulation behaviour of a transformer may require the voltage and current, that means the load, the oil temperature at different location, the gas-in-oil content, the PD activity in terms of number of PD impulses, charge per impulse or number of PD impulses per time, the weather conditions, etc. [5, 6]. The diagnosis of a high voltage cable requires the voltage and current, that means the load, the temperature of the cooling system or the temperature profile along the cable, the PD activities of the cable and the cable connections and terminations [7]. The diagnose of a high voltage circuit breaker requires beside the voltage and current of the system the number of switching operations and the switching conditions, but very often mechanical functions should be diagnosed due to their large part on the failures [8].

Not every time the information of the desired locations are available and therefore it is necessary to make a model of the apparatus, in order to evaluate the conditions on the weakest points. These models should be verified before by comparison between measurements and calculations in real experiments. The available information of all parameters from one or more monitoring systems need a weighting and a correlation between the single parameters. A simple example for this need is the harmless increasing of the oil temperature of a transformer with increasing load and/or with increasing ambient temperature as long as the temperature limits for the cooling system has not been reached.

The weighting of the different parameters needs a good knowledge on the behaviour of the electrical insulating material, which is confirmed by theoretical and experimental investigations. From these results a function can be developed which describes the performance of the electrical material under the influence of the different parameters. The combination and correlation of the different weighted function can be done by mathematical algorithms which are known as Neuronal Network, Fuzzy Logic, Knowledge Based Intelligence, etc. These algorithms can also use the results from the former tests as reference values in order to judge the actual situation by evaluation of the changes as function of the recorded parameters. With the help of a computer the recorded values can also be shown with a quick-motion camera and the engineer can judge the influences much better, because he can see the history of the parameter within a very short time. The record of former results allows the diagnosis system to adapt their algorithms and to work as a self-learning system which is able to evaluate automatically the records and to provide the diagnosis result for the engineer. It is obvious that at the beginning of the use of a diagnosis system the decision should be made by an experienced engineer, but it is also possible that in cases where a prompt reaction is necessary to save many an automatic measure by the diagnosis system takes place. This option is of interest for a later development, because at the moment the diagnosis system are understood as a system which collects recorded data from monitoring systems, weights and evaluates the information and gives a diagnosis information or result to the engineer within a time, which requires no immediate actions.

The costs of a diagnosis system should be reasonable in respect to the saved costs due to the system and in respect to the resulting costs without diagnosis system. These costs are strongly dependent on the economic factors, but the example in Figure 1 can demonstrate how the information of a diagnosis system influences the investment decision concerning maintenance and life time estimation. The performance of an apparatus decreases with the service time. In order to re-establish the performance of an equipment maintenance after fixed intervals with a constant increase of the performance is possible (thin line in Figure 1). With the knowledge of the diagnosis system the maintenance can be done when the performance has reached a given level (thick line in Figure 1) and the increase of the performance can be adjusted to the demand.

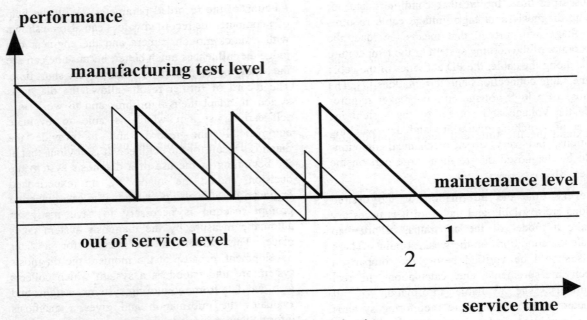

Figure 1: Performance of an apparatus as function of the service time

The demand may be influenced by technical or economical or by a mixture of both.

This example is transferable to other important factors like residual life time estimation. If the diagnosis system is able to judge the loss of insulation performance by the electrical and thermal stress of e.g. a high voltage cable a control of this cable load can prevent an unexpected outage and can extend the life time and delay a considerable investment as shown in Fig. 2.

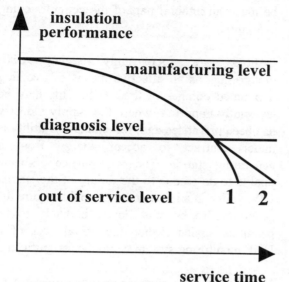

Figure 2: Insulation performance as function of the service time

Assuming the performance of the insulation follows the curve shown in Figure 2. If the diagnosis level has been reached a decision concerning the residual life time can be made. With no actions concerning the further reduction of the performance the point 1 in Figure 2 will be reached and the insulation performance is below the service level, which means that the apparatus has to be taken out of service. With actions initiated by the diagnosis system e.g. the load of the apparatus may be changed and the point 2 in Figure 2 can be reached which is a reasonable extension of the life time and saves investment costs. The shape of the curve between the point where the insulation performance crosses the diagnosis level and the point 2 where the curve crosses the out of service level is assumed and may be better judged with the knowledge of the diagnosis system and then lead to a further extension of the service time. The example takes into account that the performance of a high voltage cable insulation can not be increased by a maintenance action and therefore only the reduction of the insulation performance can be influenced by different measures.

VI CONCLUSIONS

- Tests can and should be used as part or as reference for the diagnosis system.
- A continuous monitoring of the relevant parameters is necessary for a reliable diagnosis.
- Knowledge of the insulating material, the measuring technique, the apparatus and the system are very important for the diagnosis.
- The use of a reliable diagnosis system can save costs and increases the reliability and availability of the energy supply system.
- A successful introduction of the diagnosis systems needs the support of the user of the apparatus to be diagnosed in order to collect the necessary information from the beginning and to gain experiences under service conditions.

REFERENCES

[1] R. Malewski, et. all: „Five Years of Monitoring the Impulse Test of Power Transformers with Digital Recorders and the Transfer Function Method", CIGRE 1992, paper 12-201

[2] U. Schichler: „Erfassung von Teilentladungen an polymerisolierten Kabeln bei der Vor-Ort Prüfung und im Netzbetrieb", Doctor Thesis, Universität Hannover, 1996

[3] U. Glaese, E. Gockenbach: „Water Sensor as an Integral Part of a Cable Monitoring System"; 9th ISH, Graz, 1995, paper 1027

[4] W. Rungseevijitprapa, E. Gockenbach, L. Goehlich, H. Vemmer: „Principle and Practical Experiences with a three-phase Water Monitoring System for XLPE high voltage Cables", 11th ISH, London, 1999

[5] T. Leibfried: „On-line Monitoring von Leistungstransformatoren aus der Sicht der Hersteller", Haefely Trench Symposium, Stuttgart 1998

[6] A. J. Kachler: „Diagnostic and Monitoring-Technology for large Power Transformers", CIGRE SC 12 Colloquium, Sidney, 1997

[7] G. P. Van der Wijk, E. Pultrum, H.T.F. Geene: „Development and Qualification of a new 400 kV XLPE Cable System with Integrated Sensors for Diagnostic"

[8] T.M. Chan, F. Heil, D. Kopeijtkova, P. O´Conell, J.P. Taillebois, I. Welch: „Report on the second international Survey on high voltage gas-insulated substations (GIS) service experience", CIGRE WG 23-10 Report

[9] D. Wenzel: „Teilentladungsmessungen an Transformatoren mit Verfahren der digitalen Signalverarbeitung und Mustererkennung" Doctor Thesis, Universität Hannover, 1998

UHF DIAGNOSTICS FOR GAS INSULATED SUBSTATIONS

Brian Hampton

Diagnostic Monitoring Systems Ltd., UK

Abstract

The UHF technique of detecting partial discharges in gas insulated substations (GIS), has over the past 10 years or so been developed from a laboratory curiosity into a sensitive means of giving an early warning of most impending failures in the substation. Its rapid acceptance by both utilities and switchgear manufacturers worldwide has resulted from it being a simple, sensitive and practical technique. In addition, utilities are finding it necessary to drive their plant harder and need to be assured of its reliability – even though they may have fewer staff to carry out maintenance work. The UHF technique is now seen by them as a cost effective means of avoiding failures, and providing the high quality of supply demanded by today's customers.

The principle of the UHF technique is that the current pulse which forms the partial discharge has a very short risetime, and this excites the GIS chambers into multiple resonances at frequencies of up to 1.5 GHz or so. Although the duration of the current pulse is only a few nanoseconds, these microwave resonances persist for several microseconds. They are readily picked up by UHF couplers fitted either inside the GIS chambers, or over dielectric apertures, such as glass windows, in the chamber wall. The PD signals can then be amplified, and displayed in ways which reveal the characteristic patterns of the defects which have caused them.

In addition to allowing continuous monitoring of the GIS on-line, the UHF technique is increasingly being seen as an integral part of the overpotential test when first commissioning the substation. This is not entirely new, since UHF measurements were made during the commissioning of a 420 kV substation in the UK as long ago as 1984; but recent studies through CIGRE have developed the technique significantly, and led to the recommendation that PD measurements should be made an integral part of the commissioning procedure.

Utilities are increasingly seeing UHF monitoring as an effective means of avoiding unplanned outages of their GIS. This may be especially important where the supply industry has been privatised and, for example, a utility is under contract to accept energy from a generating station. The consequence of being unable to do so due to the failure of the station GIS, which in an extreme case could require the generators to be shut down, is likely to be penalties against which the initial cost of a UHF monitoring system would be insignificant.

Utilities are also under mounting pressure to maximise their return on plant, which is a major investment and needs to be used to its limits for as long as possible. This has to be achieved with the minimum of maintenance, since this is expensive and possibly disruptive when plant is taken out of service. The utility must also provide the highest possible quality of supply to customers, especially those electronics manufacturers using processes susceptible to dips in the supply.

All this points to the need for knowing the condition of plant through on-line monitoring, so that remedial actions may be taken in time to prevent failure. Achieving this goal can mean interpreting a large amount of data, which is time consuming and requires the most important and scarce resource of all - experienced engineers. Clearly there is a need for routine data interpretation through some form of artificial intelligence, and it is reassuring to know of the many developments taking place in this field.

High Voltage Engineering Symposium, 22–27 August 1999
Conference Publication No. 467, © IEE, 1999

2. Principle of the UHF technique

The principle of the UHF technique is now well known, but as a reminder the current pulse which forms the partial discharge has a very short risetime, which recent measurements have indicated can be less than 70 ps. These pulses excite the GIS chambers into multiple resonances at frequencies of up to 1.5 GHz or more. Although the duration of the current pulse is only a few nanoseconds, the microwave resonances persist for the relatively long time of some microseconds. They may readily be picked up by UHF couplers fitted either inside the GIS chambers, or over dielectric apertures in the chamber wall. The latter are usually either the exposed edges of cast resin barriers, or glass windows; and in both cases couplers can be designed to give a perfectly acceptable output from the UHF signal which propagates through the apertures.

Whether external or internal couplers are used, the UHF signals can be amplified and displayed in ways where their characteristic patterns reveal the nature of any defect that might be present in the GIS.

3. Data interpretation

3.1 Discharge features

The features of the UHF discharge pulses that are most useful for interpretation purposes are their amplitude, point on wave, and the interval between pulses. These parameters enable typical defects such as fixed point corona, free metallic particles and floating electrodes to be identified. Other defects occur less commonly, but have their own distinctive features.

The UHF data may be displayed in any way which reveals the characteristic patterns typical of the defects causing them, as, for example, in the 3D patterns shown below. Here the pulses detected in 50 (60) consecutive cycles are shown, in their correct phase relationships over the cycle. In the 3D displays, 0 degrees corresponds to the positive-going zero of the power frequency wave, 90 degrees to the positive peak, and so on.

Corona

Three distinct phases in which corona develops from a protrusion may be seen as the voltage is raised:

Inception, where discharges occur first on the half-cycle that makes the protrusion negative with respect to the other electrode. Inception is therefore on the negative half-cycle when the protrusion is on the busbar, and on the positive half-cycle when it is on the chamber wall. The initial discharges are of very low magnitude (less than 1 pC), and are centred on one of the voltage peaks.

Streamers start at a slightly higher voltage, and appear as a regular stream of pulses on the peak of the positive half-cycle. At the same time the negative discharges become larger and more erratic. This difference between the positive and negative discharges reveals whether the protrusion is on the busbar or chamber wall. Streamer discharges will not lead to breakdown.

Leaders follow further increases of the applied voltage, and appear every few cycles as large discharges on the positive half-cycle. They propagate in steps until either they become extinguished, or reach the other electrode and cause complete breakdown. Figure 1 shows a typical display of PD from a protrusion on the busbar at this stage of the discharge process. Leaders are the precursors of breakdown, and there is always a risk of failure when they are present.

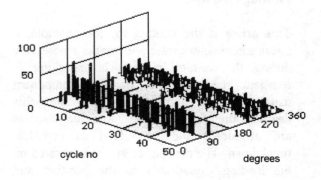

Figure 1. Busbar corona, streamers and leaders

Free Metallic Particle

A particle lying on the chamber floor becomes charged by the electric field, and if the upward force on it exceeds that due to gravity it will stand up and dance along the floor. This generates a discharge pulse each time contact is made with the floor, since the particle then assumes a new value of charge. The pulses occur randomly over the complete power frequency cycle, but their peak amplitudes follow the phase of the voltage. A typical PD pattern for a free metallic particle is shown in Figure 2. At higher voltages the particle will start to jump towards the busbar, and after several cycles may reach it. Two factors combine to make this an especially serious condition which often leads to breakdown: (i) as the particle approaches the busbar it discharges and generates a voltage transient which increases still further the stress at its tip, and (ii) breakdown can occur by leader propagation well before any space charge has had time to develop and shield the tip.

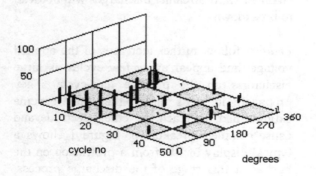

Figure 2. Free metallic particle

Floating Electrode

This arises if the contact to, for example, a stress shield deteriorates and sparks repetitively during the voltage cycle. The sparking is energetic because the floating component usually has a high capacitance, and this degrades the contact further. Metallic particles are produced, and may lead to complete breakdown. The discharges are concentrated on the leading quadrants of the positive and negative half-cycles, and their amplitude does not vary with the applied voltage.

Often the gap is asymmetrical, and sparks over at different voltages on the two half-cycles. Then a different charge is trapped on the floating component, and this gives rise to the characteristic wing-shaped patterns shown in the plan view of Figure 3. These provide a positive confirmation that the defect is a floating component.

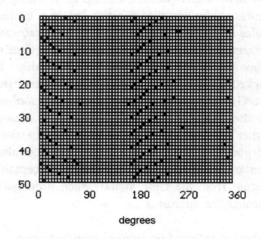

Figure 3. Floating electrode, plan view

3.2 Automatic defect classification

PD monitoring systems can generate a large amount of data, and computer-aided interpretation and classification of defects can be of great assistance when analysing this information. Interpretation is based on the analysis of statistical parameters extracted from the data, particularly those relating to the amplitude, repetition rate and point on wave of the detected UHF signals. Artificial Neural Networks are well suited to recognising the pulse patterns generated by discharges. During system training with a database of real PD patterns from known types of defect, the links between neurons in the ANNs are automatically strengthened or weakened on the basis of the required output. In practice it is found that the amount of data obtained in one second is too great to be handled by a single ANN, and it is better to use several ANNs optimised to recognise key parameters. The accuracy of this defect classification can exceed 95% [1].

A PD monitoring system is therefore able to identify the defect type, and this automated classification procedure is already being used by utilities in the UK.

4. Site test procedures for GIS

The problem of how best to undertake the site commissioning test for new GIS has been studied in depth by CIGRE JWG 33/23.12 'Insulation co-ordination of GIS: return of experience, on site tests and diagnostic techniques'. Its main conclusion [2] is that *the recommended dielectric on site test procedure is a high voltage AC test together with a sensitive partial discharge measurement. For GIS with insulation test level chosen in IEC 694, the high AC voltage test level on site has to be linked to the LIWL rather than to the PFWL of the GIS even if this results in the AC on site test level being close to the PFWL.*

The purpose of the test is to detect 'critical defects', which are defects of a size likely to affect the specified LIWL and PFWL voltages. Extensive series of tests made in a number of laboratories have shown the critical lengths of defects to be:

Free metallic particles: 2-5 mm
Protrusion from the HV conductor: 1 mm
Particle on a barrier surface: 2 mm

All these defects produce PD before breakdown occurs. An important part of the CIGRE studies has therefore been *to find the AC voltage level that will cause the critical defects to generate partial discharges of a magnitude measurable on site.* This has led to the adoption of test voltages at which the critical defects will produce PD levels of 1-10 pC. The highest permissible PD level during the test has been set at 5 pC, or the equivalent signal if the measuring system is calibrated in other units. The proposed test procedure is:

1. Conditioning with AC according to the manufacturer's recommendations.
2. A PD measurement at $0.8 \times U_t$, with a highest permissible PD level of 5 pC or equivalent *.
3. A 1 minute AC test at $U_t = 0.36$ x LIWL, or $U_t = 0.8$ x ACWL, whichever is higher **.

* Or, if PD cannot be measured, a LI test at $0.8 \times$ LIWL.
** Or, if the highest test voltage is less than $0.36 \times$ LIWL, an OSI test at $0.65 \times$ LIWL.

The UHF and acoustic techniques are the most promising for use in this test procedure, and future CIGRE work will concentrate on developing means of calibrating them.

5. Calibrating UHF measurements

The path of the signal from the defect to the UHF coupler, shown in Figure 4, involves the following stages:

Excitation of resonances in the chamber - the partial discharge in SF_6 is an extremely short current pulse, which radiates energy into the chamber and sets up many modes of microwave resonance. Studies have shown the amplitude of the microwave field to be dependent on the length of the PD current path, its position, and its orientation in the chamber.

Propagation of the signal - the UHF field strength is reduced by reflections from discontinuities such as barriers, bends, junctions and changes in the diameter of the vessel. It also undergoes dispersion, and to a small extent attenuation due to the skin effect.

UHF coupler response - the coupler acts as an antenna, producing a voltage at its output terminal in response to the UHF field incident upon it. The efficiency of this conversion, both in terms of sensitivity and bandwidth, may be measured by the technique described in the following section.

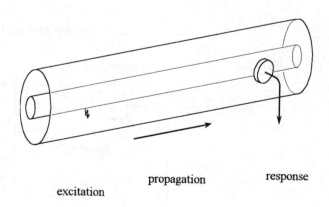

excitation   propagation   response

Figure 4. The UHF signal path

5.1 UHF coupler calibration

Even before a complete system calibration is available, it is important to be able to measure the characteristics of the UHF couplers now being fitted to GIS. This enables the couplers to be designed, both in terms of sensitivity and bandwidth, to optimise the performance of the monitoring system fitted into the substation.

A recently developed coupler calibration system [3] is shown in Figure 5. A voltage step of risetime < 50 ps is applied to the input of a gigahertz transverse electromagnetic cell (GTEM), and propagates as a step electric field towards the open end of the cell. As the field passes over the coupler aperture in the top wall of the GTEM, the incident electric field is measured using a monopole probe with a known response. The voltage signal from the probe is converted to the frequency domain using a Fast Fourier Transform (FFT), and stored as a reference. The UHF coupler is then fitted over the aperture, and the same step field applied. Having measured the voltage output from the coupler, it is again converted by a FFT.

Dividing the frequency domain signal by the reference signal previously stored allows the transfer function of the UHF coupler to be calculated. The most useful measure of the sensitivity of a coupler is its voltage output per unit incident electric field; i.e., an effective height, H_e. This quantity can be interpreted as the height of an equivalent monopole probe located on the axis of the coupler, whose output voltage is simply its height multiplied by the amplitude of the incident electric field.

Since the coupler test aperture is located at the mid-point of the 3 m long GTEM cell, a reflection-free window of 10 ns is available for the measurement before the reflection from the open end of the cell returns. Consequently, most of the coupler response to the step excitation should be contained within this 10 ns period for an accurate measurement to be obtained. This requirement has been satisfied by all the couplers tested on the system.

The signal processing used to determine the frequency response of the coupler can extend the measurement bandwidth beyond the normal limits of the test equipment. The procedure includes using identical test equipment and cables for both the incident field and coupler response measurements, and a measurement system bandwidth of 2 GHz has been achieved.

An important feature affecting the performance of a coupler is its mounting configuration, because this determines how well the coupler responds to UHF fields inside the GIS chambers. Couplers are normally located in recesses, or externally on the GIS, which makes them more remote from the UHF fields. They should therefore be tested on ports that replicate their mounting configurations on the GIS.

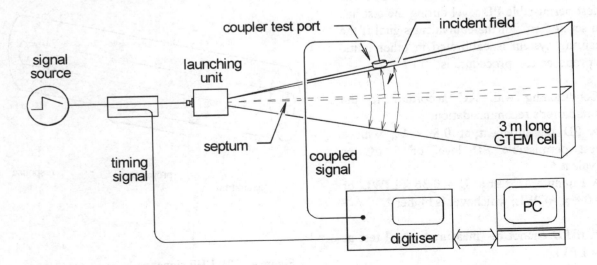

Figure 5. Block diagram of the UHF coupler calibration system

5.2 Specifying UHF Couplers

When preparing the tender for a new GIS with integral couplers, the user should specify the coupler response to suit that required by the monitoring system. A specification in terms of effective height H_e over the declared frequency range of interest has been prepared by The National Grid Company plc [4], and an example is shown in Table 1. Here the minimum effective height $H_{e\,min}$ is defined as the sensitivity that must be exceeded over at least 80% of the frequency range, and the mean effective height H_e is the sensitivity that must be exceeded by the average value of H_e over the full frequency range.

Table 1
NGC coupler specification for a 420 kV GIS

Frequency range	500-1500 MHz
Min. effective height $H_{e\,min}$	2.0 mm
Mean effective height H_e	6.0 mm

The use of two conditions rather than simply specifying H_e ensures that the couplers are broadband. If the minimum value was not specified, the average value could be achieved by using a very sensitive, but narrow band coupler. The specification requires that the calibration is carried out with the coupler mounted on a replica of the GIS mounting arrangement.

Case study 1 – a coupler design

External couplers may be used on GIS which have either glass windows, or cast resin barriers where the outer edge is exposed. The UHF signal then propagates through the dielectric and may be detected externally, although as mentioned above the apertures act as high-pass filters and the lower frequencies in the signal may be attenuated strongly.

Couplers were designed to suit the barriers of a particular GIS. The chamber flanges and a cast resin insert representing the barrier were reproduced on the calibration rig, and the coupler (Figure 6) mounted in position. The coupler had been designed to satisfy the NGC Specification, with its maximum sensitivity in the 500-1500 MHz range. The calibration

curve in Figure 7 shows that with a mean effective height of 9.2 mm over 100% of the frequency range, the NGC requirements had readily been met.

In practice, it has been found that external couplers, whether for windows or barriers, can be as sensitive as internal couplers. They are particularly convenient to use with portable monitoring equipment because they can be moved around the substation, but have also been proved very satisfactory in several permanent installations fitted to existing GIS.

Figure 6. External coupler for a barrier

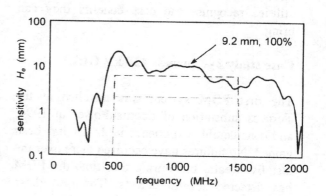

Figure 7. Coupler calibration

6. Service experience

UHF monitoring equipment is produced by some European and Japanese switchgear manufacturers, and a small number of independent companies. However since little information on the service experience of these systems has been published, the author will concentrate on the results obtained by his own company (DMS) in the hope that they will prove a useful guide to those considering using the UHF technique.

6.1 Installed PDM systems

DMS has installed on-line partial discharge monitoring (PDM) systems in a number of GIS at system voltages from 275 to 500 kV, and they are proving to be a valuable means of detecting incipient failures. At present four PDM systems are operating on GIS in the UK, two in Singapore, seven in Korea, and one each in Hong Kong, Malaysia and the USA. New PDM projects, including the survey of existing GIS using portable equipment, are currently under way in Brazil, Canada and South Africa. The UHF measurements are proving their worth by occasionally revealing defects which need to be corrected, enabling utilities to follow any development in discharges at present too small to be of concern, and (occasionally) reassuring users that their GIS is discharge free. Given in addition the significant benefits obtained from using the commissioning test procedure recommended by CIGRE, there is reason to expect that the number of UHF installations will increase significantly as utilities recognise the cost benefits they can bring.

Case study 2 – Torness 420 kV GIS

The first PDM system was installed at the Torness substation of ScottishPower in 1994, and considerable experience in its use has been gained. No failures have occurred there over the last five years, but during that time the PDM has detected several defects. Three of these were followed until they approached their critical stages, when the GIS was opened and the defects corrected. By taking this action, an in-service failure and possible disruption of supply were prevented in each case.

In the latest incident, ScottishPower monitored for several weeks a PD with a discharge pattern characteristic of a faulty busbar joint. When the discharge had grown to a level where breakdown seemed increasingly likely, the defect was located by the time-of-flight method and a decision made to take the faulty section out of service for investigation. Having opened the chamber, at first no effect of the discharge could be seen. However on removing the stress shield covering a busbar joint, it was found to be coated internally with SF_6 decomposition products. The busbar joint was then dismantled, whereupon it was seen that one of the palms had cracked between two of the bolt holes, and was on the point of fracturing completely. Had this happened while in service, the busbar would have dropped to the chamber and caused a severe fault. Metallurgical examination subsequently revealed that the crack had occurred because of flaws in the casting from which the palm was formed.

This type of fault had not been seen before, and it was reassuring to find that a PD completely enclosed by a metal shield was still able to radiate UHF energy into the chamber. In other GIS, plug and socket busbar joints have also been known to arc to an extent where the joint has melted. Laboratory simulations of a faulty joint carrying a few kiloamperes show the characteristic pattern (Figure 8) of the numerous heavy discharges typical of this defect.

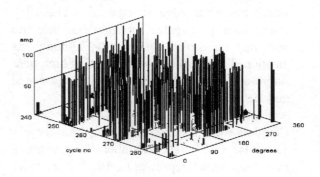

Figure 8. Discharges from a faulty joint

6.2 HV commissioning tests

Even prior to the publication of the CIGRE report of February 1998, several GIS in the UK and the Far East had been commissioned with the aid of PDM systems; indeed the very first UHF measurements were made during the commissioning of Torness GIS. Such measurements are now made much more conveniently, because PDM software dedicated to the commissioning test enables data from up to 12 UHF couplers in the test section to be displayed simultaneously, and updated every few seconds.

Should the test voltage be provided by a variable frequency resonant test set, a low-voltage signal from the test set enables the PDM to lock on to the test frequency and maintain the correct phase relationship in the data displays. It is usual for the test voltage to be raised in two or three stages, and held for some minutes at these intermediate levels before being increased to the one-minute test level. It has been an education to see the conditioning processes in action; bursts of activity while a protrusion burns off, a patch of oxide is removed, or even a dust-like particle moving away or being destroyed. Then, after a minute or so, the test section is usually completely free of discharges until the voltage is raised to the next level.

Unfortunately on occasions the defect is found to be permanent, whereupon the test may be suspended while the defect is removed, or the voltage raised with the possibility of causing a test flashover. Whatever the decision, the test engineer makes it knowing the type of defect present. Normally the GIS will be opened, but not before the defect is located. This is often done by a time-of-flight measurement of signals arriving at couplers on either side of the defect, but if the discharge is large enough then using an acoustic detector is sometimes a simpler alternative.

6.3 Sensitivity verification

A further CIGRE publication 'Sensitivity Verification for Partial Discharge Detection Systems for GIS with the UHF Method and the Acoustic Method' [5] will soon be published.

It will recommend that the following two step procedure be adopted:-

Step 1: Laboratory test

A free metallic particle (typically 3 mm long) is introduced into a test chamber, which has been fitted with two couplers reasonably close together. Having filled the chamber with SF_6, an increasing voltage is applied to the busbar until the particle moves. The PD level is then measured by both the IEC 270 and UHF methods, thereby giving a comparison between the apparent charge (IEC 270), and the UHF signal. A 3 mm long particle is recommended, because it gives an apparent charge of approximately 5 pC.

The busbar is de-energised, and fast-fronted pulses of variable amplitude injected into one of the couplers. The UHF signal at the other coupler is measured, and the pulse height increased until the amplitude of UHF signal previously produced by the particle is reached. Injecting the same pulse into a coupler in the GIS therefore has the same effect as introducing a 3 mm long particle at that point, and later will provide a convenient means of verifying the sensitivity of PD measurement in the substation.

Step 2: On-site test

Pulses of the set amplitude are injected into a coupler on site, and the sensitivity verified if the UHF signal can be detected at an adjacent coupler. This procedure needs to be repeated at other couplers in order to cover the whole site.

Case study 3 – UHF sensitivity verification

The CIGRE recommendations of [5] were followed, for what is believed to be the first time, during the factory and commissioning tests of a GIS in the Far East. They provided a model of how a PDM system should be used, and the utility and switchgear manufacturer may publish the full test details at a later date. If so, they will provide a valuable feedback to CIGRE, and be of interest to other utilities currently weighing up the benefits of the UHF technique.

However, before starting the verification tests it was most important to fit the GIS with couplers having adequate sensitivity and bandwidth, as given by the NGC Specification. For this project, DMS designed couplers to suit the existing ports of the GIS. The outline arrangement was discussed with the switchgear manufacturer, who completed the mechanical design and produced prototypes of each type. These were calibrated, and having met the Specification the coupler designs were released for manufacture. This collaborative approach is ideal, since it ensures that the couplers have the optimum electrical and mechanical design.

Verification tests

The verification tests were carried out in a short length of busbar containing two couplers, and a switchgear bay which had been assembled for its HV tests before being shipped to site. However, the tests could not be made in a shielded area, and this presented difficulties:

- The manufacturer quite properly decided it was unacceptable to introduce a free particle into equipment about to be shipped.
- At the voltage which would have been needed to lift the particle, the background noise in the IEC 270 circuit was much higher than the 5 pC to be measured.

Some alternative was needed, and so use was made of the rolling ball test cell developed some years ago in the UK. Shown in Figure 9, it consists of a gas-tight Perspex cylinder between two aluminium end plates. The lower plate carries a 2 mm diameter aluminium ball lying in a dish-shaped electrode, while the upper supports a metal sphere. When filled with SF_6 at 4-5 bar(g), and with a 0-5 kV AC supply connected across the cell, the voltage is raised until the ball hovers over the dish and generates particle type discharges.

First, the cell was placed inside an open port of the short length of busbar, and the UHF signal measured at a coupler on the second port. Then the cell was removed, and replaced by a coupler. Pulses having a risetime of less than 1 ns were injected into it, and the pulse height which gave the same amplitude of UHF signal at the second coupler was found.

Next, the cell was placed in an open chamber of the switchgear bay, and energised by leads passed through a port. UHF signals were recorded by couplers in the bay, and at the same time the apparent charge was measured using the IEC 270 method. All these measurements were made with the busbar at earth potential.

The advantages of this procedure were:

- The test rig used when calibrating the fast pulse for injection tests was an open section of busbar, without bushings.
- There was no risk of a free particle being lost in the switchgear bay.
- The IEC 270 tests in the switchgear bay were made with the busbar at earth potential, which reduced the interference level to much less than 5 pC.
- The cell gives a known and reproducible signal, and could readily be used elsewhere for comparative tests in different types of GIS.

In summary, the approach taken was straightforward in use, needed only simple rigs with no special shielding, and gave good results. It also calibrated the PDM system in terms of apparent charge, and showed that 5 pC gave signals of 50% full scale, and 25 pC gave signals of full scale.

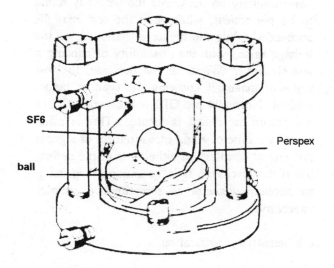

Figure 9. Rolling ball test cell

Case study 4 – portable UHF equipment

UHF measurements can readily be made using portable equipment. It is especially convenient if external couplers can be fitted over windows or exposed barriers, because they can be moved from place to place and the whole GIS surveyed. Portable equipment is also suitable for use during the commissioning of a new GIS.

The Laboratorio Central de Pesquisa e Desenvolvimento (LAC) of COPEL (Brazil) has made many measurements using a portable UHF monitor on 500 kV GIS, particularly those at the Foz do Areia (Figure 10) and Itaipu Bi-National Power Plants [6]. In the latter, many good results have been obtained from measurements undertaken in a large test rig of 500 kV chambers.

More recently, LAC performed the HV commissioning tests of the 500 kV GIS at the Salto Caxias Power Plant of COPEL.

Test voltages of up to 520 kV were applied, and LAC undertook the PD measurements using a portable UHF monitor. Attempts were also made to use the conventional IEC 270 method, but in the event they had to be abandoned because of the high level of background noise at the site.

No internal couplers had been fitted to the GIS, because it was preferred to use external couplers on the exposed edges of the barriers. Typically, the measurements were made at barriers spaced between 15 and 30 m apart. A test sequence of voltages was applied for some minutes up to the test level of 520 kV, and in most cases it was established that the GIS was in good condition. However some defects were found, and on three occasions the test was stopped while the GIS was opened to rectify them. As a result the commissioning tests were completed successfully, and without any internal flashover.

Figure 10. Making UHF measurements at Foz do Areia 500 kV GIS

7. Conclusions

Utilities and switchgear manufacturers are increasingly relying on UHF diagnostics to reveal the internal condition of their GIS, both during the HV commissioning tests and later while they are in service.

When used as recommended by CIGRE as an integral part of the commissioning tests, any defect will be detected as the test voltage is being raised to the 1 min level. If the test is then halted while the defect is located and removed, the overpotential tests will in most instances be completed without any internal flashover. The advantages of this are that it avoids dismantling perhaps long sections of the GIS in order to find the breakdown site, which the user may require; and also any secondary breakdown caused by the fast-fronted breakdown transient. More importantly, the user can be assured that the GIS will enter service free of defects and in the best possible condition.

In service, the UHF technique allows any developing discharge to be detected and located. The development can then be followed until it approaches a critical stage, when plans to correct the defect are put into operation. Experience in many countries has shown that UHF techniques are valuable aids in maintaining the reliable operation of a GIS, and they are being used with increasing confidence.

Having already proved the worth of the UHF technique, future developments are likely to be concentrated in two main areas:

- Automatically interpreting the data more comprehensively than is presently possible, so that as a routine only information will be presented to the user. This is becoming especially important for those utilities already monitoring several GIS on their systems, since their engineers do not have the time to examine in any detail the large amount of data being generated.

- Extending the condition monitoring system to include other plant in the substation. It would be especially attractive to monitor the performance of circuit breakers, where significant savings could be made by adopting condition-based maintenance systems; and there may be benefits in monitoring bushings, transformers and other plant items, as a further step towards the ultimate goal of achieving a completely 'smart' substation.

8. Acknowledgments

Grateful thanks are given to The National Grid Company plc, Scottish Power plc and COPEL for their support in the preparation of this paper.

References

[1] J. S. Pearson, O. Farish, B. F. Hampton, M. D. Judd, D. Templeton, B. M. Pryor and I. M. Welch, Partial discharge diagnostics for gas insulated substations, *IEEE Trans. Dielectrics and Electrical Insulation*, Vol. 2, No. 5, pp. 893-905, (1995).

[2] CIGRE Joint Working Group 33/23.12, Insulation co-ordination of GIS: Return of experience, on site tests and diagnostic techniques, *ELECTRA*, No. 176, pp. 67-97 (1998).

[3] M. D. Judd, O. Farish and P. F. Coventry, UHF couplers for GIS - sensitivity and specification, *10th Int. Symp. on High Voltage Engineering (ISH)*, Montreal (1997).

[4] The National Grid Company plc, *Capacitive couplers for UHF partial discharge monitoring*, Technical Guidance Note TGN(T)121 Issue 1 (1997

[5] CIGRE Task Force 15/33.03.05, 'Sensitivity Verification for Partial Discharge Detection Systems for GIS with the UHF Method and the Acoustic Method'. To be published in *ELECTRA*.

[6] M. Silva, et al. PD measurements in GIS using portable equipment, 9th Int Symp on High Voltage Engineering, Graz, Aug 1995.

classification of images. As template matching is sensitive to the relative intensities of PD patterns, alternative ways of feeding images to the recognition system were sought. A decision was made to use a low-resolution ($64q \times 32\phi$) contour version of patterns for recognition purposes, see Figure 1(b). Low-resolution patterns make recognition less susceptible to statistical variations in PD patterns and contours make recognition independent of the relative densities of PD patterns.

RESULTS

In a first example, the phase-only filter was applied to recognition of four PD patterns shown in Figure 1. This complex pattern was created by merging two different PD records. The first consisted of two cavity patterns, the second of background noise and thyristor pulses. The phase-only filter successfully recognised all four PD patterns. It can be seen in Figure 3 that during recognition of the cavity patterns a number of small false peaks appeared in the correlation plane in addition to the two dominant peaks corresponding to the two cavity patterns. The false peaks were, however, smaller than 30%. A usual procedure is to count only peaks above a certain threshold value. The threshold values must be carefully set for each template.

In a second example, an actually recorded pattern consisting of cavity, thyristor pulses and background noise was tested, see Figure 4. Recognition rates varied according to a template used. For example, recognition rates between 52% and 69% were obtained for the cavity pattern. Similar recognition rates were obtained for the background noise and the thyristor pulses patterns.

In a third example, the phase-only filter was tested to discriminate between two different cavity patterns, see Figure 5. The pattern was created by merging two different PD records. Both cavity patterns were recognised and the filter successfully discriminated between the two patterns. This can be seen in Figure 6. When the "lower" cavity served as a template, only one peak appeared in the correlation plane. The same result was obtained when the "upper" cavity served as a template.

When the "lower" cavity in Figure 5 was used as a template for recognition of four PD patterns shown in Figure 1, no dominant peaks in the correlation plane appeared, see Figure 7. The maximum height of false peaks did not exceed 25%.

In addition to the multiple patterns already discussed, thirty actually recorded single and multiple patterns (cavity, background noise, thyristor pulses, background noise + thyristor pulses) were tested. Recognition rates between 38% and 90%

were obtained for the patterns. There were no misclassifications.

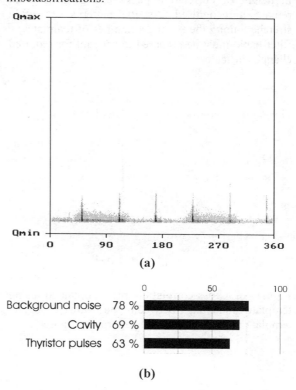

(a)

Background noise 78 %
Cavity 69 %
Thyristor pulses 63 %

(b)

Figure 4. Recognition of multiple PD pattern.

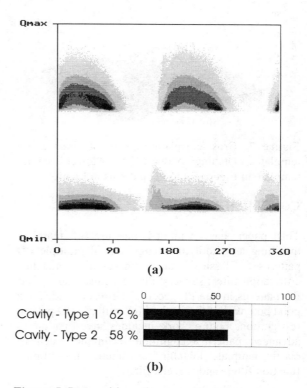

(a)

Cavity - Type 1 62 %
Cavity - Type 2 58 %

(b)

Figure 5. Recognition of two cavity patterns.

One of the major disadvantages of template matching is its sensitivity to scale changes between a template and an object to be recognised. Changes in scale as small as 10% result in a significant decrease in height of correlation peaks and patterns often remain unrecognised. As PD patterns are often stretched along the q-axis, a number of templates, a filter bank, must be prepared to account for required changes in scale.

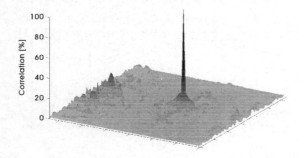

Figure 6. Cross-correlation plane obtained using template matching between the "lower" cavity template in Figure 5 and PD patterns in Figure 5.

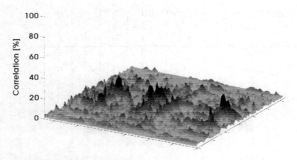

Figure 7. Cross-correlation plane obtained using template matching between the "lower" cavity template in Figure 5 and PD patterns in Figure 1.

CONCLUSIONS

This paper has examined the use of template matching as a tool for recognition of multiple PD patterns. Classical and phase-only matched correlation filters successfully recognised up to four PD patterns in one PD record. The results indicate a good potential of the method for multiple PD pattern recognition. Future work includes investigation of advanced pattern recognition techniques such as elastic template matching, synthetic discriminant function filters and wavelet filters.

ACKNOWLEDGMENT

The author wishes to thank Mr. S. Halén of Vattenfall Utveckling AB, Sweden for permission to use the PD patterns.

REFERENCES

[1]   A. Krivda, "Automated Recognition of Partial Discharges", IEEE Trans. on Dielectrics and Electrical Insulation, Vol.2, pp. 796-821, 1995.

[2]   F.H. Kreuger, E. Gulski and A. Krivda, "Classification of Partial Discharges", IEEE Trans. on Electrical Insulation, Vol.28, pp. 917-931, 1993.

[3]   E. Gulski and A. Krivda "Neural Networks as a Tool for Recognition of Partial Discharges", IEEE Trans. on Electrical Insulation, Vol.28, pp. 984-1001, 1993.

[4]   L. Satish and W.S. Zaengl, 1995, "Can Fractal Features be used for Recognising 3-D PD Patterns?", IEEE Trans. on Dielectrics and Electrical Insulation, Vol.2, pp. 352-359. 1995.

[5]   T. Hücker and H.-G. Kranz, "Requirements of Automated PD Diagnosis Systems for Fault Identification in Noisy Conditions", IEEE Trans. on Dielectrics and Electrical Insulation, Vol.2, pp. 544-556, 1995.

[6]   M. Hoof, B. Freisleben and R. Patsch, "PD Source Identification with Novel Discharge Parameters using Counterpropagation Neural Networks", IEEE Trans. on Dielectrics and Electrical Insulation, Vol.4, pp. 17-32, 1997.

[7]   A. Krivda and S. Halén, "Recognition of Partial Discharges in Generators", the 10th Int. Symp. on High Voltage Engineering, Montréal, Canada, 1997.

[8]   T. Okamoto and T. Tanaka, 1995, "Partial Discharge Pattern Recognition for Three Kinds of Model Electrodes with a Neural Network", IEE Proc. on Science, Measurement and Technology, Vol.142, pp. 75-84, 1995.

[9]   C. Cachin and H.J. Wiesmann, 1995, "PD Recognition with Knowledge-based Pre-processing and Neural Networks", IEEE Trans. on Dielectrics and Electrical Insulation, Vol.2, pp. 578-589, 1995.

[10] M. Cacciari, A. Contin and G.C. Montanari, "Use of a Mixed Weibull Distribution for the Identification of PD Phenomena", IEEE Trans. on Dielectrics and Electrical Insulation, Vol.2, pp. 614-627, 1995.

[11] J.H. Lee, D.S. Shin and T. Okamoto, "Analysis of Partial Discharge Signals from Multi Defect Insulating Systems", the 5th Int. Conf. on Properties and Applications of Dielectric Materials, Seoul, Korea, pp. 307-310, 1997.

[12] W.K. Pratt, *Digital Image Processing*, Wiley, New York, USA, 1978.

[13] A. Vander Lugt, "Signal Detection by Complex Spatial Filtering", IEEE Trans. on Information Theory, Vol.10, pp. 139-145, 1964.

[14] J.L. Horner and P.D. Gianino, "Phase-only Matched Filtering", Applied Optics, Vol.23, pp. 812-816, 1984.

WAVELET BASED PARTIAL DISCHARGE IMAGE DE-NOISING

M.Florkowski

ABB Corporate Research, Kraków, Poland

Abstract

In this paper, the application of wavelet de-noising to partial discharge (PD) phase resolved images obtained during on-line measurements on a 6kV motor is presented. The basic principles of wavelet de-noising analysis, as well as examples of hard- and soft de-noising thresholding are reported. For the purposes of decomposition, the Deabuchies wavelet and wavelet packs at different levels were applied. The efficiency of the enhancement was measured by using a mean square signal-to-noise ratio.

1. Introduction

The measurements of partial discharges, in particular on-site measurements, are immanently correlated with the issue of disturbance suppression and improvement of the signal-to-noise ratio (S/N) in order to correctly interpret the form of discharges and their source in a HV equipment under test. There are various methods applied in the time or frequency domains that are implemented either on-line in hardware (e.g. gating in the time domain, DSP filtering) or as a post-processing after the data acquisition. Usually, the tools used are digital filters (e.g. FIR, IIR, adaptive filters) or methods based on the Fourier's transform. Where the analysis of partial discharges is approached from the frequency point of view [e.g. 2, 11, 14, 15], filtration in the frequency domain takes place usually whilst a Fourier transform of the time signal is performed, and relevant spectrum elements are zeroed, and, next, a reverse transform is carried out. In the time domain, the boxcar technique which allows for the improvement of the signal-to-noise ratio proportionally to the square root obtained from an accumulation number, has several limitations with regard to the PD acquisition. Among other things, those limitations result from the situation that impulses of discharges are only quasi-repeatable (stochastic phase position), and disturbances that often accompany them, normally under industrial conditions, demonstrate a character of impulses coherent with the test voltage (e.g. thyristor pulses). Recently, the wavelet transform (WT), a scale-frequency representation of a signal, became very popular tool in signal processing. This transform constitutes a sort of remedy for limitations involved in the Short Time Fourier Transform (STFT) resolution, and it is usually applied to detection, extraction, compression and de-noising of signals.

Wavelets are also used in High Voltage Engineering, e.g. in acoustic PD measurements [1, 10], PD separation from narrowband disturbances [6], electrical treeing investigations [5], diagnostic localization of faults in power transformers [12, 13], electrical power quality [7].

2. Theory

The Equation /1/ defines a Continuous Wavelet Transform (CWT) of a signal x(t):

$$CWT(\tau,s) = \frac{1}{\sqrt{|s|}} \int_{-\infty}^{\infty} x(t)\Psi\left(\frac{t-\tau}{s}\right)dt \qquad /1/$$

where CWT is a two variable function, t - displacement, and s - scale, wheras $\Psi(t)$ is a transform function known as the "mother wavelet". A displacement which positions a window in the signal corresponds with a time information. The scale, an inverse frequency, is responsible for the width of the analyzing function, i.e. a large value corresponds with a general appearance of a signal (if s>1, signal stretching), whereas a small value refers to an analysis of details (if s<1, signal compression). Thus, a wavelet transform makes it possible to perform a signal analysis with both short high-frequency and wide low-frequency basic functions at the same time.

This feature identifies the wavelet transform which theoretically comprises an infinite set of basic functions, and distinguishes it from the Fourier transform operating with a constant set of basic functions (sinus and cosines). Owing to this particular feature, the wavelet transform can be easily applied in practice because it enables to select an adequate wavelet system for a designated issue. In the digital signal processing, a discrete wavelet transform (DWT) is mainly applied owing to its dyadic approach according to which scales and displacement are a power of two, realized as a filter bank and a pyramidal decomposition algorithm [9]. Such a family is created by wavelets satisfying the following relationship:

$$\Psi_{j,k}(x) = 2^{-j/2}\Psi\{2^{-j}(x-k)\} \qquad /2/$$

where: j, k - integers.

As a result of this processing, an approximation component **A** and a detail component **D** are obtained at each level of the decomposition; A reflects a low-frequency content of a signal x(t), and **D** - a high-frequency content. Upon filtration, a decimation (downsampling) is performed which involves the

High Voltage Engineering Symposium, 22–27 August 1999
Conference Publication No. 467, © IEE, 1999

deletion of every second sample from the **A** and **D** components in order to ensure an equal number of samples in an original x(t) signal, and totally in its transformed **cA** and **cD** components. Whilst introducing the above described processing algorithm into subsequent stages following to the **A** component, a tree-structure is obtained. The decomposition by use of wavelet packets serves to generalize about the above method; in the decomposition, both the approximation component **A** and detail component **D** undergo splitting at the subsequent levels. In 1988, Daubechies successfully implemented an orthonormal base of compactly supported wavelets. The family of Daubechies wavelets is designated as dbN in the literature, and N denotes a wavelet order [3, 8]. It is worthy to admit that 'db1', a wavelet of the first order, is also a Haar's wavelet.

Extending the one-dimensional wavelet transform into two-dimensions, an excellent tool is provided for the analysis of images. The size of an image **I** under analysis should be a multiple of 2, i.e. 2^n x 2^n (n - a natural number) that is necessary with regard to the applied dyadic algorithm of decomposition. During the first run, a low- (G_r) and highpass (H_r) filtration runs along the image rows regarded to be subsets of one-dimensional data. Next, the two separated sub-images undergo decimation and their dimension becomes 2^n x 2^{n-1}. During the second run, the columns of sub-images are scanned, and the similar low- (G_c) and highpass (H_c) filtration and decimation are realized, and as a consequence, 4 sub-images are obtained: an approximation I_1 (a lowpass filtration along the rows and columns), and three images containing details often referred in the literature as a horizontal sub-image I_2, a vertical sub-image I_3, and a diagonal sub-image I_4. Owing to the decimation, sub-images $I_{1..4}$ possess 1/4 of pixels of the original image **I**, i.e. their dimension is 2^{n-1} x 2^{n-1}. Concluding, the above transformation can be formulated as follows:

$$I_1 = G_cG_rI \qquad I_2 = H_cG_rI \qquad /3/$$
$$I_3 = G_cH_rI \qquad I_4 = H_cH_rI$$

During the next steps of the wavelet transform, an image I_1 is processed.

One of the most important applications of the wavelet transform is the de-noising of signals developed by Donoho and Johnstone [4]. While generalizing the one-dimensional wavelet method of de-noising to images, the following equation can be derived:

$$I(i,j) = O(i,j) + \sigma e(i,j) \qquad /4/$$

where: i, j = 0, ..., n-1, 'e' represents a Gaussian white noise with parameters N(0,1), 'σ' is a standard deviation, **I** is an image which is currently registered, and **O** - a de-noised image. The main purpose of de-noising is to suppress a component representing noise and to regain an original image **O**. Upon

selecting a prototype wavelet and a level P, first, an image **I** should be brought to the level P through the wavelet decomposition. With regard to components of details D_1 to D_P, a threshold operation ($TH_1...TH_P$) should be performed for each decomposition level 1 to P. The reconstruction of signal is based on both the original approximation components ($A_1..A_P$) and the details components modified by the thresholding operation. Usually, two thresholding methods are used: "soft" and "hard". "Hard thresholding" involves the zeroing of image elements located below the TH level:

$$x_{hard}(t) = \begin{cases} x(t) & \text{for } |x(t)| > TH \\ 0 & \text{for } |x(t)| \leq TH \end{cases} \qquad /4/$$

In order to avoid discontinuity of signal involved in the "hard" thresholding method, in points $x = \pm TH$, a method of "soft" thresholding has been introduced which eliminates this inconvenience. The "soft" thresholding method comprises, beside the zeroing of elements under the TH level, a displacement of this part of signal which is above the thresholding level:

$$x_{soft}(t) = \begin{cases} \text{sign}(x(t))(|x(t)|-TH) & \text{for } |x(t)| > TH \\ 0 & \text{for } |x(t)| \leq TH \end{cases} \qquad /5/$$

However, the selection of a thresholding level is still the key issue in the entire de-noising process. Besides an arbitrary selection and a choice based on a trial-and-error method, there are estimation methods to properly indicate a required thresholding level. Donoho suggested universal thresholds selected with regard to the 'n' length of the data record:

$$TH = \sigma\sqrt{2\log(n)} / \sqrt{n} \qquad /6/$$

where 'σ' is a noise standard deviation. Usually, it is impossible to 'a priori' know noise parameters, so, a standard deviation value is assumed to be σ=WFS/0.6745, where WFS is the absolute median value of the finest detail coefficients [DONO 95]. The wavelet image de-noising procedure is similar to one-dimensional case, it is based on a two-dimensional wavelet transform, and the n^2 dimension of an image, and not a 'n' length of the record, is taken into consideration while estimatively selecting an universal threshold.

3. Results

With regard to the diagnostic and monitoring analysis of partial discharges, the elimination of noises and improvement of a signal-to-noise ratio is important for the precise classification and interpretation of PD forms.

In the following part of this paper, the results are presented that pertain to the wavelet de-noising processing performed on one-dimensional PD signals in the time domain, as well as applied to phase resolved PD images.

In Fig. 1, partial discharge signals are shown that where registered in the insulation of a generator stator bar at test voltage of 15 kV. The original signal containing PD pulses and noise is illustrated in Fig.1a, while Fig.1b demonstrates a signal after de-noising with a 'db1' wavelet transformation involving a soft thresholding and an universal thresholding, selected according to the length of the data record /eq. 6/. In this case, a signal to noise rate has been enhanced by 6 dB.

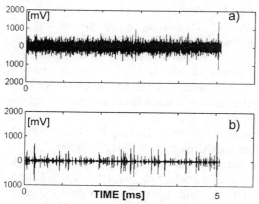

Fig. 1 A PD signal, registered in the insulation of a generator stator bar at 15 kV: a) original; b)after wavelet de-noising (db1)

The following investigations refer to the application of wavelet de-noising to PD images registered in one phase of the insulating system of a 6kV/200 kW motor winding, obtained during on-line measurements. The phase-resolved accumulated images had a size of 256 x 256 (2^8 x 2^8) pixels, i.a. 256 points on phase, 256 (+/-128) points on amplitude and two byte depth corresponding to the number of PD counts. The original image is illustrated in Fig. 2a, and in Fig. 2b, the image obtained as a result of the de-noising processing which involved a Deabuchies wavelet db2 at the level 4, and "soft" thresholding of images with diagonal details, and with thresholds for levels 1 to 4 respectively:

$$TH_{I41}= 70.0; \quad TH_{I42} = 31.6;$$
$$TH_{I43} = 34.8 \quad TH_{I44} = 1.3$$

Fig. 2 a) original PD image obtained in a 6kV/200kW motor; b) de-noised PD image, using db2, level 4, and a "soft" thresholding

As a result of de-noising, discharge groups occurring in the motor insulation system become clearly distinguishable in the image. Individual sub-images, obtained while performing subsequent steps of the wavelet decomposition at level 4 are shown in Fig. 3.

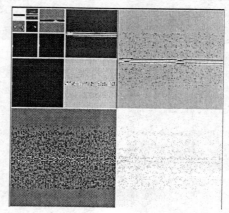

Fig. 3 Sub-images of the wavelet decomposition (db2, level 4) obtained for the image from Fig. 2a

In Fig.4, de-noising results are shown which were gained on the base of the de-noising and by use of a wavelet packet method; these results refer to the above motor case. The Deabuchies wavelet 'db2pk' at level 4 was applied, as well as "hard" (Fig. 4b) and "soft" (Fig. 4c) thresholding with a global thresholding at level TH=1. For the purpose of comparison, the results of de-noising at level 4 obtained for the Deaubuchies db4 (Fig. 4d) and Haar (Fig.4e) wavelets were listed. A mean square signal-to-noise ratio was assumed as a criterion of comparison employed to quantitatively illustrate the effectiveness of the de-noising operation, and its definition is as follows:

$$SNR_{ms} = 10 \log_{10} \frac{\sum_{i=0}^{N-1}\sum_{j=0}^{N-1} I(i,j)^2}{\sum_{i=0}^{N-1}\sum_{j=0}^{N-1}[I(i,j)-O(i,j)]^2} \quad /7/$$

where **I** represents an original image, and **O** - a de-noised image.

A comparative specification of the de-noising efficiency is tabulated in Tab. 1. To give an example: SNR_{ms} was 12 dB for an image de-noised by a wavelet 'db2' (see the example at Fig. 2b), whereas SNR_{ms} was 2.7 dB for the image de-noised by wavelet packets 'db2' at level 4 involving "soft" thresholding of TH=5.

Tab. 1

Method (applied to image Fig. 2a)	SNR_{ms} [dB]
db2, LV=4, individual SoftTH	12
db2pk, LV=4, SoftTH=1	14
db2pk, LV=4, HardTH=1	10
db2pk, LV=4, SoftTH=5	2.7
db2pk, LV=4, HardTH=5	2.3
Haar packets, LV=4,	8.4

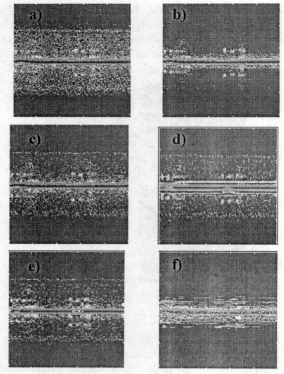

Fig. 4 Images of partial discharges registered in the insulation system of a 6kV/200 kW motor
a) original image
b) image after the de-noising operation involving wavelet packets db2, at level 4, soft thresh. TH=1
c) image after the de-noising operation involving wavelet packets db2, at level 4, hard thresh. TH=1
d) image after the de-noising operation involving wavelet packets db2, at level 4
e) image after the de-noising operation involving Haar wavelet packets db2, at level 4
f) image after the de-noising operation involving wavelet packets db2, at level 4, hard thresh. TH=5

4. Conclusions

In this paper, an useful method of improving the signal-to-noise ratio in partial discharge (PD) phase resolved images was reported. The described method is based on the wavelet de-noising processing. In a partial discharge image, registered during the on-line measurements of the insulation system in a motor, coherent PD groups were separated that could not be distinguished in the primary image. According to the tests, the fundamental issue in the image recovery using a wavelet de-noising processing seems to be the choice of the threshold value. The wavelet de-noising processing will definitely find future applications in partial discharge analyzers, beside the boxcar accumulation method, or FFT based digital filtering.

Acknowledgment

The work was partly carried out at the University of Mining and Metallurgy AGH in the Electrical Power Institute, in Kraków, Poland.

References

1. Arii K., Shibahara M., Fujii M.,"Separation of Noise from Partial Discharge Signals by Wevelet", Paper 03P07, 5th Int. Conf. on Prop. and Appl. of Diel. Materials, 1997 Seoul, Korea, pp. 232-235
2. Borsi H., Hartje M., „New Method to Reduce the Disturbance Influences on the In Situ Partial Discharge (PD)-Measurement and Monitoring", 6th ISH, New Orleans, 1989, Ref. 15.10
3. Daubechies I., Ten lectures on wavelets, CBMS, SIAM, 61, 1994
4. Donoho D.L, De-noising by soft thresholding, IEEE Trans. on Inf. Theory, Vol. 41, No. 3,
5. Fujimori S., Endoh T., Mitsuboshi K., Ishiwata K., Akasaka M., Hiruta S., „Wavelet analysis of tree developments", IEEE 5th Int. Conf. Cond. and Breakdown in Solid Diel., Leicester, England, 1995, pp.376-380
6. Hang W., Kexiong T., Deheng Z., „Extraction of Partial Discharge Signals Using Wavelet Transform", Paper 03P34, 5th Int. Conf. on Prop. and Appl. of Diel. Materials, 1997 Seoul, Korea
7. Heydt G.T., Galli A.W., Transient Power Quality Problems Analyzed Using Wavelets, IEEE Trans. on Power Del., Vol. 12, No. 2, 1996, pp.908-915
8. Lee D, Yamamoto A., Wavelet analysis: Theory and Applications, Hewlett-Packard Journal, December 1994, pp.44-54
9. Mallat S., A theory for multiresolution signal decomposition: the wavelet representation, IEEE Pattern Anal. and Machine Intell., Vol. 11, No. 7
10. Masanori M., Okano T., Nishimoto S., Kitani I., Arii K., Study on application of wavelet analysis for degradation diagnosis of partial discharge in void, IEEE 5th Int. Conf. Cond. and Breakdown in Solid Diel., Leicester, England, 1995,
11. König G., Feser K., A New Digital Filter to Reduce Periodical Noise in Partial Discharge Measurements, 6th ISH, New Orleans, 1989,
12. Satish L., Short-time Fourier and wavelet transforms for fault detection in power transformers during impulse tests, IEE Proc. Sci. Meas. Techno., Vol. 145, No. 2, 1998, pp.77-84
13. Wang Y., Li Y.M., Qiu Y., Application of wavelet analysis to the detection of transformer winding deformation, 10th ISH,Canada 1997
14. Wilson A., „Discharge Detection under Noisy Conditions", Proc. IEE, Vol. 121, No 9, 1974,
15. Włodek R., Florkowski Z., Zydroń P., „Application of Signal Theory Procedures to Partial Discharge Detection Methods", IEEE 5th Int. Conf. Conduction and Breakdown in Solid Dielectrics, Leicester, England, 1995, pp.249-253

Address of Author:
ABB Corporate Research
ul. Starowislna 13A, 31-038 Kraków, POLAND
e-mail: marek.florkowski@plcrc.mail.abb.com

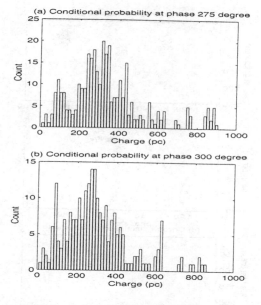

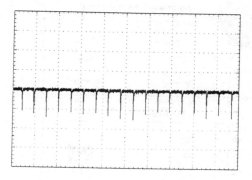

Figure 5: PD pulse oscillogram, 50mV/div (ver), $4\mu s$/div (hor)

Figure 3: Conditional distribution of charge magnitude with (a) 275 degree phase angle (b) 300 degree phase angle

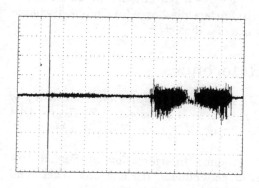

Figure 4: PD Detector's output, 1V/div (ver), $2ms$/div (hor)

ments around 270 degree phase angle.

Oscillograms of pulses developed across the measuring resistance were captured at the negative peak of the applied voltage for six different voltages in the range of 110% to 200% inception voltage. Figure 5 gives the oscillogram obtained at the applied voltage of 200% inception voltage. The periodic variation of the pulses can be clearly seen from it. The time between pulses were found to vary from $35\mu s$ at 110% inception voltage to $2\mu s$ at 200% inception voltage.

As the time interval between pulses decreased along the phase and at the same phase with different applied voltages, it can be concluded that the intra-pulse spacing decrease with increase in the instantaneous stress.

The time difference between some pulses were found to be less than the resolution time and

the dead time limits of the detector and the analyser respectively, and hence no measurement of pulses at some phase angles (near 270 degree) were observed. The output of the detector can show decrease in the pulse magnitude along the phase upto 270 degree phase due to superposition errors caused by the overlapping of pulses. However, it is difficult to comment on the discharge magnitude trend at very high applied stress as the mode of discharge may change due to the strong local field.

2.2 Inter-cluster pulses

To study the inter-cluster pulse characteristics, a set of distributions (Figure 6) were obtained by conditioning the distribution set shown in Figure 2 by the threshold value of $30\mu s$. The $\phi-q-n$ (Figure 6(a)) distribution shows a positive correlation with instantaneous stress, the charge magnitude increasing with increase in the stress.

It can be seen in the $\phi-t-n$ distribution (Figure 6(b)) that the clusters are distributed with different inter-cluster time, the largest inter-cluster time occurring at around the negative peak of the applied voltage. This is possible because of sinusoidal variation of the stress, the rate of change in the stress decrease in the phase range of 180 degree to 270 degree and same delta change in stress near 180 degree will occur at a lesser time as compared to that near 270 degree. However, the probability of finding clusters at higher stress decreases with increase in stress as then the cluster spans for larger time.

The charge distribution is found to be different from the unconditional distribution (Figure 2(c)) exhibiting dependence of charge magnitude on time characteristics of pulses.

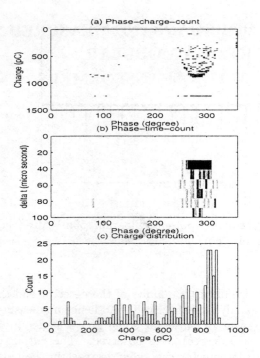

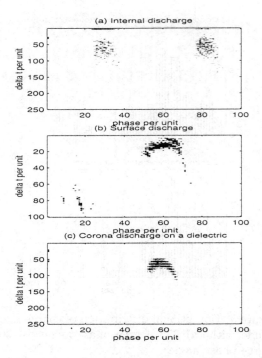

Figure 6: Corona discharge with threshold of $30\mu s$ (a) $\phi-q-n$ distribution (b) $\phi-t-n$ distribution (c) charge distribution

Figure 7: $\phi-t-n$ patterns for (a) Internal discharge (b) Surface discharge (c) Corona discharge in presence of a dielectric.

3 $\phi-t-n$ Representation

Figure 7 gives $\phi-t-n$ patterns for internal discharge (Figure 7(a)), surface discharge (Figure 7(b)) and corona discharge in presence of a dielectric (Figure 7(c)). It can be observed that the patterns are distinct and will be useful for discharge recognition. As has been discussions above, the $\phi-t-n$ can be used to explain PD pulse time characteristics. Implementation of the $\phi-t-n$ analyser is simple as only measurement of time position of the pulse is required. The detection of the pulse can be done using logic comparators which will make the $\phi-t-n$ analyser very fast overcoming the resolution limitations of conventional analysers. $\phi-t-n$ analysers are also independent of measuring instrument characteristics and experience of the user.

3.1 Conclusions

Pulse sequence analyses are shown to be useful to interpret PD distributions. Intra cluster and inter cluster time intervals indicate the stress condition around the discharge source. The characteristics of PD are represented by $\phi-t-n$ patterns which can be used for discharge source recognition. The $\phi-t-n$ analyser is simple to implement and use.

4 References

/1/ R.J. Van Brunt, *Physics and Chemistry of Partial Discharge and Corona*, IEEE Trans. Diel and Elec. Insul., VOl.1, pp. 761-784, 1994.

/2/ U. Fromm, *Interpretation of Partial Discharges at dc Voltages*, IEEE Trans. Diel. and Elec. Insul., Vol.2, pp. 761-770, 1995.

/3/ M. Hoof and R. Patsch, *Voltage- Difference Analysis, a Tool for Partial Discharge Source Identification*, Conf. Record f the 1996 IEEE International Symposium on Electrical Insulation, Montreal, Quebec, Canada, pp. 401-406, June 16-19, 1996.

/4/ R. Bartnikas and J.P. Novak, *On the Spark to Pseudoglow and Glow Transition Mechanism and Discharge Detectability*, IEEE Trans. Elec. Insul., Vol. 27, pp. 3-14, 1992.

/5/ A. Krivda, *Automated Recognition of Partial Discharges*, IEEE Trans. Diel. and Elec. Insul., Vol. 2, pp. 796-821, 1995.

/6/ S. Senthil Kumar, M.N. Narayanachar and R.S. Nema, *Response of Detectors in the Measurement of Partial Discharges - Proposal for Partial Discharge Detectors with Extended Resolution*, Fourth Workshop and Conference on EHV Technology, Bangalore, India, pp. 153-157, July 17-18, 1998.

HIERARCHICAL CLUSTER ANALYSIS OF BROADBAND MEASURED PARTIAL DISCHARGES AS PART OF A MODULAR STRUCTURED MONITORING SYSTEM FOR TRANSFORMERS

P. Werle, H. Borsi, E. Gockenbach

Schering-Institute of High Voltage Technique and Engineering
University of Hannover, Germany

ABSTRACT

In this article investigations on hierarchical cluster analysis methods concerning their ability to evaluate broadband measured Partial Discharges (PD) are presented. Therefore different cluster algorithms were tested with various data measured in the laboratory on a special modified distribution transformer and with measurements from a power transformer on-site. A comparison of the results obtained by different cluster analyses shows that these methods are suitable for a location of PD sources inside transformers as well as for a separation of these partial discharges from pulse shaped noises, e.g. corona impulses. In order to achieve a higher efficiency and performance in evaluating PD signals cluster algorithms are combined with additional pattern recognition methods in a modular structured monitoring system for transformers.

I INTRODUCTION

Transformers belong to the most important and expensive components of power transmission and distribution systems. Therefore it is of special interest to prolong their life duration while reducing the service and maintenance expenditure for these apparatus. In order to reach these aims and to give recommendations about their future life expectation it is necessary to keep the transformers under surveillance during operation which is done by using various kinds of so-called monitoring systems. However, almost all monitoring systems, that are recently available on the market, can only acquire conventionally measured values extend from the calculation of temperature distributions via the analysis of the hydrogen content of insulation liquids up to the evaluation of various transfer functions. Regarding this data only an insignificant statement about the present condition of the insulation of the transformer is possible, because due to the integral characteristic of the almost all measured values only an information about the long-term behaviour of the insulation can be given.

Partial discharge measurements can break these restrictions because they are a practical and efficient method for diagnosing the actual condition of the insulation. In opposition to other high voltage equipment where narrowband PD measurements are sufficient due to the simple construction of these apparatus, for the more complex structured transformers broadband PD measurements are more favourable [1]. These enable the location of the PD origin as well as the possibility to distinguish between partial discharges and noise impulses. This represents the two most important advantages of broadband measurements compared with narrowband PD detection methods.

Nevertheless an on-line monitoring of transformers based on broadband PD measurements is hard to realise due to the difficult evaluation of the PD signals. Therefore different methods of pattern recognition can

be used successfully although many of the used algorithms require reference impulses which are often not available. Cluster analyses may be helpful here, because they are capable to evaluate the measured impulses by separating them into different groups without any a-priori knowledge.

II HIERARCHICAL CLUSTER ANALYSIS

Cluster analysis methods use common features of objects of a regarded data set for dividing it into so-called clusters in such a way that objects belonging to any one of the clusters should be similar while objects of different clusters should be as dissimilar as possible. A simple example of this procedure is shown in Figure 1 where the cluster algorithm detects as feature for the similarity the geometric shape and therefore separates the circles, the rectangles and the triangles into three different clusters.

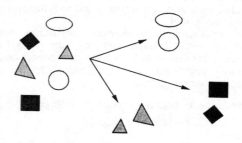

Figure 1: Cluster analysis of a simple data set

As a degree for the similarity or respectively dissimilarity between the objects, which are generally encoded as a sequence of numbers or as a vector, various kinds of distances are used like the Minkowski-distances which calculate the difference d between two objects x and y of the length n as follows:

$$d(x,y) = \left(\sum_{i=1}^{n} |x_i - y_i|^p \right)^{\frac{1}{p}} = d(y,x) \tag{1}$$

Well known distances like the City-Block distance (p=1) or the Euclidean distance (p=2) are special cases of equation (1). Beside the various types of distances also correlation coefficients are used like the Pearson Product Moment Correlation (PMC) or the Cosine Correlation (CC).

Cluster analysis methods can be divided into iterative and hierarchical algorithms while the last can be subdivided into agglomerative and diverse methods again. Iterative methods have several disadvantages like the dependency of the result on a required initialisation or the fact that the number of clusters the data should be grouped into must be given in advance in contrast to hierarchical methods which do not only

High Voltage Engineering Symposium, 22–27 August 1999
Conference Publication No. 467, © IEE, 1999

generate one separation but a sequence of so-called partitions which are accessible at the same time. Therefore these methods generate a hierarchy of partitions by means of a successive merging (agglomerative) or splitting (diverse) of clusters. Such a hierarchy can easily be represented by a dendogram as shown in Figure 2. In this case the same data as in Figure 1 are analysed and the same partition is reached after the third step which is marked in the illustration. Regarding the agglomerative methods each object represents a single cluster at the beginning. During each step two clusters are merged to a new one thus resulting finally in a global cluster after the last step. This procedure is vice versa for the diverse methods.

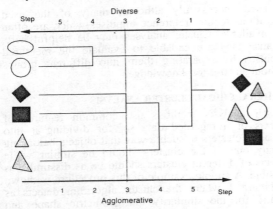

Figure 2: Dendogram for hierarchical clusters

However, diverse methods are often poor concerning their efficiency and performance and will therefore not be regarded further.

There are various agglomerative algorithms which essentially are distinguished by the way they recalculate the distances between a new cluster and the remaining objects after a merging step, which is done by the following equation:

$$d(c_n, c_o) = \alpha \cdot d(c_o, c_1) + \beta \cdot d(c_o, c_2) + \gamma \cdot d(c_1, c_2)$$
$$+ \delta \cdot |d(c_o, c_1) - d(c_o, c_2)| \qquad (2)$$

c_n: new cluster $c_{1,2}$: merged clusters
c_o: remaining clusters $\alpha, \beta, \gamma, \delta$: parameters

The parameters α, β, γ and δ for the investigated methods are described in Table 1.

Method	α	β	γ	δ
Single Linkage	0.5	0.5	0	-0.5
Complete Linkage	0.5	0.5	0	0.5
Average Linkage	$|c_1|/(|c_1|+|c_2|)$	$|c_2|/(|c_1|+|c_2|)$	0	0
Simple Average	0.5	0.5	0	0
Median	0.5	0.5	-0.25	0
Centroid	$|c_1|/(|c_1|+|c_2|)$	$|c_2|/(|c_1|+|c_2|)$	$-\alpha \cdot \beta$	0
Ward	$(|c_1|+|c_o|)/$ $(|c_1|+|c_2|+|c_o|)$	$(|c_2|+|c_o|)/$ $(|c_1|+|c_2|+|c_o|)$	$-|c_o|/$ $(|c_1|+|c_2|+|c_o|)$	0

Table 1: Parameters according to equation (2) of investigated agglomerative methods ($|c_x|$ = number of objects inside cluster c_x)

III Measurements

The described cluster analysis methods have been tested with different data acquired at the laboratory and on site. The signals from the laboratory were measured on a specially prepared distribution transformer (10kV/380V 200kVA) which was pulled out of its vessel. Furthermore, on each of the three coils 9 clamps had been mounted, one on each second of the 16 winding sections and one on the neutral point. Into the

clamps of one phase PD signals were injected which were generated by a needle plane arrangement in oil. High voltage was supplied to the tip of the needle which has a radius of 5μm while the plane electrode was connected to the mounted clamp as shown in Figure 3. At the transformer bushing the injected signals were decoupled via a 50Ω resistor and recorded by a digitiser with a sampling rate of 20MHz. For the further investigations the recorded signals were transferred to a personal computer where they are stored and processed.

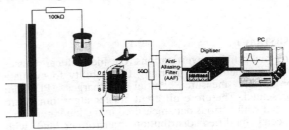

Figure 3: Measurement setup for the determination of the test patterns

Using the described measurement setup 40 partial discharges were injected into each clamp beginning at clamp number 0 close to the bushing up to clamp number 8 at the neutral. With this procedure in total 360 test patterns out of 9 different groups had been produced which were used for further investigations concerning the fault location. Each pattern has a record length of 1000 samples which is equal to 50μs.

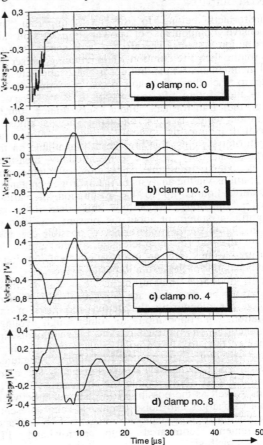

Figure 4: PD signals injected into different clamps and decoupled at the bushing
a) clamp no. 0   **b)** clamp no. 3
c) clamp no. 4   **d)** clamp no. 8

In Figure 4 four of these signals obtained from the PD injection into different clamps are shown. It is obvious that the distortion increases with the length of the coil section the injected impulse has to pass through, thus the PD signal looses more and more of its characteristic double exponential shape. The significant distortion is used by pattern recognition methods in order to locate the origin of the PD. This seems to be easy if the origins are far from each other (compare Figure 4 a and d) but it becomes more difficult if they are close together (compare Figure 4 b and c).

Measurements on-site are often superimposed by typically three different noise types. Sinusoidal noise caused by communication services as well as synchronous noise impulses caused by circuits of power electronics can be easily suppressed using various filtering techniques. Stochastically appearing pulse shaped noises caused by e.g. corona impulses are much more difficult to suppress because they are very similar to the PD signals. However, by using pattern recognition methods a separation of these noises from the partial discharges is possible [1].

In order to prove the ability of the cluster analysis algorithms to separate PD from such noises it was necessary to generate noise signals under realistic conditions. Therefore noise impulses were measured on a power transformer on site. In total 370 noise impulses were detected, which had previously been identified undoubtedly as disturbances by using several methods. For further investigations concerning the separation of pulse shaped noises from partial discharges a set of data including the 370 measured noise pulses in addition to the 360 PD signals was analysed.

IV FAULT LOCATION

In order to prove which combinations of methods and distances are the most suitable ones for an evaluation of the measured signals the following combinations were investigated :

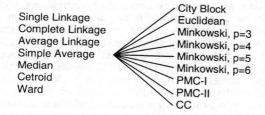

Figure 5: Investigated combinations of clustering algorithms and distances

In total 7 algorithms combined with 9 different distances have been tested, thus in total 63 combinations were investigated. For these investigations the whole data set which consists of 360 test pattern was reduced to 45 patterns, therefore 5 signals were taken from each of the 9 clamps. This procedure was performed for two reasons. On the one hand the performance of the analyses could be increased using less data thus decreasing the efforts and enabling more investigations; on the other hand it is usually more difficult for hierarchical clustering methods to analyse small data sets, because they contain less information. Therefore it is in most cases more complicated to extract the significant feature of similarity. Nevertheless, if the methods can analyse a small data set, they will in most cases be able to examine a larger one more precisely.

For each algorithm the most suitable distance, with which a minimal error for a optimal partition consisting

of 9 clusters has been achieved, is shown Table 2.

Method	Distance	Error [%]
Single Linkage	Euclidean	28.89
Complete Linkage	CC	15.56
Average Linkage	PMC-I	15.56
Simple Average	PMC-I	15.56
Median	PMC-II	15.56
Centroid	CC	15.56
Ward	PMC-I	15.56

Table 2: Best combinations of cluster methods and distances for clustering 45 PD signals into 9 clusters

All methods except the Single Linkage algorithm assign 7 objects wrong. It is obvious that no special distance or method is generally preferable. Nevertheless, if more partitions are taken into account especially the Ward method combined with the PMC-I distance seems to be more favourable. The errors of this combination for different partitions are described in Figure 6.

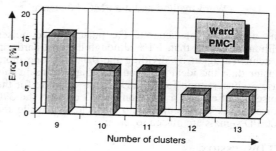

Figure 6: Errors of several partitions obtained by clustering 45 measured PD signals using the Ward method and the PMC-I distance

As described in Figure 6 the error decreases for partitions consisting of more clusters. That behaviour shows that this method can separate the PD signals well although it divides the data into more clusters than assumed, thus the signals from certain clamps are grouped into more clusters than one. A similar result is obtained if the whole data set of 360 PD impulses is investigated which is shown in Figure 7.

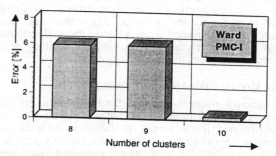

Figure 7: Errors of several partitions obtained by clustering 360 measured PD signals using the Ward method and the PMC-I distance

In this case the error also decreases with the increasing number of clusters. For a partition with 10 clusters, in which the PD signals of a certain clamp are not grouped into one cluster but into two, a minimum error of less than 0.3% is achieved.

Comparing the results shown in Figure 6 and 7 it is obvious that the total error also decreases for a larger data set as previously indicated.

V NOISE SUPPRESSION

For the separation of PD signals from pulse shaped noises the data including the 360 measured PD impulses and the 370 determined noise pulses was used. These investigations emphasised the capability of the Ward method combined with the PMC-I distance to evaluate PD signals, because with this combination the best results are obtained as demonstrated in Figure 8.

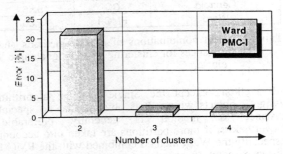

Figure 8: Errors of several partitions obtained by separating 360 measured PD signals from 370 determined noise impulses using the Ward method and the PMC-I distance

The algorithm can separate the pulse shaped noises with an error less than 1.1% although the best partition is achieved again by dividing the data into one more cluster than the ideal partition has; that means in this case 2 clusters, one for the PD signals and one for the noise impulses. For a partition with 3 clusters the 370 noise pulses are grouped into two clusters while the 360 PD signals are placed together.

VI DISCUSSION

It becomes obvious from the investigations that hierarchical cluster analysis methods have the capability to locate PD origins as well as to separate PD signals from noise impulses although they show a tendency to divide the data better into more clusters than necessary. This needs not to be a disadvantage in any case because from a comparison of all partitions generated during the analyses further conclusions can be drawn. Furthermore the algorithms are more efficient and precise if the data set increases; this predestines them for the evaluation of large data banks. Nevertheless, for PD measurements on a transformer on-site in the beginning only few data are available leading to a poor efficiency of cluster methods. That could be compensated by combining these algorithms with other pattern recognition methods resulting in a modular structured monitoring system [2].

VII MONITORING AND DIAGNOSING SYSTEM

Commonly all methods used for the analysis of PD signals show different disadvantages which may be compensated if the results of different methods are compared by an expert system which not only analyses PD patterns but also conventional measured parameters under consideration of a rule- and data base [3]. Figure 9 describes the conception of such a system which guarantees a safe and exact diagnosis of the operation situation. In order to develop such a modular structured monitoring system extensive investigations are performed at the Schering-Institute leading to both the development of various sensors for measuring several parameters and the improvement of different pattern recognition methods and feature extraction algorithms. A description of the researched basic system can be found in [4] whereas the expert system and the rule base are the subject of recent investigations which have

not been finished yet and will be subject for continuous developments due to always new discovered results.

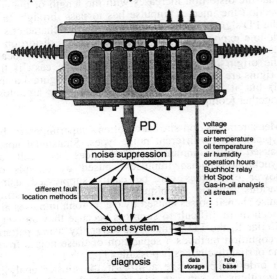

Figure 9: Modular structured monitoring system

VIII CONCLUSION

For a precise statement about the condition of the insulation of high voltage equipment Partial Discharge (PD) measurements are strictly requested. For a PD detection on transformers broadband measurements are more preferable than narrowband ones because they allow a fault location as well as a separation of partial discharges from pulse shaped noises which appear during measurements on-site. The broadband measured PD signals can be analysed using pattern recognition algorithms, although many of these methods require reference data which are often not available. Cluster analyses remove this limitation because they evaluate the signals by dividing them into groups without needing any a-priori knowledge. Investigations with different data measured in the laboratory and on a power transformer during operation show that hierarchical cluster methods are capable of performing both a fault location with a correct assignment of more than 99.7% and a separation of PD signals from pulse shaped noises with an error less than 1.1%. Although these results are achieved in cases in which the data are divided into more clusters than assumed it is possible to avoid this disadvantage by combining different algorithms in a modular structured monitoring system.

ACKNOWLEDGEMENT

The authors thank Dipl.-Ing. Delf Brüssau who performed a lot of the investigations during his work at the Schering-Institute resulting in interesting ideas and improvements.

REFERENCES

[1]   Borsi, H.
        Partial Discharge Measuring and Evaluation System based on Digital Signal Processing
        4th Volta Colloquium on PD Measurements, Como, Italy, 1997

[2]   Wenzel, D.; Borsi, H.; Gockenbach, E.
        A Measuring System based on Modern Signal Processing Methods for Partial Discharge Recognition and Localization On-Site
        IEEE, ISEI, Virginia, USA, 1998

[3]   Wenzel, D.; Borsi, H.; Gockenbach, E.
        Partial Discharge Measurement and Gas Monitoring of a Power Transformer On-Site
        7th DMMA, Bath, Australia, 1996

[4]   Wenzel, D.
        Teilentladungsmessungen an Transformatoren im Netz mit Verfahren der digitalen Signalverarbeitung und Mustererkennung
        Dissertation, Universität Hannover, 1998

PARTIAL DISCHARGE CLASSIFICATION IN GIS USING THE NARROW-BAND UHF METHOD

R. Feger, K. Feser

University of Stuttgart, Germany

R. Pietsch

ABB High-Voltage Technologies, Switzerland

Abstract: This paper discusses the use of the narrow-band ultra high frequency (UHF) partial discharge (PD) measuring technique for the recognition and classification of PD in gas insulated substations (GIS). The principle of the identification of PD signals from external disturbances or UHF signals caused by switching operations will be issued. Additionally, the interpretation of the measured signals with the use of orthogonal functions, fuzzy logic and neural networks will be reported. A genetic algorithm will be presented which can be used to determine the optimum topology of the used neural network.

Introduction

The measuring of PD in GIS using narrow as well as broad-band UHF methods is meanwhile covered widely in the literature [1,2] and commonly accepted in the industry. With the new sensitivity verification of UHF measuring systems developed by the CIGRE working group WG 15/33.03.05 its acceptance not only for monitoring of GIS but also for on-site testing will certainly gain further grow [2]. For the monitoring of GIS usually broad-band UHF systems are used since the hardware costs are lower and the frequencies with the most signal intensity changes with the type and position of the defect [3]. Hence, by using a suitable algorithm for a narrow-band system these frequencies can be determined. Furthermore, the algorithm described in this paper allows to distinguish PD signals of defects inside the GIS from those resulting from external noise or switching operations.

As former measurements have shown, the narrow-band UHF method is more sensitive to PD signals than the broad-band method [4]. Therefore, the use of a narrow-band system for on-site testing of GIS seems to be more suitable [4].

Description of the experimental set-up

The measurements reported in this paper were performed on an ABB ELK2 GIS set-up with a rated voltage of 362 kV (fig. 1). The experimental set-up contained two commercial plate sensors with a diameter of 10 cm (Sensors 2, 3) and one cone-type sensor of the same diameter (Sensor 1). The UHF spectra were measured in the frequency range 300 – 1700 MHz using a commercial spectrum analyzer and low noise / high gain UHF preamplifiers. The spectrum analyzer was connected to a PC via the GPIB interface for controlling. For the suppression of corona discharges a high pass filter with a cut-off frequency of 300 MHz was connected in series.

Additionally, conventional PD measurements according to IEC60270 were carried out.

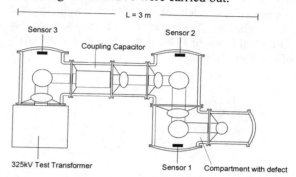

Figure 1: Experimental set-up of the 362 kV GIS with location of sensors and defect

UHF-PD-signals

Initially, some typical UHF signals of an internal defect and from external disturbances measured in the experimental set-up will be presented. The shielded laboratory environment ensured a background noise free measuring of the UHF spectra. However, on-site external disturbance caused by corona discharges, radio and television or from electrical appliances used next to the GIS can make the measurement of UHF PD signals more difficult. Figure 2 shows the UHF spectra of a particle with a length of 12 mm on a spacer as an example of an internal PD source and of an electric drill as external disturbance source coupling into the GIS.

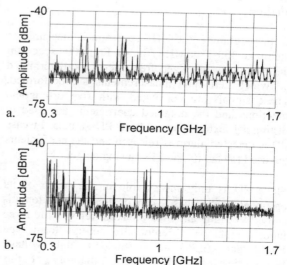

Figure 2: UHF signals measured at sensor 2
a. Particle (l = 12 mm) on spacer, 55 mm from inner conductor, Û = 250 kV, 300-1700 MHz
b. Electric drill coupling into GIS, 300-1700 MHz

High Voltage Engineering Symposium, 22–27 August 1999
Conference Publication No. 467, © IEE, 1999

Due to the commutator sparking of its motor the drill excites a UHF spectrum far up in the GHz range which makes it impossible to distinguish it in the full-span mode from the spectrum of an internal defect. Comparing the zero-span signals of the most energized frequencies clear differences between these time signals become obvious. For each internal defect type the time signal of selected frequencies (fig. 3) shows a periodical curve shape with typical characteristics, such as repetition frequency and phase relation. These distinct differences occur for all typical types of PD causing defects and all types of ambient distubances.

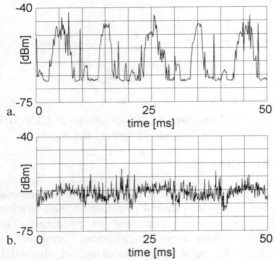

Figure 3: Time signals of selected frequencies measured at sensor 2
a. Particle (l = 12 mm) on spacer, 55 mm distance from inner conductor, Û = 250 kV, zero-span at 497 MHz
b. Electric drill coupling into GIS, zero-span at 532 MHz

PD detection using the narrow-band UHF method

At the beginning, the UHF spectrum is measured in the range 300-1700MHz (fig. 4). From this spectrum the five most energized frequencies are determined on which zero-span measurements are performed. These signals are correlated with five orthogonal functions and the obtained coefficients are used to distinguish external noise from PD signals. In case PD is detected alarm is released and the coefficients are used to classify the PD causing defect.

Since PD is only measurable in the phase of its origin all three phases can be measured automatically by using an UHF coaxial relay with three inputs. Tests on different GIS set-ups from type ELK2 and ELK3 in unshielded high voltage laboratories have proven the reliability of this procedure in a commonly disturbed location. Of course, the sensitivity of this method for a reliable PD detection is depending on the distance between the PD source and sensor and to a minor degree on the type of defect. In several test measurements performed on different test set-ups it was found that the sensitivity compared to IEC 60270 is about 1.5 pC for the most important defect, the free moving particle. For protrusions on the inner or outer conductor as well as for particles on spacers the sensitivity for one spacer between PD source and sensor is about 2.5 pC. The PD sensitivity decreases with the increasing number of intermediate spacers. In case of four intermediate spacers the sensitivity of the system reduces to 2 pC for a free moving particle and about 3.5 pC for all the other defects. These limitations are basically given by the limited sensitivity of the UHF method itself and not by the used algorithm which requires only a signal to noise ratio of the zero-span UHF signal of about 10 dBm.

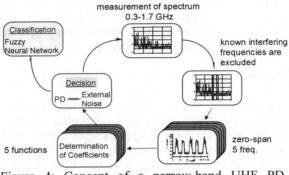

Figure 4: Concept of a narrow-band UHF PD diagnosis tool.

Interpretation of PD signals

For the interpretation and classification of PD signals usually fingerprints and their characteristics are used [3,5]. As this study has shown, a correlation of orthogonal functions and zero-span UHF signals is also appropriate for the interpretation of PD signals. Orthogonal functions have the characteristic that any time function can be synthesized by these functions with any required degree of accuracy [6, 7]. Besides the well-known Fourier transformation, Walsh-Hadamard, Haar, Slant, and Hybrid Walsh functions are used which are usually applied for pattern recognition in the field of digital image processing. Except for the Fourier all orthogonal functions consist of either square or rectangular waves so that they correspond better to the shape of a zero-span PD signal than sinusoidal curves. Another advantage of this set of orthogonal functions is the very fast computerized calculation of the coefficients since only simple multiplication has to be done. Nevertheless the number of coefficients calculated is restraint to 8 for each function which is sufficient for the required accuracy. For the decision whether the measured signal is caused by an internal defect or by external noise the coefficient values of each function are compared separately to predefined values.

To increase the system reliability the use of multiple PD identification tools based on different algorithms is advantageous [6]. The classification is done with fuzzy systems and neural networks which can make use of the coefficients of the orthogonal functions in an appropriate way.

Classification using a fuzzy system

To establish a classification system a reference data base containing a large number of measurements has to be set up. Therefore, more than 500 measurements were performed in the experimental set-up and the calculated coefficients were stored in the data base. Before setting up a fuzzy system the coefficients needed to be analyzed to find out those which are suitable for the PD defect classification. For that purpose the coefficients were normalized on the steady component which is represented by the first coefficient of each function. Then, for each defect type, the arithmetic mean value and the standard deviation of each coefficient were determined separately. Comparing these values graphically (fig. 5) only those coefficients are qualified where the curves do not overlap. From the initial 48 coefficients 16 turned out to be applicable for the classification. As further result it became evident that a reliable separation between protrusions on the inner and outer conductor is not possible in all cases without further effort. This is due to the fact that the zero-span UHF signals of protrusions on the inner and outer conductors become more and more similar for increasing PD levels.

With the remaining 16 coefficients a fuzzy system was set up according to well-known rules [7]. Since no coefficient can distinguish between all types of defects fuzzy sets for each coefficient were established and combined by logical operations for the classification.

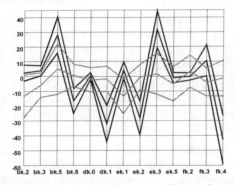

Figure 5: Non-normalized mean value and standard deviation of selected coefficients for moving particle (gray) and protrusion (black).

Classification using a neural network

For this type of classification a feed-forward-network trained with the backpropagation-momentum was used. The quality of a neural network with regard to its capability of generalization and its remaining error is depending strongly on the quality of the training and validation set and above all the chosen network topology. To determine the most suitable network topology a genetic algorithm was applied. Genetic algorithms are based on the evolutionary improvement of the characteristics of individuals by concurrence and heredity. For that purpose artificial organisms, so called individuals, are defined to describe the topology of the neural network that has to be optimized. These individuals are grouped in populations. Natural principles according to the theory of Darwin, such as heredity, crossover, mutation, and selection, are used over several generations to develop and improve the characteristics of the individuals. The selection is made according to the quality of each individual which can be understood as the capability to survive. This evolutionary process will end up into an optimum that represents the most suitable network topology.

At the beginning a so called initial population of neural networks is randomly created. While the number of neurons in the input and output layer is fixed according to the classification problem the number of hidden layers and the number of neurons in these layers is randomly selected. Experiences have shown that 40 different individuals are sufficient for the initial population. These individuals are trained with selected typical patterns and validated with all available patterns. This procedure keeps the training time short and emphasizes on the generalization of the network. After the quality of each individual was determined an intermediate population is created where successful individuals are reproduced more likely (fig. 6). On this intermediate population several different genetic operations can be applied to create a new population. The most important one is the crossover of two 'parent' individuals (fig. 7) to produce new descendents. A crossover can be an exchange of single neurons, groups of neurons, or whole layers. Crossovers are very important and frequent at the beginning of a genetic algorithm so that a wide variety of different individuals can be produced. With the increasing number of generations the probability of crossovers reduces since later in the process the displacement of weaker individuals by the successful one becomes more important than the creation of various new individuals.

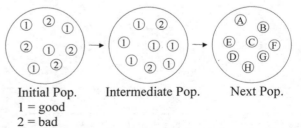

Initial Pop.   Intermediate Pop.   Next Pop.
1 = good
2 = bad

Figure 6: Principle of the evolutionary process to optimize a network topology

Furthermore, mutations can take place which change the network topology randomly by adding or removing neurons. Mutations are used to avoid that the optimization is done around a local minimum. The probability of mutations to occur during the evolution is constant. Another important genetic operation is recombination which stands for the creation of a descendent being an identical copy of a parent. Of course, this operation is used more and more frequently in the later evolution to make the most adapted individual be successful.

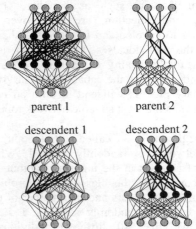

Figure 7: Crossover of groups of neurons to create new descendent individuals

The newly created population is trained and validated again and the process can be repeated. Generally, a number of 40 generations is sufficient to obtain a population which is dominated by the optimum individual representing the optimized topology of the neural network. Usually, the optimum individual occurs already after a few generations and all further generations are mainly used to displace the other individuals.

Figure 8 shows the average remaining error of all network topologies of one population as a function of the number of generations. The curve of the remaining error is approaching asymptotically the value of the remaining error of the optimum network which was 0.3% in this case. The occasional increases of the remaining error are caused by mutations which created worse topologies.

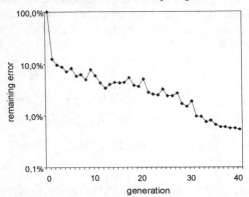

Figure 8: Average remaining error of all individuals as a function of number of generations

After reducing the network topology with the use of pruning-algorithms the characteristic of the neural network can be analyzed. By the determination of the activation of each neuron the classification result can be followed up. It also gives evidence of the relevance of input values. Thus, this information can be used for an optimization of the fuzzy system or for the definition of a neuro-fuzzy system. This will have to be part of future investigations.

Conclusions

In this paper the PD recognition and classification in GIS with the narrow-band UHF method was presented. The narrow-band UHF method is more sensitive to the detection of PD signals in GIS than the broad-band method. It can be used for on-site PD testing of GIS as well as for on-line PD monitoring. It has been demonstrated that PD signals from internal defects measured in the zero-span mode can be distinguished from external disturbances and UHF signals caused by switching operations. Using this method a classification of the defect type is possible with the restriction that protrusions on the inner conductor and the enclosure cannot be distinguished from each other without further efforts. Classification has been done using a fuzzy system and a neural network. The optimum topology of the neural network can be determined using a genetic algorithm.

References

[1] R.Kurrer, K. Feser: The Application of Ultra-High-Frequency Partial Discharge Measurements to Gas Insulated Substations, IEEE Trans. on Power Delivery, Vol.13, No. 3, July 1998

[2] G. J. Behrmann, S. Neuhold, R. Pietsch: Results of UHF measurements in a 220kV GIS Substation during on-site commissioning tests, proceedings of the 10th ISH Montreal, 1997

[3] M. D. Judd, O. Farish, B. F. Hampton: The Excitation of UHF Signals by PD in GIS, Transactions on Dielectrics and Electrical Insulation, Vol. 3, No. 2, April 1996

[4] C. Maulat, E. Fernandez, P. Almosino: Sensitivity of electrical Partial Discharge measuring methods on GIS with external disturbances, proceedings of the 10th ISH Montreal, 1997

[5] T. Hücker: UHF Partial Discharge Expert System Diagnosis, proceedings of the 10th ISH Montreal, 1997

[6] K. G. Beauchamp: Applications of Walsh and Related Functions, Microelectronics and Signal Processing, Academic Press, 1984

[7] T. Hücker: Computergestütze Teilentladungsdiagnostik unter praxisrelevanten Randbedingungen, Thesis, University of Wuppertal, 1995, in German

PRACTICAL APPLICATIONS FOR THE MACHINE INTELLIGENT PARTIAL DISCHARGE DISTURBING PULSE SUPPRESSION SYSTEM NEUROTEK II

S. Happe, H.-G. Kranz,

Bergische Universität - GH Wuppertal, Germany

Abstract

Partial Discharge (PD) measurements, especially for on-site applications, have the problem, that PD information is superimposed with electromagnetic disturbances. To solve this problem, a fast machine intelligent signal recognition system for the suppression of disturbing pulses was developed. This system distinguishes between PD and pulse shaped noise by referring to the time-resolved signal shape in real-time using a neural network hardware in a PC host.

In this contribution, practical applications for the disturbance pulse suppression will be presented. Therefore measurements with a laboratory test set-up and artificial PD sources as well as results from an on-site measurement in a Gas Insulated Switchgear (GIS) were performed.

Another application is the disturbance free determination of a PD fault location on-site.

Keywords

On-site PD measurement, VHF PD measurement, disturbing pulse suppression, Neural Network, PD fault localisation

Introduction

Noise pulses with stochastic repetition rates can be particularly troublesome and cause critical misinterpretations of partial discharge activity. All known PD measuring systems are affected by this problem. So the suppression of disturbing pulses is a key problem in Partial Discharge detection.

The wide-band time-resolved (VHF band of 20-100 MHz) signal recognition of partial discharge measurements is becoming more and more important as a method to increase the diagnosis potential of PD measurements. But in this frequency range, the measured signals are extremely influenced by any kind of noise. Compared with phase-resolved diagnosis systems, it is more profitable to investigate the single impulse shape with an adequate bandwidth as an information carrier.

With the bandwidth (about 50 MHz) of the applied measuring system it is possible to distinguish between PD impulses and pulse shaped interferences due to their different time-resolved signal shapes [3].

One possibility to process the huge amount of data of highly resolved impulse sequences, is the construction of a neural signal processing hardware, which is able to classify pulses from different sources in real-time. After the successful pulse recognition, the interference can be rejected on-line. A

system with even these abilities (NeuroTEK II) has been developed by the authors [1, 2].

This paper describes the basic functions and practical applications of the neuroprocessor slide-in-card which can co-operate with any conventional Pentium computer as a host.

Structure of the neural signal recognition system

As described before NeuroTEK consists of a neural network hardware with the neuroprocessor NLX 420 [4]. The components are:

- broadband measuring sensor (field probe or capacitor) with a cut off frequency near 50 MHz
- 200 MSample/s digital measuring system
- neuroprocessor classification hardware
- PC interface [1, 2]

The structure of NeuroTEK is described in Fig. 1.

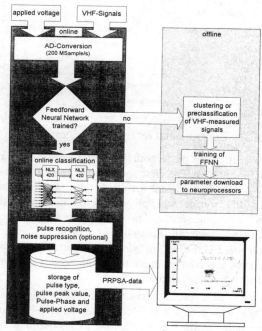

Fig. 1:   Structure of real-time signal recognition

The knowledge-base of the neural network is stored in weights and bias values. To determine them a *Preclassification Phase* has to be performed off-line in the host PC. To achieve this, captured PD pulses from different sources and disturbing pulses are investigated by an unsupervised working clustering algorithm. Afterwards a feed forward neural network (FFNN) is trained in the predetermined clusters [1, 2]. During the off-line training of the FFNN the knowledge-base is generated. The learned results are

High Voltage Engineering Symposium, 22–27 August 1999
Conference Publication No. 467, © IEE, 1999

then downloaded to the neuroprocessor hardware for real-time classification. This hardware is now able to classify on-line up to 200 pulses in one 50 Hz cycle of the applied voltage as a stand alone working system. In the *Classification Phase*, disturbance pulses can be marked or eliminated. Subsequently only the relevant PD information - represented by a phase resolved pulse sequence data set (PRPSA data) - will be transmitted to the host PC. Thereby the data flow is reduced from 200 MByte/s to about 100 kByte/s. This data can later be processed by any conventional or computer aided evaluation system.

Disturbing pulse suppression in laboratory

For testing the performance of NeuroTEK the following test set-up was used (Fig. 2).

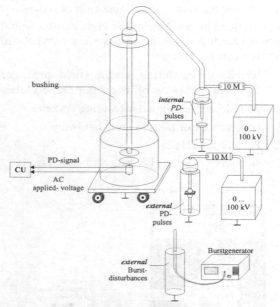

Fig. 2: Experimental set-up in laboratory

Three different pulse sources were used to generate PD like pulses. The first source represented *internal PD pulses* to simulate an UUT (Unit Under Test). This PD source was physically connected to the VHF coupling sensor. The other pulse sources were applied to generate *external burst disturbances* and *external PD pulses* to simulate disturbance from interference sources near the UUT which are coupled by earth or radiation.

A VHF field probe served as the coupling sensor which was installed in a conventional SF6-bushing. The field probe was connected to a coupling unit (CU) from where the signals were transmitted to the measuring system. The maximum bandwidth was about 57 MHz, depending on the coupling sensor, the test set up and the UUT. The minimum system bandwidth which was sufficient for the laboratory tests was about 20 MHz.

Pulses from each source were captured in the discussed frequency range and were clustered into signal classes with an automated algorithm. With at least four signal classes (PD, Burst, External PD and other noise) a neural network was trained. The de-

termined knowledge-base was transmitted to the neuroprocessor hardware. During the on-line tests, the pulses of each source were recognised very well. This was verified for the first three sources by power up and power down of the individual set-up. One PRPD pattern of this investigation demonstrates the efficiency of NeuroTEK (Fig. 3). Only the scope of internal PD activity (marked black) is the relevant information of the measured pulses. The rest are all disturbing pulses. In any other on-site investigation without disturbing pulse suppression they have the chance to cover up the relevant information!

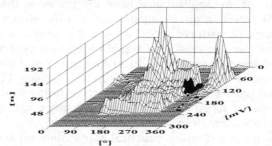

Fig. 3: PRPD pattern from lab measurement
(PD→black, disturbance→white)

Pulse recognition on-site in GIS

To evaluate the performance of NeuroTEK on-site, a measurement in a substation with a 245 kV SF6-switchgear (Type ABB ELK 12) from a major German power utility was carried out. Fig. 4 shows the block diagram of the substation.

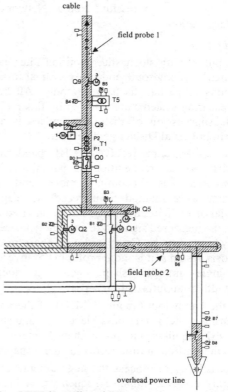

Fig. 4: VHF broadband PD measurement in a GIS

It was possible to measure at two ABB broadband field probes which were integrated into the switchgear. Despite to the upper cut off frequency of

1500 MHz it was possible to measure significant signals in the frequency range of 20-100 MHz.

The first step was to capture pulses from external interference sources. Therefore the switchgear was disconnected from the power supply. The measurement took place on field probe 1. All pulses which have been measured in this state must be disturbances. Subsequently the switchgear was switched to power. Pulses with a different signal shape compared to these previous measured pulses are with a high probability pulses from inside the switchgear (internal PD). There were no such pulses, so the switchgear in service was free of any internal partial discharges. To verify that the pulse recognition system works successfully on-site, a pulse generator was used to inject artificial PD pulses in field probe 2. These pulses were measured at field probe 1 and - of course - they had different signal shapes compared to the noise pulses which were classified before.

The preclassification was executed with these artificial PD pulses. Afterwards the knowledge-base was downloaded to the neuroprocessor hardware. Then, each measurable pulse was classified in the classification phase on-line. One result is the PRPD pattern in figure 5.

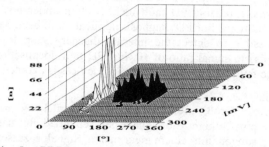

Fig. 5: PRPD pattern from an on-site measurement (PD→black, disturbance→white)

The *real external disturbing pulses* (white) can clearly be distinguished from *artificial internal PD pulses* (black). Now it is possible to suppress the external disturbance in principle and to perform a more reliable PD fault diagnosis on-site. An internal PD activity can be evaluated up to an average noise to signal ratio of 10:1.

PD fault localisation in power cable systems

For the localisation of PD defects in the laboratory, it is a well known procedure to perform pulse spacing measurements [5]. In principle, there are two concepts of measuring circuits (Fig. 6 and 7).

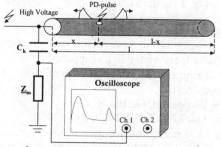

Fig. 6: pulse spacing measurement (circuit A)

A PD pulse in the cable propagates into two directions whereby each pulse moves with a propagation rate v to the ends of the cable. The value of v can be ascertained with a runtime measurement of a calibration pulse and the known length of the cable with (1). On XLPE cables v is about 171 m/µs.

$$v = \frac{2l}{t_c} \qquad (1)$$

Circuit A: If one end is not terminated with the characteristic impedance of the cable, the pulse is reflected. The first pulse, which triggers the oscilloscope, runs directly from the PD source to the beginning of the cable. The reflected pulse reaches the oscilloscope for a time Δt later (4).

1. pulse: $\qquad\qquad t_1 = \dfrac{x}{v} \qquad (2)$

2. pulse: $\qquad\qquad t_2 = \dfrac{2l - x}{v} \qquad (3)$

$$\Delta t = t_2 - t_1 = \frac{2(l - x)}{v} \Rightarrow x = l - \frac{v \cdot \Delta t}{2} \quad (4)$$

with (1) $\qquad \Rightarrow \quad x = l\left(1 - \dfrac{\Delta t}{t_c}\right) \qquad (5)$

With the difference of time Δt, the length of the cable and the calibration time t_c it is possible to evaluate the fault location x with (5).

Circuit B: If pulses are strongly deformed by dispersion it is better to measure at both cable endings.

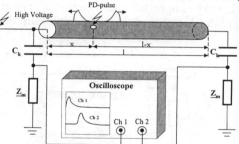

Fig. 7: pulse spacing measurement (circuit B)

Channel 1: $\qquad\qquad t_1 = \dfrac{x}{v} \qquad (6)$

Channel 2: $\qquad\qquad t_2 = \dfrac{l - x}{v} \qquad (7)$

$$\Delta t = t_2 - t_1 = \frac{l - 2x}{v} \Rightarrow x = \frac{l - v \cdot \Delta t}{2} \quad (8)$$

So it is also possible to localise the fault with equation (8). This concept needs more measuring resources than circuit A (two coupling capacitors, two measuring impedances). On-site it could also be a problem to measure at both ends of a cable because of the large distance between the ends of laid cables. This problem can be solved by connecting the ends of the cables of two neighbouring phases from a cable system. So it is possible to measure at the beginning of these two cables. Circuit A as well as circuit B can be used successfully up to a signal to noise ratio of 10:1 which is also investigated at solid type cables with a PD activity in the range of nC. With a signal to noise ratio near 1, these methods cannot work without noise suppression. At XLPE cables with a critical PD level of a few pC, it is still a challenge to localise the PD source, which is usually located at joints or terminations.

Concept for on-site PD fault localisation in XLPE cable systems

The localisation of PD defects in XLPE power cable systems on-site after laying, is a major problem because these measurements have to take place in an environment which is stressed with all known electromagnetic interferences. The biggest problem is the suppression of stochastic pulse shaped disturbances, which cause fault trigger in the discussed methods.

For the joints and terminations of medium voltage cables, the national German standard allows a maximum PD level of 20 pC at $2\,U_0$ (DIN VDE 0278). Disturbing pulses can reach amplitudes and characteristics which falsify the PD information. So a key problem is to develop a procedure for a pulse recognition which can work on-site and which is additionally able to distinguish PD and the reflected PD from pulse shaped noise. The application of NeuroTEK makes it possible to recognise any disturbances in real-time. Thereby it becomes feasible to perform a more precise PD fault localisation.

To achieve this, the described signal recognition system has to be improved. Fig. 8 shows the revised concept of the disturbance resistive fault localisation.

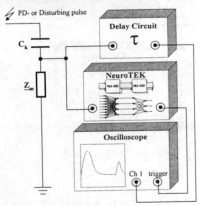

Fig. 8: Concept of fault localisation triggered by a neural pulse recognition system

This concept determines that the real-time signal recognition evaluates each measured pulse. After the recognition of a PD pulse, a trigger signal will be transmitted to the oscilloscope. The signal recognition of NeuroTEK needs about 100 µs for this evaluation. During this time, the continuous data stream has to be latched in a delay circuit. This circuit must have a delay time of at least 100 µs and a sufficient bandwidth (about 50 MHz). Such a device is not commercially available today and was developed during this investigation. The concept of the delay-line is an A/D-conversion, digital latching and

afterwards D/A-conversion. Furthermore the described NeuroTEK features have to be advanced with a trigger output.

Subsequently the oscilloscope is triggered only by clearly recognised PD pulses. With these pulses and the reflected pulses, a standard pulse spacing measurement can be performed. These measurements have the major advantage compared to other PD measurement systems for cables, that only at the beginning of the cable system one VHF measuring sensor has to be installed. After the implementation of these enhancements, a test set-up with a 20 kV XLPE cable system is currently under investigation to perform additional measurements under on-site conditions. One goal is to improve the potential to detect PD pulses at joints and terminations.

Conclusion

NeuroTEK II is a real-time signal recognition system which is able to suppress stochastic pulse shaped disturbances. PD and disturbing pulse are distinguished by evaluating time-resolved VHF measured signal shapes with a fast machine intelligent hardware. This has been confirmed with laboratory and on-site measurements. With this disturbing pulse suppression, it is now possible to perform a more reliable PD fault identification under noisy conditions. For PD diagnosis application, it becomes possible to detect the internal discharges up to a noise to signal ratio of 10. For PD fault localisation the fault trigger of pulse spacing devices is avoided.

References

[1] A. Groß, „Realzeitfähige Unterdrückung impulsförmiger Störsignale bei Teilentladungsmessungen mit einem neuronalen Signalprozessorsystem", Doctor Thesis, Bergische Universität Gesamthochschule Wuppertal, Germany, 1996

[2] H.-G. Kranz, S. Happe: „Real-Time Partial Discharge Disturbing Pulse Suppression with a Neural Network Hardware", 10th ISH '97, paper 3066, Montreal, Canada, 1997

[3] A. Groß., H.-G. Kranz: „Time Resolved Analysis of VHF Partial Discharge Measurements with Neural Networks", 9th ISH '95, paper 5611, Graz, Austria, 1995

[4] American NeuraLogix: „NLX 420 Data Sheet", Sanford, USA, 1992

[5] E. Pescke, R. v. Olshausen: „Kabelanlagen für Hoch- und Höchstspannung", Publicis MCD Verlag, Erlangen und München, 1998, p 271-272

Address of Authors
Prof. Dr.-Ing. H.-G. Kranz, Dipl.-Ing. S. Happe
Bergische Universität GH Wuppertal
Laboratorium für Hochspannungstechnik
Fuhlrottstr. 10
42119 Wuppertal, Germany,
Phone ++49 / 202 439-3027, Fax -3026,
E-Mail: kranz@uni-wuppertal.de
           happe@uni-wuppertal.de

TABLE 5- Identification rates of BP with different convergence criteria (CC)

CC	0.01	0.05	0.1	1	4
Rate (%)	95	95	95	95	88

TABLE 6- Identification rates of LVQ with different processing elements (PL) in the hidden layer

PL	2	4	8	14	20	30	40
Rate (%)	85	83	83	85	85	85	83

When the LVQ network with different training cycles is studied (the Kohonen layer consisting of 4 processing elements), the identification result is shown in Table 7. Again, the identification rates of the LVQ network are not greatly affected by the number of training cycles. The training time for the LVQ network increases when the number of training cycles increase.

TABLE 7 -Identification rates of LVQ with different training cycles (TC)

TC	200	400	1000	2000	3000	4000
Rate (%)	80	83	83	83	83	83

Table 8 shows the identification rates of the self-organizing map network with different training cycles (the training input being the first 2000 points of the spectrum), the best result is obtained when 400 training cycles are used. Both Table 7 and Table 8 indicate that the identification rate does not necessarily increase when the number of training cycles is increased.

TABLE 8- Identification rates of SORG with different training cycles (TC)

TC	200	400	1000	2000	3000	4000
Rate (%)	75	83	80	75	75	75

Finally the identification results of NNs with different training sets, from 10 sets (5 sets of AE signals from internal discharges and 5 sets from surface discharges) to 40 sets with 20 sets for each case, have been investigated (Table 9). Identification rates are consistent. In all cases, 20 training sets are enough for the neural networks to operate satisfactorily. The more training sets, the greater the NN training time.

TABLE 9 –Identification rates of NNs with different number of training sets

	BP (%)	LVQ (%)	SORG (%)
10 training sets	95	70	80
20 training sets	95	83	83
30 training sets	95	84	84
40 training sets	95	85	83

CONCLUSION

Three neural networks, the back-propagation network, the self-organising map network, and the learning vector quantization network have been used to discriminate between internal discharges and surface discharges. The results show that the power spectrum can be used to identify different PD patterns effectively, and the best results obtained when using the whole frequency spectrum as the NN input. While using part of the spectrum, the NN's learning speed was faster, but the identification rate can be lower, though it may be improved by organising NN factors. The identification rate is related to the processing elements in the hidden layer, the number of the learning cycles, and the value of the convergence criterion. The NN training time is dependent on the type of neural network selected, the number of processing elements, the convergence criterion, and the number of learning cycles. All three NNs have achieved satisfactory identification results here, and LVQ and SORG networks can obtain 100% correct identification rates. While BP has been applied to most PD pattern identification, LVQ and SORG networks, which are indicated here, may also be applied effectively.

Further work will focus on the effects of void size, shape and position, material geometry, interfaces, and electric stress on different PD patterns. A knowledge base using a neural network approach will be developed for insulation diagnosis.

ACKNOWLEDGEMENT

The support from National Grid Company plc for this work is gratefully acknowledged.

REFERENCES

1 L. Lundgard, B. Skyberg, 1990, "Acoustic Diagnosis of SF$_6$ Gas Insulated Substations", IEEE Trans. Power Delivery, 5, 1751-1759

2 R. Bozzo, F. Guastavino, 1995, "PD Detection and Localization by means of Acoustic measurements on Hydogenerator Stator Bars", IEEE Trans. Diel. and EI, 2, 660-665

3 M. Leijon, L. Ming, T. Bengtsson, O. Kristofersson, 1993, "PD Detection in Cable Terminations using Acoustic Technique", CIGRE Berlin Symp.

4 H. Suzuki, and T. Endoh, 1992, "Pattern Recognition of Partial Discharge in XLPE Cables Using a Neural Network", IEEE Trans. EI., 27, 543-549

5 R. E. James and B. T. Phung, 1995, "Development of Computer-based Measurement and their Application to PD Pattern Analysis", IEEE Trans. Diel. and EI., 2, 838-853

6. E. Gulski and A. Krivda, 1993, "Neural Network as a Tool for Recognition of Partial Discharges", IEEE Trans. EI., 28, 984-1001

7. T.R. Blackburn, B.T. Phung and R. E. James, 1993, "Neural Network Application in PD Pattern Analysis",

Conference Proceedings of International Conference on Partial Discharge, 82-83

8. H. G. Kranz, 1993, "Diagnosis of Partial Discharge Signals using Neural Network and Minimum Distance Classification", IEEE Trans. EI., 28, 1016-1024

9. H. Borsi, E. Gockenbach, D. Wenzel, 1993, "Separation of Partial Discharges from Pulse Shaped Noise Signals with the Help of Neural Networks", Conference Proceedings of International Conference on Partial Discharge, 47-48

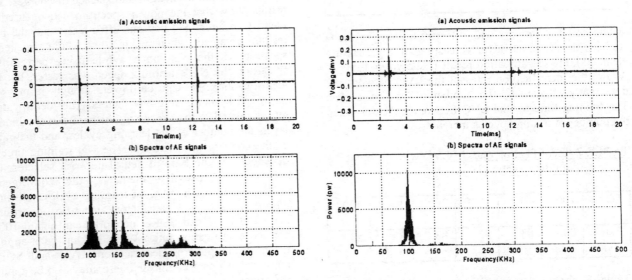

Figure 3 Acoustic emission signals and spectra from Internal discharge

Figure 4. Acoustic emission signals and spectra from surface discharge

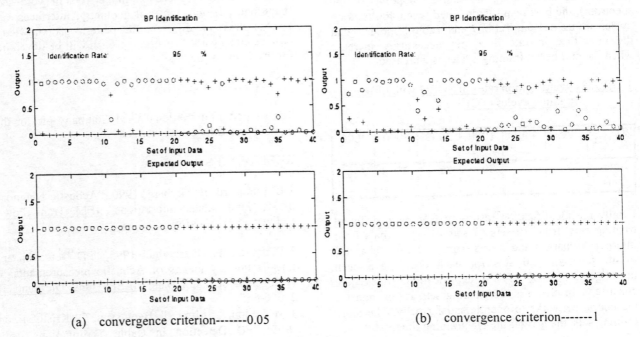

(a) convergence criterion-------0.05

(b) convergence criterion-------1

Figure 5 BP outputs with different convergence criteria

ON THE SIGNIFICANCE OF PD MEASUREMENTS IN LIQUIDS

H. DEBRUYNE, O. LESAINT

Laboratoire d'Electrostatique et de Matériaux Diélectriques (LEMD)
Centre National de la Recherche Scientifique (CNRS)
25, Avenue des Martyrs BP166, 38042 Grenoble Cedex 9 France

Abstract: **This paper presents experimental data and calculations concerning the measurement of PD's in mineral oil. The influence of the RLC measuring impedance is studied. The investigations were done using either signals of simple shapes or real PD signals. PD's were produced with a point-plane gap in air and in mineral oil. The discharge currents in air are correctly integrated by the usual RLC circuits, whereas in mineral oil the charge is strongly underestimated.**

1. Introduction

The measurement of partial discharges (PD) is a diagnostic tool used in many applications, including oil-filled apparatus such as power transformers. In such devices, different discharge types may occur (within gaseous cavities, in the liquid, etc.), for instance during tests with overvoltages. It has been previously observed that discharges in the liquid result from the initiation and propagation of streamers, either in the liquid bulk, of at a solid/liquid interface (creeping discharges). Their propagation velocity is much lower than in gases: at the inception voltage in mineral transformer oil, the velocity is usually between 0.1 and 2 km/s for streamers in the liquid bulk [1] or at a liquid/ solid interface [2]. Thus, the streamer may last up to several tens of µs, and PD measurement systems with long time constants must be used to integrate properly the associated current [3]. The total charge of the streamer is an interesting parameter since it is correlated to its physical size [1,2], and hence to its harmfulness. The purpose of this paper is to investigate the influence of measurement circuits parameters on PD recordings.

2. Behaviour of RLC circuits with signals of simple shapes.

To simulate PD's of long duration, the circuit of figure 1 was used. Charge was injected in the RLC impedance via a small capacitor C_1, with 100 V voltage steps of variable risetime t_r (measured between 10 and 90%). The voltage $u_{RLC}(t)$ was recorded with a high impedance oscilloscope probe. Such a study was previously undertaken in [4]. The parameters of RLC circuits used are given in table 1. Their main features are the resonance frequency f_0 and damping coefficient ε:

$$f_0 = \text{resonance frequency} = \frac{1}{2\pi\sqrt{LC}}$$

$$\varepsilon = \text{damping coefficient} = \frac{1}{2R}\sqrt{\frac{L}{C}}$$

Circuits 1 to 5 get a constant ε with various resonance frequency, whereas circuits 6 to 9 correspond to a fixed f_0 with variable ε.

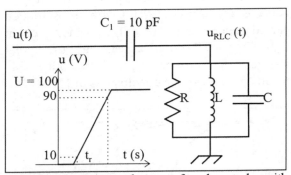

Figure 1: experimental setup for the study with signals of simple shapes

RLC circuit	R (kΩ)	L (mH)	C (nF)	f_0 (kHz)	ε
1	0.68	0.056	0.10	2 130	0.6
2	1.00	0.470	0.22	500	0.7
3	2.70	10	0.82	60	0.6
4	2.20	22	2.20	20	0.7
5	4.70	100	2.20	10	0.7
6	0.2	1.5	2.20	90	2.0
7	0.4	1.5	2.20	90	1.0
8	0.53	1.5	2.20	90	0.8
9	1.42	1.5	2.20	90	0.3

Table 1: RLC circuits characteristics

High Voltage Engineering Symposium, 22–27 August 1999
Conference Publication No. 467, © IEE, 1999

Figure 2 shows typical measurements with the circuit 3. The crest value U_{RLC} of $u_{RLC}(t)$ is maximum when t_r is small as compared to the oscillation period ($1/f_0$) of the circuit.

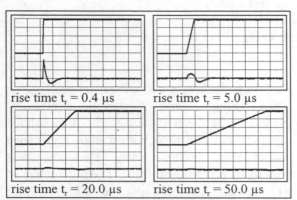

rise time $t_r = 0.4 \mu s$ rise time $t_r = 5.0 \mu s$

rise time $t_r = 20.0 \mu s$ rise time $t_r = 50.0 \mu s$

Figure 2: response of the RLC circuit 3 (10 μs/div.)
Upper trace: voltage u(t) (25 V/div.)
Lower trace: voltage $u_{RLC}(t)$ (500 mV/div.)

Figure 3 shows the ratio between the measured U_{RLC} to the maximum possible value $[C_1/(C+C_1)] \times U$, versus risetime t_r, with circuits 1 to 5 (constant ε, variable f_0). On this figure, the lines represent the response calculated analytically with the Laplace transform. To integrate properly a discharge of duration t_r, it is necessary to use a circuit with an oscillation period much longer than t_r. For instance in figure 3 ($\varepsilon \approx 0.7$), to get at least 90% of the real charge, $1/f_0$ must be larger than $\approx 50t_r$. This result agrees with the conclusions of [4].

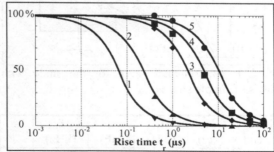

Figure 3: response of RLC circuits versus pulse rise time t_r for different resonance frequencies f_0
lines: calculated response ; points: measurements

When the damping coefficient ε is changed with a constant f_0, figure 4 shows that the response of the circuit also varies. Within the investigated range, the response of the circuits changes by a factor of nearly 2. However, these values correspond to extreme cases: with $\varepsilon = 0.3$ the circuit oscillates for some periods, and with $\varepsilon = 2$ the circuit is strongly damped

with no oscillation. Thus in practice, the relative influence of ε will be small as compared to that of the frequency f_0.

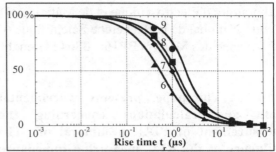

Figure 4: response of RLC circuits versus pulse rise time for different damping coefficients

3. Experiments with real PD signals.

As shown in figure 5, PD's were produced with a point-plane electrode geometry (20 mm gap), in air and in mineral transformer oil. The point tip radius of curvature r_0 was 0.8 mm in air and 10 to 40 μm in oil. With oil, the plane electrode was covered with a pressboard sheet to prevent breakdown from occurring. Typical applied voltages were 15 kV in air and 40 kV in oil.

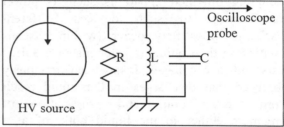

Figure 5: experimental setup for the study with real PD signals

With these conditions, stable positive and negative corona discharges were recorded in air. In oil, discharges of both polarities were also detected, but they were much less numerous (about 1 per minute), and much more scattered in amplitude and shape.

Typical shapes of transient currents corresponding to positive streamers (i.e. initiated when the point is positive) are depicted in figure 6 (air) and 8 (oil). Since the point is connected to the ground, the polarity of recordings (figures 6-9, 11-13) was inverted in such a way that a positive current corresponds to a positive streamer. These currents were recorded with a 300 MHz oscilloscope by replacing the RLC circuit by an impedance-matched 50 Ω coaxial cable directly connected to the point. As previously reported

[1,2], the shape of currents in oil is much more erratic than in air. Positive streamer currents are composed of a continuous background (some mA) with fast pulses superposed on it (figure 8).

The total discharge amplitude was measured with a RC circuit with a long time constant compared to the discharge duration (figures 7 and 9). The charge accumulates in the measuring capacitor while the discharge current flows, for about 500 ns in air, and 10 µs in oil. The average value Q_{av} of this charge was derived from 50 samples.

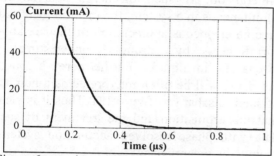

Figure 6: transient current of a positive discharge in air

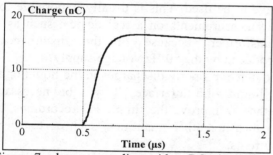

Figure 7: charge recording with a RC circuit for a positive discharge in air (R = 2.3 kΩ ; C = 7 nF)

These discharges were also measured with RLC circuits with different frequency f_0 and constant ε = 0.7. The calibration in pC was carried out using a commercial calibrator, i.e. following the principle of figure 1 with a fast risetime (t_r = 25 ns). Figure 10 shows the results of measurements versus f_0, expressed as the ratio of the measured charge to the real charge Q_{av}. The values reported are the mean of about 100 measurements.

Within the investigated range (f_0 = 20 kHz - 400 kHz), the measured charge varies from 40 to 90 % of the real charge in air, and from 5 to 40 % in oil. If we apply $1/f_0 \approx 50t_r$ to obtain 90% of Q_{av}, one should use a circuit with f_0 = 40 kHz for a 500 ns discharge in air, which

is consistent with figure 10. In oil f_0 = 3 kHz should be used.

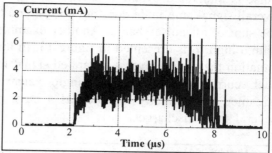

Figure 8: transient current of a positive streamer in oil

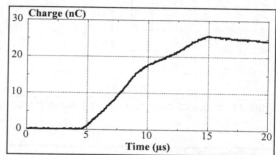

Figure 9: charge recording with a RC circuit for a positive streamer in oil (R = 10 kΩ ; C = 10.2 nF)

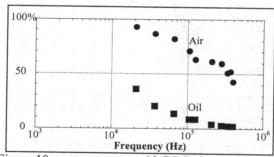

Figure 10: measurements with RLC circuit versus f_0

In these experiments, negative discharges were also detected. In air, they are much more numerous than the positive, and their charge is about 100 times smaller. However, the duration is comparable (250 ns instead of 500 ns for positive polarity), and thus their integration with different circuits is similar to that presented in figure 10 for positive discharges. In oil, the situation is quite different. As shown by figure 11, the transient current of a negative streamer is composed by a burst of fast pulses of growing intensity. Each pulse is very fast and is not accurately recorded even with the available 300 MHz bandwidth at 2Gsamples/s (the measured width was about 1 ns). The total streamer charge is the sum of all pulse charge. With the circuits used, this charge is not correctly measured. A typical response with a low f_0 is given in figure 12. Each very fast pulse is integrated by the capacitor C, but the overall signal amplitude does not correspond to the sum of all pulse charge. Another example, with a high f_0 is presented in figure 13. In this situation the circuit gives a separated signal for each current pulse. Then a single negative streamer is incorrectly recorded as a burst of several small discharges.

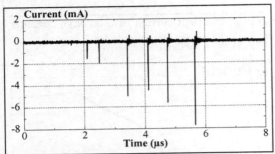

Figure 11: transient current of a negative streamer in oil

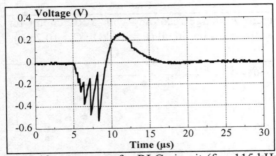

Figure 12: response of a RLC circuit ($f_0 = 115$ kHz, $\varepsilon = 0.7$) for a negative streamer in oil

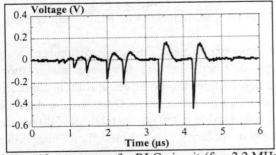

Figure 13: response of a RLC circuit ($f_0 = 2.2$ MHz, $\varepsilon = 0.7$) for a negative streamer in oil

This typical current shape (i.e. a burst of fast pulses) is frequently recorded in liquids. For instance in mineral oil at high voltage, both positive and negative streamers produce such currents [5].

4. Conclusions

In most situations, PD measurements do not give the real charge quantity transferred in a discharge, but only a fraction of this charge (the apparent charge). Thus the data obtained give a relative indication of the discharge magnitude. In addition, this study shows that with usual measurement parameters (RLC parallel circuits with 10^5 Hz$<f_0<10^6$ Hz), discharges are correctly measured in gases, but only a small fraction of the charge is recorded in liquids. In figure 10, with $f_0 = 400$ kHz, 50% of the charge will be measured in air, and only 5% in mineral oil. If such discharge sources are simultaneous in a system, the measurement will give an erroneous appreciation of their relative harmfulness by strongly underestimating discharges in liquids. Furthermore, a single discharge will be often interpreted as a number of small discharges (figure 13). Thus it is clear that the acquisition and the treatment of such data with phase resolved acquisition systems seem to be meaningless.

To integrate properly streamer currents in liquids, RLC circuits with f_0 lower than 50 kHz have to be used. This is usually impossible for PD measurements on a real device, such as a transformer, because of the disturbances produced by the 50 Hz voltage supply.

To detect properly discharges in the liquid or at a liquid/solid interface, it will be necessary either to improve the simple PD recordings, or to measure other quantities, such as transient currents.

Acknowledgements
The authors wish to express their sincere thanks to Alstom T&D and EDF DER that both supported this project.

References
[1] O. Lesaint, G. Massala, IEEE Trans. on Diel. and Elec. Insul., Vol. 5, No. 3, pp. 360-370, June 1998.
[2] P. Atten, A. Saker, IEEE Trans. on Elec. Insul., Vol. 28, No. 2, pp. 230-242, 1993.
[3] L. Lundgaard, O. Lesaint, Conf. on Elec. Insul. and Diel. Phen., pp. 596-599, 1995.
[4] R. Bartnikas, IEEE Trans. on Elec. Insul., Vol. 7, No. 1, pp. 3-8, March 1972.
[5] L. Lundgaard et al, IEEE Trans. on Diel. and Elec. Insul., Vol. 5, No. 3, pp. 388-395, June 1998.

COMPARISON OF CONVENTIONAL AND VHF PARTIAL DISCHARGE DETECTION METHODS FOR POWER TRANSFORMERS

Jan Peter van Bolhuis , Edward Gulski, Johan J. Smit, Thomas Grun*, Mark Turner*

Delft University of Technology, High Voltage Laboratory, The Netherlands
*Haefely Trench AG, Tettex Instruments Division, Switzerland

ABSTRACT

There is much experience with conventional PD measurements as a tool for transformer diagnostics. However it is not possible to use conventional PD measurements in the presence of large external disturbances. In due time several techniques for noise suppression have been devised. One such technique is VHF PD measurement. This paper compares detection sensitivity of conventional and VHF PD measurements.

1. MEASUREMENT SETUP

1.1   Test object

The transformer under test is a 50 kV/10 kV 14 MVA ONAF power transformer (Figure 1), which was withdrawn form service after 46 years. The transformer is in good condition with very low PD levels.

The transformer was energised using a 3 phase circuit. Large external disturbances coming from the mains

Figure 1: 50 kV/10 kV 14 MVA power transformer. The transformer was in good condition with low PD levels and was energized using a 3 phase circuit.

High Voltage Engineering Symposium, 22–27 August 1999
Conference Publication No. 467, © IEE, 1999

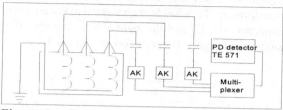

Figure 2: The measurement circuit for IEC270 measurement, showing the transformer, the coupling capacitors, the coupling devices (AKV572) and the measurement system, consisting of a multiplexer (MPR571) and a PD detector (TE571).

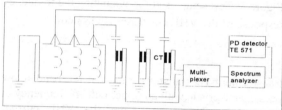

Figure 3: The measurement circuit for VHF measurements, showing the transformer, the coupling capacitor, the current transformer (CT), and the measurement system, consisting of a spectrum analyzer (HP 8590L) and a TE571 PD detector.

power were coupled onto the transformer, dominating the phase resolved PD pattern at voltages over 12 kV.

1.2   The conventional PD measurement setup

For conventional or IEC 270 measurements (Figure 2) a 1 nF/100 kV discharge free coupling capacitor is connected to each HV bushing of the transformer. The capacitors are connected to coupling devices (*Haefely Trench*, AKV 572). In order to measure the PD signal from all   three phases, a multiplexer (*Haefely Trench*, type TE 571-MPR) was used between the PD detector (*Haefely Trench*, Type TE 571) and the coupling devices. The TE-detector has a bandwidth of 40-400 kHz. The signals were recorded as well as post-processed by this equipment [1].

1.3   The VHF PD measurement setup

In the VHF measurement setup (Figure 3), 1 nF, 100 kV, discharge free coupling capacitors are connected to the HV bushings of the power transformer. To detect PD signals a VHF current transformer (*Seitz Instruments*, Figure 4) is connected to the LV part of each of the coupling capacitors [2]. The VHF current

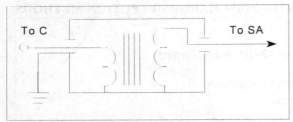

Figure 4: Electrical layout of the VHF current transformer. The left side is connected to the coupling capacitor (To C) and the right side to the spectrum analyzer (To SA). The transformer ratio is 1:1.5.

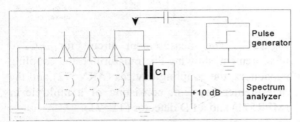

Figure 5: Test setup used for evaluating the frequency response of the VHF measurement system.

transformer has a transformer ratio of 1:1.5 and a low pass characteristic with an upper cut-off frequency at 25 MHz. The signal from the VHF current transformer was amplified using a +10 dB wide-band amplifier. To detect PD signals a spectrum analyser (*HP* 8590L or a *Tektronix* 2711) was tuned to a suitable frequency in order to achieve an optimal signal/noise ratio. In addition for phase resolved PD measurements the signal output of the spectrum analyser was connected to a PD detector (TE 571) [3, 4]. The same TE 571 PD detector was also used for post- processing. A multiplexer (TE 571-MPR) was used to switch between phases.

2. FREQUENCY RESPONSE AND SENSITIVITY OF THE VHF MEASUREMENT CIRCUIT

To evaluate the influence of the measurement circuit on the transfer of VHF signals, the frequency characteristic of the measurement setup was determined using a pulse calibrator consisting of a pulse generator (*picosecond Pulse labs*, model 2000D) and a 100 pF capacitor. Pulses of 100 pC were injected into the test circuit (Figure 5).

Figure 6 shows the differential spectrum of the signal caused by the calibration pulses. For a given configuration, a differential spectrum is computed by subtracting the background noise spectrum from the measured spectrum that contains the signal.

The pulses were then injected on the HV bushing of the transformer. The response was measured by connecting a VHF current transformer to the coupling capacitor of that particular bushing. The signals from

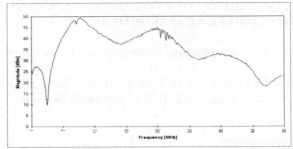

Figure 6: Differential spectrum for the calibration pulse obtained by subtracting the background noise spectrum.

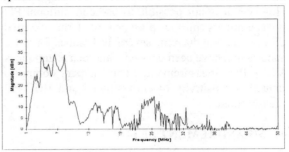

Figure 7: Differential spectrum for pulses injected into the transformer.

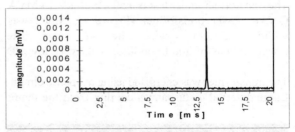

Figure 8: Phase resolved PD pattern at 4.3 MHz, 300 kHz resolution bandwidth, of 10 pC injection on the bushing of phase V, measured on phase V.

the VHF current transformer were then amplified by 10 dB and measured using a spectrum analyser (*HP*, 8590L). The differential spectrum is shown in Figure 7. From this spectrum it can be concluded that for this setup signals up to several MHz can be detected.

In order to test the sensitivity of the circuit 10 pC pulses were injected on the HV bushing of phase V. Figure 8 shows the phase resolved PD pattern of the injected pulses.

3. EXPERIMENTAL RESULTS

As mentioned above the transformer under test was in good condition and had a PD level of approximately 20 pC. To obtain higher PD levels for comparison purposes, a higher electrical stress of the main insulation was required. This was accomplished by allowing the neutral to float and grounding phase U. In this way a PD level of approximately 40 pC was achieved.

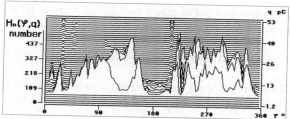

Figure 9: Phase resolved PD pattern of an IEC 270 measurement of phase V at 20 kV, without large external disturbances.

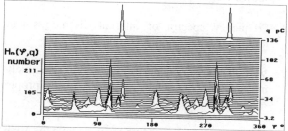

Figure 10: Phase resolved PD pattern of an IEC 270 measurement of phase V at 20 kV with large external disturbances of >136 pC present.

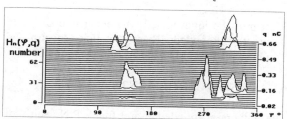

Figure 11: Phase resolved PD pattern of an IEC 270 measurement of phase V at 30 kV, with external disturbances of 660 pC magnitude. The internal PD is no longer visible.

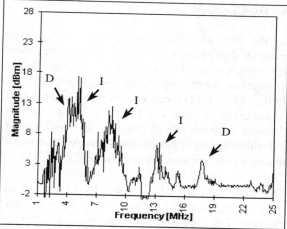

Figure 12: Differential spectrum with phase V at 16 kV. External disturbances (D) and internal PD (I) are visible.

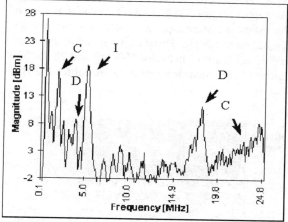

Figure 13: Differential spectrum with phase V at 16 kV and both corona on bushing V. Visible are internal PD (I) external disturbances (D) and corona (C).

3.1 IEC 270 measurements

Measurements were made with and without external disturbances present. Figure 9 shows an measurement of internal discharges without external disturbances. A PD level of approximately 40 pC is visible.

In the measurement made with external disturbances present at 20 kV (Figure 10) internal PD is much less visible. At 30 kV(Figure 11) the internal PD of 40 pC is no longer visible due to the 660 pC magnitude of the external disturbances.

3.2 VHF measurements

To investigate whether different PD sources have different PD related spectra, differential spectra were made of both internal discharges (Figure 12), and internal discharges plus corona (on purpose) (Figure 13). By comparing these spectra it is clear that the spectral content of a signal differs per discharge source.

On the basis of the differential spectrum of internal discharges (Figure 12) it was decided to investigate the discharge patterns as measured at 3, 5, 8, 13 and 18 MHz. The peaks in the differential spectrum at 3 and 18 MHz were found, by looking at the phase resolved pattern, to originate from external disturbances (D). Internal discharges (I) were visible at 5 8 and 13 MHz. Figures 14, 15 and 16 show the phase resolved PD patterns of PD measurements at 5, 8 and 13 MHz, respectively, all being made with a resolution bandwidth of 300 kHz, a linear scale with 683 µvolts per division.

The differential spectrum for internal discharges plus corona (Figure 13) shows a clear peak of internal discharges at 5 MHz. The peaks for internal discharges (I) at 8 and 13 MHz have practically disappeared.

The external disturbance (D) visible in Figure 12 at 3 MHz is also diminished in Figure 13. The external disturbance at 18 MHz are more pronounced in Figure 13 as compared to Figure 12. Corona (C) can be found in Figure 13 below 2 MHz and over 20 MHz.

4. DISCUSSION

In the IEC 270 PD measurement without large external disturbances (Figure 9) most pulses were recorded between 0-90 and 180-270 degrees. This is as expected for PD originating from internal PD.

In the IEC 270 PD measurement with large external disturbances (Figure 11) all pulses were recorded between 0-90 and 180-270 degrees. The presences of 4 distinct clusters of pulses shows however that these pulses do not originate from internal PD.

The VHF measurements (Figures 14, 15 and 16) all show a large band of pulses of varying magnitude between 0-90 and 180-270 degrees. They also all show some disturbances of small magnitude at 100 and 230 degrees.

The measurement at 5 MHz shows something more. As is visible in the differential spectrum for internal discharges (Figure 12) there is a frequency overlap of the external disturbances at 3 MHz and the internal discharges at 5 MHz. These external disturbances are visible in Figure 14 between 100-180 and 280 to 360 degrees. This measurement also resembles the IEC 270 measurement of phase V at 20 kV without large external disturbances (Figure 9).

5. CONCLUSIONS AND SUGGESTIONS

Based on this study the following conclusions can be made regarding conventional and VHF PD detection :

1. In the presence of large external disturbances IEC 270 PD measurements are not possible, as also shown earlier by others.

2. Frequency spectra of internal discharges and internal discharges plus corona differ.

3. VHF PD measurements can be used to selectively measure the PD signal by tuning to the proper frequency.

4. Field measurements should be made with the VHF PD measurement technique to gain further practical insight into the possibilities and limitations of this measurement method.

5. It must be investigated whether it is possible to create a knowledge base that links spectra and phase resolved PD patterns to typical defects.

6. ACKNOWLEDGEMENTS

The authors acknowledge and thank P.N. Seitz, of Seitz Instruments AG in Niederrohrdorf, Switzerland for the experimental support he gave.

7. REFERENCES

[1] E. Gulski, P.N. Seitz,, 1993, *Computer-Aided Registration and Analysis of Partial Discharges in High Voltage Equipments*, ISH 1993 Yokohama, 60.04

[2] M. Lauersdorf, *Ein neues Kompensationsverfahren für TE-Messungen an Transformatoren*, Hochspannungsprüftechnik und Monitoring, Haefely, Symposium 1998.

[3] J.Fuhr, 1997, *Analyse van TE-Messungen an Transformatoren im Labor und vor Ort,* Highvolt Kolloquium, beitrag 4.9

[4] J. Fuhr, U. Sunermann, R. Bähr, M. Hässig, *Vor-Ort-TE-Messungen an Grosstranformatoren*, ETG-Fachbericht 56, 253-258

[5] S. Meijer, E. Gulski, J.J. Smit, F.J. Wester, T. Grun, M. Turner, 1999, *Interpretation of Partial Discharges using Spectral Analysis*, In the ISH'99 proceedings.

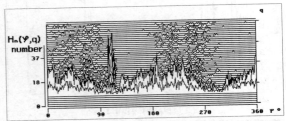

Figure 14: Phase resolved PD pattern of a VHF PD measurement of phase V (16 kV, 5 MHz). The disturbances visible at 100-180 and 280-360 degrees, come from an overlap in frequency of the disturbance peak at 3 MHz and the Internal PD peak at 5 MHz visible in Figure 11.

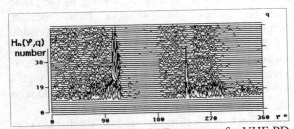

Figure 15: Phase resolved PD pattern of a VHF PD measurement of phase V (16 kV, 8 MHz).

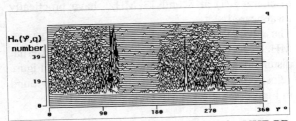

Figure 16: Phase resolved PD pattern of a VHF PD measurement of phase V (16 kV, 13 MHz).

PARTIAL DISCHARGES IN HIGH VOLTAGE DIRECT CURRENT MASS-IMPREGNATED CABLES

M J P Jeroense[†], M Bergkvist[†], P Nordberg[‡]

† ABB Corporate Research, Sweden
‡ ABB High Voltage Cables

INTRODUCTION

High Voltage Direct Current (HVDC) cables play an important role in the energy supply network. If an energy interconnection has to be established between two countries separated by a broad water, overhead lines cannot be used for obvious reasons. Two principle types of cables could then be used: an High Voltage Alternating Current (HVAC) cable or an HVDC cable. Transferring energy by a 50 Hz AC current and voltage has the disadvantage that the cable is loaded with its own capacitive current and so diminishes the capability of transferring net power. This problem can be avoided by using HVDC techniques. An HVDC cable loaded with DC current and voltage only asks for a resistive leakage current which is in the order of several μA's per kilometer cable. This number is very small compared to the typical number of several tens of Amperes per kilometer AC cable. Today's break-even point to decide between an HVAC and an HVDC cable lies roughly at 30-50 km.

Concerning HVDC cables the mass-impregnated non-draining (MIND) type has been used mostly, so far. Some experience exists with oil pressurized (1), gas pressurized cables (2) and even polymer cables through the HVDC-Light concept (3). However, the MIND cable has served as world-record keeper for the highest voltage, the highest power transfer capability and the longest length. All these record cables have been designed and manufactured successfully by ABB (4).

Understanding the physics of the MIND cable has always been the basis for success and further development. It is important to have knowledge of the dynamics of temperature, pressure and electrical field in the electrical insulation of the cable. It is known that partial discharges in the HVDC MIND cable are linked to all of these parameters. It is worthwhile, therefor, to measure partial discharges (besides measurement of temperature, pressure and fluid dynamics related parameters), as it may lead to a better understanding of the cable.

PARTAL DISCHARGES AT DC VOLTAGE

The detection principle of partial discharges under DC voltage is principally the same as under AC voltage. The interpretation of the discharge data is, however, different.

Partial discharge registration and interpretation

In the AC case two basic quantities are measured: the discharge magnitude q and the phase on the 50 Hz sine at which the discharge occurred. At DC voltage the magnitude q together with the time of occurrence t is registered (5). Although the discharge magnitude q together with the time *between* two consecutive discharges Δt is more fundamental and closer to the discharge physics (see Figure 1).

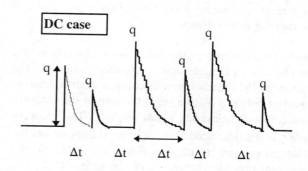

Figure 1: The two measurable basic quantities for partial discharges at DC voltage. The discharge magnitude q and the time between discharges Δt.

From these two basic quantities different related parameters and distributions may be derived, as there are: the discharge repetition rate n and the discharge repetition rate as a function of discharge magnitude n-q. More distribution types are described by Fromm (6). All these distributions can be shown as a function of measurement time also. In this article we will concentrate on the repetition rate as a function of time: $n(t)$.

Repetition rate of discharges at DC voltage

Much of the discharge physics under DC voltage can be understood with the aid of the so-called modified a,b,c-scheme (5,6,7) as shown in Figure 2.

High Voltage Engineering Symposium, 22–27 August 1999
Conference Publication No. 467, © IEE, 1999

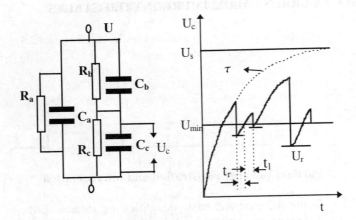

Figure 2: Modified a,b,c-scheme. **a** *stands for the healthy part of the insulation,* **b** *for the insulation in series with defect* **c**. U_c *is the voltage across the defect,* U_{min} *the minimum breakdown voltage of the defect,* U_s *is the voltage across the defect if no discharge would occur,* U_r *the residual voltage,* τ *is the time constant of the circuit,* t_r *is the recovery time of the defect and* t_l *is the time lag of the defect.*

The resistors represent the polarization currents in the sample. To calculate the repetition rate of the discharges the time between discharges must be known. The time between discharges is the sum of the recovery time t_r and the time lag t_l: $\Delta t = t_r + t_l$. The recovery time t_r is the time needed to raise the voltage across the defect from the residual voltage U_r back to the minimum breakdown voltage of the defect U_{min}. The time lag t_l describes the waiting time for a start electron. The inverse of this sum is the repetition rate n of the defect. If we set the time lag t_l to zero, then we arrive at a maximum approximation of the repetition rate and is given by (5,6)

$$n = \frac{1}{t_r} = -\frac{1}{\tau} \ln\left(\frac{U_s - U_{min}}{U_s - U_r} \right) \approx \frac{1}{\tau} \frac{U_s}{U_{min} - U_r}.$$

An explanation of the symbols can be found in the caption of Figure 2.

At first, it can be seen that the repetition rate n is in good approximation linearly proportional to the asymptotic voltage U_s across the defect, which in turn is proportional to the external voltage. The higher the external voltage is, the higher is the repetition rate.

At second, the higher the time constant τ of the insulation is, the lower is the repetition rate.

At third, the repetition rate also increases with decreasing (minimum) breakdown voltage U_{min} of the defect.

And at fourth, the closer the residual voltage U_r is to the (minimum) breakdown voltage U_{min} of the defect, the higher is the repetition rate.

This last conclusion is important. It was often thought that the residual voltage U_r was close to zero. Fromm (6), Ficker and Sikula (8) showed however, that the residual voltage is very close to the original breakdown voltage for Townsend discharges, which is the most common type of void discharges that occur under DC voltage. This is one of the reasons that the repetition rate n under DC voltage can still be quite high, despite the high resistance of the insulation and therefor the high value of the time constant τ of the insulation.

Jeroense (5) extended the calculation of the repetition rate n, which is valid for one defect, to the case of a cable in which several defects may exist. Then the total repetition rate of partial discharges that occur in several defects in the cable is given by

$$n(t) \equiv \sum_r \frac{1}{\tau_r(t)} E_r(t) D_r(t),$$

At a certain radius $r=r_l$, a certain number of defects may exist per meter cable, described by the function D. The higher the number of defects is, the higher is the total number of discharges. A field strength E exists at that radial location. The higher this field strength is, the higher is the repetition rate n for the defects at location $r=r_l$. This is in compliance with what is said earlier in the one-defect model. And finally, the higher the time constant τ of the insulation at that radial location, the lower is the repetition rate for the defects at location $r=r_l$.

These quantities, time constant τ, electric field E and the concentration of defects D depend on the radial location r in the cable insulation and are therefor described as distribution functions, τ_r, E_r and D_r.

The repetition rate n of discharges of the cable as measured, is the sum of all the independent defects in the cable. Therefor we may sum all the repetition rates across the whole insulation.

The three distribution functions are strongly coupled to parameters such as temperature T, pressure p and other fluid dynamics related quantities. As these quantities change with time, depending on the service conditions of the cable, the distribution functions depend on time as well. For that reason they are written as $\tau_r(t)$, $E_r(t)$ and $D_r(t)$ respectively.

TYPE TEST

Type test schemes for HVDC MIND are mostly based on the CIGRE document "Recommendations for Tests of Power Transmission DC Cables for a Rated Voltage up to 600 kV" (9). The most important electrical test ingredients in these recommendations are the cycle tests and the impulse tests. We shall concentrate here on the cycle tests. The CIGRE document recommends the following cycle scheme:

- 10 cycles, with positive voltage at KU_0. One cycle consists of 8 hours heating and 16 hours cooling. It is recommended to use a DC current through the conductor of the cable to heat the cable.
- 10 cycles, with negative voltage at KU_0. One cycle consists of 8 hours heating and 16 hours cooling.

- 10 cycles with polarity reversals every 4 hours at $K_{pol}U_0$. One cycle consists of 8 hours heating and 16 hours cooling. The cessation of the heating current shall coincide with the reversal of the polarity.

The factors K and K_{pol} depend mostly on the contract. Typical values are 1.8 for K_0 and 1.4 for K_{pol}. In this article we will concentrate on the cycles without polarity reversals.

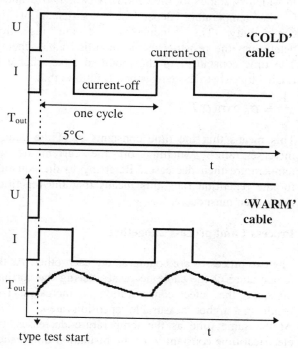

Figure 3: General type test conditions; the first two cycles in the positive cycle test. U stands for the voltage, I is the heating current and T_{out} is the controlled outer temperature of the cable.

The CIGRE document recommends also that the type test conditions match the future service conditions of the cable as close as possible. In general this means that the outer temperature of the cable during type testing is also controlled. A cable that lies at the bottom of the sea will experience a rather low and constant sheath temperature of typically 5°C, while a buried land cable experiences a changing temperature of about 20 to 30°C. The test cables under these conditions are mostly referred to as the 'cold' cable and the 'warm' cable.

The mentioned temperatures are only examples, as they depend much on sea water conditions and soil parameters. An example of voltage, heating current and outer temperatures for a 'cold' cable and a 'warm' cable is given in Figure 3.

The worst conditions for a HVDC MIND cable arise in the so-called current-off situation as there the toughest hydro-dynamical changes take place. In this article we shall therefor concentrate on this stage in the type test.

RESULTS DURING CURRENT-OFF STAGE

In the following we give some results based on partial discharge measurements on a 450 kV HVDC MIND cable. Figures 4 and 5 show the repetition rate of discharges in the magnitude class <50 pC during one of the positive cycles for a 'cold' and a 'warm' cable.

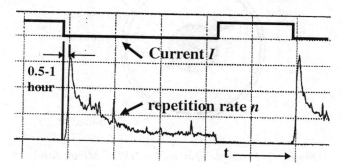

Figure 4: Measurement results of the 'cold' cable. Repetition rate n based on discharges in the magnitude class <50 pC.

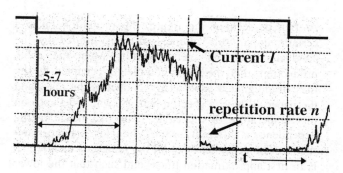

Figure 5: Measurement results of the 'warm' cable. Repetition rate n based on discharges in the magnitude class <50 pC.

The repetition rate n as a function of time shows a typical behavior for both the 'warm' and the 'cold' cable. When the heating current is switched on, the repetition rate is quite low. After switching off the current, the repetition rate increases to a maximum, where after it decreases slowly. When the current is switched on again, the repetition rate diminishes rapidly to a low value. The only difference between the 'cold' and the 'warm' cable lies in the time needed to reach the maximum in the current-off stage. This time is typically 0.5 to 1 hour for the cold cable and about 5 to 7 hours for the warm cable. The absolute numbers may differ per cable design and testing conditions. The trend, however, is clear.

EXPLANATION

The electrical insulation of the MIND cable consists of a special high voltage paper which is impregnated with oil. The paper is lapped on a metal conductor. A metal sheath is placed over the insulation. A simple sketch is given in Figure 6.

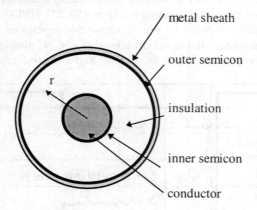

metal sheath

outer semicon

r

insulation

inner semicon

conductor

Figure 6: Principle sketch of a HVDC MIND cable up to the metal sheath. The layers outside the metal sheath, as the armoring as an example, are not shown.

Process 1: increasing repetition rate

As can be seen in the figure, the insulation takes a certain amount of volume, which is limited by the conductor and the metal sheath. After the cable production, this volume is completely occupied by the mass-impregnated paper. The conductor, the mass and the paper all have a specific density, which depends on the temperature of the cable. This magnitude of the dependency is different for the conductor, the paper and the oil. The dependency of all the materials mentioned, is in first approximation linear and such that the lower the temperature of the cable is, the smaller is the volume that the insulation occupies. The rest of the space becomes occupied by gas bubbles. This temperature dependent density of the cable materials is the major driving force in the MIND cable that may result in gaseous voids in the insulation. Generally, it can be said that the lower the temperature of the cable is, the larger may be number of gaseous voids inside the insulation.

We refer now to the equation for the repetition rate of the total number of defects in a cable. The density distribution D_r depends in first approximation, inversely on the temperature:

$$D_r \equiv F\left[\frac{1}{T_r}\right],$$

in which F is a monotone function. This means that in the current-off stage, the number of defects is growing, as the temperature of the cable decreases. The repetition rate n therefor initially increases directly after switching off the heating current.

Process 2: diminishing repetition rate

At the same time a counteracting process takes place. It should be remembered that the conductivity of the insulation depends strongly on the temperature in the following well-known way (5)

$$\sigma = \sigma_0 \exp(\alpha T),$$

in which σ_0 stands for the conductivity at $T=0°C$ and α is the temperature dependency coefficient of the conductivity. The conductivity of the insulation determines the resistors in the modified a,b,c-scheme. The time constant τ of the modified a,b,c-scheme is related linearly to these resistors. It follows that

$$\frac{1}{\tau_r} \equiv \sigma_0 \exp(\alpha T_r).$$

This means that the time constants τ_r of the defects increases after switching off the current as the temperature then decreases. Referring to the equation for the repetition rate, this means that the repetition rate n then must decrease.

Process 1 and process 2 together

The two processes are counteracting each other. As the temperature of the cable decreases during the current-off stage, the defect concentration D_r increases. This results in a higher measurable repetition rate n.

At the same time as the temperature decreases, the electrical time constant τ_r of the insulation diminishes, resulting in a lower measurable repetition rate n.

If we for simplicity set the electrical field strength E_r constant and state that the distribution functions are constant in the sense that there is no dependency on the radius r in the insulation, it follows that

$$n \equiv \exp[\alpha T(t)] \times F\left[\frac{1}{T(t)}\right].$$

This simple model is already capable to predict the essentials of the observations. It predicts the initial increase of the repetition rate n after switching off the current and it predicts the consecutive decrease of n. It also predicts the order of magnitude of the time to reach the maximum after switching off the current. Figures 7 and 8 show simulation results directly after switching off the current for a 'cold' and a 'warm' cable respectively. The actual type test temperatures $T(t)$ have been used. The simulation results may therefor be compared to the measurement results as shown in Figures 4 and 5. A simple exponential function has been chosen for the monotone function F.

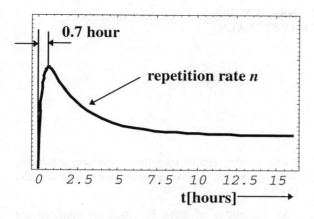

Figure 7: Simulation results of the 'cold' cable directly after switching off the heating current.

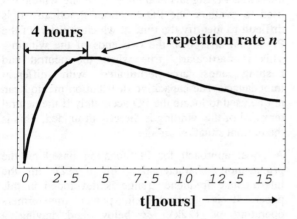

Figure 8: Simulation results of the 'warm' cable directly after switching off the heating current.

As can be seen, the times to reach the maximum in the repetition rate n, as calculated by this simple model, are already quite close to these times based on actual measurements. More information can be found in (5).

CONCLUSIONS

A HVDC MIND cable shows a quite typical and repeatable partial discharge behavior during the current-off stage in a type test cycle. After an initial increase in the level of the repetition rate n at current-off, it decreases after a certain time. The time to reach the maximum in the repetition rate is larger for a 'warm' cable than that for a 'cold' cable. A simple model has been made to explain this behavior under type test conditions. Although several assumptions and approximations have been made, the model predicts the general behavior of the repetition rate of the discharges and the time to reach the maximum value of the repetition rate after switching off the heating current. This means that the underlying principle assumptions are most probably valid. These are:

- the number of defects is increasing after switching off the heating current based on the fact that the temperature of the cable is decreasing.

- the time constant of recharging the defects is increasing after switching off the heating current, also based on the fact that the temperature of the cable is decreasing.

REFERENCES

1. Couderc D, Trinh N, Belec M, Chaaban M, Leduc J and Beauséjour Y, 1992, "Evaluation of HVDC Cables for the St-Lawrence Crossing of Hydro Quebec 500 kV Line", <u>IEEE Trans. on PD</u>, <u>7</u>, 1034-1042

2. Crabtree I, and O'Brien M, 1986, "Performance of the Cook Strait ±250 kV DC Submarine Power Cables: 1964-1985", <u>CIGRE 21-01</u>, <u>Pref. Subj. 1</u>

3. Asplund G, Eriksson K and Tollerz O, 1998, "HVDC Light, a toll for electric power transmission to distant loads", <u>SEPOPE, Salvador</u>

4. Eriksson A, Henning G, Ekenstierna B, Axelsson U and Akke M, 1994, "Development Work Concerning testing Procedures of Mass-Impregnated HVDC Cables", <u>CIGRE 21-206</u>

5. Jeroense M, 1997, "Charges and Discharges in HVDC Cables", Delft University Press, Delft, The Netherlands

6. Fromm U, 1995, "Partial Discharge and Breakdown Testing at High DC Voltage", Delft University Press, Delft, The Netherlands

7. Kreuger F, 1995, "Industrial High DC Voltage", Delft University Press, Delft, The Netherlands

8. Ficker T and Sikula J, 1984, "Spark and Glow DC-Partial-Discharges in Dielectrics", <u>Jpn. J. Appl. Phys.</u>, <u>23</u>, 1263-1264

9. CIGRE Working Group 21-02, 1980, "Recommendations for Tests of Power Transmission DC Cables for a Rated Voltage up to 600 kV", <u>Electra</u> $N^0 72$, 105-114

PARTIAL DISCHARGE LOCATION IN POWER TRANSFORMERS USING THE SPECTRA OF THE TERMINAL CURRENT SIGNALS

Z. D. Wang, P. A. Crossley and K. J. Cornick

Department of Electrical Engineering and Electronics
UMIST, Manchester, M60 1QD
United Kingdom

Abstract

Partial discharges (PD's) are a major source of insulation failure in a power transformer. The accurate location of the discharge is of crucial importance in on-site maintenance and repair. For power transformers operating at 132kV & below and having a continuous disc type winding construction, a novel approach for PD location, based on a study of how PD's propagate in a transformer winding, is presented. Experimental tests are used to show that the PD source can be located using information available in the current signals measured at the transformer terminals.

1. Introduction

If the insulation deterioration caused by partial discharge (PD) activity can be immediately detected then incipient faults can be identified and preventive maintenance undertaken. Consequently, on-line PD measurement is a powerful condition monitoring technique that is now being developed for use with power transformers [1] [2].

Techniques for the location of a PD are of major importance in on-site maintenance and repair. However in power transformers accurate location is difficult due to the complex structure of their windings. The PD pulse suffers distortion and attenuation as it travels from the site-of-origin to the measuring terminal. Consequently, the measured signal is a highly distorted representation of the original PD pulse, but it does contain useful information about the location and nature of the discharge. If this information can be extracted and analysed, the PD source may be located.

Previous research [3] [4] has suggested that within the different frequency ranges that describe the spectra of the measured PD signal, the modes of propagation in the transformer winding have different characteristics. In the lower frequency range (e.g. 0 to 0.01MHz) the component of a PD pulse that propagates in the winding is predominantly an electromagnetic wave that travels along the galvanic path. In the intermediate frequency range (e.g. 0.01 to 0.1MHz) the propagation is dominated by the characteristic resonances of the winding. In the higher frequency range (e.g. 0.1 to 10 MHz) the propagation is mainly through the capacitive elements of the winding.

Consequently, simple travelling wave and capacitive distribution methods were developed for PD location [5] [6].

The limitations of these PD location methods are mainly in their practical application. For the travelling wave method, the time delay between the measured PD signals at both ends of the winding is used to evaluate the PD location. Practically it is difficult to identify the time at which the start of the PD signal arrived at the two ends of the winding. This is particularly true when the neutral and bushing ends are terminated with different impedances. The capacitive distribution method can not be used to locate the PD accurately if the neutral terminal of the winding is directly grounded. This is the normal situation on-site.

A novel approach for PD location based on the propagation characteristics of a PD in the intermediate frequency range is developed in this paper. It is applicable to power transformers operating at 132kV & below and having a continuous disc type winding construction. The differences in the waveform of the measured current signals that result from a PD occurring at different locations will be used to indicate the location of the PD. Simulation and experimental results obtained with a continuous disc type transformer winding will be analysed. It will be seen that this approach to the location of a partial discharge results in an accurate decision.

2. Computer Simulation

A simulation model was developed to study the propagation of PD current pulses in transformer windings [7]. A lumped-element winding model was used to calculate a transfer function that describes the effect on the measured current signal as the PD instigated transient moves along the winding. The transfer function can be used to study the PD propagation patterns in different frequency ranges and with different winding constructions.

The simulation model was used to study how the PD instigated pulse propagated in a continuous disc type winding. The winding included 84 discs, a capacitance of 220pF was connected at the bushing end, and the neutral-end was grounded. A Dirac current pulse source was connected to the nodes of the equivalent network and the transfer functions from different locations of the injected PD to the

High Voltage Engineering Symposium, 22–27 August 1999
Conference Publication No. 467, © IEE, 1999

bushing and the neutral were calculated. The transfer functions, from the 1st, the 10th, the 20th, the 38th, the 54th and the 82nd disc to the bushing, are given in Fig.1 for a frequency range from 0 to 300kHz. The disc numbers were counted from the bushing end of the winding. Transfer functions from the same disc locations to the neutral, are given in Fig.2.

The characteristic resonances of the winding are clearly demonstrated in the transfer functions. The poles of the transfer functions for different PD locations are the same, i.e. the crests in the spectra occur at the same frequencies. The main effect of the PD location on the transfer function is related to the position of the zeros, i.e. the troughs in the spectra move to a frequency that depends on the PD location. In Fig.1 and Fig.2 some of the troughs are marked by "+" to help clarify the effect of the PD location. At these frequencies, some of the components included in the PD current pulse at its site-of-origin will vanish after the PD propagates through the winding. The "vanished" frequency components in the spectra of the transfer functions may help to locate the PD source.

3. Experimental Investigation

Experimental Set-up

An experimental investigation was carried out on a continuous disc type winding. A model of which was used in the simulation. The experimental set-up is given schematically in Fig.3.

A voltage generator consisting of a charging circuit and a PD simulator was connected to various discs along the winding. The PD current pulse was generated by a mercury-wetted switch short-circuiting a capacitor. The capacitor C_c simulated the equivalent capacitance of the insulation where the PD's occurred. The capacitor C_b simulated the equivalent capacitance of the series insulation. In this experiment C_c was 560pF and C_b was about 10pF.

The voltage source U was used to charge the capacitor C_c and via C_b the capacitance of the

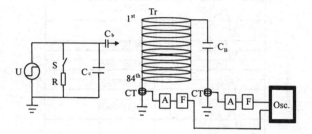

Fig.2 Transfer functions from PD locations to neutral

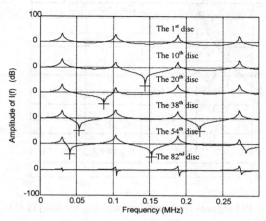

U: Voltage source, S: Mercury-wetted switch,
R: Discharge resistor, C_c,C_b: Capacitors,
Tr: Transformer winding, C_B: Bushing capacitor,
CT: PD current transducer, A: Amplifier, F: Filter,
Osc.: Oscilloscope.

Fig.3 Experimental set-up

transformer winding. The mercury-wetted switch S was open during the charging time, after the completion of the charging process switch S was closed. C_c, C_b, and the winding were discharged through the switch and its series resistor R. Since the mercury-wetted switch closes in a few nanoseconds, the duration of the discharge current pulse was dominated by the value of the discharge resistor. In this experiment it had a value of 50 Ohms and a 100ns PD pulse was generated.

The PD current pulse was injected into different discs along the winding, and the current responses measured at the bushing and neutral terminals. The measurements were obtained using wide-band current transducers (CT's) with a bandpass frequency response centred at about 40kHz, 250kHz and 400kHz respectively. Each CT had a -6dB frequency band of about 100kHz.

Experimental Results

The PD current pulse was first injected into the 1st disc, then the 10th and then the 20th, the 38th, the 54th and the 82nd disc. The current signals measured at the bushing using a 40kHz CT for each of these tests are given in Fig.4. Similarly, the current signals measured at the neutral by the 40kHz CT are given in Fig.5.

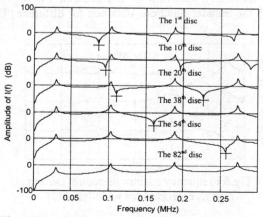

Fig.1 Transfer functions from PD locations to bushing

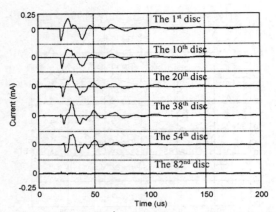

Fig.4 PD signals measured at bushing by 40kHz CT

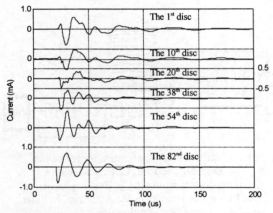

Fig.5 PD signals measured at neutral by 40kHz CT

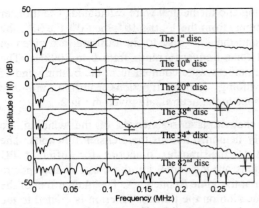

Fig.6 Frequency spectra of bushing current signals

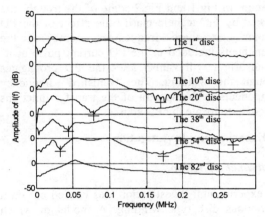

Fig.7 Frequency spectra of neutral current signals

The difference in the *terminal signals* resulting from a PD at different locations can be observed. When a PD occurs near the neutral-end, for example at the 82nd disc, most of the current will flow to the neutral ground, therefore the current propagating to the bushing will be small see 82nd disc in Fig.4. The PD signals measured at the neutral, see Fig.5, do not support the view that the more distant the PD source, the smaller the measured signals. This is due to the fact that the passband of the CT (0-100kHz with 40kHz CT) falls in the range where the transfer functions from the PD locations to the neutral vary as in Fig.2. It can be observed that for a PD source near the middle of the winding the PD signals measured at the neutral have a smaller magnitude than those observed for a PD source near the ends.

Spectrum analysis can give a more sensitive demonstration of the effect on the measured current signals of the PD location. The spectra that describe the PD signals measured at the bushing, see Fig.4, are given in Fig.6, and the spectra of the PD signals measured at the neutral, see Fig.5, are given in Fig.7. As previous, the zeros of the transfer functions are marked by "+" in Fig.6 and Fig.7. When compared to the simulation results, see Fig.1 and Fig.2, the zeros in the spectra occur at similar frequencies. The results are surprisingly close particularly when one considers the influence of the CT, i.e. for a 40kHz

CT frequencies components above approximately 100kHz are severely attenuated.

By multiplying the spectra of the measured signals with the inverse frequency characteristics of the CT, the spectra with a "perfect" CT can be derived. For example the spectra of the PD signals measured at the neutral, with a perfect CT, are given in Fig.8. Comparing with Fig.7, the effect of the CT is reduced and the spectra have close agreement with the simulation results in Fig.2.

4. Discussion of Results

The easiest way to locate a PD source is to compare the frequency spectrum of a measured PD signal with the transfer functions obtained for different PD

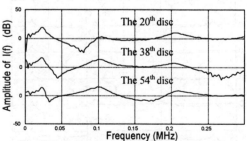

Fig.8 Frequency spectra of neutral current signals corrected for wide-band CT

locations using the computer model. The location of the PD source is obtained when the frequency position of the measured zero is approximately equal to the simulation result obtained for a PD at a particular disc. By this means the neutral signals have a better sensitivity than the bushing signals.

If one looks at the experimental results in Fig.8, the frequency of the first zero for a PD at the 20th disc is 81.25kHz, whilst the first zero with the 38th disc is at 43.75kHz, and the first zero with the 54th disc is at 34.38kHz. The simulation results in Fig.2 show that the frequency of the first zero for a PD at the 20th disc is 87.5kHz, whilst the first zero with the 38th disc is at 53.13kHz, and the first zero with the 54th disc is at 40.63kHz. Comparing the frequency of the zeros observed with the experimental results against those obtained using the simulation, the PD location can be determined, but in some cases it may be difficult to identify the exact location by only comparing the frequency of the first zero. The frequency position of the second zero may help with the identification. In the experimental results the frequency of the second zero for a PD at the 38th disc is 268.75kHz, whilst the second zero at the 54th disc is at 168.75kHz. The simulation results in Fig.2 show that the second zero for a PD at the 38th disc is at 218.75kHz, whilst the second zero with the 54th disc is at 153.13kHz.

5. Conclusions

A novel approach for PD location in power transformers with a continuous disc type winding construction was proposed. It makes use of the PD propagation characteristics in the intermediate frequency range. It is shown that only the zeros of the transfer functions change as the location of the PD moves along the winding. By comparing the zeros in the measured PD current signals with the zeros in the simulation results, the PD can be located accurately.

Acknowledgements

The first author thanks the United Kingdom "Overseas Research Students Awards Scheme" for financially supporting her PhD study. This work is a part of the research project (Project No. 59637200) supported by the National Natural Science Foundation of China and the Northeast China Electricity Company. Thanks are given to Ms. X.H. Chen and Mr. G.C. Fan of the high voltage laboratory of Shenyang Transformers, China for their assistance with the experimental tests. All the authors acknowledge the support of the Manchester Centre for Electrical Energy, UMIST.

References

[1] C. Bengtsson, "Status and Trends in Transformer Monitoring", IEEE Transactions on Power Deliverly, Vol.11, No.3, July 1996, pp.1379-1384.

[2] G. C. Stone, "Partial Discharge – Part VII: Practical Techniques for Measuring PD in Operating Equipment", IEEE Electrical Insulation Magazine, July/August 1991-Vol.7, No.4, pp.9-19.

[3] A. T. Thoeng, "Detection and Localization of Partial Discharges in Power Transformers", Conf. Publications No.94, Part 1, IEE Diagnostic Testing of HV Power Apparatus in Service, March 6-8, 1973, London.

[4] J. Fuhr, M. Haessig, P. Boss, ec al, "Detection and Location of Internal Defects in the Insulation of Power Transformers", IEEE transactions on Electrical Insulation, vol. 28, No. 6, December 1993, pp. 1057-1067.

[5] T. Bertula, V. Palva, E. Talvio, "Partial Discharge Measurement on Oil-paper Insulated Transformers", Report 12-7, CIGRE, 1968.

[6] R. E. James, B. T. Phung, Q. Su, "Application of Digital Filtering Techniques to the Determination of Partial Discharge Location in Transformers", vol. 24, No.4, August 1989, pp.657-668.

[7] Z. D. Wang, P. A. Crossley, K. J. Cornick, "A Simulation Model for Propagation of Partial Discharge Pulses in Power Transformers", Proceedings of the International Conference on Power System Technology (PowerCon-98), August, 18-21, 1998, Beijing, China, pp.151-155.

FIELD EXPERIENCES FROM ON-SITE PARTIAL DISCHARGE DETECTION IN POWER PLANTS

J.P. Zondervan[*], A.J.M. Pemen[**], J.J. Smit[*], W. de Leeuw[***]

[*] High Voltage Laboratory, Delft University of Technology, The Netherlands
[**] High Voltage & EMC group, Eindhoven University of Technology, The Netherlands
[***] KEMA Transmission & Distribution, The Netherlands

ABSTRACT

In this report recent experiences are discussed with on-line PD measurements using the VHF PD detection technique. The discussion is focussed on the great care that should be taken into account with on-line PD detection, regarding the following topics:
- Localization of PD sources
- Comparison of PD results
- Selecting measurement frequencies

1. INTRODUCTION

Partial discharge (PD) measurements are frequently in use as a tool to get insight in the condition of the stator insulation of turbogenerators. Much effort is taken in developing on-line PD measurement techniques to obtain this insight during regular operation of the generators.

This contribution reports on experiences with the formerly introduced VHF PD detection technique [1,4]. By discussing several cases it will be shown that this technique is exceptional suitable for on-line PD detection on turbogenerators in order to identify defects or monitor the condition of the stator winding. Nevertheless, as for the analysis and the interpretation of the PD results, great care has to be taken, in order to justify the conclusions.

The topics that are treated in this report are:
- localization of PD, i.e. determining if the PD originate from the generator or from the transformer
- comparison of PD results, i.e. when trying to identify a PD source by comparison with other results, make sure that a sound comparison is made
- selecting measurement frequencies, i.e. the measurement frequency affects the recorded PD results. Thus, make sure a suitable frequency is selected.

2. SENSORS AND MEASUREMENT TECHNIQUES

When measuring PD on turbogenerators during regular operation, two difficulties arise:

(i) From the power plant as well as from the rotor excitation, system interferences may occur in the measuring circuit. Especially at frequencies up to 5 MHz intense interference is present.
(ii) Due to the complex propagation of PD pulses through a stator winding, cross talk takes place between the three phases [1].

For the purpose of suppressing interference, we use a spectrum analyser in the 'zero-span' mode as a tuned filter. The spectrum analyser is tuned to a frequency where PD dominate and no interference and cross talk is present. Experiences show that the best results are obtained at tuning frequencies between 10 and 30 MHz [2,3], however in some cases higher frequencies may also serve as suitable tuning frequencies [4]. The output of the spectrum analyser can be connected to a modern PD-detector (Haefely TE-571), to provide a digital registration of all PD quantities. Since various PD sources might give different frequency responses at the generator terminals, for each phase a series of PD patterns is recorded at various tuning frequencies, the so called frequency scan. To detect the PD signals during regular operation, various sensors can be used, such as Rogowski coils or capacitive sensors [5]. A representation of the measuring setup is shown in figure 1.

3. FIELD EXPERIENCES

3.1 Localization of PD in a large power plant

During regular on-line measurements on a 650 MW generator, at times an abnormal PD pattern was recorded. As the step-up transformer of the power plant showed a rapid increase of gas-in-oil concentrations, it was assumed that the abnormal discharges originate from the step-up transformer instead of the generator. Especially the C_2H_6 (ethane) and C_3H_8 (propane) levels showed a rapid increase, which is an indication for the presence of a hot-spot inside the transformer [6]. To gain more information, it was decided to do additional on-line PD measurements. Normally, these measurements are done by means of permanently installed Rogowski coils around the

High Voltage Engineering Symposium, 22–27 August 1999
Conference Publication No. 467, © IEE, 1999

HV terminals of the generator. For the additional measurements, also a pick-up loop near the 21 kV bushings of the transformer was used.

A partial discharge inside the generator manifests itself at the terminals and then propagates along the isolated phase bus (IPB) in the direction of the step-up transformer. An example is given in figure 2, where a PD in the generator first shows up at the Rogowski coil at the generator. After a transit time of 115 ns the signal also shows up at the pick-up loop near the step-up transformer.

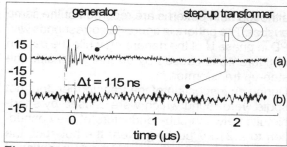

Figure 2　PD inside the generator, observed by means of:
(a) a Rogowski coil around the high-voltage terminal of phase U of the generator, and
(b) a pick-up loop near the 21 kV bushing of phase U of the step-up transformer.

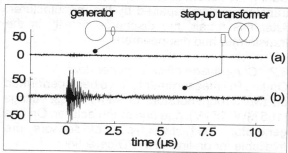

Figure 3　PD inside the step-up transformer, observed by means of:
(a) a Rogowski coil around the high-voltage terminal of phase U of the generator, and
(b) a pick-up loop near the 21 kV bushing of phase U of the step-up transformer.

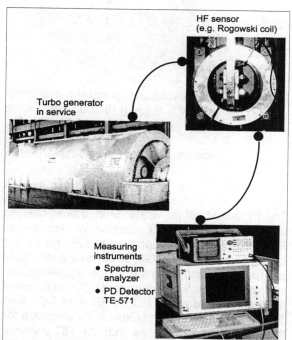

Figure 1　Representation of the measuring setup as used for the on-line PD detection technique. The PD detector with spectrum analyser records the discharges from the stator insulation of the turbo generator through HF couplers (e.g. Rogowski coils around HV bushings)

Figure 3 shows a signal which first shows up at the transformer side. Due to damping of the IPB or to poor coupling of the signal into the IPB, it can not be seen at the generator side. About 50% of the recorded signals were of this type and seem to originate from the step-up transformer. To verify whether these signals are indeed caused by PD inside the transformer, the PD patterns were recorded. Therefore both the outputs of the Rogowski coil near the generator and the pick-up loop near the step-up transformer were connected to a two-channel spectrum analyser. The tuning frequency was 20 MHz for both channels and the patterns were recorded during 64 cycles of the 50 Hz voltage. The results are shown in figure 4.

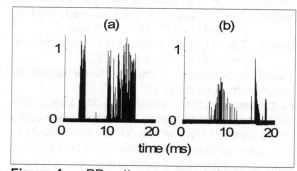

Figure 4　PD patterns measured with
(a) a Rogowski coil near the high-voltage terminal of phase U of the generator, and
(b) a pick-up loop near the 21 kV terminal of the step-up transformer. Both patterns are recorded simultaneously by means of a 2-channel spectrum analyser, tuned to 20 MHz.

Figure 4 shows two completely different patterns,

although both patterns are recorded at the same busbar. The pattern in figure 4a corresponds with PD in phase U of the generator, while the pattern of figure 4b is caused by discharges inside the step-up transformer.

During an inspection of the transformer, four hot spots were found which were caused by short-circuits between sub-conductors of the connection to a 21 kV bushing. Near the hot spot, the insulating paper was burned. This degradation of the paper, in combination with the local high temperature, was responsible for the increased gas-in-oil concentrations. The short-circuits led to intermittent sparking. This sparking resulted in signals similar to partial discharges, which were detected with the pick-up loop near the 21 kV terminals of the transformer.

During the next maintenance period, the step-up transformer was repaired by the manufacturer. After this repair, the discharge activity in the transformer had disappeared.

3.2 Comparison of sister generators

On-line PD measurements have been performed on two turbogenerators of same class, ratings (165 MW/ 15 kV), age and manufacturer. On one generator, TG1, two types of sensors are available for on-line VHF PD detection:
1) A HF-capacitor housed in a support insulator of the IPB (S1), and
2) A ring capacitor, a metal ring installed inside the IPB (S2).

The other generator, TG2 is equipped with only sensor S2.

Measurements were performed on all phases of both generators, using the available sensors.

Figure 5 shows the results of two measurements:
upper: The $H_n(\varphi,q)$ distribution measured at phase U of TG1 using S1, and
lower: The $H_n(\varphi,q)$ distribution measured at phase U of TG2 using S2.

Visual comparison shows a large similarity between the two patterns.

Furthermore, all measurements are statistically processed and compared with each other using group analysis [7]. The result is shown in figure 6. The tree structure illustrates the relationship between the individual measurements, by means of a dissimilarity percentage scale. Similar PD results connect at a low dissimilarity level, different PD results connect at relatively high dissimilarity levels. In order to make a comparison with some kind of reference, on-line PD results from a 650 MW/ 21 kV generator (Gen. 3) are added to the statistical analysis. Figure 6 shows that all measurements from TG1 and TG2 cluster with each other, whereas the results from Gen. 3 form a separate cluster. This concludes that TG1 and TG2 produce similar results for the VHF PD detection technique.

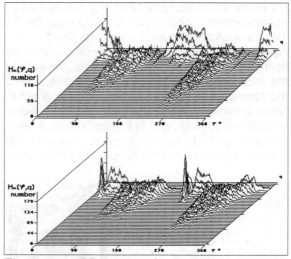

Figure 5   PD results of two sister generators:
upper: $H_n(\varphi,q)$-distribution from phase U of TG1 using S1, $f_o = 47$ MHz
lower: $H_n(\varphi,q)$-distribution from phase U of TG2 using S2, $f_o = 15$ MHz

The PD results of generator Gen. 3 are obtained with the same PD detection technique. Furthermore, its condition should quite match the condition of TG1 and TG2, as all three generators are more or less of the same age. However, the statistical analysis as presented in figure 6 shows a difference between PD patterns of TG1 and TG2 and PD patterns of Gen. 3. One reason for this occurrence might be that the PD patterns respond different to the winding configurations. Gen. 3 has a different winding configuration than TG1 and TG2, which means that there is a matter of different pulse propagation characteristics. As a result, a similar insulation condition can give a different response at the coupler's site, when the measuring frequencies are not matched with each other.

Anyhow, this example emphasizes the care that should be taken when comparing PD results of different machines (or different phases) with each other. Due to the complexity of the on-line measuring technique and the interpretation of PD patterns, results might not always agree with each other, although expected. In the following section, the discussion will be focussed on the influence of the measuring frequency on the recorded PD results.

3.3 Frequency scan

As PD signals endure complex propagation processes when travelling through the stator winding, they might give different frequency responses at the generator terminals. Furthermore, the coupler's transfer functions strongly influence the measured responses. As a result, measuring on one single frequency might not reveal the right response of the PD activity that holds information about the insulation condition.

For this reason, a PD measurement on one phase should be preceded by performing a frequency scan, i.e. a series of PD patterns at various measuring frequencies within a certain range (e.g. from 10 to 40 MHz, in steps of 2 MHz). After performing the frequency scan, suitable measuring frequencies are selected by examining the patterns for high PD response and low interference and cross talk response. Figure 7 illustrates part of a frequency scan (in practice, a scan consists of 30 measuring frequencies). As can be seen from this figure, the responses are quite different at the various measuring frequencies. For this example, the measuring frequency of 32 MHz would be regarded as suitable, as at this frequency the PD activity from the phase under test is recorded, and the pattern is little disturbed by interference and cross-talk.

4. CONCLUSIONS

This report discusses the utilization of the VHF PD detection technique for on-line PD tests on generators in power plants. It has been experienced that the selection of the measuring frequencies plays a key role when collecting significant PD responses. As the measuring frequency affects the localization of PD sources and comparison with other PD results, the selection should be performed with the greatest care. A frequency scan may contribute to a sound selection.

5. REFERENCES

[1] A.J.M. Pemen, P.C.T. van der Laan, P.T.M. Vaessen, *Partial discharge monitoring of turbine generators; laboratory and live measurements*, Proc. CEIDP, Oct. 1994.

[2] A.J.M. Pemen, P.C.T. van der Laan, *Pitfalls of PD measurements on statorwindings of turbine generators*, 7th. DMMA, Sep.1996

[3] J.P. Zondervan, E. Gulski, J.J. Smit, R. Brooks, *Use of VHF detection technique for PD pattern analysis on turbogenerators*, 1998 IEEE ISEI, Sept. 1998

[4] J.P. Zondervan, E. Gulski, J.J. Smit, R. Brooks, *PD Pattern Analysis of On-line Measurements on Rotating Machines*, Proc. CEIDP, Oct. 1997

[5] A.J.M. Pemen, W. de Leeuw, P.C.T. van der Laan, *On-line PD monitoring of statorwindings; comparison of different sensors*, 10th ISH, August 1997

[6] G.P. Krikke, *Trends in gas-in-olie analyse bij vermogenstransformatoren*, KEMA report 44190-KET/BVA 96-7001, Apr.'96 (in Dutch)

[7] A. Krivda, *Recognition of Discharges, Discrimination and Classification*, Delft University Press, 1995

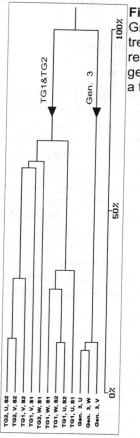

Figure 6
Group analysis using the tree method of on-line PD results from two sister generators (TG1&TG2) and a third generator (Gen. 3)

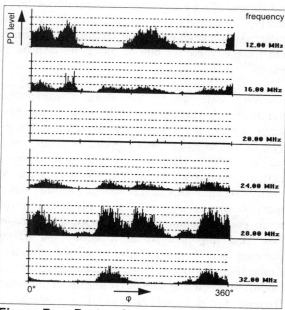

Figure 7 Part of a frequency scan as measured on one phase of a 165 MW/ 15 kV turbogenerator

STUDY ON MECHANISMS AND PROPAGATION CHARACTERISTICS OF PARTIAL DISCHARGES IN SF$_6$ GIS

M HIKITA, T HORIUCHI, Z H TIAN, H SUZUKI *

Kyushu Institute of Technology, Japan
Takaoka Electric Mfg. Co., Ltd, Japan*

ABSTRACT

It is imperative to study partial discharge behavior in SF$_6$ GIS in order to carry out reliable insulation diagnosis and conditioning evaluation of practical gas-insulated system. This paper deals with investigation on propagation characteristics of electromagnetic waves owing to partial discharges in a SF$_6$ GIS model as well as the PD mechanisms, by utilizing various electrical measuring and optical detecting techniques. Leader discharges prior to breakdown can be readily discriminated and the attenuation characteristics of electromagnetic waves within the GIS has been established. Suggestion is also put forward to set up a comprehensive PD detection and insulation performance evaluation system for reliable and efficient application to practical GIS.

INTRODUCTION

Nowadays, compressed SF$_6$ gas-insulated switchgears and substations (GIS) have been widely used in many electric power networks owing to a lot of advantages such as compact structure and high reliability as well as almost maintenance-free characteristics, and so on [1~2]. In addition, 275 kV gas-insulated transmission line is drawing much attention as one of the attractive transmission methods for high electric power transfer into large city centres due to increasing demand for power [3].

However, some defects like metallic particles or protrusion or contamination on the surface of an epoxy spacer in GIS would be a cause of partial discharges (PD) which can lead to final dielectric breakdown of the equipment [4~6]. Detection of PD which precede dielectric breakdown is a very effective means for quality control of GIS so that serious accidents during operation can be avoided. Thus, PD measurements during commissioning tests and service of GIS are very useful and important so as to ensure the reliable operation and minimize damage of the apparatus. Furthermore, on-line PD continuous detection and condition monitoring are also valuable for insulation diagnosis and risk evaluation of GIS. Nevertheless, there are still some problems concerning PD mechanisms, measuring techniques and especially the assessment of the measured results from GIS due to complicated attenuation and reflection as well as much interference in situ [4~6].

In this paper, we investigated electromagnetic waves (EMW) emitted from PD and their propagation characteristics in a 66 kV GIS model by several internal sensors which were built into the insulating spacers of the GIS. Moreover, we also measured simultaneous current pulse waveforms and light emission as well as discharge images of PD activities using several measuring and detecting techniques, such as ultra-wide-band (UWB) circuit (DC~800 MHz) [6], photomultiplier tubes (PMT) and intensified charge-coupled device (ICCD) camera, respectively, in order to research different mechanisms of PD and their development characteristics in the GIS.

EXPERIMENTAL SETUP

Fig. 1 schematically illustrates the experimental setup with the 66 kV SF$_6$ GIS model used. A needle-plane electrode system (the tip radius of the needle electrode r = 0.5 mm, gap length between electrodes g = 10 mm) was set to simulate metallic protrusion at a high potential conductor and produce PD signals within the GIS model. Power frequency ac high voltage (60 Hz) was applied to the electrode system of the GIS through a HV ceramic bushing for all experiments. SF$_6$ gas pressure inside the GIS can be changed for experimental purposes. In this study, the pressure P = 0.2 MPa was specifically chosen so as to readily discriminate streamer type and leader type PD in compressed SF$_6$ gas. The internal sensors A~E consist of O-ring couplers or EMI probes built into the insulating spacers of the GIS and are used for specially measuring EMW signals from PD. A digitizing oscilloscope (1.5 GHz, 5 GSa/s) and a spectrum analyzer (100 kHz ~ 6 GHz) were utilized to carry out various measurements. Furthermore, two PM tubes with red and blue filters were separately adopted for optical detection of light emission generated by PD in SF$_6$ gas. In addition, an ICCD camera was also used for observing developing images of PD in SF$_6$ gas directly.

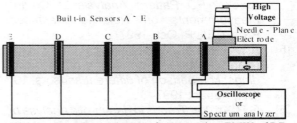

Fig. 1   Experimental setup for measuring EMW of PD in SF$_6$ gas using built-in sensors

High Voltage Engineering Symposium, 22–27 August 1999
Conference Publication No. 467, © IEE, 1999

EXPERIMENTAL RESULTS AND ANALYSIS OF PD MECHANISMS

Fig. 2 shows the breakdown voltage (BDV) and PD inception voltage (PDIV) characteristics of the needle-plane electrode system in the SF_6 GIS model. On the other hand, Figs. 3 (a) and (b) display simultaneously measured current waveforms of streamer and leader type discharges, along with PD light emission signals and images, respectively. When streamer type discharges occur, PD image looks like short bush and only smaller magnitudes of PD current pulses as well as light emission pulses from two kinds of PMT are also obtained. It is clearly seen from Fig. 3 (b) that a filamentary discharge emerges with much larger magnitudes of PD current pulses (about 4 times) as well as light emission pulses from two kinds of PMT with two peaks. It is also found that the magnitude ratio of red/blue components of two PM tubes was about 1/5 for streamer type PD, but the ratio was 1/2 for leader type discharge, which means that leader discharge has much stronger infrared component than that of streamer PD [2, 5].

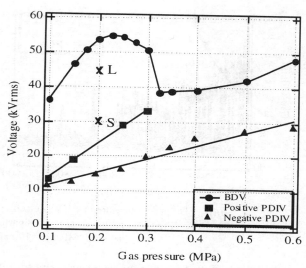

Fig. 2 PDIV and BDV characteristics as a function of gas pressure in SF_6 gas
($g = 10$ mm, $r = 0.5$ mm, room temperature)

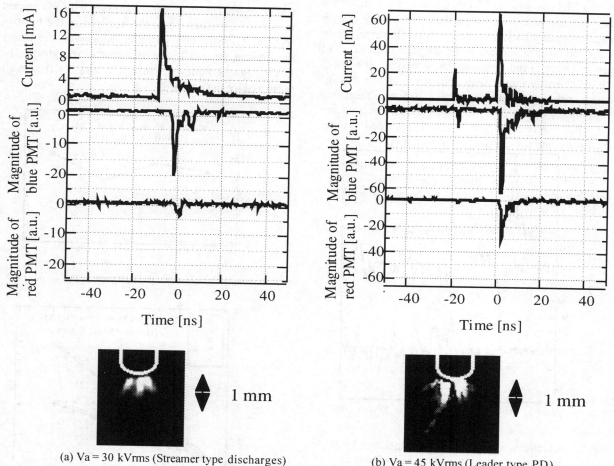

(a) Va = 30 kVrms (Streamer type discharges)   (b) Va = 45 kVrms (Leader type PD)
Fig. 3   Waveforms of PD current pulses and light emission in SF_6 gas
($P = 0.2$ MPa, $g = 10$ mm, $r = 0.5$ mm, room temperature)
Blue and red PMTs means the transparent wavelengths in
the ranges of 320~540 nm and 580~760 nm, respectively.

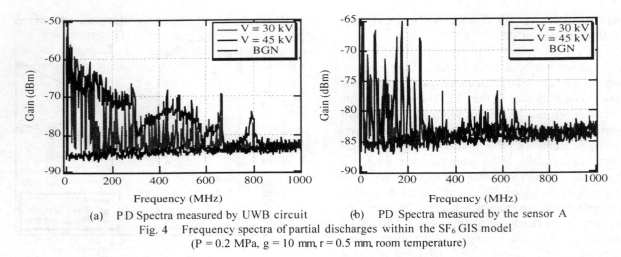

(a) PD Spectra measured by UWB circuit   (b) PD Spectra measured by the sensor A

Fig. 4   Frequency spectra of partial discharges within the SF₆ GIS model
(P = 0.2 MPa, g = 10 mm, r = 0.5 mm, room temperature)

Figs. 4 (a) and (b) illustrate frequency spectra of streamer and leader type partial discharges measured at both 30 and 45 kVrms separately with resolution frequency bandwidth RFB = 100 kHz using UWB circuit and the sensor A, respectively. Comparing frequency components in the ranges of 0～200 and 400 ～660 MHz as well as 750~820 MHz, a few difference can be apparently distinguished, especially larger amplitudes at 45 kVrms in Fig. 4(a). However, when using the built-in sensor, it is somewhat difficult to tell any difference in the frequency range above 200 MHz of the frequency spectra at both voltages as shown in Fig. 4 (b), although there are a few different frequency components below 200 MHz. These results indicate that the UWB electrical method provided higher sensitivity and more valuable information.

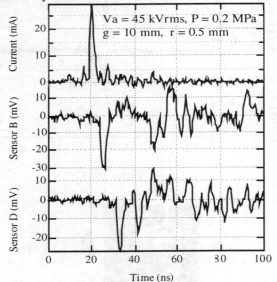

Fig. 5   Waveforms of single PD current pulse and EMW measured with the built-in sensors B and D.

Fig. 5 shows a typical example of single PD current pulse and the corresponding EMW waveforms measured with built-in sensors B and D at the same time. It is obvious from Fig. 5 that time delay exists

because of different propagation distances. The time delay measurements by the built-in sensors allow us to estimate the propagation velocity of EMW owing to PD in the GIS, as shown in Fig. 6, which is a little bit slower than the speed of light. Fig. 7 shows relation between PD current and EMW values measured with the built-in sensors. It is clear that the magnitude of EMW signal almost linearly increases with PD current in the GIS, irrespective of the position of the built-in sensors along the GIS chamber. Moreover, the nearer the internal sensors to the PD source, the greater the EMW signals from the sensors, except for the result measured by the sensor E.

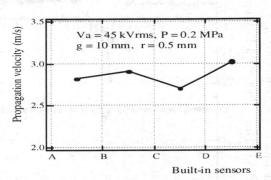

Fig. 6   Propagation velocity of EMW of PD in the GIS

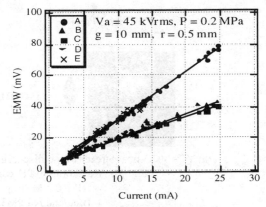

Fig. 7 Relation between PD current and EMW values (mV_pp) measured with the built-in sensors.

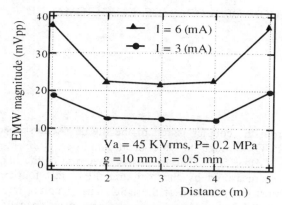

Fig. 8 Distance characteristics of EMW measured with the built-in sensors for different PD currents

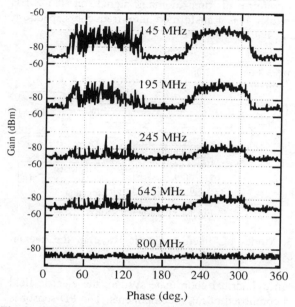

Fig. 9 POW at different centre frequency (Sensor A) (Va = 45 kVrms, P = 0.2 MPa, g = 10 mm, r = 0.5 mm)

Fig. 8 shows the propagation characteristics of EMW from PD when I = 3 and 6 mA, respectively. It is noted that EMW signal of PD decreases gradually up to the 4th spacer, while it increases at the 5th sensor because of serious reflection at this far end.

Fig. 9 shows the point-on-wave (POW) of PD at several selected centre frequencies (CF) with RFB = 100 kHz. The result indicates greater magnitudes in both positive and negative half cycles at lower CF values and larger attenuation for higher frequency components. Fig. 10 summarizes the damp characteristics of each frequency component of EMW signals measured by the built-in sensors. From the results, the attenuation rate is estimated as 2.1 and 2.3 dBm/m for 145 and 245 MHz, respectively. If it is assumed that GIS is a loss-free waveguide, then magnitudes of EMW signals will only reduce at spacers; the reduction rate of EMW signals due to spacers is 0.5~0.6 dBm/spacer for 145 and 245 MHz components. On the other hand, the magnitudes of

EMW signals at 195 and 645 MHz do not exhibit good relationship with distance. The result may be caused by complicated phenomena of reflection at the far end of the GIS and resultant standing wave occurring at specific frequency. As for 800 MHz POW, little effective signal appears by the built-in sensor.

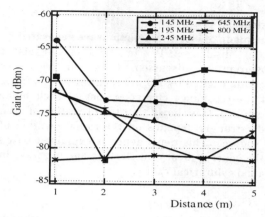

Fig. 10 Distance dependency of EMW at several CFs (Va = 45 kV, P = 0.2 MPa, g = 10 mm, r = 0.5 mm)

CONCLUSIONS

We investigated the mechanisms and propagation characteristics of partial discharges in a SF_6 GIS model by utilizing various electrical measuring and optical detecting techniques. Leader discharges prior to breakdown and streamers could be readily distinguished by these methods. Moreover, we studied attenuation characteristics of electromagnetic waves due to PD within the GIS using several special internal sensors built in the insulating spacers of the GIS. It was found that the propagation speed of EMW of PD in the GIS is close to, but slightly lower than, the speed of light. It was also established that larger PD currents generally produce bigger EMW signals, and larger attenuation rate of higher frequency component of PD is mainly caused by the insulating spacers within the GIS.

REFERENCE

1. M. E. Holmberg et al., 1998, "Electric charge and field at lift-off for metallic particles in GIS", IEEE ISEI, USA, June 1998, 53-56.
2. H. Okubo et al., 1998, "Discrimination of streamer / leader type PD in SF6 gas based on discharge mechanism", 8th ISGD, USA, June 1998, Paper 18.
3. T. Egawa et al., 1997, "Partial discharge detection for a long distance GIL", 10th ISH, Canada, 1997, p. 163.
4. J. S. Pearson et al., 1991, "A continuous UHF monitor for gas-insulated substations", IEEE Trans. EI, 26, No. 3, 469-478.
5. A. G. Sellars et al., 1995, "UHF detection of leader discharges in SF6", IEEE Trans. DEI, 2, 143-153.
6. M. Hikita et al., 1996, "Phase dependence of PD current pulse waveform and its frequency characteristics in SF6 gas", IEEE ISEI, Canada, June 1996, 103-106.

RESONANCE CHARACTERISTICS AND IDENTIFICATION OF MODES OF ELECTROMAGNETIC WAVE EXCITED BY PARTIAL DISCHARGES IN GIS

H.Muto, M.Doi, H.Fujii and M.Kamei

Mitsubishi Electric Corporation, Japan

Abstract

The mode of electromagnetic wave generated by partial discharge(PD) in GIS(Gas Insulated Substation) is discussed. Peaks in frequency spectrum excited by PD are identified in its mode and the resonance characteristics of each mode are studied. It is shown that the excitation of each mode strongly depends on the radial position of PD source. This dependence is explained based on the general theory on electromagnetic wave in co-axial cylindrical cavity.

1. INTRODUCTION

Measuring PD is very important for predictive maintenance for GIS. For the detection of PD in GIS, electromagnetic wave of very wide frequency band up to 1.5GHz has been utilized recent years.

There are three categories of electromagnetic wave which can be excited and propagate in the coaxial cylindrical structure like GIS: TEM, TE, TM modes. Peaks in measured frequency spectrum are originated in different mode of electromagnetic wave. Therefore the characterization and identification of mode are indispensable to understand the PD phenomena in GIS. Many report have been presented on the attenuation characteristics in frequency domain[1,2], and in time domain for TEM mode[3]. However, there has been less analysis on TE or TM modes[4]. As we will show later, TE mode is dominant above the cut-off frequency of TE11 mode. In this paper, we try to identify the mode of electromagnetic wave excited by actual PD in GIS, and discuss the resonant characteristics of each mode.

2. EXPERIMENT

2.1 GIS configuration

Experimental setup is shown in Fig.1. The GIS which we used is composed of four tanks separated by cone type spacer and at the right end of the tank a PT(potential transformer) for applying high voltage is set. The diameter of the inner conductor is 130mm, and that of the grounded tank is 550mm. As the PD source, an aluminum wire with the length of 20mm and the diameter of 0.2mm was used. We fixed the PD source on the inner conductor or the inner wall of the tank.

2.2 PD sensor

As a PD sensor we used internal UHF coupler placed in the flange L2. The coupler is a disk type electrode facing to the inner conductor. The signal

from the coupler is fed to the digital oscilloscope (DSO) through a wide band amplifier (100kHz-1.3GHz, 25dB). The frequency spectrum of the PD was measured by the real time FFT (fast fourior transform) function of the DSO. The sampling rate was 2.5Gsample/sec giving the frequency span of 1.25GHz and the resolution of 0.5MHz. Each waveform of time domain signal on the UHF coupler was transformed into frequency spectrum and averaged 50 times.

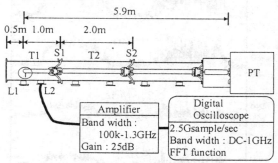

Fig.1 Experimental setup

3. Modes of Electromagnetic wave in the coaxial cylindrical structure

In cylindrical coordinate system, the electric field is composed of r,θ, z components. The PD sensor is set as its surface is parallel to the inner wall of the tank. Since the electric field on metal is perpendicular to its surface, the main detectable component of the electric field is Er. In this section the r component of the electric field for each mode is described.

(1) TEM mode

$$E_r = C/r, \quad (1)$$

where C is a constant.

(2) TE mode

The z component of electric field of TE mode is zero. The r component is given by

$$E_r = jZ_H \frac{m\beta_g}{k_c^2 r}\{AJ_m(k_c r) + BN_m(k_c r)\}\cos m\theta e^{-j\beta_g z}, \quad (2)$$

where m and n are integers(one or over).

(3) TM mode

The z component of magnetic field of TM mode is zero. The r component is given by

$$E_r = -j\frac{\beta_g}{k_c}\{AJ'_m(k_c r) + BN'_m(k_c r)\}\sin m\theta e^{-j\beta_g z}, \quad (3)$$

where m and n are integers,(m is zero or over, n is one or over.) Jm is the Bessel function, Nm is the Neumann function, kc is the wave number at the cut-

High Voltage Engineering Symposium, 22–27 August 1999
Conference Publication No. 467, © IEE, 1999

For comparison and correlation between time and frequency domain beside the depolarisation current and recovery voltage also the dissipation factor (tan δ) was measured in the range 10 mHz to 100 Hz. This allowed to verify the status of the cable samples without an influence of the parameters intentionally varied in the time domain.

The parameters charging time (tc) and measurement time (tm) for the depolarisation current resp. recovery voltage measurements were varied between 300 sec and 1800 sec; the discharging time (ts) was constant 1 sec (fig. 1). The charging voltage resp. the ac voltage (rms) was in both cases 1, 6, 12, 24 kV (0,1 U_0 - 2 U_0).

In addition, all service-aged cables were investigated by destructive test methods. The breakdown voltage was obtained by using a voltage test presented in fig. 2. The water tree analysis was done microscopically with 30 slices à 200 μm thickness.

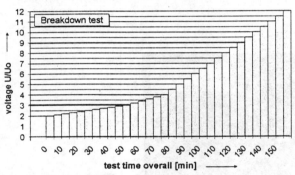

Fig. 2: *Breakdown test procedure*

Measurement equipment

The dissipation factor was obtained using our high voltage dielectric test system containing basically a high voltage source with variable frequency and a capacitance/tan δ measurement part [1, 6]. The depolarisation current resp. recovery voltage measurements were done by a self-developed diagnostic system using a Keithley 617-electrometer [1, 6].

3. Results and discussion

All achieved results are represented in table 1. An example of the depolarisation current for different charging periods is shown in fig. 3. Fig. 4 to 6 show the results achieved for the service-aged PE/XLPE cables. The sequence of the cables is arranged with respect to their breakdown voltage resp. their dissipation factor.

Fig. 4 represents the three measurements of the dissipation factor (here at 0.1 Hz) according to our experimental procedure (fig. 1). It characterises the dielectric properties of the cables without an influence of the parameters intentionally varied in the time domain measurements (fig. 5 and fig. 6). Obviously, in case of the cables with a low amount of water trees (C6 - C8, C12) except the third measurement of cable C6, where the tan δ increased for unknown reasons, the starting point for time domain measurements with different charging periods was constant. In case of the stronger water-treed cables (C5, C9 - C11) the dissipation factor increased from measurement to measurement (fig. 4, tab. 1). This was probably due to a to short preconditioning period prior to the first dissipation factor measurement (>

100h / 50Hz at U_0)). This period was not long enough to concentrate the humidity inside the water tree structures. This has to be considered when discussing the influence of the charging period as represented in fig. 3, resp. fig. 5 and fig. 6.

However, in general, fig. 4 to 6 show evidently, that there is a correlation between time and frequency domain responses. Strong water-treed cables (C5, C9 - C11) give higher responses than unaged or slightly water-treed cables (C6 - C8, C12), and usually break down at lower voltages (tab. 1; for correlation of the dielectric response to water-tree-analysis based on more measurements see [7]). However, the differentiation between different aged cables is in time domain not as good as in frequency domain, e.g. there is no differentiation in the time domain response in case of the cables C8 - C11 (fig. 5, fig. 6) despite their different water tree deterioration and breakdown voltage (tab.1, further results [7]). A longer charging period increases the absolute value of the response only slightly. However, the classification of different service-aged cables with respect to the absolute value is almost not affected.

Due to a higher probability of polarising charges during a longer charging period, the relaxation of the depolarisation current decreases slower as can be seen clearly in fig. 3 for the unaged cable. Nevertheless, the relaxation resp. the slope n of the depolarisation current (calculated with the approach $i_d(10s-100s) = A*t^{-n}$) seems not to correlate to the water tree deterioration. For example cable C11 is stronger water-treed (U_{bd} = 5,5) than cable C7 (U_{bd} > 14), but the relaxation of C7 is slower than of C5 (further results tab.1). In fact, this confirms the results published in [2]. Otherwise measurements of cables with a high amount of crosslinking byproducts show at certain voltages (> 0.5U_0) a strongly influence to the relaxation behaviour (tab. 1, C3). Therefore, an evaluation of the relaxation behaviour of the depolarisation current as diagnosis criterion can not be recommended.

However, due to the same reason as in the case of the relaxation behaviour, a longer charging period increases also the 'Non-Linearity' of the dielectric response as mostly visible in case of cable C6 and C12 (fig. 5 / 6, tab.1). Furthermore fig. 4 - fig. 6 show clearly, that the 'Non-Linearity' is much stronger in the frequency domain than in the time domain. However, the 'Non-Linearity' correlates less to the water tree deterioration than the absolute value of the dielectric response and is also affected by the amount of crosslinking byproducts.

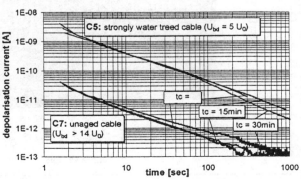

Fig. 3: *Depolarisation current normalised to 100pF (at 2U_0) parameter: charging period*

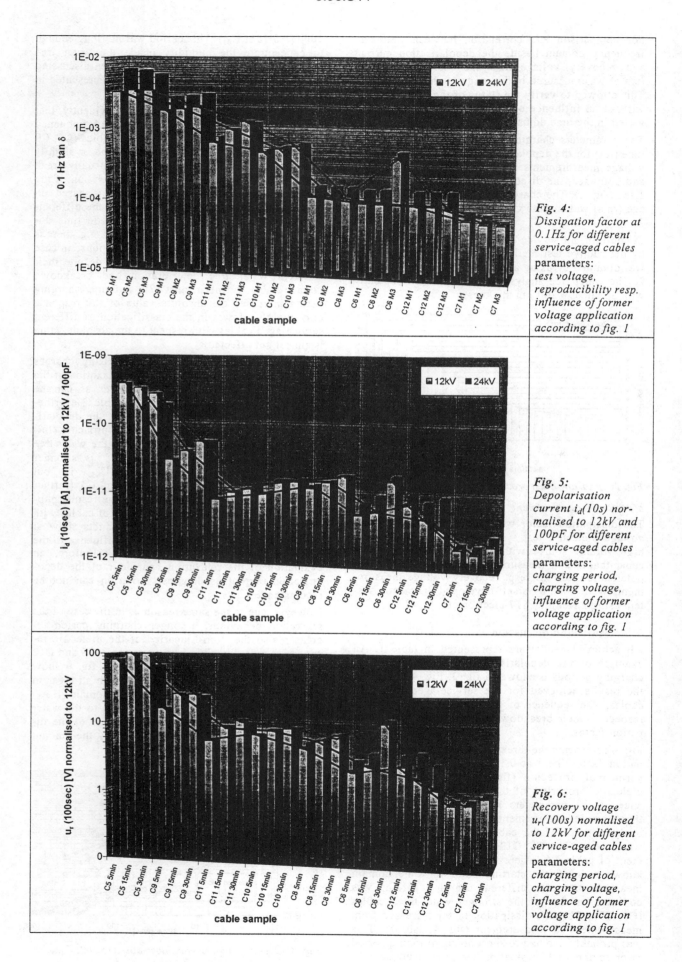

Fig. 4:
Dissipation factor at 0.1Hz for different service-aged cables
parameters:
test voltage, reproducibility resp. influence of former voltage application according to fig. 1

Fig. 5:
Depolarisation current $i_d(10s)$ normalised to 12kV and 100pF for different service-aged cables
parameters:
charging period, charging voltage, influence of former voltage application according to fig. 1

Fig. 6:
Recovery voltage $u_r(100s)$ normalised to 12kV for different service-aged cables
parameters:
charging period, charging voltage, influence of former voltage application according to fig. 1

off frequency, A,B are constants, β_g is a wave number in the cavity.

The r component of electric field for TE and TM mode changes periodically when the PD source is rotated around the inner conductor in θ direction. The period is π/m. So by measuring the distribution of Er, the mode of electromagnetic wave can be identified.

4. Results

4.1 Dependence of frequency spectrum on the angle between the PD source and the sensor

Here results of rotating the PD source around the inner conductor are described. Figure 2 shows the result of the dependence of frequency spectrum on the angle between the PD source and the sensor. Fig.2(a) is the data when the PD source is placed on the inner conductor. The cut-off frequency of each mode is also shown in Fig.2(a).

Below the fc_{TE11}(the cut-off frequency for TE11 mode), the peaks in frequency spectrum have no dependence on the angleθ. They are certainly TEM modes. Above the fc_{TE11} the intensity of the peaks changes periodically along the θ axis.

Between fc_{TE11} and fc_{TE21}, only TEM and TE11 mode can be excited. Because each spectrum peak in the frequency range has two nodes in θ direction, they are TE11 mode. The distribution with two nodes is also observed around the area of 740MHz. It shows that the TE11 mode is excited in higher frequency region. The peaks around 650MHz are TE21 mode since they have four nodes in θ direction.

TE31 or TE41 mode seems not to be excited although TEM, TE11, and TE21 mode are clearly observed. If exist, when rotating the PD source around the inner conductor in θ direction, six or eight nodes would appear. But spectrum peaks with such number of node can't be found in Fig.2(a). It shows that when the PD source is on the inner conductor, the excitation of higher order mode like TE31 and TE41 mode is very weak.

Figure 2(b) shows the dependence of frequency spectrum on the angleθ when the PD source is on the inner wall of the tank. The amplitude of the spectra of TEM mode seen in the frequency below fc_{TE11} is small comparing to the case that the PD source is on the inner conductor. On the other hand, above the frequency of fc_{TE31} or fc_{TE41}, frequency peaks with six or eight nodes in θ direction are clearly observed. They are TE31 or TE41 mode. When the PD source is on the inner wall of the tank, higher order mode can be excited easily comparing to the PD at the inner conductor surface.

In these circumstances, the mode of each spectrum can be identified by observing the dependence on the angle between the PD source and the sensor. And the amplitude of the spectrum depends on the position of the PD source (the inner conductor side or the inner wall of the tank side). In theory there is no cut-off frequency for TEM mode, so the electromagnetic wave of TEM mode can propagate regardless of its frequency. However in actual PD, the TEM mode can hardly be observed above the fc_{TE11}. The reason is thought to be the localization of the PD source. Because of the localization, the electric field doesn't have the same

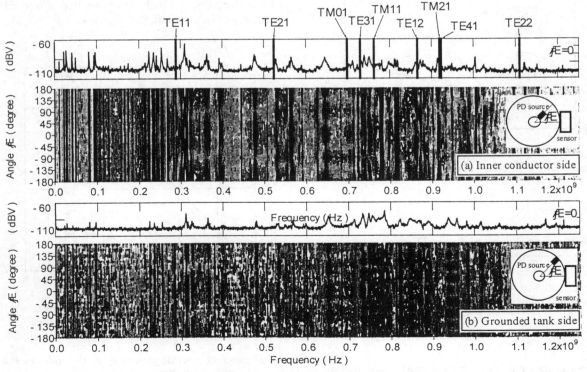

Fig.2 Dependence of frequency spectrum on the angle between the partial discharge source and the sensor.

amplitude around the inner conductor. So the wave of TEM mode (whose electric field has the same amplitude around the inner conductor) is hardly observed above the fc_{TE11}, and the amplitude of TE mode (whose electric field doesn't have the same amplitude around the inner conductor) becomes larger than that of TEM mode.

4.2 Dependence of the amplitude of spectra on the position of the PD source along radial direction

We made another experiment changing the position of the PD source along radial direction. The result is shown in Fig.3. The intensity of the ac electric field, which determines PD magnitude, changes as $1/r$. Consequently depending the position of PD source, the magnitude of the PD may change even if a given voltage is applied to the inner conductor. To eliminate the effect of PD magnitude variation, the amplitude of each spectrum is normalized by the amplitude of TE21 mode. This normalization is justified by the small dependence of TE21 mode on radial position as will be shown in theoretical calculation(Fig.5).

Fig.3 shows that the amplitude of lower order mode such as TEM and TE11 decreases when the PD source moves toward the grounded tank side, while the amplitude of higher order mode such as TE31

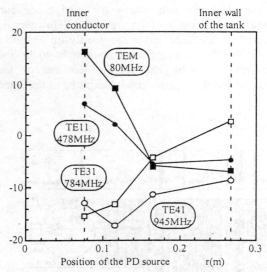

Fig.3 Radial distribution of each mode along r axis and TE41 increases.

5. Discussion

5.1 Resonance characteristics of TE mode

In the experiment TEm1(m:1-4) mode is observed. Considering the GIS as a resonant cavity with the length l, the resonant frequency of TE mode is determined in eq.(4).

$$f_{mn,p} = \sqrt{f_{cTEmn}^2 + \left(\frac{pv}{2l}\right)^2} \quad (4)$$

wherev is the velocity of light. Eq.(4) shows that an interval between resonant frequencies becomes

small for the mode with high cut-off frequency. This characteristic appears in the experimental result shown in Fig.2.

Equation (4) is rewritten as follows.

$$\sqrt{f_{mn,p}^2 - f_{cTEmn}^2} = p\frac{v}{2l} \quad (5)$$

The left side of the eq.(5) can be calculated from the resonant frequency that is experimentally observed and the cut-off frequency that is theoretically calculated. By plotting the value of the left side of the eq.(5) to the index p, the fundamental frequency of the resonance in z direction can be determined. Figure 4 is the result of the plotting. All points for each mode make a straight line. The slope is approximately 100 MHz. Suppose that this is a resonance of 1/2 wavelength, the length of the cavity would be 1.5m. Suppose that this is a resonance of 1/4 wavelength, the length of the cavity would be 0.75m. In this experiment the length of the tank T1 is 1.5m, the length of the inner conductor inside the tank T1 is 1m. And the spacer recesses into the tank T1. Moreover the end of the left side of the tank is not coaxial but cylindrical. Therefore it is difficult to determine the fundamental frequency in the tank T1 by its structure. However if the reflection point is the spacer S2, which is expected as the reflection point next to S1, the length of the cavity would be 3m. That is too large to regard as the reflection point of the resonance. Consequently the resonant frequency of TE mode measured in this experiment is the result of resonance inside tank T1.

Resonant frequencies appeared just above the cut-off frequency calculated theoretically. So the peaks in frequency spectrum can be estimated from the cut-off frequency which is calculated from the diameters of inner conductor and tank.

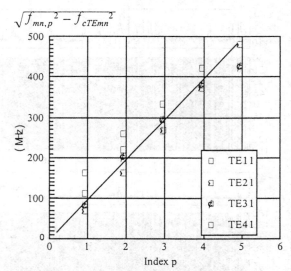

Fig.4 Relation between resonant frequencies of each mode and resonant index of z direction

5.2 Dependence of the power of each mode on the position along radial direction

In this section dependence of the intensity of each

mode on the position along radial direction will be discussed. The distribution of electromagnetic field build by resonance is determined not by the position of the PD source but by the structure of the cavity. On the other hand, the intensity of a mode is determined by the position of the PD source in the cavity. In general, when the PD source is placed on the position where the amplitude of the electric field of a given mode is large, the mode is excited strongly. Intensity variation of a mode is to be estimated from the distribution of the electric field of the mode under consideration.

Figure 5 is the Er distribution of TE modes along radial direction calculated from eq.(2). In the case of m=1, the intensity of the electric field is stronger at the inner conductor than at the inner wall of the tank. As the index m becomes larger, the intensity of the electric field at the inner conductor becomes weaker. This means that for large m, excitation of the TEm1 mode is easier at the tank wall than at the inner conductor. This characteristic agrees with Fig.3. To confirm this characteristic quantitatively, the ratio of the electric field at the inner conductor to that at the wall of the tank is shown in Fig.6. The ratio of the peak intensity of frequency spectra measured for the corresponding two positions is also plotted. The ratio becomes smaller as the index of the mode becomes larger. For TEM and TE11 mode the ratio is +10dB, while for TE41 mode the ratio is -20dB. The measured value(●) from spectrum peak shows relatively good agreement with the theoretical value (the bar graph). Thus the dependence of the amplitude of each mode on the PD position along radial direction can be estimated from the distribution of the theoretical electric field.

The above results means that depending on the position of the PD source, the sensitivity changes if detecting with a given frequency range. The TEM and TE11 modes are suitable for the detection of the PD at the inner conductor, while higher order mode such as TE31 and TE41 are suitable for the PD at the inner wall of the tank.

incorporated in GIS could be at the inner conductor surface, tank surface, and in-between when it is on the spacer. Therefore it is very important to be able to have knowledge on its radial location without opening GIS tanks. The discussion above showed that the radial position of PD source can be estimated from its frequency spectrum with consideration on its mode.

6. Conclusion

Resonance characteristics of electromagnetic wave generated by partial discharges in GIS were investigated. Main results are summarized as follows.

(1) Mode of peaks in frequency spectrum was identified up to 1 GHz.

(2) TEM mode is little observed by actual PD above the cut-off frequency of TE11 mode.

(3) Depending on the radial position of PD source, the amplitude of each mode changes according to the electric field distribution expected from the theory of electromagnetic wave in co-axial cylindrical cavity. PD source at inner conductor excites TEM and TE11 mode strongly while PD at tank wall excites TE31 and TE41 modes.

References

[1] H.Muto et al.:"Frequency Spectrum due to Standing Waves Excited by Partial Discharges in a GIS", 10th Int. Symp. on High Volt. Eng., 4, 179-183 (1997)

[2] R.Kurrer et al.:"Attenuation Measurement of Ultra-High-Frequency Partial Discharge Signals in Gas Insulated Substations", 10th Int. Symp. on High Volt. Eng., 2, 161-165 (1997)

[3] K.Mizuno et al.:"Investigations of PD Pulse Propagation Characteristics in GIS", IEEE Trans. on PS, 12, 2 (1997)

[4] M.D.Judd et al.:"The Excitation of UHF Signals by Partial Discharges in GIS", IEEE Trans. on DEI, 3, 213-228 (1996)

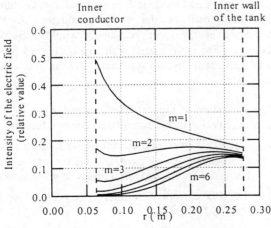

TEm1 mode⌐Er/ ZH Ø E= 0)

Fig.5 Distribution of radial electric field along r axis.

Metallic particles which may possibly be

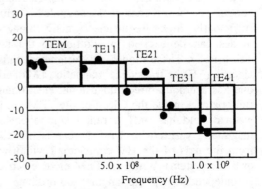

● : Ratio of the amplitude of the spectrum (measured)

Bar graph : Ratio of the amplitude of the electric field (theoritical)

Fig.6 Comparison of peak intensity in frequency spectrum between PD at the inner conductor and the inner wall of the tank

TRANSFER FUNCTIONS FOR UHF PARTIAL DISCHARGE SIGNALS IN GIS

M D Judd and O Farish

University of Strathclyde, UK

Abstract

The principles of electromagnetic radiation by partial discharges (PD) in gas insulated substations (GIS) are described. Calculated transfer functions are presented, showing that transverse electric modes are predominant for radial PD currents in the coaxial geometry. By using actual PD current pulses from a free moving particle together with the measured sensitivity of a UHF coupler, the entire transfer function for the PD detection problem is modelled. The relationship between UHF signal energy at the coupler output and the electronic component of the PD pulse is investigated using the model, revealing a significant degree of correlation.

1. Introduction

The theory of UHF signals excited by partial discharges (PD) in gas insulated substations (GIS) has been described in detail previously [1]. In this paper, we present examples of calculated GIS transfer functions and explore the implications for UHF PD detection. The transfer of energy from PD to distant coupler is undoubtedly a complex process. However, a proper understanding of the transfer functions will assist with the interpretation of UHF signals when used in conjunction with improved measurements of typical PD current pulses in SF_6 and information about the defect that can be gained from phase resolved measurements.

2. Theory

Essentially, the PD is a current pulse that occurs somewhere in the insulating region between the high voltage (HV) conductor and the metal cladding. The conductors form a coaxial "waveguide" in which, at sufficiently high frequencies, many higher-order modes can be excited in addition to the familiar TEM mode. These modes are designated transverse electric (TE) or transverse magnetic (TM) and each one has a unique three-dimensional electromagnetic field pattern inside the GIS. For the TEM mode, the electric field has only a radial component whose magnitude varies with $1/r$, where r is the distance from the axis of the HV conductor (cylindrical co-ordinate system). While the TEM mode electric field is independent of the angular co-ordinate ϕ, the fields of higher-order modes have amplitudes that are modulated over the GIS cross-section by $cos(n\phi)$, where n is the integer in the designation of a mode as TE_{nm} or TM_{nm}. In addition, radial variations of the field patterns in higher-order modes involve Bessel functions and their derivatives.

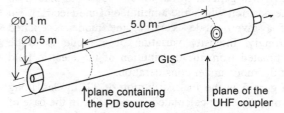

Fig. 1 Relative position of PD source and coupler used in the transfer function calculations.

To gain a qualitative understanding of how the modes are excited, consider the PD source as a short burst of current flowing in a radial direction from $r = r_1$ to $r = r_2$. Let the distance ℓ over which this current flows be small, $\ell = \Delta r = r_2 - r_1$. In simple terms, the current will excite a particular mode in proportion to the potential difference V which that mode would impress across ℓ. Thus if a mode satisfies the condition of equation (1) it will not be excited at all by the radial component of PD current.

$$V = \int_{r_1}^{r_2} \mathbf{E} \cdot d\mathbf{r} = 0 \qquad (1)$$

Conversely, the PD will excite maximum signal amplitude in that mode if it occurs at a radius where the integral in (1) is a maximum. If ℓ is small, the integral in (1) can be approximated by

$$\ell E_r \qquad (2)$$

where E_r is the radial component of $\mathbf{E}$. Now it is clear that a small radial PD will excite a mode in proportion to the magnitude of its radial electric field at the PD position. The TEM mode can be used to illustrate this principle. In a GIS with conductor radii $a = 0.05$ m and $b = 0.25$ m (see Fig. 1), the electric field of the TEM mode is 5 times greater at the inner conductor than at the outer conductor. A PD occurring at the inner conductor would therefore excite a TEM mode signal with an amplitude 5 times that generated by an identical PD at the outer conductor.

The excitation of higher-order modes by PD in GIS is primarily due to the combination of two factors:

1. The PD is small and cannot excite fields that are independent of ϕ over the GIS cross-section because of the propagation delay at velocity c.

2. The PD current i exhibits a very high rate of change (di/dt).

Higher-order modes are only a mathematical approximation to reality, representing three-

High Voltage Engineering Symposium, 22–27 August 1999
Conference Publication No. 467, © IEE, 1999

dimensional fields by summing repetitive field patterns in the spatial domain. This is analogous to the way in which a periodic signal in the time-domain can be represented by a Fourier series. If one considers the early stages of radiation from a PD, the time taken for the transient field to spread out from the defect to the surface of a sphere of radius r_0 is $t = r_0/c$. When t is small, this field clearly cannot be represented by the TEM mode alone, which is non-zero over the whole cross-section of the GIS. Thus any TEM mode contribution must be offset by higher-order modes in the exact proportions required to ensure that the fields cancel everywhere except inside the volume defined by r_0. The overall transfer function $G(\omega)$ between a PD source and the UHF electric field at a distant coupler is the sum of the transfer functions of the individual modes. This leads to a complex relationship that is best investigated using mathematical simulation software.

3. GIS transfer function

Consider the section of GIS shown in Fig. 1, in which a PD source and UHF coupler (both at $\phi = 0$) are separated by a distance $z = 5.0$ m. The cut-off frequencies f_c of the higher-order modes can be calculated from the conductor diameters. All those for which $f_c < 2000$ MHz are listed in Table 1.

Fig. 2 shows two examples of the GIS transfer function $|G(\omega)|$ calculated using the theoretical approach described in [1]. The PD source is identical in both cases, but in Fig. 2(a) it is located at the outer conductor while in Fig. 2(b) it is located at the inner conductor. $G(\omega)$ relates the electric field at the coupler to the PD current flowing at the site of a defect and has units of Vm^{-1}/A. Signal reflections and ohmic losses have not been included. At frequencies below TE_{11} cut-off only the TEM mode propagates. In this region the transfer function is independent of ω, and if the graphs are compared the magnitude ratio is seen to be 5, corresponding to the ratio of the conductor radii. At higher frequencies, as each new mode begins to propagate, there is a step in the transfer function followed by a series of ripples. In practice, these steps will not be so abrupt because, although signal attenuation is generally quite low, it does increase sharply for each mode at frequencies close to cut-off. The ripples are introduced because the phase of each mode changes rapidly at frequencies just above cut-off.

$G(\omega)$ is dominated by the TE_{n1} modes, particularly when the PD source is at the outer conductor. In Fig. 2(a), their contribution increases with n, while the TE_{n2} and TE_{n3} modes are visible only as minor ripples just above their cut-off frequencies. When the

Table 1 - *List of modes used in the simulation and their cut-off frequencies (MHz) in the 400 kV GIS.*

TE_{11}	325.4	TE_{61}	1431.6	TE_{32}	1519.9	TM_{01}	728.3	TM_{51}	1674.1	TM_{32}	1884.5
TE_{21}	579.2	TE_{71}	1637.1	TE_{42}	1769.9	TM_{11}	808.4	TM_{61}	1896.3		
TE_{31}	801.4	TE_{81}	1841.2	TE_{13}	1609.5	TM_{21}	996.6	TM_{01}	1485.9	TEM	n/a
TE_{41}	1014.8	TE_{12}	946.8	TE_{23}	1822.6	TM_{31}	1220.4	TM_{12}	1537.4		
TE_{51}	1224.4	TE_{22}	1239.6			TM_{41}	1448.7	TM_{22}	1680.3		

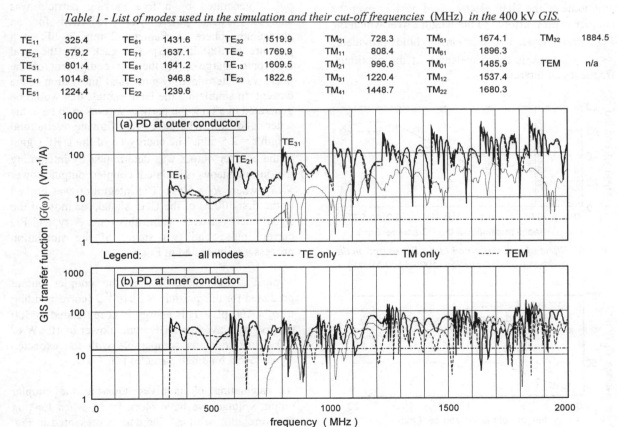

Fig. 2 *Transfer function $|G(\omega)|$ for a GIS, relating the electric field at the coupler to the PD current. In these examples the PD source is a 10 mm protrusion. In (a) the protrusion is on the outer conductor, while in (b) it is on the inner conductor. Contributions from the TEM, TE and TM type modes to the overall transfer function are also shown.*

PD source is at the inner conductor, the TE mode contribution does not increase so markedly with n. The explanation lies in the fact that for a given value of n, the $\cos(n\phi)$ variation of the fields results in a greater rate of change with distance at the surface of the inner conductor so that fewer modes are required to represent the same source field.

How should the complex information presented in Fig. 2 be interpreted? To evaluate variations in the transfer function, consider the frequency band 500 - 1500 MHz, which is typical of broadband UHF PD detection systems. Taking the average the value of $|G|$ over this range, we obtain a value of 65 Vm^{-1}/A for PD at the outer conductor and 45 Vm^{-1}/A for PD at the inner conductor. More generally, if the transfer function is calculated at 10 mm intervals as the PD source is moved between the conductors, we find that its average value varies as shown in Fig. 3. This variation does not imply that the energy radiated by the PD source varies with its position. Remember that the coupler only extracts a small portion of the energy travelling along the GIS. In this example, the PD source will generate a maximum electric field at the coupler if it is located at $r = 0.225$ m.

The effect of the size of the PD source can be seen in Fig. 4. If all other parameters remain constant, the amplitude of the UHF electric field passing over the coupler is proportional to the length ℓ over which the PD current flows. This relationship holds provided $\ell \ll \lambda$, where λ is the wavelength at the maximum frequency of interest.

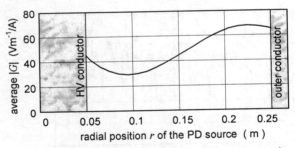

Fig. 3   Effect of radial position of the PD source on the average transfer function (500 - 1500 MHz).

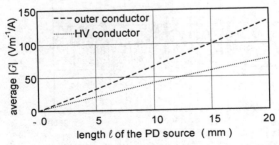

Fig. 4   Effect of PD source size on the average transfer function (500 - 1500 MHz), showing results for the PD occurring at the inner and outer conductors of the GIS.

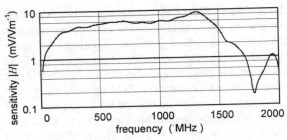

Fig. 5   Frequency response of the UHF coupler.

4. A study using the transfer function

Until now we have only been concerned with signal transfer from PD current to electric field at the coupler. To complete the transfer function for UHF PD detection we need to know the sensitivity $H(\omega)$ of the UHF coupler, which relates the coupler output voltage to the incident electric field. Calibration results for actual couplers [2] can be used. In this case we will use real data from an internal disk-type coupler. The magnitude response $|H(\omega)|$ for the coupler is shown in Fig. 5.

As input data to the simulation, broadband measurements of real PD pulses can be used. To illustrate the use of the transfer function to simulate PD detection, a randomly selected set of 64 current pulses generated by a free moving particle was analysed. These pulses were recorded for an aluminium sphere of diameter 2 mm in SF$_6$ at a pressure of 380 kPa [3]. For each PD, the total electronic charge q_e in the fast component of the pulse was determined by numerical integration of the current. In simulating the UHF signals that would be generated in a GIS, the particle was taken to be at the outer conductor, with the current flowing over a total length $\ell = 2.5$ mm. The energy U_c of the UHF signal at the coupler output was determined by integrating the instantaneous simulated coupler output power V^2/R_L, where $R_L = 50 \, \Omega$. The integration was limited to the first 500 ns of the UHF signal, as most of the signal energy arrives within this time. A typical PD current pulse and two stages of the simulation process are illustrated in Fig. 6.

Typical signal energy U_c at the coupler output predicted for this particle is 5×10^{-15} J (corresponding to $q_e \approx 100$ pC). This energy is accumulated in 500 ns, giving an average UHF signal power of 10 nW or -50 dBm, which is comparable with the expected signal levels for small particles [4].

As calculating U_c involves squaring the coupler output voltage, we have plotted $\sqrt{U_c}$ when looking for correlation with q_e. The data is presented in Fig. 7, which shows results for PD pulses of both polarities. Note that the polarity of q_e relates to the charge acquired by the particle from the measurement electrode.

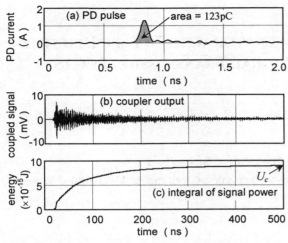

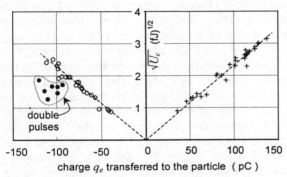

Fig. 6 (a) *Measured PD current pulse,* (b) *calculated coupler output voltage into 50Ω and* (c) *integrating signal power to determine the energy received at 500 ns.*

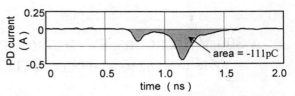

Fig. 8 *Example of a negative PD pulse with precursor.*

Fig. 7 *Correlation between q_e and $\sqrt{U_c}$.*

5. Discussion

On the whole, correlation between q_e and $\sqrt{U_c}$ is quite good. An analogous result has been derived for VHF signals in GIS [5]. Previous experiments [1] have shown that the UHF signal is very sensitive to the rate of change of PD current (di/dt), so this correlation is an indication of the similarity of PD pulses with different amplitudes. A significant deviation occurs for some of the larger negative pulses shown in Fig. 7. An examination of the time-domain records revealed that these were all double pulses, in which a larger PD pulse followed a smaller precursor. The contribution from both pulses was included in the calculations for q_e. The reason for the lower U_c in these cases is the reduced di/dt associated with these pulses, one of which is shown in Fig. 8. This observation is a natural consequence of electromagnetic radiation arising from the acceleration and deceleration of charges [6], remembering that

$$\frac{di}{dt} = \frac{d^2 q_e}{dt^2} \qquad (3)$$

Herein lies the fundamental difference between conventional and UHF PD measurements. In the conventional method, PD current is essentially integrated to obtain the apparent charge. The UHF method is primarily sensitive to charge acceleration at the defect site and therefore requires a different approach to the issue of calibration.

6. Conclusions

Calculated transfer functions for UHF PD detection in GIS show that transverse electric modes are the dominant modes excited by a radial PD current. The amplitude of the UHF signals is primarily dependent on the acceleration and deceleration of charge in the PD. The shape of the PD current pulse therefore plays a critical role in determining UHF signals. However, the physics of SF_6 ionisation can lead to quite consistent pulse shapes, in which case a linear relationship between the electronic component of the PD charge and the square root of the coupled UHF signal energy can be demonstrated.

Further experimental work is required to address issues of risk assessment using UHF PD detection. An appreciation of the underlying theoretical principles is important if the potential of the technique is to be fully realised.

7. Acknowledgement

This work is funded by an EPSRC research grant.

8. References

[1] M D Judd, O Farish and B F Hampton, "Excitation of UHF signals by partial discharges in GIS", IEEE Trans. DEI, Vol. 3, No. 2, pp. 213-228, April 1996

[2] M D Judd and O Farish, "A pulsed GTEM system for UHF sensor calibration", IEEE Trans. Instrumentation and Measurement, June 1998

[3] M D Judd and O Farish, "High bandwidth measurement of partial discharge current pulses", Conf. Record IEEE Int. Symp. on Electrical Insulation (Washington), Vol. 2, pp. 436-439, 1998

[4] R Kurrer and K Feser, "Attenuation measurements of UHF partial discharge signals in GIS", Proc. 10th ISH (Montreal), Vol. 2, pp. 161-164, 1997

[5] H-D Schlemper, A Vogel and K Feser, "Calibration and sensitivity of VHF partial discharge detection in GIS", Proc. 9th ISH (Graz), Vol. 5, paper 5619, 1995

[6] G S Smith, "On the interpretation for radiation from simple current distributions", IEEE Antennas and Propagation Magazine, Vol. 40, No. 3, pp. 9-14, June 1998

A 300 V Mercury Switch Pulse Generator with 70 Psec Risetime for Investigation of UHF PD Signal Transmission In GIS

S. M. Neuhold, H.R. Benedickter[*], M. L. Schmatz[*]

FKH Zurich, Voltastrasse 9, CH-8044 Zurich, Switzerland
[*]ETH Zurich Lab. for Electromagn. Fields and Microwave Electronics, Gloriastrasse 35, CH-8092 Zurich Switzerland;
email: neuhold@fkh.ch

Abstract

A Mercury switch pulse generator for the purpose of high frequency characterization of **G**as **I**nsulated **S**witchgear (GIS) is presented. The compact device is well matched to 50 Ω and operates with a DC voltage of 300 V. It consists of a charged cable, a fast mercury switch relay and supporting passive elements.

With this combination, a 150 V pulse may be applied to a GIS **P**artial **D**ischarge (PD) sensor with a 70 ps rise time and a 50 ns pulse length, having a maximum repetition rate of more than 100 Hz.

The robust device is light-weight and easy to handle. Due to its modular design, it may be directly connected to a GIS PD sensor. It may be used for the simulation of PD discharges in GIS, for impulse propagation speed measurements and for the sensitivity check of GIS PD sensors together with UHF PD measurement systems.

Keywords: mercury switch, picosecond, pulse generator, Gas insulated switchgear, GIS, Partial discharges, PD, Ultra high frequency, UHF, SF6

Introduction

The spectrum of PD signals in GIS contains relevant frequency components above 3 GHz [1, 2]. To check the sensitivity of installed PD sensors in GIS, artificial pulses are injected into one PD sensor and measured at the other sensors [3, 4]. A calibration of the UHF-method is not possible [5]. Up to now, pulse generators were used with rise time of 200 psec – 400 psec [1, 6, 7]. According to formula (1), this results in a frequency spectrum limitation of 0.9 GHz to 1.4 GHz [8].

Formula 1: $$f_{3dB} = 0.35/ t_r$$

When injecting a charge Q into a GIS via PD sensor, the capacity C_1 of the PD sensor to the center conductor determines the pulse amplitude of the artificial PD signal [4]. For some sensor designs, the value of C_1 is small (0.15 pF). To apply a charge of 15 pC, a pulse amplitude of 100 V is necessary. This results in a charging voltage of 200 V for a cable pulse generator design [4]. Many pulse generators have a maximum output voltage in the range of only 5 V – 50 V [1, 9]. The attenuation of propagating PD-Signals in GIS increases considerably towards higher frequencies [7, 10]. This fact allows a rough localization of the PD source via the measurement of the power spectrum of the PD signal. On the other hand, some constructive parts of the GIS (circuit breakers, disconnectors) lead to a high damping of high frequency signals [7]. To study the frequency damping characteristics of a GIS design and to get information for possible PD source localization, it is important to use a pulse generator with high amplitude (to reach the most distant sensor) and a spectrum as similar as possible to a real PD-source. The pulse generator should be connected as close as possible to the PD sensor to reduce signal distortion and skin effect damping [11]. With standard pulse generators, this is very difficult due to their weight and/or size.

What would be needed is a small, light-weight pulse generator with very short rise time (< 100 psec), high signal amplitude (> 200 V) and a maximal repetition rate of approximately 100 Hz. It should be possible to connect the switch very close to the PD sensor and to synchronize it with 50 Hz / 60 Hz or with an arbitrary frequency (e. g. of a resonant high voltage testset) up to 100 Hz.

Design

The realized pulse generator consists of a mercury switch relay, connecting a DC charged cable to the output (Fig. 1). All components are well matched to 50 Ω. Due to the mercury wetted contacts, no bouncing occurs. The closing and opening of the mercury switch is controlled with a coil. Between charging cable and DC supply, a matching circuit M_2 is connected to eliminate reflections and to control the charging time.

High Voltage Engineering Symposium, 22–27 August 1999
Conference Publication No. 467, © IEE, 1999

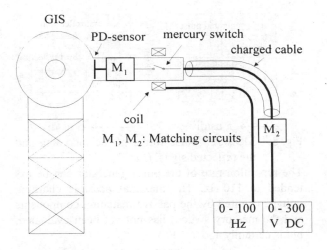

Fig. 1: Design of the pulse generator

The mercury switch, together with a matching circuit M_1 can be connected directly to a GIS sensor (Fig. 2). The charging cable with the matching circuit M_2 and the supply cable for the coil may be connected via long cables to the coil driver unit and the charging unit. This has the advantage, that the mercury switch of the pulse generator is connected as close as possible to the PD-sensor. Due to the small size of all components, the device (mercury switch with matching unit) can easily be connected directly to PD sensors, even at hardly accessible locations of the GIS with limited space or at high locations.

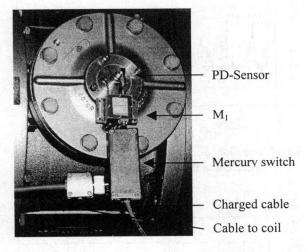

Fig. 2: Mercury switch with matching unit connected to a PD sensor

Fig. 3 shows the pulse generator circuit. The charging cable is connected via a charging resistor R_3 at a DC voltage level of U_1. Closing the mercury switch by energizing the coil, a pulse with an amplitude of $\frac{1}{2}$ * U_1 is applied to the PD sensor. The resistor R_1 to ground between mercury switch and PD sensor discharges the PD sensor with a time constant several orders of magnitude higher than the risetime and duration of the pulse.

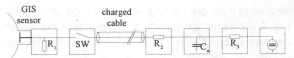

Fig. 3: Pulse generator circuit

Performance

Fig. 4 shows the measured rise time of the pulse generator at a charging voltage of 300 V and a repetition rate of 100 Hz. To measure the risetime (10% - 90% of the peak amplitude), a 6 GHz sampling oscilloscope with internal delay line together with a 40 dB attenuator was used (Tektronix TDS 820). The rise time of the oscilloscope was calculated (Formula 1) to measure 58 psec.

Due to the fact, that the measured rise time (t_r) of approximately 70 psec is very close to the rise time of the oscilloscope (t_{Osz}), the effective rise time (t_r') of the pulse generator can be estimated to approximately 35 psec [9,12].

Formula 2:
$$t_r' = \sqrt{t_r^2 - t_{Osz}^2}$$

(Assuming fist order spectrum decay at high bandwidth limit)

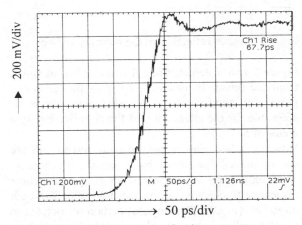

Fig. 4: Measured rise time of pulse generator

In Fig. 5, the pulse shape is shown, with a length of the charging cable of 1 meter. The length of the pulse is a function of the capacitance C_a in Figure 3.

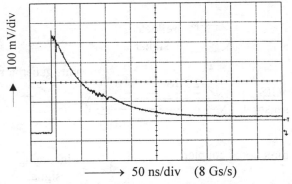

Fig. 5: Pulse shape with a charging cable of 1 meter.

The very fast rise time of the realized pulse generator is a direct result of the optimized source impedance match to 50 Ω. The mercury switch is mounted in a coaxial system minimizing any unwanted reflections. Fig. 6 shows the measured source match of the closed mercury relais. Up to 3 GHz, the measured reflection coefficient was better than −17 dB. This guarantees that any energy reflected by an unmatched sensor is dissipated in the source and no significant re-reflections disturb the measurement. This is not the case with traditional designs where no cable termination is used on the source side of the pulse generator. No matching resistor is needed after the switch. This results in higher signal amplitude and makes the design independent of the load reflection coefficient. Even with a PD sensor other than 50 Ω impedance, no multiple pulse reflections are generated. This is very important because such multiple injected pulses could not be separated from a GIS reflection measured on another PD sensor.

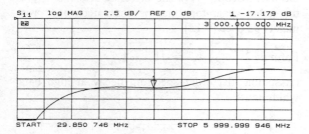

Fig. 6:   Source reflection of the mercury switch

Figure 7 and 8 demonstrate, that the reflected signal from the sensor is almost as large as the generated pulse. The good match of the pulse generator is responsible for the attenuation of the re-reflection by a factor of 5.

However, plate sensors used in wide spread GIS are not optimized for broad band signal transmission. Thus only parts of the spectrum at typical sensor resonance frequencies are in fact transmitted in both directions (Fig.9). This imposes certain restrictions to the PD scaling with plate type sensors. Nevertheless, better sensor designs with other shapes (conical, stripline) are already available an will preferably be used in the future.

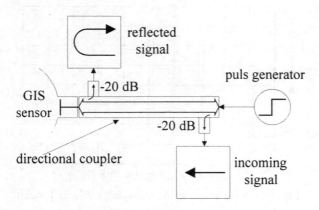

Fig. 7:   Measurement setup for the determination of the pulse reflection at the sensor

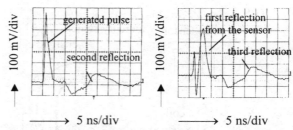

Fig. 8:   Measured pulse forms of the incoming and the reflected signal (Fig. 7)

The repetition rate of the pulse generator can be extended to 110 Hz. The maximal possible charging voltage (limited by the passive matching components and the mercury switch) has not yet been evaluated, but is certainly above 300 V.

Simulation of PD Discharges

The pulse generator design was tested on a 170 kV GIS substation having 12 PD sensors. At one PD sensor (capacity of sensor disk to center conductor: 0.15 pF), pulses of 100 V were applied, resulting in a 15 pC charge injection. On the adjacent sensor, the signals were measured with a wide bandwidth (0.1 - 1.8 GHz) measurement system. The frequency spectrum measured (Fig. 9) shows high signal energy between 1.3 GHz and 1.7 GHz. Therefore it is essential for this type of sensor to have a pulse generator with a frequency spectrum up to 2 GHz.

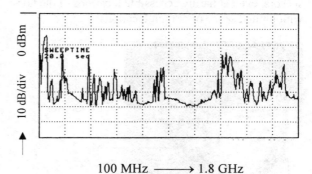

100 MHz ⟶ 1.8 GHz

Fig. 9: Injection of pulse in a PD sensor: measured spectrum at the adjacent sensor

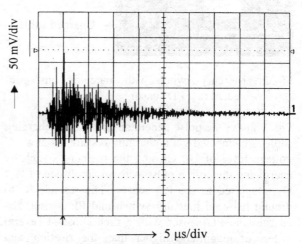

⟶ 5 μs/div

Fig.10:   Sensitivity check of PD detection in GIS

Given a certain layout of PD sensor placement, it is proposed to perform a sensitivity check of the PD measurement to determine the sensitivity for all locations in the GIS [3]. To scale the PD sensitivity an artificial PD pulse is injected via one PD sensor and measured at another PD sensor. Figure 10 shows the measured artificial PD pulse at a PD sensor which is injected at the adjacent PD sensor. With a coupling capacity of the sensor disk to the center conductor of 0.15 pF and a pulse amplitude of 33 V, a charge of 5 pC is injected.

Simulation of PD source localization

To simulate a PD source localization procedure, impulse propagation speed measurements have been carried out on a GIS in service. Fig. 11 shows the experimental setup, fig 12 presents the measured waveforms. With a wide band (0.1 – 1.8 GHz) measurement system, a source localization accuracy of approx. 30 cm was obtained using an oscilloscope with 600 MHz bandwidth.

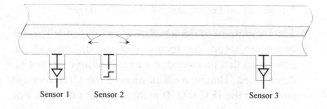

Fig. 11: Measurement setup for simulation of localization measurement. PD like pulses injected in sensor 2 and registered at sensors 1 and 3

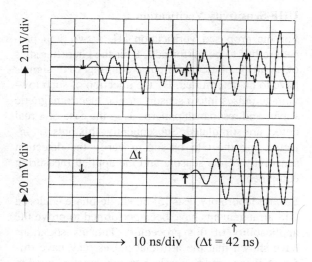

$\longrightarrow$ 10 ns/div   ($\Delta t = 42$ ns)

Fig. 12: Simulation of PD source localization: Waveforms

Conclusions

A modular mercury switch pulse generator for GIS HF characterization was realized. A rise time of < 70 psec and a pulse length of approximately 50 nsec were measured with a DC input voltage of 300 V and a pulse repetition rate of 100 Hz. The circuit is well matched to 50 Ω, therefore multiple reflections can be excluded with this design. This implies that the trans-

ferred pulse into the GIS is much more reliable compared to earlier designs.

Due to the fact, that the measurement rise time of approximately 70 psec is very close to the rise time of the oscilloscope, the effective rise time of the pulse generator it is supposed to be even shorter than 70 psec.

References

[1] R. Kurrer, R.Feger, K. Feser: The Application of UHF-Impulses for Testing UHF PD Measurement Systems applied to Gas-Insulated Substations. CIGRE 1997, Joint TF 15/33.03.05 ...... IWD Feser 4

[2] G. Wanninger: Apparent Charge Measurement in GIS by Modern Diagnostic Methods. ETEP Vol. 7, No. 4, July/August 1997

[3] CIGRE 1998 TF 15/33.03.05 IWD 73: Sensitivity Verification for Partial Discharge Detection on GIS with the UHF and the Acoustic Method.

[4] Albiez, M.: Teilentladungsmessung an SF6-isolierten Schaltanlagen. ETH Zurich, Switzerland, 1992, Thesis Nr. 9694.

[5] Sellars, A.G., MacGregor, S.J., Farish, O.: Calibrating the UHF Technique of PD Detection using a PD simulator. IEEE Trans. on Diel. and El. Ins., Vol. 2, No. 1, 1995, pp. 46-52.

[6] S. Meijer, E. Gulski and W.R. Rutgers: Evaluation of Partial Discharge Measurements in SF6 Gas Insulated Systems. 10th International Symposium on High Voltage Engineering, Montreal 1997

[7] Behrmann, G.J.; Neuhold, S., Pietsch, R.: Results of UHF measurements in a 220 kV Substation during on-site commissioning tests. 10th International Symposium on High Voltage Engineering, Montreal 1997

[8] Schuon, E., Wolf, H.: Nachrichtenmesstechnik. Springer Verlag 1981

[9] J.R.Andrews: Picosecond pulse generators using microminiature mercury switches. NBSIR 74-377, Boulder, Colorado.

[10] Hitoshi Okubo, Toshihiro Hoshino, and Thoshihiro Takahashi, Masayuki Hikita, Akinobu Miyazaki.: Insulating Design an On-Site Testing Method for a Long Distance, Gas Insulated Transmission Line (GIL). IEEE 1998 -Vol. 14, No. 6

[11] W. Bächthold: Lineare Elemente der Höchstfrequenztechnik, vdf verlag / Zürich 1994

[12] G. E. Valley and H. Wallman: Vacuum Tube Amplifiers, pages 77-79, McGraw-Hill Book Company, Inc., New York, 1948.

APPLICATION OF THE CIGRE-SENSITIVITY VERIFICATION FOR UHF PD DETECTION IN THREE-PHASE GIS

M. Knapp, R. Feger, K. Feser

University of Stuttgart, Germany

A. Breuer

ABB Calor Emag Schaltanlagen AG, Germany

ABSTRACT

This paper reports results with the new sensitivity verification for ultra high frequency (UHF) partial discharge (PD) detection applied to three-phase SF_6 gas-insulated substations (GIS) for rated voltages up to 170 kV. The procedure was developed by CIGRE WG 15/33.03.05 [1]. The aim is to examine the practical use of the proposed procedure and to investigate its extension to an unconventional sensor type.

UHF PD measurements were carried out on a typical GIS section via potential grading electrodes. Derived results demonstrate the ability of these already built-in devices to serve as UHF sensors, to replace additional sensors sufficiently, and to make further design efforts unnecessary, if UHF diagnostics are required. Measurements, the experimental setup, and the adequate test-vessel are described.

The sensitivity of the new sensor is compared to the sensitivity of the IEC 60270 method and to other UHF sensors. Results concerning the detection of characteristic defects demonstrate the sensitivity of the new sensor arrangement.

INTRODUCTION

One- and three-phase SF_6-encapsulated substations with operating voltages up to 170 kV are proven to be reliable devices. Although designs differ in many aspects, both types of GIS increase the reliability of supply due to common advantages. On the other hand, the reduction in dielctric strength caused by irregularities like free moving particles or fixed protrusions remains for both. Defects of the insulating system of typical GIS equipment cause PD which can be measured in different frequency ranges and therefore in entirely different ways.

PD Detection for Diagnostics in GIS

As introduction a comparison between the conventional PD measurement (according to IEC 60270) and the UHF method is performed to explain the logical necessity of a sensitivity verification for the UHF method.

IEC 60270 Measurement. Environmental noise limits the sensitivity of these conventional PD measurements. Especially in case of incomplete encapsulation, predominating disturbances are able to suppress evident results. Furthermore, the evaluated apparent charge depends on the PD location, so that easily comparable results are prevented in case of non-symmetrical cross-sections (e.g. three-phase commonly enclosed GIS).

UHF Measurement. In contrast to the IEC 60270 measurement, noise insensitive narrow-banded UHF measurements demonstrate the characteristic spectra obtained by the relevant PD [2].

Since the discharge currents of the defects have rise times in the range of a few ten picoseconds, electromagnetic waves of several modes with frequency contents up to more than 2 GHz are excited. Reflections at various discontinuities of a GIS cause standing waves and complex resonance patterns.

The magnitude of detectable signals depends strongly on the setup, the location of the defect and the sensor, so that an overall transfer function can not be determined. Hence, a calibration of the UHF method similar to the IEC 60270 method is not possible. For this reason, a new sensitivity verification is proposed by the CIGRE WG 15/33.03.05. It will ensure that typical defects can be detected by an UHF measuring equipment.

UHF Sensitivity Verification

For the proposed verification [1], a two step procedure has to be performed. First the characteristics of an artificial electrical pulse emitting UHF signals have to be determined. By the injection of such low-voltage pulses into a sensor, similar electromagnetic waves and resonance patterns as in the case of a real defect are stimulated. Subsequently, this pulse is injected into the GIS on-site to verify the detection sensitivity of the sensors and the applied measuring equipment.

Several laboratory tests [3] on single-phase enclosed GIS arrangements have been performed to prove the applicability of this procedure. The investigations have shown that the artificial pulses may have different shapes and that only a few parameters need to be considered. To excite waves in the whole measuring bandwith, the rise time of the pulse has to be less than one nanosecond. Further parameters with minor importance are the time to half value and the repetition rate. As a main parameter, the magnitude of the pulse has to be determined depending on the GIS design, on sensors and on applied measuring devices.

High Voltage Engineering Symposium, 22–27 August 1999
Conference Publication No. 467, © IEE, 1999

To determine the necessary voltage level, a defect is placed close to a sensor in a compartment of the laboratory setup. Since free moving particles have turned out to be the most important defect type in GIS, this was chosen for basic comparison measurements. The value of 5 pC for the apparent charge is considered to be the sufficient threshold [2]. As soon as the related PD signal is 5 pC, the UHF signal has to be measured at another sensor located in a compartment nearby. This UHF signal ought to be compared with the one from the artificial pulse injected into the sensor close to the defect. The pulse amplitude can be varied until the UHF signal of the pulse equals the magnitude of the defect signal with an accepted tolerance of 20%.

For the on-site test, artificial pulses with the same magnitude and shape are injected into a first sensor. If the UHF signal can be measured at a second sensor, the sensitivity verification is successful for the GIS section between the two sensors. Of course, the same type of equipment has to be used as during the laboratory test.

MEASUREMENTS AND COMPARISON

Description of the Experimental Setup

GIS. The considered setup consists mainly of a three-phase commonly enclosed GIS section. It is fed by a typical non-encapsulated test circuit for single-phase high voltages up to $U_{rms} = 100$ kV (*Fig. 1*). A SF$_6$ outdoor bushing connects the low-situated inner conductor L1 to adjustable high voltage, both uppermost conductors (L2 and L3) are grounded. Additionally, the particularly developed

setup of GIS-components *(Fig. 2)* supports observation of moving particles.

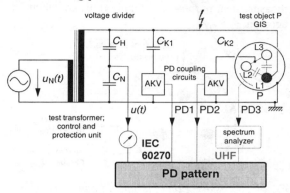

Fig. 1: Equivalent circuit diagram

Measurements according to IEC 60270 use an external coupling capacitor C_{K1} (PD1, *Fig. 1*) and reflect phase resolved PD characteristics and magnitudes of the apparent charge q in pC.

Unconventional UHF Sensor. As an important component of the considered GIS type, each separating gas barrier insulator includes auxiliary potential grading electrodes *(Fig. 3)* of a certain geometry which support stress control. During the entire rated operating condition, screws connect each poured-in electrode to the grounded enclosure and ensure their peculiar purpose: to equalize and homogenize the electrostatic field distribution and to relieve boundaries between epoxy dielectric insulator material and aluminium. Even if one screw is removed, these properties still remain. The generated loop retains zero potential, but gains an accessible tapping point in the middle of its frame.

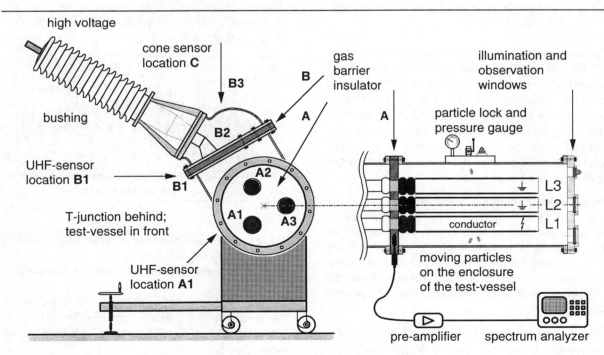

Fig. 2: Setup and test-vessel (three-phase GIS-components); insulator, tapping point and measuring devices

From now on, it serves as an UHF antenna - directly expanded into the pipe - and can be contacted for measurement purposes, if simple rules are taken into account.

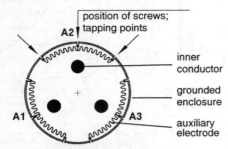

Fig. 3:   *Cross-section of a built-in gas barrier insulator (illustrated in principle)*

Transfer Function. Transfer functions are recorded by the use of a tracking generator. They indicate the complexity of stimulated resonance patterns which are composed of electromagnetic waves between the sensors. Normalization eliminates the frequency response error in the test-setup.

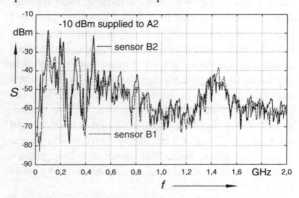

Fig. 4:   *Transfer functions measured at the sensors B2 and B1 (Fig. 2)*

The output power of -10 dBm is supplied to sensor A2 on top of the vertically positioned insulator. Stimulated resonances exist within the performed frequency range up to 2 GHz *(Fig. 4)*. The exemplarily chosen electrodes B1 and B2 allow sensitive measurements at sufficient accuracy due to geometrical characteristics. To support later zero span measurements, a pronounced resonant frequency can be chosen with low expenditure.

Free Moving Particles. Critical particles are usually shaped like thin metallic shavings (e.g. aluminium), rarely longer than 10 mm in practice. Affected by the electric field, particles exchange charge with the enclosure they lie on. If the electric force increases sufficiently to equalize the gravitational field, the adhesion as well as the frictional force, the particle lifts off.

Comparison with a Cone-shaped Sensor. For comparative purposes, a suitable designed cone sen-

sor is located in C (behind the bushing and near the sensor B1, *Fig. 2*). It receives slightly improved resonance patterns excited by the identical moving particle used as PD source. The length of the considered bouncing particle is $l_P = 9$ mm and its diameter $d_P = 0.5$ mm. It leads to an apparent charge of $q \approx 40$ pC at an applied voltage $U_{rms} = 65$ kV. In case of non-movement of the particle, q is 0.5 pC due to environmental noise. The background noise spectrum is supposed to complete further comparative figures uniformly.

The auxiliary electrodes are almost as sensitive as the additionally manufactured and optimized cone sensor (*Fig. 5*). The measurements are performed in the frequency domain within 0.3 and 1.8 GHz. The deviation of the amplitudes S can be neglected, except for frequencies of about 1.2 GHz.

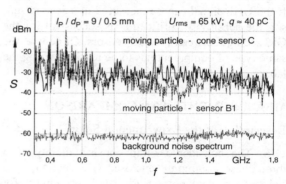

Fig. 5:   *Spectra of a moving particle; $U_{rms} = 65\ kV$*
*   *$l_P / d_P = 9 / 0.5$ mm*
*   *sensor B1 and adjacent cone sensor C*

To verify the sensitivity of the UHF method [1], the particle's size has to be reduced until the apparent charge is about 5 pC. With respect to the given arrangement, a particle of $l_P / d_P = 3.8 / 0.22$ mm fulfils this requirement.

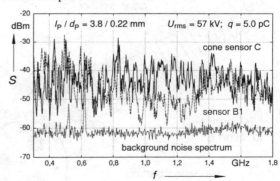

Fig. 6:   *Spectra of a moving particle; $U_{rms} = 57\ kV$*
*   *$l_P / d_P = 3.8 / 0.22$ mm*
*   *sensor B1 and adjacent cone sensor C*

In spite of decreasing amplitudes, the detection of the small moving particle corresponding to a low PD-value of 5 pC is still possible with both sensor types *(Fig. 6)*.

the UHF band has to be considered, which is highly influenced by the gas. Therefore the UHF method with all its advantages may be attractive also in gas mixture insulated GIL. If N_2-SF_6 gas mixtures are used for insulation purposes it has to be examined whether the UHF method can be applied for diagnostics. It has to be investigated for all SF_6 percentages of interest if the magnitude of the excited electromagnetic waves is sufficient for PD measurements.

3. Experimental Setup

3.1 General Arrangement

The test arrangement consists of components of a real 420 kV type GIS. A metal-enclosed test transformer with a SF_6 insulation is used for voltage supply. The test transformer with a maximum voltage of 510 kV is connected with a coupling capacitor by a 5kΩ resistor (Fig. 1). The capacitor is used for measurements with the conventional method and consists of a 3 m duct with an incorporated measuring electrode. It is also used to make PRPD patterns. The 5kΩ resistor protects the transformer from overvoltages created by travelling waves in case of breakdown. The coupling capacitor is also connected with a transfer vessel containing the UHF sensor. This transfer vessel is linked with a test vessel which contains the defects (free moving particles and fixed protrusions).

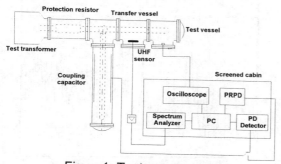

Figure1: Test arrangement

3.2 Equipment for measurements in the frequency domain

Figure 1 shows the arrangement for measurements in the frequency and in the time domain. The UHF sensor consists of a disc electrode with a diameter of 160 mm like it is used for capacitive high voltage dividers.

The UHF signals are put into a 20 dB amplifier. This amplifier feeds a spectrum analyzer with a frequency range of 1.8 GHz where the spectra are recorded. The foreign noise level of the whole UHF measuring equipment is -65 dB

without defects. The coupling capacitor is connected with a measuring impedance to measure the apparent charge according to IEC 60270. The signals of the measuring impedance are fed into the broadband amplifier of a PD measuring device. This device is used for measuring the applied voltage as well as the apparent charge. The foreign noise level here is 0.8 pC. The measuring data of the spectrum analyzer and of the conventional PD measuring device are fed into a PC where they are processed. To avoid electromagnetic interference all measuring devices are placed in a screened cabin.

3.3 Equipment for measurements in the time domain

In the time domain the PD currents caused by the incorporated defects are measured. Figure 1 shows the measuring arrangement. The current impulses are recorded with a high-speed oscilloscope. The measuring data of the oscilloscope are read into the PC and are processed. The measuring equipment for the time domain is placed in the screened cabin as well.

4. Measurements

4.1 General conditions

Measurements were carried out with N_2-SF_6 gas mixtures with a SF_6 percentage of 0.1%, 1 % and 20 % as well as with pure N_2 and pure SF_6. The gas pressure was always 0.5 MPa (abs). Free moving particles and fixed defects were inserted to cause realistic PD. A needle with a tip radius of 20 μm fastened at the outer conductor represented the fixed defect. The distance between the tip and the inner conductor was 170 mm. In the case of free moving particles a particle bowl was installed at the outer conductor. The free aluminium particle was of cylindrical shape with a length of 5 mm and a diameter of 0.5 mm.

In order to reach similar conditions during the various experiments with the fixed defect first the inception voltage (U_{inc}) for every case was determined. Then the measurements happened at a test voltage (U_{test}) 10 % above inception level,

$$U_{test} = 1.1 \cdot U_{inc}$$

in the time domain as well as in the frequency domain. Table 1 presents the inception voltages

Frequency spectra for different defects and gas mixtures

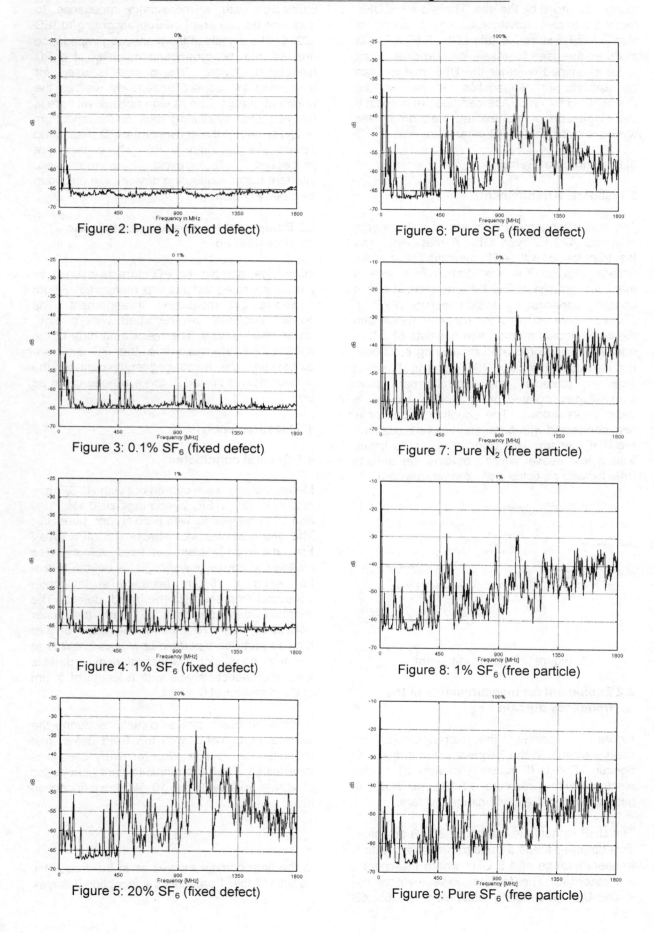

Figure 2: Pure N₂ (fixed defect)

Figure 6: Pure SF₆ (fixed defect)

Figure 3: 0.1% SF₆ (fixed defect)

Figure 7: Pure N₂ (free particle)

Figure 4: 1% SF₆ (fixed defect)

Figure 8: 1% SF₆ (free particle)

Figure 5: 20% SF₆ (fixed defect)

Figure 9: Pure SF₆ (free particle)

and the applied test voltages for the fixed defect.

SF$_6$-percentage	0	0.1	1	20	100
U$_{inc}$ [kV]	46	56	81	98	120
U$_{test}$ [kV]	51	62	90	107	132

Table 1: Inception and test voltages

The free moving particles started their motion at a voltage of 210 kV. Therefore all measurements with free particles both in the time domain and in the frequency domain were carried out at a voltage of 230 kV.

4.2 Frequency domain

All measured spectra were recorded for a duration of 15 minutes.

Figures 2-6 show the spectra for fixed defects. Obviously there are no spectral components in the UHF range when pure N$_2$ is used. With 0.1% first signals in the UHF range appear, with 1% there are much more spectral components and their magnitudes are higher as well. With 20% the spectrum is the same as it is with pure SF$_6$. One can easily see that already very low percentages of SF$_6$ cause signals in the UHF range whereas such signals do not appear in pure N$_2$. To compare the spectra of the various gases at a fixed voltage the same measurements were made at 150 kV for all gases. The spectra with 1%, 20% and pure SF$_6$ are almost equal and look like figure 5 or 6. With 0.1% a spectrum like figure 3 appeared and with pure N$_2$ the spectrum is the same as in figure 2.

The spectra recorded with free moving particles in pure N$_2$, 1% SF$_6$ and pure SF$_6$ are almost equal (Fig. 7-9). The spectra recorded in 0.1% and 20% SF$_6$ are not presented because they are also like those in figures 7-9. These figures indicate that the electromagnetic waves excited by PD with free moving particles are independent of the applied gas.

4.3 Time domain

In all cases the current was recorded 50 times. In order to consider the current pulses causing the UHF spectra each time the pulse with the shortest rise time was selected.

All the current pulses except that of the 0.1% mixture and pure N$_2$ have rise times below 500

ps in case of fixed defects. In 0.1% mixtures the rise time is below 2 ns, in pure N$_2$ the maximum rise time is about 50 ns. The currents recorded with the free moving particle have all the same course. They have maximum rise times below 500 ps.

5. Conclusion

In case of free moving particles the UHF spectrum is independent of the applied gas and has the highest amplitudes in the UHF range. With fixed protrusions there are spectral components in the UHF range even in a mixture with a SF$_6$ percentage of only 0.1%. In pure N$_2$ there is no UHF spectrum. The rise times of the PD currents confirm this. But spectra with especially high amplitudes in the UHF range are given for mixtures with an SF$_6$ content of 20% and more.

References

[1] Diessner, A.; Koch, H.; Kynast, E.; Schuette, A.: Progress in High Voltage Testing of Gas Insulated Transmission Lines. 10th Int. Sympos. on High Voltage (ISH), Montreal, 1997

[2] CIGRE-Rep. 15/23-01 (WG 15.03): Diagnostic Methods for GIS Insulating Systems. CIGRE, 1992

[3] Albiez, M.; Leijon, M.: PD-Measurement in GIS with Electric Field Sensor and Acoustic Sensor. 7th Int. Sympos. on High Voltage (ISH), Dresden, 1991

[4] Kurrer, R.; Feser, K.; Herbst, I.: Calculation of Resonant Frequencies in GIS for UHF Partial Discharge Detection. Proc. Of the 7th Int. Sympos. on Gaseous Dielectrics, Knoxville, paper 70, 1994

[5] Hampton, B.; Meats, R. J.: Diagnostic Measurements at UHF in GIS. IEE Proc. C-135 no. 2, pp. 137 - 144, 1988

[6] Egawa, T.; Miyazaki, A.; Takinami, N.; Miyashita, M.; Kamei, M.; Sakuma, S.: Partial Discharge Detection for a Long Distance GIL. 10th Int. Sympos. on High Voltage (ISH), Montreal, 1997

Address of author:

Günther Schöffner
Lehrstuhl für Hochspannungs- und Anlagentechnik
Technische Universität München
Arcisstraße 21

D-80333 München

COMPARISON AMONG PD DETECTION METHODS
FOR GIS ON -SITE TESTING

L.De Maria*, E.Colombo *   W.Koltunowicz**

*ENEL RICERCA, Italy   **CESI, Italy

ABSTRACT - **Electrical, optical and acoustic methods for Partial Discharge detection in Gas Insulated Substation (GIS) are simultaneously compared in laboratory and their application during on-site testing is discussed. The sensitivity of these methods to detect and recognize different GIS defects is verified. The most common defects of critical sizes and very low apparent charge values are considered.**

I. INTRODUCTION

Effective on-site tests are requested to verify that the GIS is still defect free after transportation, final assembly and mechanical testing on site. The procedure for on-site GIS testing proposed by CIGRE [1] indicates, as the best solution, the application of power frequency voltage together with very sensitive Partial Discharge (PD) measurements. Voltage test levels higher than the ones stated in IEC-Standards [2,3] are proposed both for short duration AC test and for PD measurements. In Table 1 the above mentioned values are referred to GIS of 420kV rated voltage.

TABLE.1
On site voltage test levels for 420 kV GIS

	Voltage test levels by IEC	*Voltage test levels by CIGRE*
Short duration power frequency level	416 kV	513 kV
PD measurement.	267kV	410 kV

According to the new procedure the maximum permissible PD value should not exceed 5 pC when measured with standard PD method (IEC 270). This involves the application of very sensitive PD methods. Since the standard method is very difficult to apply on-site, the use of alternative methods should be

High Voltage Engineering Symposium, 22–27 August 1999
Conference Publication No. 467, © IEE, 1999

envisaged. In the paper the sensitivity of ultra high frequency method (UHF), optical and acoustic methods, to detect and recognize different GIS defects is analyzed.

II. TEST SET-UP AND CONSIDERED DEFECTS

The measurements were performed on the 420 kV GIS module with the conductor outer diameter of 90 mm and the welded aluminium enclosure inner diameter of 480 mm. The module was filled with SF_6 gas at a pressure of 0.45 MPa and was supplied with power frequency voltage. The test set-up is shown in Fig.1.

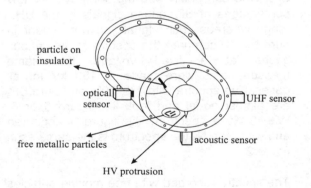

Fig.1. Test set-up

Three sensors, one for each method, were located on the GIS enclosure. Optical and UHF sensors were inserted into optical accesses already available for arc location. Acoustical sensor was externally attached to the bottom part of the enclosure.

Various defects were simulated on the basis of the following criteria:
- common presence of the defect after assembling on-site
- critical size and location of the defect that cause the dielectric strength to drop below the AC and LI withstand voltage on-site test levels
- low apparent charge value generated by the defect.

In particular, the following two defect types were simulated:

- Free moving metallic particles as the most common defects being critical at power frequency voltage [4,5]. Copper and aluminium wires of 3 mm length, with diameters of 0.25 and 0.38 mm respectively, were positioned on the bottom of the enclosure under the HV conductor. Five wires of each material at time were tested and the voltage was increased up to particle "lift-off" PD inception value.

- Fixed defects that are critical under LI voltage [4,5]. Particle attached to the surface of the insulator and sharp protrusions on HV conductor were tested. In particular, a copper wire with a length of 30 mm and a diameter of 0.2 mm was positioned on the surface of the insulator, close to the maximum voltage gradient [6]. HV protrusion, steel needles with a 70 microns curvature radius tip and of different lengths ranging from 1 to 10 mm were fixed to the conductor, as shown in Fig.1.

All the simulated defects produce apparent charge lower than 5 pC when measured with the standard method. The PD values in relation to defect type and to voltage applied at inception level are presented in Table 2.

TABLE 2
Defect apparent charges

Defect type	Inception PD voltage [kV]	PD value [pC]
Free moving particles (*)	90-100	2-3
Particle on insulator	125	2
HV protrusion - 2 mm	175	2

(*) - same results were obtained for copper and aluminium particles

III. PD DETECTION METHODS

UHF, acoustic and optical methods were used to measure electric field, pressure wave and light emission phenomena associated with PD activity at defect location. The measured signals were analysed in real time or by accumulating PD activity in relation to the phase of the test voltage for longer time periods. By means of the second procedure higher sensitivity and better recognition of defect type were obtained.

In the paper the following measuring PD signals are considered:

- UHF signal captured by internally located sensor is amplified and sent to portable measuring equipment where time domain analysis is performed. The signal amplitude over several consecutive voltage cycles and

correlated to the phase of the voltage is 3-D displayed in per-cent of the full-scale[7].

- Acoustic signal is measured with ultrasonic sensor operating in the 20-100kHz range, commonly used for the measurements on GIS [8,9]. The accumulation procedure is used to analyse the data.

- Optical signal is measured by photomultiplier tube, using photon counting technique [10,11]. The signal is obtained by storing signals coming from the optical system. The resulting pattern is an histogram whose amplitude is related to the total number of photons (counts per seconds) collected by the optical detector.

IV. TEST RESULTS

PD measurements were carried out with all the methods simultaneously and their comparison was performed for free moving particles and fixed defect type separately.

The sensitivity of the methods was evaluated as PD inception voltage level (Table.3 and Fig.2) and was extimated as the ratio between the amplitude of the PD signal and the background noise level measured at U=0kV (Table.4).

The defect recognition was performed by evaluating the correlation of the pattern with the phase and polarity of the voltage applied.

TABLE 3
Sensitivity of different methods to free moving particles and to particle on insulator

Defect type	Voltage PD inception value [kV]		
	UHF	Acoustic	Optical
Free moving particles	90-100	90-100	90-100
Particle on insulator	110	110	65

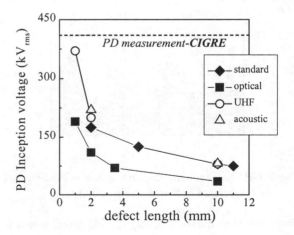

Fig.2: PD inception voltage of different methods as a function of the length of the HV protrusion.

TABLE 4
Signal to noise ratio at PD inception level for different methods and defects types

Defect type	UHF [%of full scale]	Optical [cps/cps]	Acoustic [mV/mV]
free particles	55-60	4-6	8-10
particle on insulator	55-60	4-6	1.5
HV (2 mm) protrusion	50-60	5-6	1.5

Free moving particles
All the techniques have detected free particles movements at the same voltage level (Table.3), which is the " lift-off" voltage for the considered particles.

The pattern produced by moving particles was easily recognisable and similar for all the methods (Fig.3).

(a)

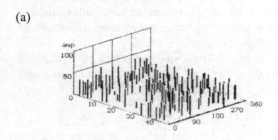

(b)

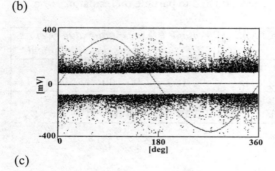

(c)

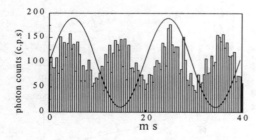

Fig.3: UHF (a), acoustic (b), and optical (c) PD signals related to free particle activity.

With real time measurements, the PD signal seems to have no correlation with the phase of the applied voltage. As an example, the UHF PD random signals acquired over 50 consecutive voltage cycles, are shown in Fig.3. When the accumulating procedure is applied,

as for acoustic and optical techniques, the correlation with the phase of the voltage is noted on both polarities, but particle activity is also observed along the whole voltage period. At inception level, PD signals have the same amplitudes on the positive and negative semi-cycles. The sensitivity of the measurements was higher for acoustic technique, as in this case the direct mechanical impacts of the particles on the aluminium enclosure were measured. A lower signal to noise ratio was obtained for UHF and optical methods, since they measure the electrical charge released by the particles at their impact on the enclosure.

Fixed defects
The sensitivity of the various methods is presented in Table.3, for particle on insulator, and in Fig.2 for HV protrusions of different lengths.

(a)

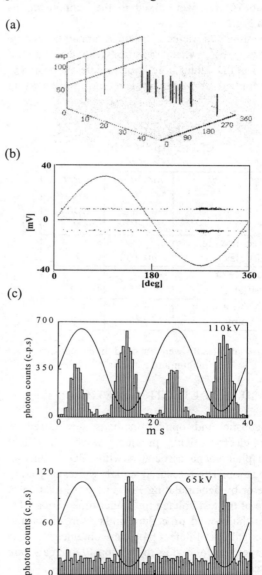

(b)

(c)

Fig.4: UHF(a), acoustic (b) and optical (c) PD signals related to particle on insulator.

Optical technique turns out to be the most sensitive one and the inception voltages for both type of fixed defects result even lower than those for the standard method. In Fig.2, the voltage level for PD measurements proposed by CIGRE is also indicated (dashed line) and even the 1 mm HV protrusion was detected with optical and UHF methods below this proposed voltage value. This defect length is critical as it can still lower the dielectric strength below the LI level during on-site tests. The acoustic measurements were not performed, being strongly disturbed by corona effect on HV connections.

As far as the defect pattern is concerned, both real time and accumulating techniques highlight the presence of the defect. The inception of the defect took place on the negative polarity of the voltage applied, but, at higher voltage levels, the activity was observed on both polarities. An example of the measurements performed with all the methods and related to the particle on insulator is presented in Fig.4. Measurements are referred to U=110 kV, which was the PD inception level for acoustic and UHF techniques. In the figure the pattern of optical signal for its inception level (U=65kV) is also reported. The signal to noise ratio, when compared to free particles, was much lower for acoustic and equal for the UHF and optical methods (see Table.4).

V. CONCLUSIONS

UHF, optical and acoustic techniques can be recommended for PD measurements on-site. With all these methods the detection and recognition of GIS defects of low apparent charges is possible, thus fulfilling the requirements of the new CIGRE testing on-site procedure.

VI. ACKNOWLEDGEMENT

The authors would like to thank M. Boldrin of Nuova Magrini Galileo (Italy) for his contribution in the research activity.

Address of Author:
L. De Maria, ENEL RICERCA
via Reggio Emilia 39
20090 SEGRATE (MI), ITALY
E-mail address: 0826DEMA@cise.it

VII. REFERENCES

[1] CIGRE WG 33/23.12,:"Insulation Co-ordination of GIS: Return of Experience, On-site Tests and Diagnostic Techniques." Electra 176, February 1998

[2] IEC-Publication 517:" Gas Insulated, Metal Enclosed Switchgear for 72.5 kV and above".

[3] IEC-Publication 694: "Common Clauses for High-voltage Switchgear and Controlgear Standards".

[4] A.Bargigia, W.Koltunowicz, A.Pigini,:"Detection of Partial Discharges in Gas Insulated Substations", IEEE Transactions on Power Delivery T-PWRD,July 1992, pp.1239-1249.

[5] CIGRE WG 15-03,:"Effects of Particles on GIS Insulation and Evaluation of Relevant Diagnostic Tools" CIGRE' Session 1994, Report 15-103.

[6] E.Colombo, W.Koltunowicz, A.Pigini, G.Tronconi: "Long Term Performance of GIS in Relation to the Quality of Spacers", in Proceedings of 10th ISH in Montreal, 1997.

[7] M.D.Judd,O.Farish and B.F.Hampton, "Excitation of UHF Signals by Partial Discharges in GIS",IEEE Transactions on Dielectrics and Electrical Insulation, Vol.3, N.2, pp.213-228, April 1996.

[8] E.Colombo, W.Koltunowicz,A.Pigini,:"Sensitivity of Electrical and Acoustic Methods for GIS Diagnostics with Particular Reference to On-site Testing" in Proceeding of CIGRE Symp.Berlin,1993, paper 130-13.

[9] L.E.Lundgaard, M.Runde, B.Skyberg,:"Acoustic Diagnosis of Gas Insulated Substations: a Theoretical and Experimental Basis", IEEE Transactions.on Power Delivery T-PWRD,November 1992, pp.1751-1759.

[10] B.Cox:"Partial Discharge Detection in GIS by an Optical Technique", in Proceeding of International Symposium on GIS, Toronto, 1985, pp.341-349.

[11] L.De Maria, A.Martinelli, E.Paganini, U.Perini, E.Colombo, W.Koltunowicz: "Non-invasive Optical System for Partial Discharge Detection in Gas Insulated Systems" (in italian), Elettroottica '98, Matera, May 1998, pp.396-400.

RELEVANCE OF THE CHARGING PERIOD TO THE DIELECTRIC RESPONSE BASED DIAGNOSIS OF PE/XLPE CABLES

M. Kuschel   W. Kalkner

Technical University of Berlin / Federal Republic of Germany

Abstract

The measurement of the so-called dielectric response is one of the approaches for non-destructive insulation diagnosis. In frequency domain the dielectric response can be obtained by capacitance and dissipation factor measurements, in time domain by depolarisation and polarisation current or recovery voltage measurements.

The paper presents dielectric response measurements of different aged and unaged XLPE-insulated medium voltage cables in dependence of the required charging period prior to the time domain measurement. Furthermore, the results are compared to results of destructive test methods.

Keywords: diagnosis, charging period, dielectric response, non-linearity, relaxation

1. Introduction

A lot of research efforts and activities is directed towards a better understanding of ageing phenomena and the finding of tools for insulation diagnosis and remaining life estimation techniques as published in e.g. [1-7].

Some of the testing procedures applied today base on measurements of the 'dielectric response' being a direct function of polarisation and conduction processes, mainly influenced by the structure of the material and by the amount and type of ageing.

In time domain the dielectric response measurement requires an application of a dc voltage to the test object for a certain time. This so-called charging period varies dependent on the applicator between 5 min and 30 min, the applied dc voltage between $0.1U_0$ and $2U_0$ [1-5]. Also the evaluation criteria differ from each other, some recommend to take the absolute value of the response, others believe in the non-linearity with respect to the applied voltage. Others consider the relaxation behaviour as an evaluation possibility.

However, these quantities are related to the charging period, the relevance of which is investigated and presented in this paper. Furthermore, the results are compared to results of destructive test methods. Finally, suitable diagnosis procedures are discussed and recommended.

2. Experimental techniques and procedures

Test objects

The following 12/20kV cable samples of different German manufacturers with an active length of between 0.5m and 2m were investigated:

Cable samples C1 and C2

Two different new XLPE-Copolymer cables (extruded outer semicon layer; produced 1997; thermal conditioned for 1000h at 90°C, low amount of crosslinking-byproducts)

Cable samples C3 and C4

Two different 1 year laboratory aged (VDE long duration test [8]) XLPE-Copolymer cables (extruded outer semicon layer; produced 1996; C3 high and C4 low amount of crosslinking-byproducts)

Cable samples C5 - C12

Eight different water tree service-aged PE- and XLPE-Homopolymer cables (graphite insulation screen; produced 1972 - 1979; former the investigation the cables were stored in water in order to prevent a reduction of water content due to drying)

Each cable sample was prepared with a guard-ring arrangement, therefore no creepage current influenced the measurements. Furthermore, also a metallic sheath against capacitive coupling from the high voltage electrode was realised.

Experimental procedure

The experimental procedure is shown schematically in fig. 1. Prior to the diagnostic measurements all samples were checked by a partial discharge (PD) measurement at 50 Hz and a measurement of the guard-ring resistance. Only samples with no respectively low PD-level at 24 kV (< 50 pC) and very high guard-ring resistance (> 0.1 TΩ) were accepted for further investigation. This guaranteed the absence of influences on the dielectric response as proved in some extra experiments.

Prior to the measurements and during the intervals between the measurements the cables were energised for approx. 50h with their nominal service voltage. This preconditioning assured that in case of water tree aged cables the humidity was concentrated inside the water tree structures due to dielectrophoresis. Furthermore this preconditioning prevented an influence of former measurements and simulated on-site conditions.

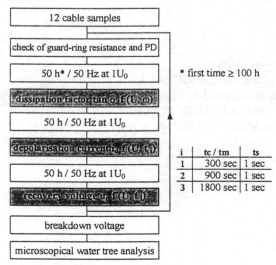

Fig. 1: Schematic diagram of the experimental procedure

High Voltage Engineering Symposium, 22–27 August 1999
Conference Publication No. 467, © IEE, 1999

Tab. 1: *Overview and correlation of results*

cable sample		C1	C2	C3	C4	C5	C9	C11	C10	C8	C6	C12	C7
absolute values of tan δ 0,1Hz, id (10s) [pA], ur (100s) [V] measured at U0													
tan δ 0,1Hz	M1	1,99E-4	1,70E-4	5,23E-3	1,85E-4	3,3E-03	1,8E-03	6,7E-04	5,0E-04	1,2E-04	1,1E-04	1,02E-4	6,6E-05
	M2	2,21E-4	1,88E-4	2,16E-3	2,07E-4	3,5E-03	1,9E-03	1,0E-03	5,6E-04	1,3E-04	1,1E-04	1,01E-4	6,1E-05
	M3	3,03E-4	2,66E-4	1,84E-3	2,34E-4	3,5E-03	2,1E-03	1,3E-03	6,1E-04	1,2E-04	4,4E-04	9,54E-5	6,0E-05
id [pA] (10s) nor. to 100pF	tc = 300s	2,35E-11	1,62E-11	4,02E-10	1,39E-11	3,9E-10	3,1E-11	8,4E-12	1,1E-11	1,3E-11	5,7E-12	8,68E-12	2,0E-12
	900s	2,55E-11	2,69E-11	3,23E-10	2,1E-11	3,4E-10	4,2E-11	1,1E-11	1,6E-11	1,7E-11	6,6E-12	5,37E-12	1,8E-12
	1800s	4,27E-11	2,73E-11	3,34E-10	2,6E-11	2,8E-10	5,8E-11	1,2E-11	1,8E-11	2,1E-11	1,8E-11	3,93E-12	2,3E-12
Slope n of id	tc = 300s	1,01	1,17	0,66	0,88	0,91	1,76	2,07	1,45	1,19	1,11	0,89	1,01
	900s	0,98	1,11	0,72	0,74	0,95	1,80	2,17	1,55	1,20	1,09	0,82	1,04
	1800s	0,99	1,17	0,88	0,76	1,00	1,77	1,95	1,51	1,19	0,98	0,87	0,89
ur [V] (100s)	tc = 300s	8,00	6,00	180,94	5,77	80,04	16,24	5,65	4,14	4,81	2,33	2,77	0,83
	900s	9,73	10,61	117,97	8,74	103,97	26,77	9,21	8,46	6,13	2,77	1,91	0,85
	1800s	12,29	10,17	127,74	10,16	100,85	26,39	11,38	8,82	7,23	1,01	1,35	1,08
Ubd		-	-	-	-	5,0	5,5	5,5	9,0	> 6,5	> 8	8,5	>14
ΔDoNL-2 relative change of the Degree of Non-Linearity (2U0 to 1U0)													
tan δ 0,1Hz	M1	0,34	-0,05	2,79	1,11	1,05	0,48	0,1	0	0,24	0,31	0,33	0,13
	M2	0,27	-0,11	2,42	0,84	0,86	0,48	-0,1	-0,02	0,21	0,31	0,34	0,10
	M3	0,02	-0,12	3,34	0,90	0,64	0,46	-0,11	-0,01	0,23	0,10	0,43	0,10
id (10s)	tc = 300s	-0,13	-0,05		0,12	-0,54	-0,14	-0,17	-0,09	0,16	0,24	-0,15	0,03
	900s	0,07	0,10	0,59	0,22	-0,43	-0,15	-0,08	-0,13	0,09	0,44	0,44	0,04
	1800s	0,04	0,07	1,26	0,17	-0,33	0,10	-0,07	-0,02	0,08	0,40	1,35	0,10
ur (100s)	tc = 300s	-0,13	-0,08	0,52	2,86	-0,25	-0,08	-0,08	-0,11	0,06	0,22	0,27	-0,03
	900s	0,05	0,00	0,19	3,20	-0,39	-0,09	-0,08	-0,09	-0,05	0,23	1,04	-0,03
	1800s	0,16	0,01	0,41	3,14	-0,39	-0,10	-0,21	-0,07	-0,05	0,17	0,97	0,01

4. Conclusions

According to the results obtained up to now the following conclusions can be drawn:

- Heavily water tree-deteriorated XLPE cables give in frequency as well as in time domain higher dielectric responses than unaged or slightly water-treed XLPE cables. The absolute value of the dielectric response is a suitable and a recommendable diagnosis criterion.

- Nevertheless, the absolute value of the response is also strongly affected by the polymer matrix and the amount of crosslinking byproducts and other additives of the insulation.

- A longer charging period prior to dielectric response measurement increases the absolute value of the response slightly; however, the classification of different service-aged cables with respect to the absolute value is almost not affected.

- Also the relaxation time or the slope of the dielectric response in case of the depolarisation current is affected by the charging period, usually the longer the charging period the lower the relaxation time and the slope of the response; however, the relaxation time or the slope of the depolarisation current is not correlated to the water tree deterioration and depends at certain voltages (> 0.5U0) on the amount of crosslinking byproducts and other additives in the insulation.

- The Non-Linearity of the dielectric response increases also with a longer charging period, however the Non-Linearity is less sensitive to water tree deterioration than the absolute value of the response; the Non-Linearity depends strongly of the amount of crosslinking byproducts and other additives in the insulation.

For an optimisation of the test period in the time domain based diagnosis further investigation of the behaviour of the dielectric response for charging periods < 5min are needed and are currently done.

5. References

[1] M. Kuschel, W. Kalkner: "Dielectric response measurements in time and frequency domain of different XLPE Homo- and Copolymer insulated medium voltage cables", Submitted to IEE Proceedings Science, Measurement and Technology

[2] T. Heizmann: "Ein Verfahren zur Bestimmung des Alterungszustandes von verlegten polymerisolierten Mittelspannungskabeln", Diss. ETH Nr. 105858, ETH Zurich 1994

[3] M. Schacht: "Cable diagnosis on Medium Voltage Cables in public utility networks. Description of a return voltage test set for on site tests and report about experience and results of tests", Nord-IS, Bergen 1996

[4] H.-G. Kranz, D. Steinbrink, F. Merschel:" IRC-Analysis - A new test procedure for laid XLPE-Cables", 10th ISH, Montreal 1997

[5] S. Hvidsten, E. Ildstad, B. Holmgren, P. Werelius: "Correlation between AC breakdown strength and low frequency dielectric loss of water tree aged XLPE cables", IEEE Trans. on Pow. Del., vol.13, no.1, pp. 40-45, 1998

[6] M. Kuschel, W. Kalkner: "Nichtlinearität der dielektrischen Antwort von water-tree-geschädigten Mittelspannungskabeln im Zeit- und Frequenzbereich", 43rd International Scientific Colloquium, Ilmenau 1998

[7] M. Kuschel, W. Kalkner: "Time and frequency domain based non-destructive diagnosis in comparison to destructive diagnosis of service-aged PE/XLPE insulated cables", JiCable, Paris-Versailles 1999

[8] Standard DIN VDE 0276-620

[9] M. Kuschel, T. Kumm, W. Kalkner: "Vergleich der 'dielektrischen Antwort' von laborgealterten VPE-Homo- und Copolymer-isolierten Mittelspannungskabeln", ETG-Fachtagung: Einfluß von Grenzflächen auf die Lebensdauer elektrischer Isolierungen, Bad Nauheim 1999

Acknowledgement:
The authors acknowledge the financial support by DFG (Deutsche Forschungsgemeinschaft).

Address of the authors:
Dipl.-Ing. M. Kuschel, Prof. Dr.-Ing. W. Kalkner
TU Berlin, Institut für Elektrische Energietechnik, FG Hochspannungstechnik, Sekr. HT 3, Einsteinufer 11, 10587 Berlin
WWW: http://ihs.ee.tu-berlin.de   Tel.: + 49 30 - 314 23470
Email : kuschel@ihs.ee.tu-berlin.de   Fax : + 49 30 - 314 21442

DISCHARGE CURRENT METHOD - A TEST PROCEDURE FOR PLASTIC-INSULATED MEDIUM-VOLTAGE CABLES

Michael MUHR, Rudolf WOSCHITZ [1]

Institut für Hochspannungstechnik mit Versuchsanstalt (IHV)

Technische Universität Graz

Inffeldgasse 18, A-8010 Graz, Austria

1 GENERAL

The diagnostic method for medium voltage cables based on the method of discharge current is an integral measurement technique, which means that the dielectric properties of an insulating material can only be measured in general, singular vacancies cannot be detected. The non-linear correlation between charging voltage and discharge current is a criterion for the state of ageing of the insulating material and gives information about its condition.

2 DETERMINATION OF THE AGEING OF PLASTIC-INSULATED CABLES

Fig.1 gives a brief survey of the test procedures being used in order to assess the dielectric properties of plastic-insulated cables.

During voltage tests the cable is exposed to a prescribed testing field strength.

If the remainig electrical field strength is lower than the testing field strength a breakdown occurs. In order to avoid this problem a modern diagnostic method can be used to test a cable. Investigations of conduction and polarisation effects in insulating materials are able to give further information about the ageing process of plastic-insulated cables.

As a consequence the criterion of assessing a cable get much more differentiated for the diagnostic method reckons the state of ageing of a cable and can give assessions about remaining lifetime.

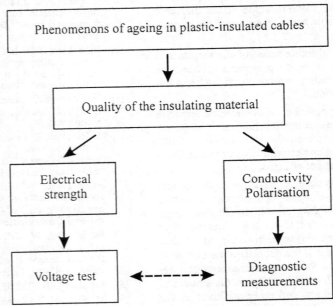

fig.1:   Test procedure to assess the dielectric properties of plastic-insulated cables

Fig.2 gives a survey of test procedures for plastic-insulated cables. Measuring methods being known

---

[1] The authors are members of the international working group „Insulation Diagnostics" (WGID).

High Voltage Engineering Symposium, 22–27 August 1999
Conference Publication No. 467, © IEE, 1999

at present can be split up into two groups which are methods in frequency and time domain and shall not

be discussed any further.

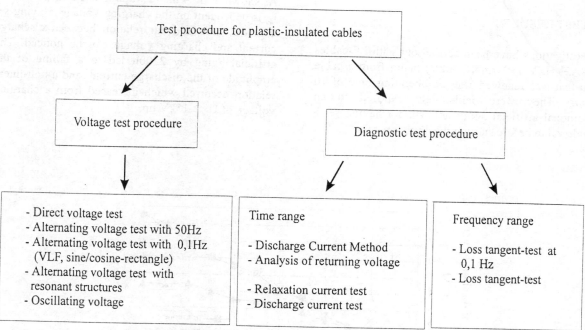

fig.2:   Survey of test procedures for plastic-insulated cables

3 COURSE OF MEASURES

The course of measures according to the method of discharge current can be divided into three parts (fig.3):

1. *Period of charge:* The cable is charged with direct voltage. The level of charging voltage and the duration of formation activate different mechanisms of polarisation and conduction being of great importance.

2. *Period of self-discharge:* The cable is disconnected from the direct voltage source and

from every discharge device. According to the condition of the dielectric of the cable inner discharges occur.

Therefore the potential cannot be maintained, which consequently leads to a falling of voltage. The slope of self-discharge voltage S_d being a tangent at the beginning of the phase of self-discharge serves as one diagnostic parameter.

3. *Period of discharge:* After a defined period of self-discharge the residual charge of the cable is measured. The maximum of discharge current $\hat{i}_{dc}$ being the second diagnostic parameter is evaluated.

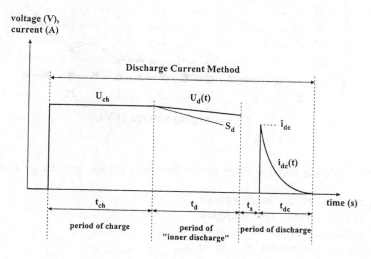

fig. 3: measuring procedure of discharge current method

4 TEST RESULTS

Investigations have been carried out with PE-cables (E-A2YHSY 1x95rm 18/30kV) being from a cable run that was renewed after an operating time of 10 years. They were deliberately exposed to an additional artificial aging not without having set a guide value before starting the aging process.

As shown in fig.4 the maximum of discharge current $\hat{i}_{dc}$ is dependent on the charging voltage. Having set a guide value a linear relation between discharge current and charging voltage can be noticed. The artificial aging by 2200h led to a falling of the amplitude of the discharge current, and a non-linear relation occurred which increased from a charging voltage of $U_L \geq 15kV$ up.

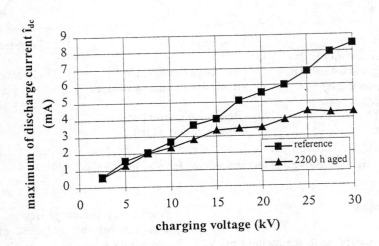

fig.4:   relation between maximum of discharge current $\hat{i}_{dc}$ and charging voltage

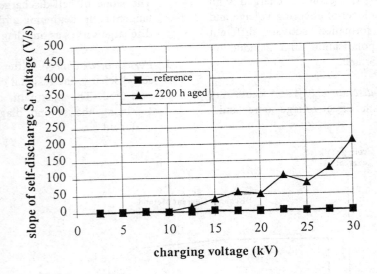

fig.5:   relation between slope of self-discharge voltage and charging voltage

Fig. 5 shows the slope of the self-discharge voltage S_d for the reference and for 2200h aged samples. This implies that the artificial aging process causes a diminution of the dielectric properties and inner discharges during the process of self-discharge.

5   CONCLUSIONS

Characteristic Value of Diagnosis CV:

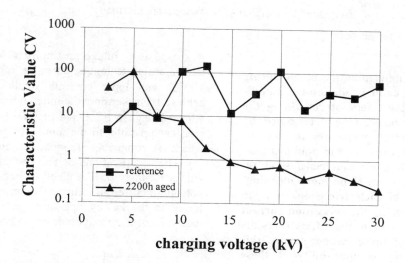

fig.6:   relation between characteristic value CV and the charging voltage

In order to get a better meaningfulness of the state of the dielectric cable you should not consider the parameters of diagnosis ($\hat{\imath}_{dc}$ and S_d) separate but combine them to a characteristic value CV.

Fig.6 shows the relation between characteristic value CV and the charging voltage for the PE-cable (reference, 2200h aging).

As a first guideline three ranges for the characteristic value CV having the following limits can be given to assess the state of the dielectric of medium-voltage (5) cables:

1. CV>10:   the cable shows no remarkable signs
             of aging
2. 1<CV<10: some signs of aging occur, further
             changes can be observed during the
             next inspection intervals
3. CV<1:   the cable shows signs of progressive (5)
             aging, increasing the danger of
             failure caused by breakdowns

6   SCRIPT

(1) Muhr M., Strobl R., Woschitz R.: Discharge
    Current Method for aged-insulated medium-
    voltage cables. 5th Höfler´s Days of High Voltage
    Technique, Portoroz 1996;
(2) Muhr M., Woschitz R.: Discharge Current
    Method to examine the ageing process of plastic-
    insulated cables. Contribution N[O]40 to

the conference at Ilmenau, IWK, September
    1995
(3) Heizmann T.: A method of determining the
    state of ageing of installed polymere-insulated
    medium-voltage cables. Dissertation, ETH-
    Zürich, 1994
(4) Porzel R., Sturm M.: Use of dielectric diagnosis
    to examine artificially aged high voltage
    insulation or insulation aged in service.
    Contribution N[O]40 to a conference. IWK,
    Ilmenau, September 1995
(5) Biasiutti G.: Examination of plastic-cables by
    using direct voltage. Bulletin SEV/VSE 78
    (1987)

Authors

O. Univ.-Prof. Dr. Michael MUHR
Ao. Univ.-Prof. Dr. Rudolf WOSCHITZ
Institut für Hochspannungstechnik mit Versuchsanstalt
(IVH), Technische Universität Graz
Inffeldgasse 18
A-8010 Graz, Austria
Tel. ++43/316/873-7401
Fax ++43/316/873-7900
E-mail: NAME@hspt.tu-graz.ac.at

CORRELATION BETWEEN RETURN VOLTAGE AND RELAXATION CURRENT MEASUREMENTS ON XLPE MEDIUM VOLTAGE CABLES

G. Hoff, H.-G. Kranz

Bergische Universität - GH Wuppertal, Germany

Abstract

The destruction free determination of the ageing status of a polymeric cable insulation is a significant issue for the assessment of remaining life time. The ageing status can be determined by measuring the dielectric properties in time or frequency domain. Because of the nonlinear behaviour of aged polymeric materials, time resolved measurements seem to be more meaningful. In time domain the dielectric properties can be obtained by measuring the relaxation current or the return voltage. Due to the different results and interpretations of those measurements, this paper will discuss the comparability of these procedures. Furthermore, influences of electromagnetic disturbances will be discussed.

Introduction

In the literature are a lot of measurement systems and methods described, to access the deterioration and the ageing status of polymer cables. These methods base on the fact, that ageing in a polymer changes the electrical, physical, mechanical and morphological properties of the insulation [1]. Dielectric properties are characterised by the time dependent polarisation P(t) (eq. 1).

$$P(t) = \varepsilon_0 (\varepsilon_r - 1) \cdot E(t), \quad \Delta P(t) = \int_0^t f(t - \tau) \cdot E(\tau) d\tau \quad (1)$$

where E(t) is the applied electrical field and ε the permittivity. Therefore it is taken into account that deterioration and ageing change the insulation properties as well as the dielectric response f(t) [2]. The task of diagnosis is to determine the dielectric response. This is performed by measuring the relaxation current or the return voltage directly. Considering that especially aged cables show a nonlinear behaviour [3, 4], the calculation of the dielectric response, based on frequency domain measurements, cannot be successful.

Measurement System

In order to achieve optimal comparability of measured relaxation current and return voltage data, a special measurement system was developed for this investigation. The principles are shown in Fig. 1. R_C are charging and R_D are discharging resistors. The internal resistances R_{MU} and R_{MI} represent the impedance of the voltage or current measurement circuit, which influence is discussed more detailed in this paper. The current preamplifier has bias and offset current as well as current noise lower 1pA. To allow a maximum input impedance an electrometer

was used as a voltage preamplifier. With a specialised guarding technique an input impedance higher 100TΩ and input currents smaller 2fA can be achieved. Furthermore the internal impedance can be adjusted (e.g. in the range of GΩ) by connecting a resistance parallel to the inputs, to get informations which are comparable to measurements presented in [5]. The specifications of the system were verified with a Keithley 6517A electrometer. The amplified voltage or current is being A/D converted and transferred to the PC by optical fibre. On the PC the measured data is visualised, stored and evaluated.

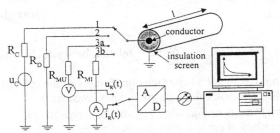

Fig. 1: Concept of the dielectric time domain measurement system

Measured data presented in this paper were recorded by using measurement conditions which are reasonable for on-site ageing diagnosis:

- According to results published in [6] all specimen were charged with a dc voltage u_C of less than 20% of the rated voltages. Only this can guarantee a destruction free monitoring.
- The charging time t_C, discharging time t_D and the measuring time t_M were set by the experience of more than 500 on-site investigations.
  Parameters : (u_C/[kV], t_C/[s], t_D/[s], t_M/[s])

Fig. 2 shows the voltage and current shapes during the respective measurement.

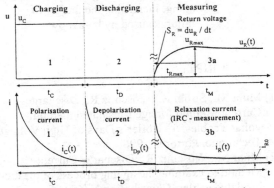

Fig. 2: Measurement parameters for dielectric response investigations in time domain

High Voltage Engineering Symposium, 22–27 August 1999
Conference Publication No. 467, © IEE, 1999

Evaluation parameters :
- Return voltage $u_R(t)$: S_R, t_{Rmax}, u_{Rmax}
- Relaxation current $i_R(t)$: according to *IRC[1]-Analysis* [1]

By performing dielectric measurements in the laboratory, naturally the length of the cable specimen is limited in the range of 1-20m. The relaxation currents of these samples are in the range of 5...500pA. To obtain plausible measuring data, not only the accuracy and the input circuitry of the preamplifiers has to be optimised, but also the polarisation of the measurement system itself has to be taken into account. This was achieved by minimising potential differences on all devices in the measurement circuit during the charging time t_C. Consequently the used equipment is much more complex than illustrated in Fig. 1.

Cable samples

As mentioned above the test length of the cables was limited. The intention of this paper is not performing an ageing diagnosis on different cable samples. The main aim is a comparison of the potential of the measured current and voltage data. Especially the influences of disturbances on the different procedures are investigated. Consequently the used samples were not specially treated. But all measurements which are compared in this paper were performed under the same conditions, e.g. the same temperature and measurement parameters. To avoid influences of mechanical relaxation the samples were not moved or otherwise mechanically stressed during the measurement series.

Test sample : l=4m, C_0'=400pF/m, R_0'=6.8TΩm, U_{Rated}=12/20kV, cross-section 500mm^2, manufactured in 1979.

C_0' is the cable capacity and R_0' the dc conductivity of the cable. The sample was in service for about 16 years and was replaced due to critical ageing.

Equivalent circuit

In order to compare the results of voltage and current measurements it is necessary to discuss an extended equivalent circuit of the interconnection of sample and measurement system. The general consensus is, that the physics inside an aged polymer acts like a current source. A first approach to model the relaxation behaviour is the classical Debye response shown in Fig. 3.

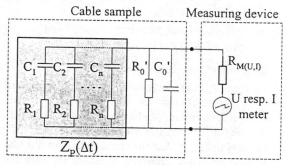

Fig. 3: RC-based equivalent circuit [4]

R_i and C_i represent discrete and independent polarisation mechanisms. $R_{M(U,I)}$ is the very important impedance of the measurement circuit. The value of $R_{M(U,I)}$ to perform correct voltage or current measurements strongly depends on the impedance of the cable sample, e.g. the values of R_0, C_0, R_i and C_i. In order to simplify the circuit, the polarisation characteristics are represented by only one *time dependent* impedance $Z_P(\Delta t)$ (grey shaded part in Fig. 3).

The measured current $i_R(t)$ is described as the superposition of several relaxation current components (eq. 2), which have a strong correlation to the internal polarisation processes [6].

$$i_R(t) = i_{R0} + \sum_{i=1}^{n} a_i \cdot \exp\left(-\frac{t}{\tau_i}\right) \quad (2)$$

In [7] it is shown, that the *IRC-Analysis* is able to separate the components (a_i, τ_i) of the measured relaxation current.

Solving the differential equations of the equivalent circuit (Fig. 3), enables the calculation of the values R_i and C_i. Using Fourier transform and considering a measuring time 1s<t_M<1000s and a cable length in the range of 1m<l<500m, allows an estimation of the limiting values of $Z_P(\Delta t)$. This estimation based on the analysis of a major number of *IRC-measurements* on aged cable samples according to VDE 0276 and on-site measurements of service aged cables up to 1000m length.

	Z_P	Z_{C0}	R_0
Δt_M=1s, l=1m	100GΩ	1MΩ	5TΩ
Δt_M=1000s, l=500m	20TΩ	500GΩ	10GΩ

Tab. 1: Limiting values of the impedances

Consequently in case of current measurements the measuring device should short-circuit the cable. But by using $R_M \approx M\Omega$ the relaxation current also flows through the *ampere meter* and the real dielectric response is measured. In case of return voltage measurements with $R_{MU} \gg R_0$, the physical relaxation current $i_R(t)$ charges also C_0 and the voltage response can be calculated as follows:

$$u_R(t) + C_0 \cdot (R_{MU} \| R_0) \frac{\partial u_R(t)}{\partial t} = (R_{MU} \| R_0) \cdot i_R(t) \quad (3)$$

Therefore return voltage measurements with R_{MU} less than R_0 or $Z_P(\Delta t)$ cause non interpretable results. The evaluation parameters t_{Rmax}, u_{Rmax} and S_R are dominated by the behaviour of R_{MU}. This can also be confirmed by an analysis of the self-discharge time τ_0 of the cable insulation, which is independent of the construction and the polymer cable length. With ε_r=2.3 and $\kappa \approx (10^{-14}-10^{-15})/\Omega m$, the relaxation time is in the range of hours.

$$\tau_0 = R_0' \cdot C_0' = \varepsilon_0 \cdot \varepsilon_r / \kappa \approx 2000...20000s$$
$$\tau_0 \approx 0.5...5.5h \quad (4)$$

From this it is assumed that often presented return voltage diagrams with t_{Rmax} in the range of seconds or minutes result from an insufficient proportioning of the input impedance R_{MU}.

---

[1] Isothermal Relaxation Current

The RC-based model (Fig. 3) is very useful to discuss the influence of the measuring system on the measured values, but it is not able to describe the total behaviour of the insulation material. Especially the nonlinear behaviour of heavily water treed insulation and the quasi direct current i_{R0} of old insulation cannot be explained by this model. Therefore it is useful to consider $Z_P(\Delta t)$ as a *Black Box*, which can imply impedances and sources (e.g. a dc source with i_{R0}). But this extension of the model cannot explain the nonlinear behaviour. A new approach is to introduce the response of the sample as a time dependent current source $i_D(t)$. In this case the measured *IRC-current* $i_R(t)$ is identical with $i_D(t)$. By performing voltage measurements with sufficiently high values of R_{MU}, the depolarisation current $i_D(t)$ divides into a part charging C_0 and a part flowing through the unknown impedance $Z_P(\Delta t)$. Therefore eq. 3 is no longer valid. This might explain the experience that it is not at any rate possible to use eq. 3 for the correct calculating of return voltages from relaxation current measurements and vice versa.

Measuring results

To compare the results of return voltage and relaxation current measurements it is necessary to check whether the measured data is reproducible. The measurable dielectric characteristics of an insulation strongly depend on the previous history of electrical, mechanical, chemical and temperature treatment of the material. In case of on-site measurements this is a minor problem, because the cables are buried for years hence there exists no mechanical stress. Furthermore they are service stressed with the rated voltage and with operating temperatures due to the current flow. Solely it must be taken into account, that the actual temperature of the insulation is in the normal range and constant during the *IRC-measurement*. By performing laboratory measurements, such conditions cannot be guaranteed and due to short samples the gathered results are scattered. Measuring $i_R(t)$ or $u_R(t)$ one after another show reproducible results. Four *IRC-measurements* of the same test sample from one day in the $i \cdot t$ vs. $\log(t/s)$ diagram are plotted in Fig. 4. This diagram was introduced as the *IRC-plot* to characterise the dielectric properties of an insulation and to perform ageing diagnosis [1].

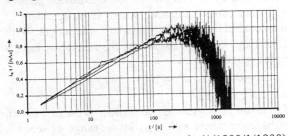

Fig. 4: *IRC-plots* from test sample (1/1800/1/1800)

It can be seen that the 4 wave-shapes are very similar and one can assume that these measurements are comparable.

The measured return voltages (Fig. 5) seem to be more different, but the deviation about 1V after 1800s can be caused by an equivalent relaxation current of 1pA. Furthermore Fig. 5 shows calculated return voltage based on the data presented in Fig. 4 by using eq. 3. The numbers beside the diagram show the measuring order. The initial steepness of the measured and calculated slopes seem to be comparable, but u_{Rmax} or t_{Rmax} cannot be compared because u_{Rmax} was not reached, not even after 5h (compare eq. 4). By translating the current data about 2pA, the measured and calculated voltages are in the same tolerance band.

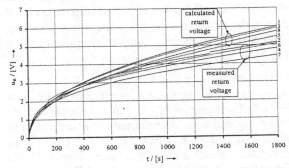

Fig. 5: Measured and calculated return voltages $R_{MU} \to \infty$, (1/1800/1/1800)

Fig. 6 shows measured and calculated return voltage data of the same cable with an electrometer input impedance $R_{MU}=6.8 G\Omega$. This demonstrates that now the value of t_{Rmax} is in the range of seconds. In case of XLPE insulations this is only the consequence of the proportioning of the voltage amplifier. The return voltage peak is also shifted to longer times by an increasing cable length. So it can be concluded that these evaluation parameters (voltage slope S_R, voltage peak u_{Rmax} and time to peak t_{Rmax}), are not only influenced by the physics of the dielectric.

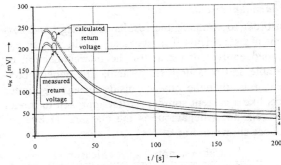

Fig. 6: Measured and calculated return voltages $R_{MU}=6.8 G\Omega$, (1/1800/1/1800)

Comparison of diagnosis criteria

To perform ageing diagnosis the most significant issue is interpretation of the measured values. Nowadays the nonlinear behaviour of aged dielectric can be used as criterion in all procedures. A significant disadvantage is that this behaviour can only be recognised by using charging voltages u_C with the risk of predamage [6]. Furthermore it is shown, that

conventional transient voltage shape is schematically shown in Fig.1.

From a practical point of view the front time (charging of the cable capacity) of the transient voltage is chosen in the order of 10 seconds, which keeps the power demand as low as in the case of 0.1 Hz VLF voltages. The tail time (discharging of the cable capacity) is chosen in the order of 10 milliseconds, which corresponds to the duration of a power frequency half cycle.

The conditions for PD ignition are much better and more realistic applying power frequency stress than VLF voltages. Consequently, the PD detection is performed during the tail time of the applied transient voltage. This conditions for PD ignition can further be improved, if the test voltage decreases not periodically but oscillating (Fig. 1). Due to this the effective voltage magnitude during the tail time is further increased.

Taking into account the statistics of PD ignition a sequence of such transient voltage shots is applied. Moreover, the detection and also the location of PD faults is performed (i.e. complex analysing) while discharging the cable capacity. The presented procedure is therefore called COMPLEX DISCHARGE ANALYZING (CDA-procedure) and the developed test voltage is called CDA voltage.

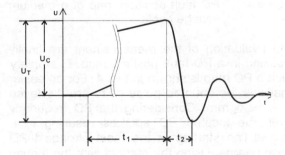

Figure 1: Voltage shape of the CDA test voltage

t_1	Front time
t_2	Tail time
U_C	Charging voltage
U_T	Test voltage

III. CDA - PD MEASUREMENT

The condition for PD ignition under CDA test voltages can be explained as follows: According to the well known equivalent circuit for PD faults the average capacitive current flowing through an assumed imperfection, such as a gas-filled void, is proportional to the steepness of the applied test voltage. The slope is relatively slow during the long front time t_1 of the CDA voltage. If within this time interval of the increasing test voltage finally the critical electrical strength for ionisation processes inside the void is exceeded, then the average discharge current is limited by the previous mentioned low capacitive current. Thus, only so-called micro-discharges may appear, characterised by current pulses of extremely low magnitudes, which cannot be detected by usual PD measuring circuits according to IEC 270 [6].

Much different conditions arise, if the test object capacity is suddenly discharged during the tail time t_2. Because the tail time is about 1000 times shorter than the front time, the voltage steepness increases accordingly and even so the average current flowing through the gas-filled void. This results into the appearance of well detectable PD pulses.

In this context it has also to be noted, that the PD inception voltage under CDA stresses is remarkable lower than those under other transient test voltages, such as standardised switching impulses (SI). The reason for this is, that during the long front time of the CDA voltage charge-carriers may have already been produced by micro-discharges, as mentioned previously. Hence, the time lag in generation of primary electrons, as it is usual for SI, can be neglected [8]. Consequently, PD pulses may appear during the tail time of the CDA voltage without any significant time-delay, if the critical strength for ionisation processes is exceeded.

Compared to other predictive diagnosis tools for power cables the CDA test procedure is distinguished by the following advantages:

1. Low voltage stress – a test level of $2 \times U_0$ is sufficient for PD ignition
2. Short time consumption – at most 5 CDA voltage impulses are recommended for each test voltage level
3. Low power demand – using a slow charging time
4. Low weight => no transportation problems
5. Network-independent power supply possible – caused by the low power demand
6. Location of the PD faults – the system uses wideband PD measurement technology for impulse reflectometry and an automatic PD location software is implemented

IV. PARTIAL DISCHARGE LOCATION CHART FOR MV CABLE SYSTEMS

As an example, in Fig. 2 results of on-site PD tests for an extruded medium voltage cable, tested with CDA voltages, are presented. Only the tail region of a single shot is recorded. Additionally to the integrated current pulses, which represent the standardised PD quantity "apparent charge" according to IEC 270 [6], the wideband amplified PD pulses are displayed. If the pulses are zoomed in the time domain, the PD location can be evaluated using the well known pulse reflectometry method, as evident in Fig. 3.

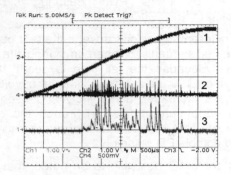

Figure 2:   Recorded CDA Voltage and PD signals at a test shot

    1 Test voltage at t_2 (discharging the cable capacity)

    2 Wideband PD signal for fault location

    3 Integrated PD signal / "Apparent Charge"

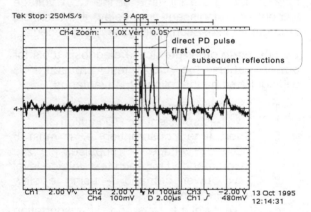

Figure 3:   PD fault location with the pulse reflectometry method applying CDA voltage

When a PD impulse arises, starting at the location of the PD source two travelling waves will travel along the cable. The first impulse will directly go to the measuring end, where an high sophisticated wideband signal conditioner picks it up. The other part of the impulse travels to the far end of the cable, where it will reflect and travel back to the measuring end as well. In consideration of the propagation velocity and the time difference between the original impulse and the reflected one, the location of the PD source is evaluated automatically by a complex software system.

The received PD impulses will be digitised by a high speed analogue-digital converter. After

that, according to the noise situation, different digital filters can be applied to suppress harmonic interference's. The recorded and pre-processed PD-signals will be analysed by a software with regard to the previously described reflectometry method.

A data base with all relevant information of the tested power cable is implemented in the system. Therefore, the detected PD impulses, the PD level and the PD repetition rate can be displayed in a chart as a function of their location together with all available information's like the test voltage levels, the cable type and structure, the absolute length of the cable, the joint positions etc. .

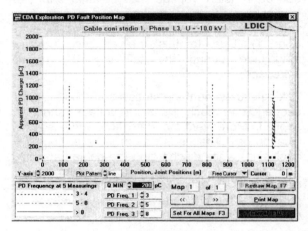

Figure 4:   PD fault position map of a medium voltage cable

The evaluation of the measurement are finally resulting in a PD fault position map. Exemplary such a PD map is shown in Fig. 4. For an easier analyses the counted pulse pairs are displayed in a histogram. Considering the PD frequency level, the located PD sources resulting in a vertical line starting at the most strongest PD pulse charge. Using the cursors with the mouse in a snap-to-point mode the PD fault position can easily be measured in the chart.

Such as, this Complex Discharge Analysing System permits extremely comfortable and cost-effectively the discovery of impending defect joints and therefore a maintenance corresponding to the actual condition of the cable system.

V.   PD COUPLING SENSORS FOR HV CABLE SYSTEMS

In case of an on-site PD diagnosis on HV and EHV cable systems two fundamental problems have to be solved [12]. First, the strong high frequency signal attenuation characteristic of the HV cable reduces strongly the sensitivity below an acceptable level [13;14;15] and secondly the on-site detection system must be able to distinguish between different PD sources as well as between real PD impulses and equivalent noise signals.

On the other hand, strictly controlled production environments combined with complete and sensitive factory tests and the robust design of the HV cables justifies the assumption that the tested cable lengths are free of service-relevant defects. Statistically, most problems of XLPE-insulated EHV cable systems are caused by their accessories (joints and terminations) which have to be mounted on-site [16]. Therefore, a sensitive on-site PD test of the accessories can be considered necessary and sufficient, because - as mentioned above - the cable length itself has been PD-tested before.

A successful method with respect to both of these aspects is the usage of two directional coupler sensors placed inside or outside the housing of the accessory.

The directional coupler principle is well known and easy to apply in homogenous dielectrics [12;17]. High voltage cables use a semicon layer for electrostatic field smoothing. The dielectric constant of such a semicon layer is complex and frequency dependent [18] making broadband directional coupler design for this application more sophisticated.

To avoid any degradation of the high voltage performance of the cables the directional coupler is installed between the outer semicon layer and the metallic cable sheath. Due to the semicon layer the directional coupler design has to consider a non-homogenous and frequency dependent partially conducting dielectric. A conductor is placed on the semicon layer and by adjusting the design of the sensor the directivity is optimised in the desired frequency range.

As shown in figure 5 two sensors are installed to the left and to the right of a cable joint.

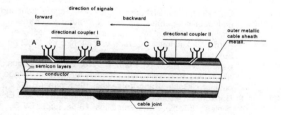

Figure 5:   Principal set-up of two DCS at a cable joint

Signals propagating 'forward' (as indicated in fig.5) are coupled to port A and C, signals propagating 'backward' couple to port B and D. PD in the joint will couple to port C and B. The sensors can be part of the joint or can be

mounted during joint installation or later beside the joint.

Time-gating is used to avoid misinterpretation due to reflections (e.g. from the test set-up terminations). The time-gating parameters are derived from the length of the joint and the propagation velocity. As no parameters depend on engineering judgement, an exact discrimination between noise and PD from the joint is achieved [12].

VI. PD-SIGNAL EVALUATION UNIT - DIRECTIONAL COUPLER DETECTOR

To evaluate the PD-signals coupled by the sensors the Directional Coupler Detector (DCD) electronically determines PD in a joint and discriminates any noise signal. Figure 6 shows the DCD. It has four input ports for the DCS signals, a PD counter unit and gated outputs to connect the PD visualisation system.

Figure 6:   Front-Panel of the DCD

After the system amplifies, filters and carries out the time-gating for each detected impulse it calculates the propagation direction and the source location.

	A	B	C	D
Noise from left	X		X	
PD in the joint		X	X	
Noise from right		X		X

Table 1:   Combination of the measuring signal to the direction of the impulses

According to the source-output-function shown in table 1 a source will be assigned to every detected impulse. Sources are: 'From the right', 'From the left' and 'PD from the joint'. Depending on the assignment results the signals are switched to the appropriate output. This allows to visualise the phase resolved distribution diagram (PD pattern) of each separated signal source.

VII. APPLICATION OF THE DIRECTIONAL COUPLER TECHNOLOGY

The figure 7 shows the principle set-up of the measurement system with sensors, evaluation unit (DCD) for PD location and a PD-pattern recording system. The test object in this case was a 110 kV XLPE-cable with water terminations and a silicon joint.

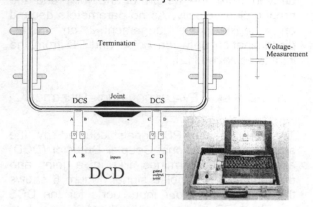

Figure 7: Measurement system with DCS and DCD

The PD signals are amplified and evaluated by the DCD as shown in figure 7. The output signals of each separated, different sources are digitised and transmitted to a computer, which also simultaneously controls all functions of the DCD. The PD-measurement system LDS-5 handles the signal-processing, the continuos recording and the analysis of the PD-data.

The magnitude of the apparent charge, the phase position and the event-time of each PD-signal will be recorded and stored together with the actual value of the voltage with a very high resolution. In this way a continuos monitoring and phase resolved analysis of the PD events can be performed. By using high sophisticated software tools the stored PD-data can be post-processed, displayed on the computer monitor or printed.

The separation abilities of the system allow to get single PD-pattern for each PD-location. Without this capability the patterns are usually mixed, because PD-pulses from different sources are overlapping one pattern and moreover disturbances lead to unrecognisable phase resolved PD diagrams.[12]

VIII. SUMMARY

The paper deals with the new developed diagnostic tools for non-destructive PD test of MV and HV power cables under on-site conditions.

For MV cables the cost-effectively CDA test procedure has been developed. It bases on the detection and location of PD faults if the cable capacity is discharged according to a power frequency half cycle. In order to keep the power demand low, the cable is charged by a slowly rising voltage comparable to the VLF test voltage.

If compared to alternative procedures for on-site testing of MV power cables the here presented method is characterised by the following main benefit:

The PD inception voltage under CDA stress is much lower than under other alternative voltages. This results into a comparatively low test voltage level which seems sufficient for recognition of dangerous imperfections in the cable insulation. Furthermore, PD events will ignite during the first subsequent CDA shots with a high probability. This permits a very short time for testing the cable.

Using two Directional Coupler Sensors, one on each side of a joint, placed above the semicon layer, a sensitive PD measurement system for HV cable systems has been built. On-site tests proved that this system allows an exact discrimination between noise and PD from the fault. The abilities of the system to plot individual PD-patterns, while simultaneously strong PD and noise from other sources interfere, were demonstrated.

The fields of applications of the Directional Coupling Sensors Technology on HV cables include:

- Type and unit acceptance tests of prefabricated accessories and silicone rubber stress cones in unshielded test rooms. Due to its accurate PD - noise discrimination abilities no shielded rooms are necessary.
- After laying tests.
- Permanent PD monitoring of cable systems.

References

[1] Grönefeld, P., v. Olshausen, Selle, F.: „Fehlererkennung und Isolationsgefährdung bei der Prüfung water tree-haltiger VPE-Kabel mit Spannungen unterschiedlicher Form". Elektrizitätswirtschaft, Vol. 84, No. 13, 1985, pp. 501-506.

[2] Kalkner, W., Bach, R., Plath, R., Wei, Z.: „Investigation of alternative after laying test methods for medium voltage cables". JiCable, Paris-Versailles, 1991, paper B 3.2.

[3] Dorison, E., Aucourt, C.: „After laying test of HV and EHV cables". JiCable, Paris-Versailles, 19-84, paper BV-5

[4] Lefèvre, A., Legros, W., Salvador, W.: „Dielectric test with oscillating discharge on synthetic insulation cables". CIRED, Paris, 1989, pp.270-273.

[5] Farneti, F., Ombello, F., Bertani, E., Mosca, W.: „Generation of oscillating waves for after-laying test of HV extruded cable links". CIGRE Session, Paris, 1990, paper 21-10.

[6] „Partial discharge measurements". IEC Publication 270, 1981.

[7] Lemke, E., Röding, R., Weißenberg, W.: „On-site testing of extruded power cables by PD measurements at SI voltages". CIGRE Symposium, Vienna, 1987, paper 1020-02.

[8]   Friese, G., Lemke, E.: „PD phenomena in polyethylene under AC and impulse stresses with respect to on-site diagnosis tests of extruded cables". 7th ISH, Dresden, 1991, paper 75.07.

[9]   Plath, R.: „„Oscillating voltages' als Prüfspannung zur Vor-Ort-Prüfung und TE-Messung kunststoffisolierter Kabel". Doctoral Thesis, TU Berlin, 1994

[10]   Strehl, T.: "Measurement and Location of Partial Discharges During On-Site Testing of XLPE Cables with Oscillating Voltages", International Symposium on High Voltage, ISH 1995, Graz, Austria.

[11]   Lemke, E., Schmiegel, P.: „Complex Discharge Analyzing (CDA) - an alternative procedure for diagnosis tests on HV power apparatus of extremely high capacity". 9th ISH, Graz / Austria, 1995, paper 5617

[12]   Pommerenke D.,Strehl T., Kalkner W.: "Directional Coupler Sensor for Partial Discharge Recognition on High Voltage Cable Systems", International Symp. on High Voltage, ISH 1997 Montreal, Canada

[13]   T. Heizmann, W.S. Zaengl, "Impulsausbreitung in Hochspannungskabeln", III Int. Kabelkonferenz, Budapest 1989

[14]   U.Schichler, H.Borsi, E.Gockenbach, "Teilentladungs (TE)-Messungen an Hochspannungskabeln unter Vor-Ort-Bedingungen", ETG Fachbericht 47, 1994

[15]   E.Pultrum, "On-site testing of cable systems after laying, monitoring with HF partial discharge detection", IEE Two day Colloquim on Supertension (66-500KV) Polymeric Cables and their Accessories, IEE Digest No. 1995/210, 21. Nov. 1995

[16]   C.G. Henningsen, K. Polster, B.A. Fruth, D.W. Gross, "Experience with an On-line Monitoring System for 400 KV XLPE Cables", Proc. of the 1996 IEEE Power Engineering Society Transmission and Distribution Conference, Sept. 1996, pp. 515-520

[17]   D.Pommerenke, I.Krage, W.Kalkner, E.Lemke,P. Schmiegel, "On-site PD measurement on high voltage cable accessories using integrated sensors", International Symposium on High Voltage, ISH 1995, Graz, Austria.

[18]   B.M.Oliver, "Directional Electromagnetic Couplers", Proc. IRE 42 (1954) 1686-1692

Author(s)

LEMKE DIAGNOSTICS GmbH, GERMANY

Prof. Dr.-Ing. E. Lemke
Dipl.-Ing. T. Strehl
Dipl.-Ing. D. Rußwurm

Radeburger Str. 47
01468 Volkersdorf
Germany

Voice ++49 35207 863 0
Fax ++49 35207 863 11
mailto: info@ldic.de

ON-SITE PD DIAGNOSTICS OF POWER CABLES USING OSCILLATING WAVE TEST SYSTEM

Edward Gulski, Johan J. Smit, Paul N. Seitz*, Jacco C. Smit and Mark Turner**

Delft University of Technology, High Voltage Laboratory, The Netherlands
*Seitz Instruments &Co, Switzerland
**Haefely Trench AG, Tettex Instruments Division, Switzerland

ABSTRACT

This paper discusses the results of partial discharge measurements in medium voltage power cables using oscillating test voltages. In particular, the sensitivity and the reproducibility of this new method for on-site testing are discussed. Based on laboratory and on-site tests the usefulness of this PD measuring technique for practical applications is presented.

1. INTRODUCTION

The detection, location and recognition of partial discharges (PD) at an early stage of possible insulation failure in medium voltage is of great importance for maintenance purposes. As a result, maintenance actions can be planned more precisely to prevent unexpected discontinuities in operation of the cable network.

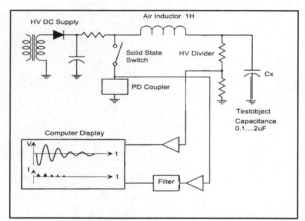

Figure 1: Schematic diagram of measuring circuit for on-site PD detection and location in power cables (<50 kV).

To obtain a sensitive picture of discharging faults in power cables the PD should be ignited, detected and located at power frequencies which are comparable to operating conditions at 50 or 60 Hz. In this way realistic magnitudes in [pC] and reproducible patterns of discharges in a power cable can be obtained. PD measurements during service as well as on-site continuous energizing at 50(60)Hz of MV cables are not economically realistic for on-site inspections. Different energizing methods have been introduced

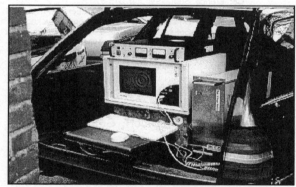

Figure 2: On-site testing of a 3 km long 10 kV power cable system.

Figure 3: 50 kV HVDC power supply and OWTS control data acquisition unit during on-site test.

and employed during recent years [1-4]. Therefore based on the assumption that sensitive detection of critical PD sites occurs by a method most similar to 50 Hz energizing conditions, a method as introduced in [5] for on-site PD diagnosis of MV cables will be discussed in this paper, (figure 1).

2. OSCILLATING WAVE TEST SYSTEM (OWTS®)

A compact, low weight solution is used to generate HV oscillating waves with a duration of a few tens of cycles of AC voltage at frequencies up to a few hundreds of Hz. The application of this voltage to the power cable under test and analysis of PD in the cable sample represents a new advance in the adaptation of the latest digital technology for HV insulation diagnosis (figures 2-3).

This method is used to energize, measure and locate

High Voltage Engineering Symposium, 22–27 August 1999
Conference Publication No. 467, © IEE, 1999

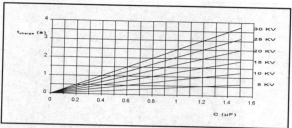

Figure 4: Charging times needed to charge different cable samples capacitances to a specific voltage level.

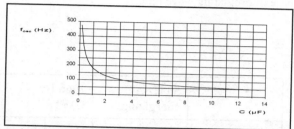

Figure 5: Oscillating wave frequency as function of the cable capacitance.

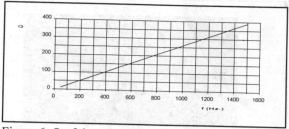

Figure 6: Q of the resonance circuit as function of the oscillating wave frequency.

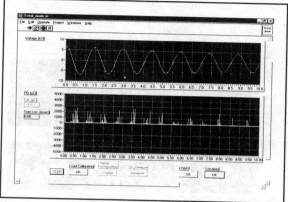

Figure 7: Example of PD pulses as measured during 12 kV oscillating wave on a 12 kV XLPE power cable, having internal discharges.

on-site partial discharges in power cables in accordance with IEC 60270 recommendations.
The system consists of a digitally controlled flexible power supply to charge long cable lengths at power frequencies of a few hundreds of Hz, fast digital recording and statistical evaluation system for discharges.

2.1 Basic theory of oscillating waves

With this method, the cable sample is charged with a DC power supply over a period of just a few seconds to the usual service voltage (figure 4).
Then a specially designed solid state switch connects an air-core inductor to the cable sample in a closure time of <1 µs (figure 1).
Now series of voltage cycles starts oscillating with the resonant frequency of the circuit:

$$f = 1/(2\pi \cdot \sqrt{L \cdot C})$$

where L represents the fixed inductance of the air core and C represents the capacitance of the cable sample (figure 5). The air core inductor has a low loss factor and design, so that the resonant frequency lies close to the range of power frequency of the service voltage: 50 Hz to 1 kHz. Due to the fact that the insulation of power cables usually has a relative low dissipation factor, the Q of the resonant circuit remains high depending upon cable: 30 to more than 100 (figure 6). As a result, a slowly decaying oscillating waveform (decay time 0.3 to 1 second) of test voltage is applied to energize the cable sample. During tens of power frequency cycles the PD signals are initiated in a way similar to 50 (60) Hz inception conditions. All of these PD pulses are measured using a fast digitizer.

2.2 PD detection

The advantage of a high Q circuit, is that PD can be measured on-site for a series of undisturbed sinusoidal cycles of the test voltage. For this purpose a special PD detection circuit has been used providing sensitive detection of discharge signals. Due to the fact that the switched DC power supply produces disturbances during charging of the cable sample, the PD circuit is therefore inhibited during this time. During the oscillating phase of the test cycle the DC power supply is disconnected to provide sensitive PD measurement (figure 7).
Since the oscillating frequency represents the AC conditions of the power line frequency, the measurement bandwidth of the PD circuit has been chosen in accordance with IEC 60270 recommendations. For the purpose of location by travelling waves, the bandwidth is increased up to 10 MHz figure 8). In combination with a 100 MHz digitizer and depending upon cable type e.g. XLPE or paper-oil the detection and location of PD pulses remains sensitive for cable length of few kilometres.

2.3 PD evaluation

The PD signals which are ignited during one or more oscillating waves and are detected by the system can be processed for two purposes.

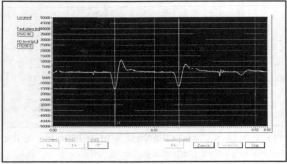

Figure 8: Example of PD site location after 12 kV oscillating wave charging a 10 kV paper-oil power cable.

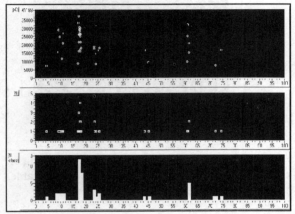

Figure 9: Example of statistical evaluation obtained after oscillating waves applied to a 840 m long 10 kV paper-oil power cable:

top: PD magnitudes versus location in the cable,

middle: PD intensity versus location in the cable,

bottom: PD intensity (5 m classes) versus location in the cable.

Firstly, each of the PD pulses can be analysed for reflections using travelling wave analysis (figure 8). Statistical evaluation of PD signals obtained after several oscillating waves can be used to evaluate the location of discharge sites in the power cable (figure 9).

Secondly, values of capacitance C and tan δ can be calculated based upon the oscillating wave time and frequency characteristics (figure 10).

Thirdly, after several oscillating waves the whole discharge sequence can be resolved into a phase-resolved PD pattern. In this way patterns can be obtained which are similar to those recognized under 50 (60) Hz conditions (figure 12).

For recognition purposes such PD phase-resolved patterns can additionally be processed using statistical discrimination and classification tools [6-7]. As a result, PD databases with reference to typical degradation examples in cables or cable accessories can be created for maintenance purposes [8].

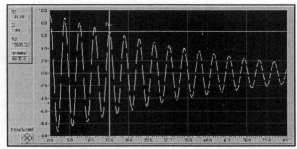

Figure 10: Example of calculation of C and tan δ on the base of oscillating wave time.

Defect	50 Hz AC energizing		1066 Hz OWTS	
	U_{inc} [kV]	PD [pC]	U_{inc} [kV]	PD [pC]
Bad contact between semicon and the stress cone	2.6	25	3	30
Bad adjustment of the stress cone	15	20	13	40
Internal cavities	13	150	14	70

Figure 11: Comparison of PD magnitudes and inception voltages obtained for the same internal defects in cable accessories by different charging voltages: 50 Hz (AC) power frequency, and oscillating wave voltages of 1066 Hz.

3. EXPERIMENTAL RESULTS

3.1 Measurement of PD magnitude

Laboratory experiments have been performed to provide insight into the inception conditions and measurable PD amplitudes [Gulski et al, (5)]. A series of PD measurements were performed on realistic internal defects made in 6 kV plastic insulated cable accessories. In particular, the same samples have been energised with initially 50 Hz and then again using 1066 Hz oscillating wave voltages. In figure 11, results of inception voltages and measurable in [pC]/[nC] PD levels are summarised. It follows from this figure that in comparison with 50 Hz (AC) the PD level as well as the PD inception voltages are in the same range. As a result, taking into account the stochastic behaviour of internal discharges no significant difference has been

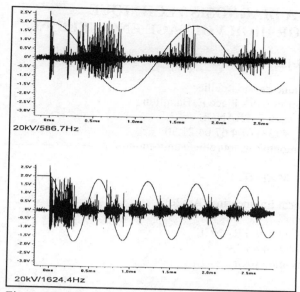

Figure 12: Comparison of phase-resolved PD patterns obtained by OWTS for the same internal defect for different wave frequencies.

observed.

Also the comparison of phase-resolved PD patterns measured at the same voltage level of different oscillating wave frequencies shows that the frequency applied does not seriously influence the shape of PD patterns (figure 12).

3.2 On-site experimental results

Initial examples of field application of the oscillating wave method have been discussed in [5] and systematic application of OWTS® is given in this proceedings, see [9].

4. CONCLUSIONS AND SUGGESTIONS

In this study, a new on-site method for PD detection and location in power cables has been applied in the field. Based on these results, the following conclusions can be drawn.

1. It is possible to use the oscillating wave method OWTS® to test power cables under AC conditions, enabling the ignition of discharges at particular sites and sensitive detection of PD signals.
2. Based on the travelling wave principle of PD signals in power cables, the location of particular discharge sites can be traced back.

5. OUTLOOK

As mentioned earlier in this report, the following investigations should be performed in the near future to investigate the full potential of this new method.

1. Using a laboratory set-up representative cable

samples with well defined typical faults have to be investigated systematically.
2. Based on feedback from on-site tests, investigation has to be made of sensitivity in locating and recognizing single & multiple PD sources in power cables in practice.
3. Using advanced statistical discrimination and classification tools a database consisting of typical degradation examples in cable accessories has to be developed as far as possible.

6. ACKNOWLEDGMENT

The authors would like to acknowledge Energie Noord West in Alkmaar, The Netherlands for experimental support in performing on-site measurements and fruitful discussions about PD on-site testing.

7. REFERENCES

[1] Plath, R., 1994, "Oscillating Voltages als Prüfspannung zur Vor-Ort-Prüfung und TE-Messung kunstoffisolierter Kabel", Band 1, Verlag Berlin

[2] E. Lemke, P. Schmiegel, H. Elze, D. Russwurm, " Procedure for Evaluation of Dielectric Properties Based on Complex Discharge Analyzing" , 1996 IEEE International Symposium on EI, Montreal, Canada

[3] E. Pultrum, E. Hetzel, " VLF Discharge Detection as a Diagnostic Tool for MV Cables", IEEE PES 1997 Summer Meeting, 20-24 July 1997., Berlin

[4] J.T. Holboll, H. Edin, " PD-Detection vs. Loss Measurement at HV with Variable Frequencies" 10th ISH, Montreal, Canada, 1997

[5] E. Gulski, J. J. Smit, P. N. Seitz and J.C. Smit, "Partial Discharge Measurements On-site Using Oscillating Impulse Test System", 1998 IEEE International Symposium on EI, Washington, USA

[6] E. Gulski, P. Seitz, "Computer-aided registration and analysis of partial discharges in high voltage equipment", 8th ISH, Yokohama, Japan, 1993

[7] E. Gulski, H.P. Burger, A. Zielonka, R. Brooks, Classification of Deflects in HV Components by Fractal Analysis, 1996 CEIDP, San Francisco, USA

[8] E. Gulski, J. J. Smit and R. Brooks, PD Databases for Diagnosis support of HV Components, 1998 IEEE International Symposium on EI, Washington, USA

[9] F. Wester, E. Gulski, J.J. Smit, "Experiences form on-site PD measurements using oscillating wave test system", 11th ISH, London, UK, 1999

THE THERMAL STEP METHOD : A DIAGNOSIS TECHNIQUE FOR INSULATORS AND COMPONENTS FOR HIGH VOLTAGE ENGINEERING

S. AGNEL, P. NOTINGHER jr., A. TOUREILLE

Laboratoire d'Electrotechnique de Montpellier
Université Montpellier II - Case Courrier 079 - Place E. Bataillon
34095 MONTPELLIER Cedex 5, FRANCE
Ph. + (33) (0)4.67.14.34.85, Fax + (33) (0)4.67.04.21.30
E-mail agnel@univ-montp2.fr, petru@univ-montp2.fr, toureille@univ-montp2.fr

J. CASTELLON, S. MALRIEU

Advanced Metrology for Electrical Engineering (Am2e)
Cap Alpha
34940 MONTPELLIER Cedex 9, FRANCE
Ph./Fax + (33) (0)4.67.14.45.73
E-mail am2e@univ-montp2.fr

1. INTRODUCTION

It is well known that the electrical properties of insulating materials used in high voltage applications are altered in time. This phenomenon, called « ageing », is strongly related to the external factors acting upon the materials (electric field, temperature etc.) and to the manufacturing process. Over the last decades, considerable efforts have been made to understand and to prevent ageing.

An insulating material is supposed electrically neutral ; however, electric charges can penetrate within the material. It has been shown that the accumulation of electric charges (*space charge*) within insulators, by creating a supplementary internal field (*remnant electric field*), is one of the main causes of ageing. Indeed, it seems that the more an insulation stores space charge, the more its ageing is accelerated by the increase of the global electric field (sum of the applied field and the remnant field). These charges can cause significant field distortion to affect the insulator performance and to reduce its lifetime.

The *thermal step method* (TSM) [1-3] is a sensitive method for measuring remnant electric field and space charge density. The principle of the TSM consists of applying a thermal step to an insulating material placed in short circuit and of measuring a current response, which is related to the space charge distribution within the material. The response is due to the variation of the sample's capacitance. The TSM is a non-destructive measurement method : the space charge located in the bulk of the sample is not removed, and repeated measurements are possible. An electrical survey of the sample can therefore be carried out in order to predict the beginning of the electrical ageing.

The thermal step method is directly applicable both to material samples and to electrical engineering components with insulating thickness up to 20 mm (cables [4], electrical machines windings [5], circuit boards [6], capacitor films [7]etc.).

In this study, two applications of the TSM to high voltage power cables are presented. We focus on the appearance and the evolution of the space charge within cable insulation, as well as on the effect of the extrusion process on the radial anisotropy of a high voltage cable.

2. THE THERMAL STEP METHOD APPLIED TO POWER CABLES

Let us consider a short-circuited cable sample containing space charge (*Fig. 1*). In the absence of the thermal step, the system consisting of the cable, the outer semicon electrode, the core and the conducting wire is electrostatically equilibrated; consequently, image charges will appear on the semicon layers.

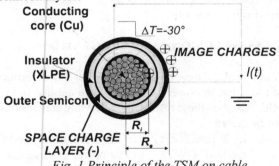

Fig. 1 Principle of the TSM on cable

If a thermal step ΔT is applied on the outer semicon layer (sharp cooling or heating), the electrostatic balance of the system is changed. This is due to the contraction (or expansion) of the cable's insulator, which causes a slight and reversible movement of the space charge within the sample, and to the variation of the electric permittivitty of the insulator with the temperature. As the system tends to rebalance, the image charges on the electrodes are redistributed. Hence, charge transport occurs from an electrode to the other in the external circuit. This corresponds to an external current, given by [1] :

$$I(t) = -\alpha C \int_{R_e}^{R_i} E(r) \frac{\partial \Delta T(r,t)}{\partial t} dr \quad (1)$$

where α is a constant of material related to the sample's contraction (or expansion) and to the variation of its permittivitty with the temperature, C is the capacitance of the sample, $E(r)$ is the radial remaining electric field distribution in the sample, and $\Delta T(r,t)$ is the relative temperature distribution in the sample : $\Delta T(r,t) = T(r,t) - T_0$ (T_0 is the temperature of the sample before applying the thermal step).

The current is recorded and then digitally processed using deconvolution algorithms [1-3] in order to find the remaining electric field distribution in the radial direction $E(r)$. Then, by applying the Poisson equation :

$$\rho(r) = \varepsilon \frac{\partial E(r)}{\partial r}, \quad (2)$$

where ε is the permittivitty of the sample's insulator, the space charge density distribution within the cable can be calculated (*Fig. 2*).

Fig. 2 Space charge density in the radial direction

The experimental set up in the case of the outer cooling is presented in *Fig. 3*. The thermal step is created by a cold liquid (-10°C) circulating within a heat exchanger in contact with the outer region of the cable. The core and the outer semicon are connected through a current amplifier. The latter is connected to a computer, which records the current and does the numerical data processing. After performing a measurement, the cable can be reheated using a warm liquid, and the experiment can be repeated.

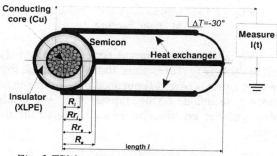

Fig. 3 TSM measurement by outer cooling

It is also possible to apply the thermal step by inner heating (*Fig. 4*). In this case, a strong current (some kA) is applied for heating the core. After the heating time (a few seconds), the core is connected to the ground and the external electrode of the cable is connected to the current amplifier. Hence, the cable response is measured. This technique presents the advantage of being applicable for on-site measurements.

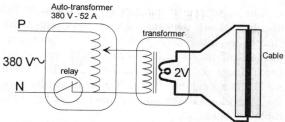

Fig. 4 The thermal step applied by inner heating

3. SPACE CHARGE EVOLUTION IN A POWER CABLE

This example of application of the *TSM* presents the evolution of the space charge within a cable insulation with the time of submission to electrical stress.

The tested cable is a 20 kV XLPE-insulated power cable, with an insulating thickness of 6 mm. A first space charge measurement has been performed by outer cooling before applying the electrical stress. The cable has then been submitted to a DC voltage of 41.6 kV at room temperature (25°C). The high voltage, applied to the cable core, corresponds to an electrical stress of 10 kV/mm near the conducting core and 5 kV/mm near the external semicon layer. Space charge measurements by outer cooling have been carried out after 4 h and 22 h of application of the DC HV stress. The *TSM* signals obtained are presented in *Fig. 5*.

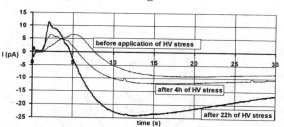

Fig. 5 TSM signals for the cable submitted to HV DC stress

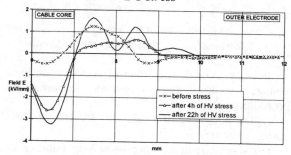

Fig. 6 Remnant electric field

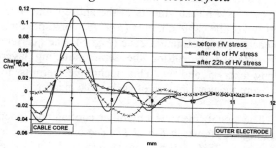

Fig. 7 Space charge density

The experimental currents (*Fig. 5*) lead to the distributions of remnant electric field and space charge presented in *Fig. 6* and *7*. A continuous increase of the remnant electric field and of the space charge density with the time of application of the HV stress can be noticed. The high sensitivity of the *TSM* allows to see the beginning of ageing of the cable insulation.

However, the reduced values for remnant electric field and space charge density do not reveal an ageing level of the cable which could dangerously affect its electrical properties. This result is in accordance with the short time of submission of the cable to HV stress and with the low polarising temperature. A critical ageing of the cable occurs for longer periods of DC stress.

4. RADIAL AND ANGULAR STUDY OF XLPE IN A HIGH VOLTAGE CABLE

The study of the radial and angular components of the charge distribution is important for cable manufacturing. In fact, the polyethylene extrusion does not respect completely the cylindrical symmetry of the cable, creating double refraction zones (medium residual stress zones and physical and chemical heterogeneity zones - *Fig. 8*).

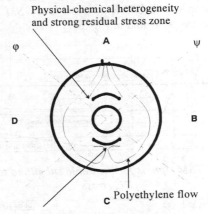

Fig. 8 Residual stress zones in an extruded cable

In order to study the effect of extrusion on high voltage cables, the following experimental procedure was applied :

(*i*) DC polarising of a cable (60 kV DC, at 70°C during 20 hours) ;

(*ii*) short circuiting after polarising (storage at room temperature) ;

(*iii*) *TSM* measurements of the entire cable (*Fig. 10*) and of the cable quarters (*Fig. 11*) corresponding to each double refraction zone (*Fig. 8*). The quarters are obtained by cutting the semiconducting layer along axis φ and ψ. This technique permits to measure the contribution of each cable quarter.

The signals measured by outer cooling and inner heating are shown in *Fig. 9-11*. The outer cooling is done by a -35°C real thermal step (a cold liquid at a temperature of -10°C is applied on the outer region),

while the inner heating is done by the Joule effect of a strong current (5 kA) passing through the cable core during 3 seconds. The latter gives a thermal step of only +1.2°C. The magnitude of the *TSM* current obtained by inner heating (*Fig. 11*) is about 30 times weaker than in the case of outer cooling (*Fig. 10*). The explanation is related to the amplitude of the thermal step, which is 30 times weaker in the case of inner heating.

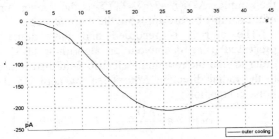

Fig. 9 TSM signal obtained for the entire cable by outer cooling

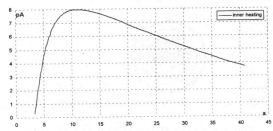

Fig. 10 TSM signal obtained for the entire cable by inner heating

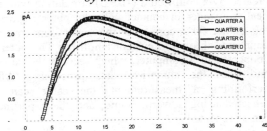

Fig. 11 TSM signals obtained for each quarter by inner heating

The inner heating current corresponding to the entire cable (*Fig. 10*) is equal to the sum of the four quarter currents (*Fig. 11*) obtained by the same inner heating technique. This proves that we really measured the contribution of each quarter of the cable.

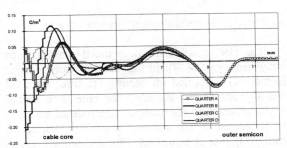

Fig. 12 Space charge distribution in each quarter (thermal step applied by inner heating)

The digital processing of the measured signals lead to the space charge distributions presented in *Fig. 12*. Slightly different profiles are observed for the space charges, especially near the extrusion point. The results reveal an increase of defects near the entrance (quarter A) and near the turbulence of the PE flow (quarter C). On the contrary, quarters B and D are similar.

The origin of observed charges is related to the structure of PE. The results show that the manufacturing process, in particular the cooling time of PE during solidification, is an important parameter to create defects and traps.

5. CONCLUSIONS

In this work, we have presented a diagnostic technique based on space charge measurements: the thermal step method (*TSM*). The *TSM* can be used for detecting ageing of high voltage components and for analysing the structure of insulating materials, both directly after manufacturing and after service.

Applications of the *TSM* directly on power cables have been brought in focus. The sensitivity of the TSM allow to detect very low levels of defects and traps (below 1 mC/m^3), and we have shown that the very beginning of the ageing phenomenon can be observed. Two methods of applying the thermal step on cables have been presented: the outer cooling (by a liquid circulating in a heat exchanger adjoined to the outer region of the cable) and the inner heating (the thermal step is generated by a strong current which heats the cable core). The outer cooling is suitable for detailed analysis of short cable samples, whilst the inner heating can be used for on-site measurements and for cable control just after manufacturing.

Similar applications to those presented on cable have been carried out for electric machines windings [5] and circuit boards [6] and are possible for a wide variety of components used in electrical engineering.

6. REFERENCES

[1] A. Toureille, Rév. Gén. d'Electricité, No. 8, 1991, pp. 15-20.

[2] A. Cherifi, M. Abou-Dakka, A. Toureille, IEEE TDEI, Vol. 27, No. 6, 1992, pp. 1152-1158.

[3] M. Abou-Dakka, S. S. Bamji, A. T. Bulinski, IEEE TDEI, Vol. 4, No. 3, 1997, pp. 314-320.

[4] J. Santana, J. Berdala, A. Toureille, Proc. JICABLE 1995, pp. 685-687.

[5] J. Castellon, J. Bouquart, J.P. Reboul and A. Toureille, Proc. IEEE CEIDP 1997, pp. 447-450.

[6] S. Malrieu, P. Notingher jr., F. Pacreau and A. Toureille, Proc. IEEE CEIDP Minneapolis, 1996, pp. 88-91.

[7] S. Agnel, A. Toureille, C. Le Gressus, Proc. IEEE CEIDP San Francisco 1996, pp. 190-193.

ACKNOWLEDGEMENTS

The authors would like to acknowledge Ms. Régine Clavreul (research division of Electricité de France) for supplying the sample used for the experiments described at §3.

XIPD TESTING OF A 275 KV CABLE JOINT INSULATOR

J.M. Braun , N. Fujimoto and S. Rizzetto

Ontario Hydro Technologies, Toronto, Canada

K. Watanabe, H. Niinobe and N. Kikuta

Fujikura Ltd., Tokyo, Japan

Abstract

Diagnostic tests were performed on a 275 kV prefabricated cable joint insulator using a new test technology, the X-ray Induced Partial Discharge (XIPD) test system developed at Ontario Hydro Technologies. Special shields were designed to facilitate the evaluation of PD characteristics of the large epoxy casting. The test chamber was PD free to 600 kV; discharges from controlled voids inserted near the conductor were readily detected. A PD sensitivity of 0.08 pC was achieved during testing at up to 600 kV. The cable joint insulator remained PD free under X-ray irradiation to the target test voltage of 420 kV.

Introduction

Transmission class cables insulated with polymeric insulation offer several potential advantages over conventional oil-filled paper cables. They have little adverse environmental impact, give virtually maintenance-free operation and have a significantly lower life-time cost. Crosslinked polyethylene (XLPE) cables have been used extensively in Japan and Europe at voltages up to 500 kV. Unlike oil/paper insulations, XLPE cables are very vulnerable to partial discharge degradation; in particular, joints and terminations demand careful manufacture and installation.

This paper describes tests performed on a 275 kV cable joint insulator using a new test technology, the X-ray Induced Partial Discharge (XIPD) test system, at Ontario Hydro Technologies' laboratories [1]. The purpose of the test was to evaluate the partial discharge (PD) characteristics of the large epoxy casting used in Fujikura's prefabricated cable joint. The present tests were intended to test the epoxy casting in isolation. That is, the insulator was to be tested without the other components which normally comprise the completed cable joint. In this manner, the quality of the insulator casting itself could be evaluated without the possibility of interference from other components and the various interfaces between the components.

X-Ray Enhanced Void Discharge

Partial discharges in voids are gas discharge events; initiation occurs only when the electric field stress exceeds the minimum field required for discharge and that a free electron becomes available which can initiate a discharge event. In practice, because of the scarcity of a free electron, the true inception level can be easily exceeded without discharges occurring. Sustained PD would only be possible at higher electric stress where other mechanisms of electron production are favored. Once sustained discharges are initiated, free electrons become abundant and discharges are often able to continue at voltages below the apparent inception level. As a result, the threshold at which sustained discharges cease to exist (discharge extinction voltage) is often much lower than the apparent inception voltage.

The application of X rays during electrical PD measurements is intended to artificially increase the electron production rate in the void. As a result, sustained partial discharges can be observed at much lower applied voltages with void stresses which approach the minimum critical value. In addition to the ability to induce sustained discharges, the X rays provide a unique opportunity to locate the site of the discharge by modulating PD activity with the ionizing beam. Void location within opaque systems can be implemented by using a narrow X-ray beam which systematically scans the entire sample (while energized) while correlating the PD measurements with the irradiated area of the sample. For development purposes, the ability to locate a discharge source allows more precise investigations of PD and failure mechanisms in solid insulation systems.

Test Configuration

The XIPD system was originally designed for gas-insulated substation (GIS) components using SF6 gas as the main insulation medium. The testing of a cable joint insulator (CJI) raised two issues which needed to be addressed:

- when tested without the cable joint's other components, is the electric stress within the epoxy representative of end-use conditions?

- how do we test a CJI in an SF6 environment when the required stresses would be higher than normally experienced in GIS equipment?

To address these issues, a series of electric field calculations were performed. The calculations were used to develop a set of stress shields for adapting the CJI to the XIPD system. From the calculations, we demonstrated that the CJI could be tested to the target voltage of 400 kV without any dielectric problems due to excessive stress in the SF6 gas. In

High Voltage Engineering Symposium, 22–27 August 1999
Conference Publication No. 467, © IEE, 1999

addition, the calculations showed that the electric stresses within the CJI using these stress shields were similar to those in a completed cable joint. In the high stress regions, the stresses were nearly identical and stresses throughout other parts of the CJI were representative.

Shield Design for CJI

A series of shields were constructed according to the stress calculations performed on the CJI. The insulator assembly was placed in an aluminum cradle supported by insulating supports which keep the CJI assembly centered in the test cell. The insulating supports facilitate the measurement of PD by detecting the PD currents through a measurement impedance between the CJI ground and the test cell ground. The external semiconducting layer (ground potential) was mated to the aluminum cradle by a layer of semi-conducting foam to ensure good, uniform contact. The upper, exposed portion of the CJI was also covered (via the semiconducting foam layer) with copper foil to ensure good contact with ground potential over the entire circumference. The shields and the overall CJI assembly are shown in Figure 1.

XIPD Test System

The XIPD system is integrated into OHT's High Voltage Laboratory complex. An X-ray shielded room adjoins the high voltage control room and the high voltage bay. The X-ray room is constructed from solid poured concrete and lined with lead. The excitation voltage source is a Haefely 800 kV, 2400 kVA AC resonant system. The overall XIPD test system is shown in Figure 2 . The connection to the test cell includes a specially constructed filter resistor which effectively reduces the impact of any externally-generated electrical noise. After the CJI was installed in the test chamber, voltage measurements made in the test chamber confirmed that the filter resistor had no significant impact on power frequency voltage applied to the insulator.

The X-ray generator is a 100 kW output unit capable of voltages to 150 kV and currents to 1200 mA. X-ray exposure duration can be controlled from 3 ms to 5 s. A rectangular X-ray beam collimator can control the beam pattern from fully closed to a 12 degree angular flood. The collimator control is switchable from either manual, remote manual or complete computer control. X-ray beam positioning is accomplished by two methods depending on the resolution required and the sampling area. Manual positioning for large areas is provided with an overhead tube crane.

The partial discharge (PD) detection system front end consists of 1-1000 MHz bandwidth amplifiers which feed a Pulse Height Analyzer (PHA). The

Figure 1 – CJI assembled with shields on metallic cradle cushioned by a layer of semiconducting foam

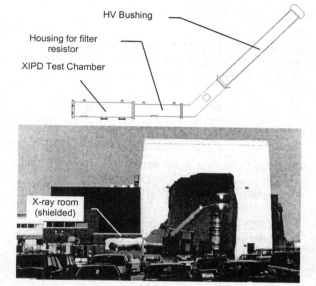

Figure 2 Overall XIPD test system. The actual test cell is show in the upper diagram. The lower composite photograph shows the placement of the test cell within the laboratory.

PHA is of a proprietary design and is capable of responding to 1-2 ns wide pulses in real time. For each detected pulse, the analyzer can detect the pulse polarity, measure amplitude (spanning three orders of magnitude) and determine its relative occurrence relative to the phase of the high voltage excitation voltage reference. The PHA has a maximum pulse throughput of 50 kHz. PD pulse confirmation is accomplished by simultaneously viewing the signal with a Tektronix 7104 oscilloscope.

The XIPD test cell has been equipped to make partial discharge measurements by various means. These include:

- direct measurement of the PD current in the insulator ground lead

- direct measurement of the PD current in the HV conductor, via high speed fiberoptic link

- capacitive couplers (sensors) installed in the test cell similar to those used for diagnostic measurements in gas-insulated substations.

For the tests on the CJI, the direct measurement of the PD current in the ground lead was used as the primary measurement method. The maximum sensitivity of measurement was estimated to be about 0.5 mV. Measurements with an artificially introduced void in this configuration have shown that the PD pulse measured has a half-magnitude width of about 8 ns. If this value is used as being representative of PD measured on the CJI, the 0.5 mV sensitivity corresponds to a charge sensitivity of about 0.08 pC.

XIPD Tests

In order to meet the practical and technical requirements, several test sequences were performed on the CJI. These included:

1) Standard Hipot Test - to ensure that the test configuration has been properly installed and is capable of withstanding full test voltage

2) Hipot test with X-ray irradiation - to ensure that the use of X rays do not compromise the withstand capabilities of the test cell

3) XIPD Tests - Full sequence XIPD measurements at various voltages. Tests were also performed where X ray exposure was limited to a small portion of the CJI at one time, in order to increase the effective X-ray intensity at the insulator. Exposure sites were systematically changed to cover the entire insulator

4) PD measurements with an artificially-introduced void to demonstrate the integrity of the measurement system.

Test Sequence 1 - Standard Hipot test

No flashovers were observed up to 450 kV for an SF6 pressure of 46 psig and a background PD level of 1.5 pC (with occasional bursts of 30-100 pC).

Test Sequence 2 - Hipot Test with X rays

No flashovers were observed up to 450 kV; the voltage raised to the desired level and the insulator exposed to X rays for a short (1-2 sec) duration

Test Sequence 3 - XIPD Test

No confirmed PD was detected in any of the measurements. The voltage was initially raised at approx. 100 kV steps up to 420 kV with XIPD measurements performed at each step. A systematic

Figure 3 - Grid Pattern used for xray irradiation of CJI. The photograph shows the guidelines on the XIPD test cell used to define the grid.

set of measurements were then performed at 420 kV with the area of X ray exposure limited to a small area as defined by a grid pattern as shown in Figure 3. Tests were performed at an SF6 pressure of 45 psig and X ray settings of 100 kV @ 600 mA and 1 sec exposure.

Test Sequence 4 - Measurement Verification

Small samples containing a void of known size (0.2, 0.29 and 1.48 mm in dia.) were introduced into the test chamber to verify that PD measurements can be successfully made on a CJI geometry installed in the XIPD system. The samples were made of clear epoxy with imbedded electrodes at a spacing of approx. 5mm. Two separate locations were used: the first was directly between the HV conductor and the CJI shield near the end of the insulator. The second location was between the HV part of the CJI and the ground shield. The second location was chosen to ensure that X ray can adequately penetrate the thicker portions of the insulator material.

For all samples and locations, partial discharges were easily and clearly measured using the same measurement scheme used for testing the CJI. An example of a measurement is shown in Figure 4 . An additional benefit of these tests was that the most appropriate X-ray level for testing the CJI could be determined. Most subsequent measurements were performed with an X ray setting of 90 kV/300 mA.

Conclusions and Recommendations

From the tests performed on the CJI, we conclude that the CJI is discharge free to the target test voltage of 420 kV at a sensitivity of 0.08 pC. Any detected signal was attributed to noise and not related to the CJI itself.

The techniques used for interfacing the CJI to an SF6-insulated system worked well. With the stress shields designed by OHT, the CJI was successfully tested up to 450 kV without problems. This voltage corresponds to maximum electrical stresses more than double those experienced by 550 kV class GIS at the power frequency withstand level.

It is recommended that testing be extended to include the elastomeric stress cones and epoxy/rubber interfaces, areas of considerable interest in joint performance. This approach using the existing test cell can be recommended with confidence as testing of high stress designs has now been proven with the current test sequence.

References

1) J.M. Braun, S. Rizzetto, N. Fujimoto, G.L. Ford, T. Molony and J.P. Meehan, X-ray Induced Partial Discharge Testing of Full-Size EHV Insulators, IEEE PES T&D Conference, Los Angeles, CA, Sept. 1996

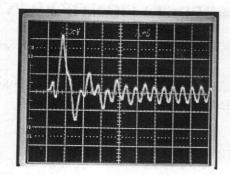

a) typical PD waveform

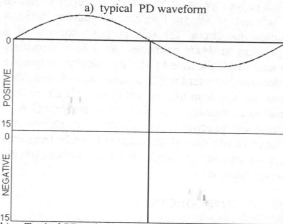

b)   Typical PD pulse height/phase distribution The number of pulses in each bin is represented by the grey tone

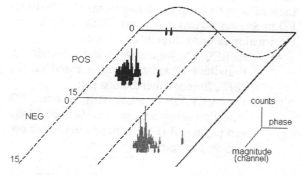

c)   Distribution of b) above displayed in 3D format. The number of pulses in each bin is represented by the height of the image

Figure 4 – Sample measurement from a void in a small insulator inserted into CJI configuration. The distributions show PD pulses accumulated over a 5 sec interval ans sorted by phase and magnitude. Magnitudes are shown independently for both positive and negative polarity pulses and are sorted in 16 logarithmically spaced channels. The magnitudes show in the figure correspond to approximately 26-70 pC.

INTERPRETATION OF PD IN GIS USING SPECTRAL ANALYSIS

Sander Meijer, Edward Gulski, Johan J. Smit, Frank J. Wester, Thomas Grun* and Mark Turner*

Delft University of Technology, The Netherlands
*Haefely Trench AG, Switzerland

ABSTRACT

This paper reports on VHF/UHF detection of partial discharges in SF_6 gas insulated substations (GIS). A measuring circuit for this purpose consists of a spectrum analyser connected to a high frequency sensor inside the tested GIS. In particular the use of a spectrum analyser for PD measurements is described and its applicability is validated. Using several examples measured on GIS, the interpretation of frequency spectra is explained. It is shown that characteristic key values obtained from the frequency spectra can be very useful for trend analysis of PD measurements.

1. INTRODUCTION

Most breakdowns of high voltage (HV) equipment are preceded by partial discharge (PD) activity [1]. Therefore, detection of PD activity can give an early indication of faults in insulating materials. To detect PD in SF_6 gas insulated systems (GIS), two kinds of electrical measuring circuits are in use:

A) the IEC 270 measuring system: broadband (hundreds of kHz) detection based on the IEC 270 recommendations [2];

B) the VHF/UHF measuring system: PD detection using narrow or wide band filters (up to several GHz) which is in use for on-site/on-line tests [3].

During on-site PD tests external disturbances can affect considerably the measurements, especially in the case of an IEC 270 measuring circuit. To solve this problem and to achieve a sensitive measuring circuit for on-site/on-line tests a VHF/UHF measuring circuit has been introduced [4]. It has been shown that this detection system has about the same sensitivity as the IEC 270 system, regarding PD magnitude [7] and PD patterns [5]. However, the advantage of effective noise suppression must be weighed against the missing [pC] indication for the measured signals.

In general, a VHF/UHF measuring circuit consists of a VHF/UHF coupler mounted inside the GIS. These couplers pick up the by partial discharges excited electromagnetic waves. The processing equipment for these signals is shown in figure 1 and consists of

① a spectrum analyser (SA) and
② a computer containing software for controlling the SA, data storage and analysis of measured frequency spectra.

The investigations described in this paper demonstrate the suitability of a SA for PD detection. The VHF/UHF measuring system and the analysis of obtained frequency spectra is explained with reference to various GIS test setups. Finally a detailed example of PD analysis on a 420 kV GIS substation concludes this contribution.

2. VERIFICATION TESTS TO CHECK THE PD DETECTING CAPABILITY OF A SA

PD are localised breakdown phenomena accompanied by current pulses of a few nanosecond duration. Frequencies up to several hundreds of MHz can be expected, for which a SA is an appropriate analysis tool.

The block diagram of a (radio frequency scanning) SA is shown in figure 2. First the input signal (f_{in}) is passed through a low-pass filter with a cut-off frequency of 1.8 GHz. This signal is then passed to the 1st mixer together with the signal from the local oscillator (f_{lo}). The mixer is used for frequency translation of the input signal. It can be shown that the input signal is converted into two output signals, one at the up-conversion band, where $f_u = f_{in} + f_{lo}$, and one at the down-conversion band, where $f_d = f_{in} - f_{lo}$, Couch (7).

Example:
See figure 2. Assume f_{lo}=2300 MHz, then the frequency component of 200 MHz of an input signal

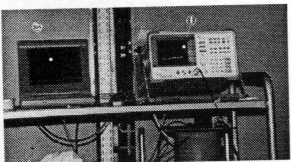

Figure 1: *VHF/UHF measuring system consisting of ① a spectrum analyser (HP 8590L) and ② a personal computer to control the spectrum analyser, for data storage and for data analysis.*

High Voltage Engineering Symposium, 22–27 August 1999
Conference Publication No. 467, © IEE, 1999

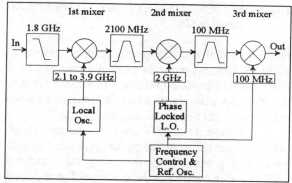

Figure 2: *Part of the block diagram of a spectrum analyser.*

is converted into signals at 2100 MHz and 2500 MHz. A bandpass filter selects the 2100 MHz signal for further processing. A second and third mixer convert the signal further producing a frequency-shifted output corresponding only to the 200 MHz spectral component (with a bandwidth of 10 MHz) of the input signal.

By sweeping the frequency of the local oscillator between 2.1 and 3.8 GHz all frequencies between 0 and 1.8 GHz in the input signal can be detected. It takes a specific amount of time, called sweep time, to sweep from 2.1 to 3.8 GHz. The selection of a specific sweep time can significantly affect the accuracy of the measurement. Increasing the sweep time increases the available measurement time and the SA can use a smaller frequency step size to step through the frequency spectrum. Hence a more accurate measurement can be made.

Principally, proper use of a SA is based on a continuous non-changing input signal. However, it is known that a PD process can not be seen as a continuous input and it is in most cases characterized by a stochastic behaviour. Several important aspects regarding proper setting of the SA are studied [6]. Important settings include: sweep time, measurement time and acquisition mode. The tests described in this section were carried out using a fast pulse generator, Picosecond Pulse Lab model 2000D. The generated pulse shape is shown in figure 3. The repetition rate of the pulses can be varied between single shot and 1 MHz.

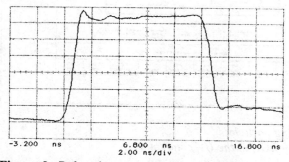

Figure 3: *Pulse shape generated by the Picoseond Pulse Lab 2000D pulse generator.*

PD processes can have repetition rates between a few Hz and several hundreds of Hz. To test the percentage of detected PD pulses from a PD source with a low repetition rate, 50 single shot pulses were injected and the number of detected pulses counted. As previously mentioned, the sweep time of the SA can have a large influence on the accuracy of the measurement. The sweep time of the SA was therefore varied between 36 ms (minimum value when measuring the full frequency span) and 10 s to determine the optimum sweep time. Table 1 summarizes the percentage of processed pulses as function of the sweep time. This table shows that by setting the SA to a sweep time of at least 5 s, 90 % of all injected pulses are captured.

Table 1: *Percentage of processed pulses out of 50 randomly injected pulses as function of the sweep time.*

Sweep time:	Percentage of processed pulses:
36 ms	40 %
100 ms	70 %
500 ms	80 %
1 s	80 %
5 s	90 %
10 s	90 %

The repetition rate of PD pulses can vary between a few Hz up to several hundreds of Hz. This means that the frequency spectrum may change during or between sweeps. However, for further analysis of frequency spectra an unchanging (stable) spectrum is necessary. To obtain a stable frequency spectrum the results of several sweeps can be processed in two ways:

1. Hold max: process a stable frequency spectrum from the maximum amplitudes at each frequency as measured during the series of sweeps.

2. Averaging: process a stable frequency spectrum by averaging the measured amplitudes at each frequency over the series of sweeps.

Table 2 shows the measurement times required to obtain a stable frequency spectrum as a function of the repetition rate of the pulse generator.

The previous experiments show that a measurement consisting of 20 sweeps of 5 s generates enough information to obtain stable frequency spectra, irrespective of the PD source and the processing of the frequency spectrum. All measurements described hereafter have been obtained accordingly.

Table 2: *Measurement time required to build up a stable frequency spectrum as function of the repetition rate of the pulse generator.*

Repetition rate:	Stable spectrum using max hold after:	Stable spectrum using averaging after:
1 Hz	100 s	60 s
10 Hz	50 s	50 s
100 Hz	50 s	50 s

3.   PD MEASUREMENTS ON GIS

A GIS is a totally encapsulated system. Consequently external disturbance coupling into the GIS is often low. In the following example spectral analysis is used for PD detection and trend analysis in a 420 kV GIS.

3.1   PD EVALUATION BY SPECTRAL ANALYSIS

Figure 4a and 4b show frequency spectra measured in a 420 kV GIS test setup. In particular resulting from a free moving particle (4a) and a protrusion fixed to the HV conductor (4b). Clear differences between both defects can be observed in the spectra.

Figure 4c and 4d show frequency spectra obtained from the same faults, however these spectra have been processed by averaging the measured amplitudes. In contrast to the spectra of figure 4a and 4b, the processing result in spectra with very stable peaks. In a similar way, processing using hold max also results in very stable spectra.

Peaks in a frequency spectrum can indicate PD activity so a more detailed analysis of these peaks is necessary. This can be done in the zero span mode of the SA [5]. Figure 5a shows a resulting PD pattern measured at 541 MHz during 10 minutes. Unfortunately, no indication of the magnitude in *[pC]* of the PD pulses can be given. However, methods to provide at least a sensitivity calibration of different GIS couplers are described in [7].

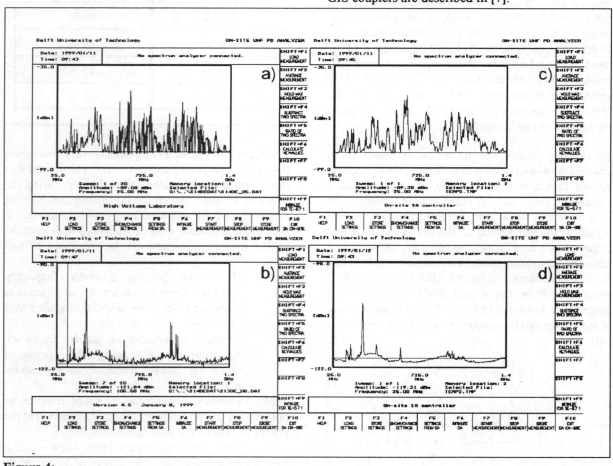

Figure 4:
a)   *One sweep of 5 s in the obtained from a free moving particle;*
b)   *One sweep of 5 s in the obtained from a protrusion fixed to the HV conductor;*
c)   *20 sweeps of 5 s from a free moving particle;*
d)   *20 sweeps of 5 s from a protrusion fixed to the HV conductor;*

3.2 PD EVALUATION BY APPLYING KEY-VALUES

To monitor changes in the PD process, periodic measurements have been done. Four key-values have been calculated from the spectra:

1) average amplitude in the full measured frequency span
2) average amplitude between 500 and 700 MHz
3) total energy in a the full measured frequency span
4) total energy between 500 and 700 MHz.

These values are displayed on a graph in figure 6.

Following the trends of all key-values in figure 6, an increased level on the last day of measurements can be observed. This could indicate an increase of the PD magnitude or a change in the pattern which can indicate a dangerous situation. Therefore another PD pattern was measured, as shown in figure 5b. Comparison of figures 5a and 5b shows no significant change in the PD pattern. Moreover, the increase in the trend of figure 6 is caused by external disturbances as can be seen in figure 5b. Nevertheless this example shows that the VHF/UHF method is a very sensitive means to detect even small changes in the measured signals.

4. CONCLUSIONS

In this paper:

1) using proper settings for sweep time and number of sweeps the application of a SA for PD measurements on GIS is validated
2) different frequency spectra can be measured

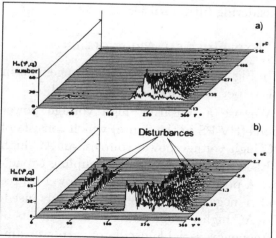

Figure 5: *3-dimensional PD pattern measured at a measuring frequency of 541 MHz*
a) *observed and unchanged during observations 1-6*
b) *observed and unchanged during observations 7-8. The influence of external disturbances in clearly visible.*

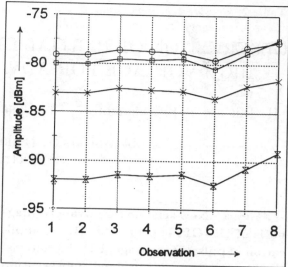

Figure 6: *Trend analysis observed in a 420 kV GIS over a period of 6 months. Starting at the top:*
1) *average amplitude between 200 and 1500 MHz (○)*
2) *average amplitude between 500 and 700 MHz (□)*
3) *total energy between 200 and 1500 MHz (×)*
4) *total energy between 500 and 700 MHz (Ⅹ).*

   for different defects
3) specific key-values to describe frequency spectra were calculated which can be effectively used for trend analysis
4) it was shown that based on this trend analysis possible changes in PD processes can be observed.

REFERENCES

[1] R. Baumgartner, B. Fruth, W. Lanz and K. Petterson, *PD-Part IX: PD in GIS - Fundamental Considerations*, IEEE Electrical Insulation Magazine, Vol. 7, No. 6, 1991.

[2] IEC, *Partial discharge measurements*, 2nd edition, IEC Publication 270, 1981.

[3] H.-D. Schlemper, R. Kurrer and K. Feser, *Sensitivity of on-site partial discharge detection in GIS*, ISH Yokohama, Vol. 3, pp. 157-160, 1993.

[4] B.F. Hampton, R.J. Meats, *Diagnostic measurements at UHF in gas insulated substations*, IEE Proceedings, Vol. 135, Pt. C, No. 2, pp. 137-144, 1988.

[5] S. Meijer, E. Gulski, J.J. Smit, R. Brooks, *Comparison of Conventional and VHF/UHF Partial Discharge Detection Methods for SF6 GIS*, ISH Montreal, Vol. 4, 1997.

[6] S. Meijer, W.R. Rutgers and J.J. Smit, *Acquisition of partial discharges in SF6 insulation*, CEIDP, 1996.

[7] Task Force 15/33.03.05, *Partial Discharge Detection System for GIS: Sensitivity Verification for the UHF Method and the Acoustic Method*

OBSERVATION OF PARTIAL DISCHARGE IN SF$_6$ ON PULSATING HIGH VOLTAGE FOR ADVANCED DIAGNOSIS OF DC-GIS

H. Takeno   S. Nakamoto   Y. Ohsawa   K. Arai[†]   Y. Kato[‡]   T. Hakari[‡]

Kobe University, †Fukui University of Technology,
‡Kansai Electric Co., Ltd., JAPAN

Abstract—**New scheme for advanced diagnosis of DC-GIS is proposed. By an application of pulsating voltage, PD characteristics about 'phase' is obtained. Observation of basic characteristics of PD is performed and results are presented.**

1 INTRODUCTION

GIS and other gas insulated power equipments are widely used because of its advantages, such as stability and saving space. They have one crucial disadvantage, that is, degradation of equipments is not found easily because inside of tanks sealed hermetically is not visible in operation. To overcome such a disadvantage, lots of studies for diagnosis of degradation of GIS are reported. Various signals induced by partial discharge(denoted by PD in the following) are often used to know start of degradation. To examine characteristics of PD pulse against phase of applied voltage advanced informations, such as left time to breakdown, are also obtained[1].

On one hand, DC transmission is expected to be employed for large and long power transmission. In such a situation, DC-GIS is also expected to be used, and some diagnostic techniques are necessary to maintain equipments. Signal sensors for AC-GIS are commonly able to be used, but there is a problem in analyzing system because there is no information about 'phase' in DC equipments. In order to obtain advanced informations, new technique which is applicable to DC equipments should be developed.

The authors propose new scheme for advanced diagnosis of degradation of DC-GIS. We superpose AC voltage on DC voltage in working equipments and force to give 'phase'. The equipments are resulted in working with 'pulsating voltage' during diagnosis. To examine characteristics of PD against given phase, we can expect advanced informations such as level of degradation. This scheme has a merit that we could employ the same kind of techniques as those for AC-GIS. The problem in principle of this scheme is that the characteristics of PD will be different from those on AC voltage because polarity of voltage does not change. It would be also different from those on DC voltage.

In order to examine a possibility of this new scheme for advanced diagnosis of degradation of DC-GIS, we carried out observation of fundamental characteristics of PD in pulsating voltage. In this paper, we present results of experimental study. In the next section, we explain the experimental arrangement. In section 3, we present the experimental data and its analysis. In the last section, we summarize the contents of the paper with considering future subjects.

2 EXPERIMENTAL ARRANGEMENT

In Fig. 1, the experimental arrangement in this paper is schematically presented.

We use a pulsating high voltage power source(HV-PS in the figure) which consists of DC high voltage power sources, an AC high voltage power source, and a coupling capacitor. A positive DC high voltage power source and a negative one are alternatively used. The AC high voltage power source consists of a frequency variable oscillator, a power amplifier, and a high voltage transformer. The power amplifier and the transformer can be used in the frequency range of 45 Hz ~ 1 kHz, thus we can superpose AC voltage of that frequency range.

The electrodes for PD are pin to plane type

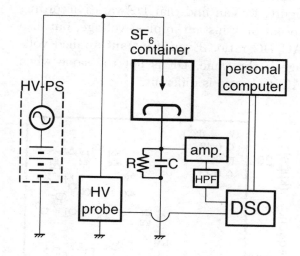

Fig. 1: Experimental arrangement for PD measurement on pulsating voltage.

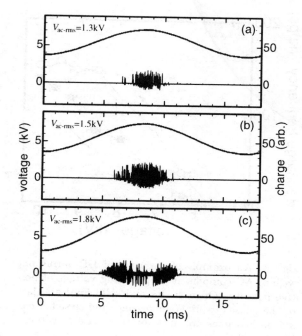

Fig. 2: Examples of waveforms of applied pulsating voltage and PD pulse.

and the gap length between them is 10 mm. We suppose PD induced by metal particles in GIS. The whole electrodes are contained in SF_6 gas container and the pressure of the gas is 1 atm.

The applied voltage is measured by a high voltage probe(HV probe). The charge quantity of PD pulse is measured by detecting voltage of a capacitor(C: 30 pF) connected in series. The pulse signals are amplified(amp.) and low frequency components are filtered out(HPF). Both measured signals are recorded by a digital storage oscilloscope(DSO), and transferred to a computer for data analysis.

3 EXPERIMENTAL RESULTS

<3·1> Partial discharge induction depending on phase of AC component

At first, relation between phase and appearance of PD pulse on pulsating voltage is observed.

Figure 2 shows a few examples of waveforms of applied pulsating voltage and PD pulse. Voltage of DC component is positive and constant($V_{dc} = 5.5$ kV) for all figures. As for AC component, the frequency is constant($f = 60$ Hz), but the amplitude(V_{ac}) is varied as shown in the figure. In Fig. 2, we can clearly find that PD is induced depending on phase of AC component. The PD pulses are found around peak voltages, and the range of the phase where pulses exist

becomes wider as V_{ac} increases. This fact is consistent with the idea that PD is induced when instant voltage is high enough.

When we look at the figure (c), we find that the pulse height around peak voltage(corresponding to phase of about 90°) is smaller compared with that in other phases. This phenomenon is similar with that observed in AC alone application[2].

We also examine the case of negative polarity application($V_{dc} < 0$). The phenomenon that PD is induced according to the phase of AC component is also observed. On the other hand, weakening of discharge around phase of 90° is not observed in negative polarity, This is also similar with the case of AC alone application[2].

<3·2> Observation of discharge frequency

Next, we evaluate 'discharge frequency'(denoted by DF in the following). The definition of 'DF' in this paper is the number of discharge pulse during unit time period. This corresponds to 'n' in usual ϕ-q-n characteristics of PD in AC-GIS. For example, when we find 9 pulses during 200 μs, (DF) = 9 pulses/200 μs = 45 kpps(kilo − pulse per second). In the following, DF is evaluated for short period, and 'applied voltage' means the averaged value of

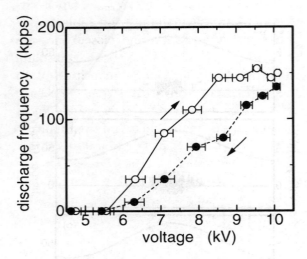

Fig. 3: An example of variation of DF during one cycle of AC component on pulsating voltage of positive polarity. Open circles and solid lines are data for the phase of voltage-increasing, and closed circles and dashed lines are data for the phase of voltage-decreasing.

instant voltage during corresponding period.

In Fig. 3, DF versus applied voltage is presented. The data are those for one cycle of AC component. The voltage condition is $V_{dc} = 6.3\,\mathrm{kV}$(positive polarity), $V_{ac} = 2.7\,\mathrm{kV_{rms}}$, and $f = 180\,\mathrm{Hz}$. The data that are in the phase of voltage-increasing($i.e.$ $-90° \sim 90°$) are indicated by open circles and solid lines, and closed circles and broken lines mean the data in the phase of voltage-decreasing($90° \sim 270°$). This figure clearly indicates that DF in the phase of voltage-increasing is different from that in the phase of voltage-decreasing. This phenomenon is found wide range of voltage conditions. On negative polarity application, the phenomenon is not clear. At least, the difference between DF for voltage increasing and that for voltage-decreasing is much smaller compared with that on positive polarity application.

We further examine DF by changing the condition of high voltage. Figure 4 shows variation of DF versus instant applied voltage on positive polarity application. In order to clarify the variation, only data in the phase of voltage-increasing are plotted. In these data, the ratio of V_{ac} to V_{dc}(denoted by 'AC/DC ratio' in the following) is varied. The conditions are selected by keeping peak voltages constant. Frequency of AC component is constant($f = 180\,\mathrm{Hz}$). From this figure, we can find that DF varies according to not only instant applied voltage, but also AC/DC ratio. Even if instant applied voltage is the same, DF is not the same when AC/DC ratio is different.

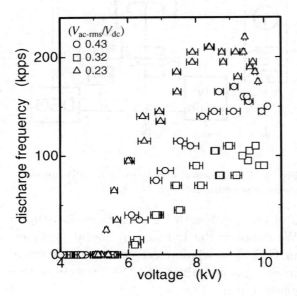

Fig. 4: Variation of DF due to change of AC/DC ratio on positive polarity application.

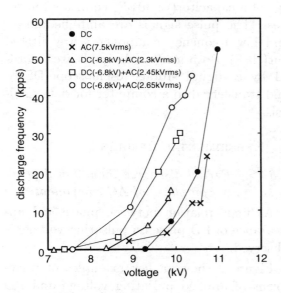

Fig. 5: Variation of DF due to the condition of applied high voltage on negative polarity application. Absolute values are taken for voltage(abscissa).

In Fig. 5, we also present the same kind of data of DF on negative polarity application. Note that absolute values are taken for abscissa(voltage). Open symbols in the figure distinguish just the difference of V_{ac}, and V_{dc}

is constant. The different values of DF due to different AC/DC ratio are also found on negative polarity application. In the figure, data on DC alone application and AC alone application are also shown, and they are also different from data on pulsating application.

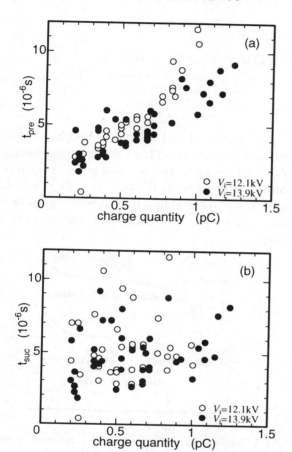

Fig. 6: Time intervals of PD pulse versus charge quantity. (a) is for predecessor time and (b) is for successor time. Open circles are data for instant voltage $V_i = 12.1 \pm 0.4$ kV, and closed circles are those for $V_i = 13.9 \pm 0.2$ kV. Absolute values are taken for charge quantity.

<3·3> *Correlation between charge quantity and time intervals of PD pulses*

Beyer *et al.* report new scheme to diagnose degradation of DC capacitor by utilizing PD signal[3]. They examine time to the discharge successor(t_{suc}) and time to its predecessor(t_{pre}). They analyze the correlation between charge quantity(q) and these time intervals. We apply the same examination to PD in SF$_6$ gas on the application of pulsating voltage.

Figure 6 shows example of analysis. The condition of voltage application is $V_{dc} =$

-12.4 kV and $V_{ac} = 1.2$ kV$_{rms}$. Open circles are data when the instant applied voltage is about -12.1 kV and closed ones are those on $V_i \cong -13.9$ kV. Note that absolute values are taken for charge quantities(q). In Fig. 6(a), indicating correlation between q and t_{pre}, clear positive correlation and almost linear relation is found between two quantity. The increasing rate for open circles is larger compared with that for closed circles. On one hand, according to Fig. 6(b), we cannot find clear correlation between q and t_{suc}.

The linear relation between q and t_{pre} and no correlation between q and t_{suc} are confirmed for various voltage condition of negative polarity. The results are partially similar with that reported by Beyer *et al.*[3].

4 Summary

In order to examine a possibility of our proposal to develop new scheme for advanced diagnosis of DC-GIS, basic studies are performed. The characteristics of PD in SF$_6$ gas on pulsating voltage are observed, and followings are confirmed. 1) PD is induced corresponding to phase of AC component. 2) DF is evaluated for various kinds of voltage condition. The values depend on AC/DC ratio as well as instant voltage. Whether voltage is increasing or decreasing also affects DF on positive polarity. 3) Almost linear relation is found between q and t_{pre}, but correlation with t_{suc} is not found.

In future, in order to confirm the characteristics of PD on pulsating voltage, same kind of observation in usual pressure($3 \sim 6$ atm) should be done. The most important future subject is clarification of the relation between PD signals and level and/or kind of insulation defects. If it is clarified, design of diagnosis system, which includes a controller of AC voltage superposing system, is not difficult.

References

[1] T. Kato, *et al.*, Proc. 4th ICPADM 574 (1994).

[2] M. Hikita, *et al.*, Trans. IEE of Japan Vol.115-B 1215 (1995).

[3] J. Beyer, *et al.*, Proc. of 10th ISH, No.3358 (1997).

Diagnostics of Surge Arresters Using High Voltage Gradient ZnO Elements for the 800kV GIS by UHF Partial Discharge Methods

S.Shirakawa   S.Tanaka   S.Yamada   S.Watahiki   S. Kondo   T.Kato   W-P.Son
(HITACHI, Ltd.)   (Hyosung Industries CO, Ltd)
Japan   Korea

Abstract

It is important to perform partial discharge tests in order to evaluate the soundness of the insulation parts of surge arresters. Normally, surge arresters have been evaluated by ERA methods in factory tests. In this case, newly developed surge arresters for the 800kV GIS using high voltage gradient ZnO elements have been diagnosed by UHF partial discharge methods. This paper describes test results of surge arresters for the 800kV GIS by partial discharge tests.

Keywords : Surge Arrester, Zinc Oxide Elements,
UHF PD Diagnostics Techniques

1. Introduction

For the partial discharge test the power frequency voltage applied to the complete arrester be increased up its rated voltage and after less than 10 s decreased to 1.05 times its continuous operating voltage. At that voltage the partial discharge level according to IEC 270 shall be measured. When required, the partial discharge in the arrester energized at 1.05 times its continuous operating voltage shall not exceed 10 pC according to IEC 60099-4 A.3/CD, 1998 or JEC-2373-1998 "Gas Insulated Metal-Enclosed Surge Arresters". Satisfactory absence from partial discharges and contact noise shall be checked on each unit by any sensitive method adopted by the manufacture. In this paper, newly developed surge arresters for the 800kV GIS using high voltage gradient ZnO elements have been diagnosed by UHF partial discharge methods.

2. Tested Surge Arresters for the 800kV GIS

Recently, SF₆ gas insulated tank type surge arresters using about 2 times high voltage gradient ZnO element have spotlight because of reducing the stacking height of ZnO elements

to half value of the nominal height and GIS original layouts /1/. These basic characteristics of ZnO elements are expressed in relation of voltage (V) − current (I) as shown in Fig.1. ZnO elements are able to obtain reasonable reference voltage by controlling the size of ZnO grains.

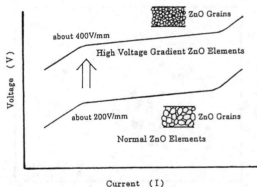

Fig.1 Voltage (V) −Current (I) Characteristics and Microstructure of Zinc Oxide Elements

We have newly developed a vertical mounting type surge arresters using high voltage gradient ZnO elements for 765kV power systems. By reducing the height of ZnO elements

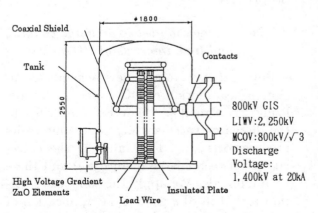

800kV GIS
LIWV:2, 250kV
MCOV:800kV/√3
Discharge
Voltage:
1, 400kV at 20kA

Fig.2 Composition of Surge Arresters using High Voltage Gradient ZnO Elements Connected to 800kV GIS

High Voltage Engineering Symposium, 22–27 August 1999
Conference Publication No. 467, © IEE, 1999

elements connected in series, these arresters can obtain the compact design as shown in Fig.2. These arresters are connectable to the 800kV GIS at a lower floor place. Fig.3 shows the internal connection between two columns of ZnO elements. The shape of these ZnO elements is doughnut type. A simple construction of surge arresters (1,400kV at 20kA) for the 800kV GIS (LIWV 2,250kV) has been established. Fig.4 shows a test circuit of line discharge tests. Fig.5 shows a waveform of line discharge test results (line discharge class 4) on IEC 60099-4 "Metal-oxide surge arresters without gaps for a.c". High voltage gradient ZnO elements are resistible for 18 discharge operations divided into six groups of three operations.

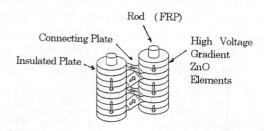

Fig.3 Internal Connection of ZnO Elements

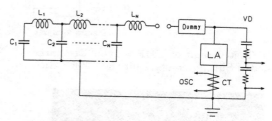

Line Discharge Class 4
· Surge Impedance : 0.8 Ur (=0.8x576kV= 460 Ω)
· Charging Voltage : 2.6 Ur (=2.6x576kV=1498kV)
· Virtual Duration of Peak : 2.8ms

Fig.4 Test Circuit of Line Discharge Test

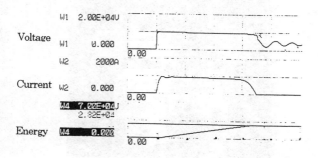

1ms/div.

Fig.5 Tested Waveform of Line Discharge Test
(Line Discharge Class 4)

3. Partial Discharge Methods

Normally, surge arresters have been evaluated by ERA (Electrical Research Association, England) methods. In this case, newly developed surge arresters for the 800kV GIS using high voltage gradient ZnO elements have been diagnosed by UHF (Ultra High Frequency) partial discharge methods.

3.1 UHF Methods

The UHF partial discharge method enables defects to be found because it has the necessary high sensitivity, and partial discharge of pC level can readily be detected on site in large sections GIS. The technique has the added advantage that measurements may be made equally well in GIS at any system voltage. UHF measurements can be made from 300 MHz upwards, but if any interference from air corona is being fed into the GIS this can be much reduced by filtering out signals below 500 MHz. Signals of the partial discharge in the GIS appear in region of 750 − 1,500 MHz bands.

A big issue for the on site test is related to calibration matter. For the UHF partial discharge method, no calibration exists at present /2/. In this case, we have calibrated detected signals between ERA methods and UHF methods.

(a) Calibration by the External Input 10 pC

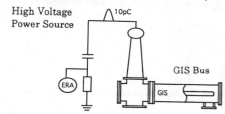

Fig.6 Tested GIS Model without Particles

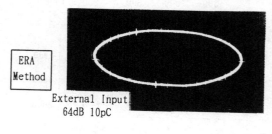

ERA Method

External Input 64dB 10pC

UHF Method

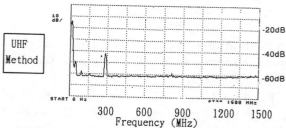

Fig.7 Calibration of ERA and UHF Methods
without Particles

(b) Calibration by Metallic Particles in GIS

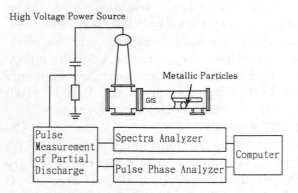

Fig.8 Tested GIS Model with Metallic Particles

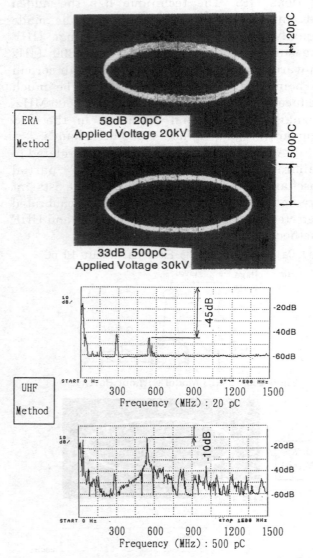

Fig.9 Calibration of ERA and UHF Methods by Partial Discharge with Metallic Particles in the GIS

Fig.6 and Fig.7 shows outputs signals of ERA and UHF methods in the GIS without metallic particles in case of the external signal input 10 pC.

On the otherhand, Fig.8 and Fig.9 show partial discharge signal outputs of ERA and UHF methods in the GIS with metallic particles. In this case, calibrations between ERA and UHF methods have been performed by the comparison of the above signal output.

3. 2 Evaluation of Surge Arresters for the 800kV GIS by UHF Partial Discharge Methods

Fig.10 shows a detection test circuit of a UHF partial discharge method of surge arresters for the 800kV GIS. UHF sensors for partial discharge detection are installed in a hand hall on the GIS apart from the surge arrester. Fig.11 shows an outside view of the tested surge arrester for the 800kV GIS. Fig.12 shows output waveforms of the surge arrester for the 800kV GIS by UHF partial discharge methods.

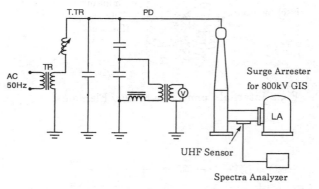

Fig.10 Test Circuit of UHF Partial Discharge Method of Surge Arresters for the 800kV GIS

Fig.11 Outside View of UHF Partial Discharge Method of Surge Arresters using High Voltage Gradient ZnO Elements for the 800kV GIS

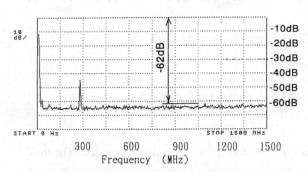

(a) Back Ground Noize

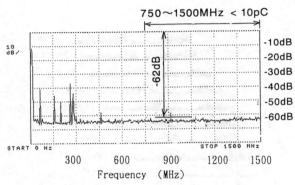

(b) Partial Discharge Voltage Noize
    at 485kV (=800kV/√3 ×1.05)

Fig.12 UHF Sensor Output of UHF Partial Discharge Method of Surge Arresters for the 800kV GIS

3 pC of the surge arrester for the 800kV GIS can be obtained at MCOV 800 kV /√3 ×1.05. This measured value clears the specified value 10 pC according to IEC 60099-4 A.3/CD, 1998 and JEC-2373-1998.

4. Conclusion

Surge arresters using high voltage gradient ZnO elements for the 800kV GIS can be diagnosed by UHF partial discharge methods in factory tests.

References
/1/ S. Shirakawa, I. Ejiri, S. Watahiki, N. Iimura, S. Yamada, S. Kondo : " Application of High Voltage Gradient Zinc Oxide Elements to SF6 Gas Insulated Surge Arresters for 22kV — 765kV Power Systems ", **IEEE** SM 1998, Preprint Number PE-424-PWRD-0-05-1998
/2/ CIGRE WG33/23-12 : "Insulation Co-ordination of GIS : return of experience, on site tests and diagnostic techniques ", ELECTRA No. 176 Feb. 1998
/3/ S. Okabe, E. Zaima, T. Yamagiwa, T. Ishikawa : " Partial Discharge Characteristics and Diagnostics Technology for Metallic Particles in Gas Insulated Switchgear ", **T. IEE** Japan, Vol. 115-B, No. 10, 1995

There is a sensibility curve /3/ reported by Dr. Okabe et al. about the relation of the detection level and the coulomb of the partial discharge under UHF methods. Data of Fig.9 have been plotted on this sensibility curve. Fig.13 have been obtained as the calibration curve of partial discharge tests at this test equipment. From this Fig.13, the partial discharge

Address of author
Shingo Shirakawa, IEEE Senior Member
Kokubu Works, HITACHI, Ltd.
1-1, Kokubu-cho 1 chome, Hitachi-shi, Ibaraki-ken, 316-8501 Japan
E-mail : s-shirakawa@cm.kokubu.hitachi.co.jp

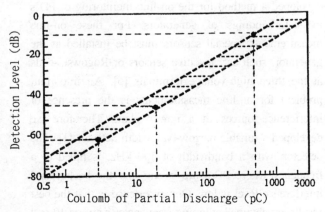

Fig.13 Relation of UHF Sensor Output Detection Level and Coulomb of Partial Discharge

ON-LINE PARTIAL DISCHARGE MONITORING OF HV COMPONENTS

A.J.M. Pemen[*], W.R. Rutgers, T.J.M. van Rijn, Y.H. Fu

KEMA T&D Power, Arnhem
[*]now with the Eindhoven University of Technology, High-Voltage & EMC group
THE NETHERLANDS

Abstract - Partial discharge (PD) measurements are an important tool to assess the condition of the insulation of high-voltage systems. Preferably, these measurements are done on-line during regular operation. This paper gives an overview of developments and experiences with on-line PD measurements in The Netherlands on turbine generators, transformers and switch-gear.

Introduction

Power utilities strive to produce and transport electrical energy with a high availability and at low costs. To achieve this, two aspects are of major importance: (i) minimal costs for maintenance, e.g. by applying condition based maintenance, and (ii) a reduction of in-service failures by early detection of possible failures. This can only be realized when effective tools are available to assess the condition of important components, such as generators, transformers, GIS, cables and their accessories and switch-gear.

Partial discharge (PD) measurements are an important tool to assess the quality of high-voltage insulation. Already more than 35 years, KEMA has experience with off-line PD measurements, carried out during commissioning tests and maintenance periods. However PD measurements are preferably done on-line during regular operation. The advantages of on-line measurements are: (i) they take place under realistic stresses, (ii) early detection of possible failures, (iii) permanent monitoring of the condition, and (iv) inspection is possible at every wanted moment (e.g. before instead of during a revision, before and after repair).

Therefore, KEMA developed various techniques for on-line PD measurements on generators [1], transformers [2], GIS [3], cables, cable terminations and joints [4] and medium-voltage sub-stations [5]. This paper gives an overview of experiences with these techniques in the form of three cases. In the first case, large discharges in a 570 MW power plant are located. The second case describes the on-line PD detection in a

machine transformer and a compensation coil. The third case gives an example of how on-line PD measurements can be used in the condition assessment of a 10 kV switch-bay.

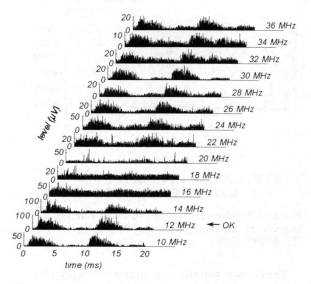

Fig. 1 *PD-patterns for a large turbine generator, measured at various tuning frequencies*

Case 1: PD monitoring of a 570 MW power plant

By order of the Dutch power utilities, KEMA developed a method for the on-line monitoring of PD's in statorwindings of generators. For these on-line measurements, special sensors must be installed at the generator, such as capacitive sensors or Rogowski-coils at the three high-voltage terminals [6]. An important problem for on-line measurements is the presence of interference-sources in a power plant. Therefore we developed a tunable narrow-bandwidth detector [1]. The detector, with a bandwidth of 300 kHz, is tuned to a frequency where PD's are strong and the background interference level is low. Our experience is that the best results are obtained at tuning frequencies between 10 and 30 MHz. We developed a 3-channel detector with two outputs for each channel:

High Voltage Engineering Symposium, 22–27 August 1999
Conference Publication No. 467, © IEE, 1999

• a pulse output that can be connected to a laptop PC, which provides the digital registration of the PD-patterns and has built-in tools for recognition and classification of these patterns. Since various PD-sources might give different frequency-responses at the generator terminals, for each phase a series of patterns is recorded at various tuning frequencies. An example is given in Fig. 1.

• a level output to continuously record the PD-level, e.g. on a display in the control room. Now the detector is tuned to a fixed, optimal frequency. In the case of Fig. 1 this optimal frequency is 12 MHz, since at this frequency: (i) the largest signal is recorded, (ii) no interference can be recognized (for instance as at 20 MHz), and (iii) little cross-talk from other phases takes place (in contrast to 34 MHz).

During regular measurements on a 570 MW generator, a large PD-level was recorded in phase W. The discharges in phase W are about a factor 3 larger than for the other two phases.

These measurements were done with capacitive sensors in the isolated-phase-bus. For each phase, two sensors are installed, one close to the generator and one close to the machine transformer. For further analysis it was decided to measure the waveshapes of the large PD's. A result is given in Fig. 2. As can be seen, the signal first shows up at the sensor near the machine transformer, which proves that the large PD's do not come from the generator. In the case that the PD's would come from the transformer, the transit-time between both signals has to be equal to the distance between the two sensors times the speed of light (or 11.5 m * 3.3 ns/m = 38 ns). However, the transit-time is much shorter (12 ns), which is only possible if the PD originates between the two sensors. Thus we come to the conclusion that the large PD's occur in the generator-circuit-breaker.

Therefore it was decided to locate the PD-source inside the circuit-breaker, by means of a HF-probe [5]. This was done during regular operation of the power plant, and the result is given in Fig. 3. It can be seen that the PD originates from the bleeding-resistors at the top of the circuit-breaker.

Now that the origin of the discharges is known, appropriate actions can be undertaken. After consultation with an expert on generator-circuit-breakers, it was concluded that the PD's will, on a short term, not lead to serious problems. So it was decided to keep the unit in operation until the next planned maintenance overhaul. During this maintenance period, a visual inspection of

the circuit-breaker will be performed to find, and repair, the actual cause of the discharges.

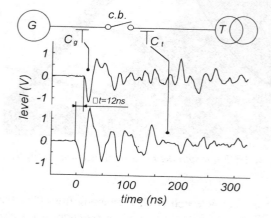

Fig. 2 *A PD in phase W, measured simultaneously with the two sensors in this phase. From the 12 ns transit-time it could be determined that this PD originates from the generator-circuit-breaker*

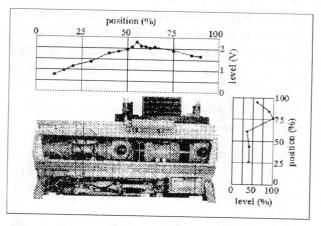

Fig. 3 *PD-levels near the generator-circuit-breaker in phase W of a 570 MW power plant*

Case 2: On-line PD measurements on a machine transformer and a 3-phase compensation reactor

PD measurements on power transformers are performed to determine the dielectric condition of the HV insulation. For on-line detection of PD's a broad-band UHF antenna has been constructed that can be installed inside the transformer during operation (Fig. 4). The antenna is mounted on an oil valve and is pushed into the transformer tank to a position aligned with the inside wall.

Fig. 4 *UHF antenna mounted on a machine transformer*

For detection of UHF emission from PD either a commercial spectrum analyzer (HP 8590B, 0-1.8 GHz bandwidth up to 1 MHz) or a broadband receiver (bandwidth 90 MHz) and pulsshaper made by KEMA Technical Services have been used. The center frequency of the latter can be chosen between 500 and 1000 MHz (adjustable in steps of 10 MHz). With both instruments phase resolved PD patterns can be measured. In this case the spectrum analyzer is operated in the zero span mode or POW (Point- on- Wave) mode combined with the hold-max mode. The KEMA receiver is connected to a portable PC (Field-works) with a fast digitizer and software to measure the time and amplitude of the UHF pulses after the pulsshaper.

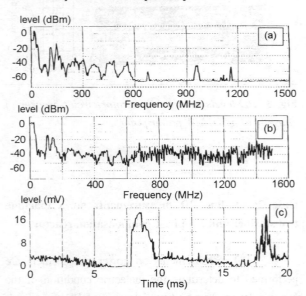

Fig. 5 *Frequency spectra measured on a compensation reactor: (a) background, (b) under operating voltage, (c) POW at 1403 MHz*

Figure 5 shows the frequency spectrum of the background (no HV) and PD spectrum (reactor under operating voltage), measured on the antenna mounted on

a 3-phase compensation reactor. Generally speaking a strong narrow-band signal around 750 or 950 MHz is always present in the background spectrum which comes from mobile telephone transmitters. The background signal is normally high up to 200 MHz and sometimes, as in this case, up to 500 MHz. However above 500 MHz there is always a band were the background level is below the noise level of the PD detector. The frequency spectrum of PD's in oil and SF_6 gas is broad, up to frequencies above 1 GHz (Fig. 5b), while the background corona discharges on the high-voltage line have much lower frequency components (upto 200 MHz). In Fig. 5c the POW signal is shown for a detection frequency of 1402 MHz. This phase-resolved PD pattern shows a high PD level (order 500 pC) that is fluctuating in time and nature. After maintenance the level was measured to be below 20 pC.

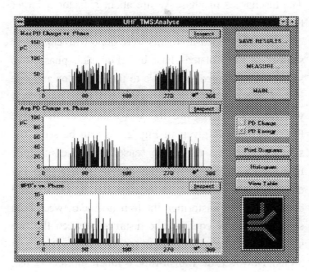

Fig. 6 *Phase resolved PD pattern of a 570 MVA machine transformer, detection frequency 920 MHz. The weak discharges are harmless for the transformer insulation*

PD measurements have been performed on a number of HV transformers in the Dutch 150 kV and 380 kV grid. The spectrum analyzer is always used during the first measurements to select a good detection frequency. If two sensors can be installed, a pulse is injected for proving the sensitivity of the detection system. Phase-resolved PD patterns are then recorded with the broadband detection system. An example is shown in Fig. 6. It shows Q_{max}, Q_{mean} and the number of PD's as a function of the phase angle. A weak discharge activity is observed, which is harmless for the transformer insulation.

Case 3: 10 kV switch-bay

Condition based maintenance is introduced in sub-station maintenance practice to reduce total maintenance cost, extend service life or reduce and avoid costs of failures. To asses the condition of the HV insulation and estimate the trend of degradation, PD measurements are used as part of the inspection program.

In this case an example is given of PD measurements by use of a HF-probe [5] during inspection of a 10 kV metal enclosed sub-station containing 80 (three-phase) switch-disconnectors. The PD sensor with a capacitive probe can measure E-M emission from PD up to a frequency of 5 MHz. It has been used to find and localize the position of partial discharges. If the signal is high the electrical output of the sensor is displayed on a transient recorder to measure the phase angle of the pulses. The phase resolved discharge pattern is stored in a database.

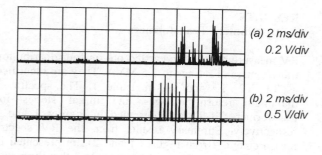

(a) 2 ms/div
0.2 V/div

(b) 2 ms/div
0.5 V/div

Fig. 7 *PD patterns measured with a HF-probe near a cable termination, during 20 ms or one period. (a) Non-audible PD's. (b) Audible PD's, crack in cable mantle.*

PD signals were measured near the cable-terminations at several different places. In particular the terminations for indoor use only show problems due to pollution and moisture or condense. Surface tracks from discharges were found on 11 locations, often combined with condense and salt deposition.

Figure 7 gives two examples of measured PD patterns of (a) a non audible and of (b) an audible discharge in a cable termination. In the latter case further inspection showed a crack in the insulation of the cable mantle. From the results of PD measurements and other inspection results, all performed under service conditions, a list of urgent conditions and of not urgent but degraded conditions of components in a sub-station was made.

Conclusions

Measuring techniques for on-line PD measurements on turbine generators, power transformers and switch-gear has been developed and used on-site for assessment of the condition of HV insulation. Information on the location and nature of PD, used in combination with other inspection results, are very useful to detect failures in an early state of development and to plan maintenance actions. In future, when more historic data are available, the information can be used to predict the rate of degradation of the HV insulation and to estimate remaining life-time.

Acknowledgment - this research was undertaken by order of Dutch electricity generation and distribution companies. Their support in the measurement campaigns is gratefully acknowledged.

References

[1] A.J.M. Pemen, P.C.T. van der Laan, 1998, *"On-line detection of PD's in statorwindings of large turbine generators"*, IEE Colloquium on discharges in large machines, London.

[2] W.R. Rutgers, Y.H. Fu, 1997, *"UHF PD detection in a Power Transformer"*, Proc. 10th ISH, Montreal, pp. 219-222

[3] S. Meijer, E. Gulski, W.R. Rutgers, 1997, *"Evaluation of Partial Discharge Measurements in SF$_6$ Gas Insulated Systems"*, Proc. 10th ISH, Montreal, pp. 469-473.

[4] E. Pultrum, E.F. Steennis, M.J.M. van Riet, 1996, *"Test after laying, diagnostic testing using partial discharge testing at site"*, CIGRE session 1996, Paris

[5] Lemke PD detector type LDP 5.

[6] A.J.M. Pemen, W. de Leeuw, P.C.T. van der Laan, 1997, *"On-line PD-monitoring of statorwindings; comparison of different sensors"*, Proc. 10th ISH, Montreal, pp.105-108.

Addresses of authors:

KEMA T&D Power,
P.O. Box 9035, 6800 ET Arnhem, The Netherlands
phone: +31 26 356 3244, email: w.r.rutgers@kema.nl

Eindhoven University of Technology,
High-Voltage and EMC group
P.O.Box 513, 5600 MB Eindhoven, The Netherlands
phone: +31 40 2474492, email: a.j.m.pemen@ele.tue.nl

FIELD EXPERIENCE WITH RETURN VOLTAGE MEASUREMENTS FOR ASSESSING INSULATION CONDITION IN TRANSFORMERS AND ITS COMPARISON WITH ACCELERATED AGED INSULATION SAMPLES

Dr. T K Saha, Prof. M Darveniza, Z T Yao

Department of Computer Science & Electrical Engineering, University of Queensland, Australia

INTRODUCTION

Large proportions of power transformers within electric utilities around the world are approaching the end of their design life. Many, perhaps most, seem to be operating satisfactorily. However, insulation degradation continues to be a major concern for these aged transformers. Insulation materials degrade at higher temperatures in the presence of oxygen and moisture. The degradation from thermal stress affect electrical, chemical and mechanical properties [1] [2]. Utility engineers use many modern techniques to assess the insulation condition of aged transformers. Among them Dissolved Gas Analysis (DGA), Degree of Polymerisation (DP) measurement and furan analysis by the High Performance Liquid Chromatography (HPLC) are frequently used [1,2]. In recent years there has been growing interest in the condition assessment of transformer insulation by the Return Voltage Method (RVM). In our previous research project, we have extensively used the return voltage method for analysing the condition of aged cellulose insulation. The results from these experiments have been presented elsewhere [3][4][5]. In our recent research project, accelerated ageing experiments were completed under air and nitrogen environments over the temperature range 115-145°C. The results have been described in an IEEE paper, which has been accepted for publication [6]. In this research we have also investigated a number of power and distribution transformers of different age with the return voltage measurement technique. Results from two groups of measurements (one from the distribution and the other from power transformers) will be presented in this paper. A comparison of results with those obtained from the accelerated ageing experiments will be also described.

RETURN VOLTAGE METHOD

When a direct voltage is applied to a dielectric for a long period of time, and it is then short circuited for a short period, after opening the short circuit, the charge bounded by the polarization will turn into free charges i.e, a voltage will build up between the electrodes on the dielectric. This phenomenon is called the return voltage. The theory of return voltage measurements has been presented elsewhere [3]. The charging voltage used in this measurement was 1000 volt DC. When a transformer has two windings, only one measurement was carried out between the primary and secondary winding. While for a three winding transformer, two different RV measurements were carried out. The first was carried out for the bulk insulation between the primary and the secondary windings and the second RV measurement was carried out for the bulk insulation between the secondary and the tertiary windings. As interfacial polarization is predominant at longer time constants, the spectrum of the return voltage was investigated by changing the charging and discharging time over a range of times greater than 0.5 second until the peak value of the maximum return voltage was obtained. Then the spectra of maximum return voltage and the initial slope were plotted versus the central time constant (the time at which the return voltage is maximum). The peak value of the maximum return voltage (from the return voltage spectrum) and the corresponding initial slope (from the initial slope spectrum), along with central time constant (from either of the spectrum), are the parameters used to assess the insulation condition from the return voltage measurements [3]. Details of the transformers tested are shown in Table 1.

RESULTS
(a) Distribution Transformers

RV measurements were carried out on four distribution transformers of ratings to 1.5 MVA. They are identified as T1, T2, T3 and T4 in Table 1. The spectra of maximum return voltages and initial slopes for transformers T1-T4 are presented in Figures 1 and 2 respectively. Summary results from the RV spectra measurements (from Figures 1 and 2) are presented in Table 2. It can be observed from Table 2 that there was a considerable variation in peak maximum return voltages for different transformers. Two transformers (T2 and T3) of same rating and age (1955) produced almost identical return voltages. Figure 1 and Table 2 show that the variation of the central time constant for different transformers is very significant. For transformer T4, the central time constant is 164 seconds (still rising), for transformers T2 and T3 it is only 48 and 47 seconds respectively, while for transformer T1 it becomes only 0.4 second. In general, the central time constant decreases sharply with ageing. In particular, for the oldest transformer T1, the central time constant dropped sharply. Table 2 shows that the variation of the initial slopes for different transformers is also very significant. From the results it is also clear that peak maximum return voltage and initial slope have some relationship with the size of the transformer. However, it is hard to compare the initial slope result of the 1 MVA transformer to those of 1.5 MVA transformers without any normalisation. It is interesting to compare T2, T3 and T4 as they are of same rating with different ages. T2 and T3 showed almost identical results in all respects. However, T4 shows a lower peak maximum return voltage, lower initial slope and longer central time constant compared with those of T3. All three of this group had moisture contents in the similar range

High Voltage Engineering Symposium, 22–27 August 1999
Conference Publication No. 467, © IEE, 1999

(24-28 PPM). Finally, it can be concluded that transformer T1 shows high level of insulation degradation. Similarly T2 and T3 in the 1.5 MVA group have significant insulation degradation compared to T4.

(b) POWER TRANSFORMERS
Return Voltage Measurements on a 30 MVA Transformer

The rating of this transformer (T5) is 30MVA between primary and secondary, 10MVA between primary and tertiary and 10MVA between secondary and tertiary. The voltage rating is primary =132kV, secondary = 33kV and tertiary = 11kV. The year of manufacture was 1959. The transformer was in service for 25 years and then it was not in operation over the last ten years. Two different RV measurements were carried out for this transformer. The first was carried out for the bulk insulation between the primary and the secondary windings and the second RV measurement was carried out for the bulk insulation between the secondary and the tertiary windings. The ambient temperature during RV measurements was 31°C. The RV parameters obtained from return voltage and initial slope spectra, as shown in Figures 3 and 4, are given in Table 3. Among the RVM parameters, the initial slopes in both the measurements are much higher than the normal initial slope of similar size transformer as found previously. From the RV parameters and spectra in Figures 3 and 4, the insulation paper has shown a certain degree of degradation. The paper has been contaminated by ageing products. The oil moisture content was low (10 PPM). Since the initial slopes are very high and the central time constants are relatively low (compared to the usual few hundred seconds for new insulation), the general paper insulation condition in these windings display evidence of some degradation.

Comparing the RV parameters and spectra in Test-2 with those of Test-1, the central time constant of 30 s is smaller and the initial slope of 69 V/s is higher in Test 2. This indicates that the general oil-paper insulation condition between secondary and tertiary shows more evidence of degradation than the condition of the insulation between primary and secondary windings. If the predominant cause of degradation is moisture, then it can be removed if the transformer is refurbished. It is expected that after refurbishment, which includes moisture removal by vapour cleaning or even by hot oil circulation and vacuum moisture removal, the general paper insulation condition will be improved significantly. Of course, the lowered moisture content would be reflected in improved RVM central time constants and initial slopes (as would other electrical parameters like insulation resistance and dissipation factor). Clearly the insulation between the secondary and the tertiary would benefit even more from moisture removal and refurbishment. From the RVM spectra in Test-2, it is observed that the insulation between secondary and tertiary winding is associated with more degradation and with higher percentages of ageing product than the insulation between primary and secondary. Based on the measurement results described in Test-1 and Test-2, refurbishment of this transformer is highly recommended, and particular attention should

be given to moisture removal. It would be of value to repeat the RVM measurements after refurbishment.

Return Voltage Measurements on a 145 MVA Transformer

The rating of the transformer (T6) is 145 MVA and voltage rating is 13.8/295 kV. The year of manufacture was 1971. The transformer was in operation for more than 25 years. After a failure, the windings of the transformer were replaced by new ones and the old oil was replaced with new oil. The RV measurement was carried out for the bulk insulation between the primary and the secondary windings. The ambient temperature during the measurements was 25°C. The following RV parameters were obtained from return voltage and initial slope spectra (graphs not shown here). Peak maximum return voltage = 81.5 V, central time constant = 1093 seconds, and initial slope = 1.70 V/s. In the graph of maximum return voltage versus central time constant spectra, the return voltage maximum value continued to rise. In fact, the return voltage was still rising after 1000 seconds charging. As the trend was clearly observed, after 1000 seconds charging, the measurement was finally stopped. This transformer has not been in operation after rewinding. The moisture content in the paper is likely to be evenly distributed if equilibrium is reached. This does not show an evidence of any degradation within the transformer.

Among the RVM parameters, the initial slope at peak maximum return voltage is 1.7 V/s. This is much smaller than the normal initial slope (as measured from a number of new and old transformers). From the RV parameters, the insulation paper has shown no degree of degradation. The oil moisture content is measured and is only 8 PPM. Since the initial slope is very low and the central time constant is high (comparable to the usual several hundred seconds for a new insulation), the general paper insulation condition in these windings shows no evidence of degradation. From the RVM parameters, it is observed that the insulation between primary and secondary winding is not associated with any degradation (as expected). It would be of great interest to compare the results with those of similar sized and aged transformers, as well as with measurements of this transformer after several years in service.

COMPARISON WITH ACCELERATED AGEING RESULTS

In our present research project, accelerated ageing experiments were carried out under air and nitrogen environments over the temperature range 115-145°C. It is observed that the variation of the central time constant for different ageing times is very significant for air ageing. Initially up to 14 days at 145°C ageing, the central time constant is many hundreds of seconds, while it becomes only a few seconds after 21 days of ageing in the presence of air. In general, the central time constant decreases sharply with respect to ageing time. In particular, after 21 days, the central time constant drops sharply. At 135°C up to 12 days ageing, the central time constant is several hundred seconds, while it becomes only a few seconds after 37 days of ageing in the presence of air. There was also a clear trend for the initial slopes to increase with time. In contrast, it was found that the central time constant was unaffected of ageing in nitrogen and the initial slope also changed very little with time. The most degraded samples (similar two conductors side by side forming the specimen) from a 25 year old transformer showed central time constants in the range of 20 to 35 s[3]. A good replication of the insulation condition in the aged transformer was observed for high temperature aged samples of the insulated conductor for 21 days of ageing at 145°C, for 37 days of ageing at 135°C, for 67 days of ageing at 125°C and for 125 days of ageing at 115°C in presence of air.

Transformer T6 (145 MVA) had central time constant of 1093 seconds. This is comparable to unaged samples. However, transformer T1 had a central time constant of 0.4 second. This time constant was even smaller than the corresponding times for 21 days of ageing at 145°C, for 37 days of ageing at 135°C, for 67 days of ageing at 125°C and for 125 days of ageing at 115°C in presence of air. This transformer had significant degradation in its insulation system. By comparison, transformers T2, T3 and T5 have central time constants in the range 30-60 seconds. This group of transformers has shown some degree of degradation and still can be used after proper refurbishment. Our experience with the accelerated ageing of paper wrapped insulated conductors has been reported previously [3,4]. These findings were somewhat different than those found with the full size transformers. For paper wrapped insulated conductors, both the initial slopes and central time constants were found to be very sensitive to ageing. The samples were of exactly identical in geometry. So the sample geometry had no impact on the results. The geometric capacitance and the insulation resistance were the same for all samples during unaged state. The experiments were conducted under controlled conditions. However, the transformers were of different ratings and have different geometric insulation resistance and capacitance at the unaged state. As a result, the return voltages and corresponding initial slopes were all different for different ratings. This phenomenon needs to be carefully investigated. We shall investigate this in our next project.

CONCLUSIONS
RV measurements have been applied to a number of power and distribution transformers.

The RV parameters are found to vary significantly and consistently with the ageing condition of insulation systems. Among RV parameters, central time constant has been found to be most sensitive to ageing and is independent of rating of transformers. However, the peak maximum return voltage and initial slope are dependent on the geometry of the insulation, and hence it was difficult to compare these for different ratings of transformers. For the same rating transformer, the initial slope was found to be very sensitive to ageing as well. Peak maximum return voltage has been found to increase with rating (perhaps due to the effect of different insulation resistance). This phenomenon will be investigated more carefully in our next research project. The initial slope and central time constant can therefore be used for assessing the condition of aged insulation in power transformers. A good replication of ageing condition was achieved by the ageing experiments in the presence of air.

REFERENCES
[1] D. H.Shroff & A. W. Stannett, "A Review of Paper Ageing in Power Transformers," IEE Proc., Vol. 132, pt. c, no. 6, pp. 312-319, Nov. 1985.
[2] J. Unsworth & F. Mitchell, "Degradation of Electrical Insulation Paper Monitored Using High Performance Liquid Chromatography," Proceedings of the 2nd International Conference on Properties and Applications of Dielectric Materials (ICPADM), pp. 337-340, Beijing, China, Sept. 12-16, 1988.
[3] T. K. Saha, M.Darveniza, D. J. T. Hill, T. T. Le, "Electrical & Chemical Diagnostics of Transformers Insulation, Part A: Aged Transformers Samples," IEEE Transactions on Power Delivery, Vol. 12, no. 4, pp. 1547-1554, October 1997.
[4] T. K. Saha, M.Darveniza, D. J. T. Hill, T. T. Le, "Electrical & Chemical Diagnostics of Transformers Insulation, Part B: Accelerated Aged Insulation Samples," IEEE Transactions on Power Delivery, Vol. 12, no. 4, pp. 1555-1561, October 1997.
[5] M. Darveniza, T. K. Saha, D. J. T. Hill, T. T. Le, "Investigation Into Effective Methods For Assessing the Condition of Insulation In Aged Power Transformers," IEEE Transactions on Power Delivery, Vol. 13, no. 4, pp. 1214-1223, October 1998.
[6] T. K. Saha, M.Darveniza, D. J. T. Hill, Z. T Yao, G. Yeung, "Investigating the Effects of Oxidation and Thermal Degradation on Electrical and Chemical Properties of Power Transformers," Paper accepted for the IEEE Transactions on Power Delivery, October 1997. (In Press).

Table 1: Details of the transformers investigated

Identity	Capacity (MVA)	HV/LV (kV)	Year of Manuf.
T1	1	10.5/0.415	1952
T2	1.5	33/11	1955
T3	1.5	33/11	1955
T4	1.5	33/11	1966
T5	30/10	132/33/11	1959
T6	145	295/13.8	1971 *

*New oil and new windings installed in 1998.

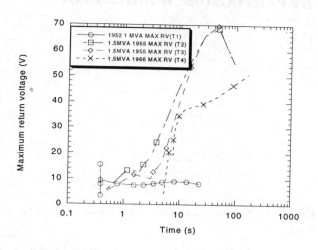

Figure 1: Spectra of maximum return voltage of transformers T1-T4

Table 2: Summary results of RV measurements from transformers T1-T4

Trans-for-Mer	Peak Max. Ret. Volt. (volt)	Cent. Time Cons. (s)	Init. Slope (volt /s)	Oil Water PPM
T1	15.5	0.4	18	40
T2	69	48	69	24
T3	70	47	64	28
T4	51	164	16	24

Table 3: Summary results of RV measurements from transformer T5

Trans-for-mer	Peak Max. Ret. Volt. (volt)	Cent. Time Const (s)	Init. Slope (volt/s)	Oil Water PPM
T5 (p-s) Test 1	99	58	31	10
T5 (s-t) Test 2	100	30	69	10

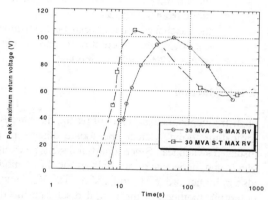

Figure 3: Spectra of maximum return voltage of transformer T5

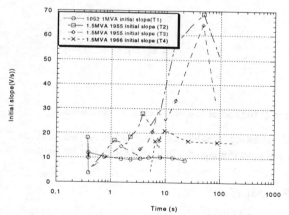

Figure 2: Spectra of initial slopes of transformers T1-T4

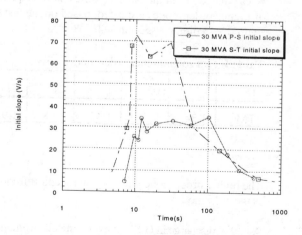

Figure 4: Spectra of initial slope of transformer T5

TEM- AND TE-MODE WAVES EXCITED BY PARTIAL DISCHARGES IN GIS

M.C. Zhang, H. Li

Xi'an Jiaotong University, Xi'an, China

Abstract

The possibility of using information from the TEM- and TE-modes for the detection and location of partial discharges has been investigated. The propagation and attenuation of the E and H components of the different modes excited radially by a line current in coaxial cylinder is described. To give the verification of it, experiments have been performed with simulated partial discharge currents injected radially in a GIS test facility. It is shown that:

- both the TEM-mode and TE_{11}-mode are excited by partial discharges in GIS
- inductive sensor allows to distinguish TEM-mode from TE-mode
- the discharge location can be determined by employing the different propagation velocities of the TEM- and TE_{11}-modes

1. Introduction

High-frequency measurement of partial discharges is an important tool to detect and locate defects in a gas insulated switchgear. Since a partial discharge excites an electromagnetic wave, not only in the TEM mode but also in higher order TE and TM modes, VHF and UHF detecting methods of the partial discharge in GIS are widely used /1//2/.

In GIS, most parts can be considered as a coaxial transmission line. According to the dimension of the GIS, the cutoff frequency of various higher order modes can be calculated /3/. In the table 1, some cutoff frequencies of higher order modes are listed, based on the test GIS setup (r_i=45 mm, r_o=155 mm) /4/.

Table I: f_c of some higher order modes (GHz)

f_c	01	02	11	21	31
TM	1.34	2.71	1.43	1.68	1.99

The higher order mode with the lowest cutoff frequency is the TE_{11} mode.

A partial discharge in GIS can excite the electromagnetic wave in two ways which are shown in figure 1: in radial direction (a), or in axial direction (b). The radial directed partial discharge can excite an electromagnetic wave not only in TEM and TM modes, but also in TE modes.

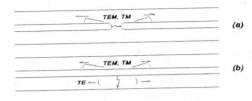

Figure 1: wave excited by axial (a) and radial (b) discharge

As Compared with TEM mode, the TE mode has its own properties which are help to partial discharge detection in GIS. However, the previous work /1//2/ was interested in the signal detection, no matter it is TEM or TE mode. This paper describes the possibility of using information from the TEM and TE mode for detection and location of partial discharge in GIS.

2. TE mode analysis

- Excitation

The TE mode excitation by the radial directed discharge in a coaxial waveguide can be described as shown in figure 2. The current flowing from the inner-conductor to enclosure in radial direction at z=0 and $\varphi=\varphi_0$ is defined as:

$$\vec{J}_{prim} = I(\omega)\,\delta(\varphi-\varphi_0)\,\delta(z)\,\vec{r} \qquad r_i \le r \le r_o \qquad (1)$$

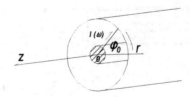

Figure 2: the radial discharge channel distribution

Under existing conditions, after some distance, the Maxwell's electromagnetic field equations can be written as:

$$\nabla \times \vec{E} = -j\omega\mu\vec{H}$$
$$\nabla \times \vec{H} = \vec{J}_{prim} + j\omega\varepsilon\vec{E} \qquad (2)$$

Combining equation 1 with equation 2, the axial component of the magnetic field H can be written as:

$$(\nabla^2 + k_0^2)H_z = \frac{1}{r}I(\omega)\,\delta'(\varphi-\varphi_0)\,\delta(z) \qquad (3)$$

According to the expression of the TE mode, under the

$I(\omega)$ excitation, the $\mathbf{H_z}$ can be composed of a series of TE modes /5/:

$$H_z \sum_m^\infty \sum_n^\infty A_{nm}(z,\omega) D_{nm}(k_{nm}r) \sin(n\varphi) \qquad (4)$$

Based on the orthogonal property and no-variation of the phase of $I(\omega)$ along $\mathbf{r}$, the expression of $\mathbf{H_z}$ is simplified as:

$$H_z(r,\varphi,z,\omega) = \frac{I(\omega)}{2\pi} \sum_n^\infty (\frac{F_{n1} n}{j\beta_{n1} B_{n1}} E)$$
$$E = \cos(n\varphi_0) D_{n1}(k_{n1}r) \sin(n\varphi) \exp(-j\beta_{n1}z) \qquad (5)$$

In equation 5, F_{n1}, $D_{n1}(k_{n1}r)$, B_{n1} and β_{n1} are represented respectively as:

$$F_{n1} = \int_{r_i}^{r_o} D_{n1}(k_{n1}r) \, dr$$
$$D_{n1}(k_{n1}r) = J_n(k_{n1}r) + D_n N_n(k_{n1}r)$$
$$\beta_{n1}^2 = \omega^2 \varepsilon\mu - k_{n1}^2 \qquad (6)$$
$$B_{n1} = \frac{(k_{n1}^2 r_o^2 - n^2) D_{n1}^2(k_{n1}r_o) - (k_{n1}^2 r_i^2 - n^2) D_{n1}^2(k_{n1}r_i)}{2k_{n1}^2}$$

- Propagation

In a lossless transmission line, the TEM mode wave signal propagates in velocity v_c. For higher order modes, the characteristics of the signal propagation is depended on the frequency. Here, the propagation of the higher order modes wave is discussed for a modulated Gaussian pulse excitation at z=0 point.

$$g(t) = C \exp(-(\frac{2(t-t_0)}{T})^2) \cos(\omega_0 t) \qquad (7)$$

In the direction of z axis, the propagation of signal g(t) in the frequency domain can be expressed as:

$$G(z,\omega) = G(\omega) \exp(j\omega t - j\beta(\omega)z) \qquad (8)$$

$\beta(\omega)$ is a phase constant and is a function of the frequency. Expanding $\beta(\omega)$ in a Taylar series at ω_0 and neglecting the $\beta'''(\omega_0)$ as well as higher order terms, we can get the Gaussian pulse after a distance z:

$$g(z,t) = A \exp(-\frac{(t-t_0-\beta'(\omega_0)z)^2}{\frac{T^2}{4} + 16(\frac{\beta''(\omega_0)z}{T})^2}) \cos(\omega_0(t-\frac{z}{v_p}))$$
$$A = \frac{CT}{(T^4 + (8\beta''(\omega_0)z)^2)^{\frac{1}{4}}} \qquad (9)$$

Combining equation 5 with equation 8, in the TE mode, H_z can be expressed in time domain under the modulated Gaussian pulse excitation as:

$$H_z(r,\varphi,z,t) = \sum_n^\infty AB \exp(-\frac{(t-t_0-\beta'_{n1}(\omega_0)z)^2}{\frac{T^2}{4} + 16(\frac{\beta''_{n1}(\omega_0)z}{T})^2}).$$
$$\qquad (10)$$
$$B = \frac{F_{n1} n\cos(n\varphi_0)\sin(n\varphi)D_{n1}(k_{n1}r)\cos(\omega_0(t-\frac{z}{v_{pn1}}))}{4\pi^2\beta_{n1}(\omega_0)B_{n1}}$$

How many TE modes are superimposed, depends on the modulation frequency ω_0 and the cutoff frequency ω_{cn1} of the different modes. In this equation the group velocity is defined as:

$$v_{gn1} = \frac{1}{\beta'_{n1}(\omega_0)} = \frac{1}{\sqrt{\mu\varepsilon}} \frac{\sqrt{\omega_0^2 - \omega_{cn1}^2}}{\omega_0} \qquad (11)$$

For different modes, the group velocity v_{gn1} is different and it is less than the velocity of the TEM mode. The same as other higher order modes, the TE mode signal will be deformed during its propagation. The deformation of the signal is caused by the different mode split and by the group dispersion.

- Attenuation

Unlike the TEM mode signal attenuation, the attenuation of higher order modes is caused not only by the skin effect, but also by the group dispersion, even though the signal is only in one mode. From equation 9, the attenuation of one higher order mode can be expressed as:

$$\alpha_g = 20\log\frac{T}{(T^4 + (8\beta''(\omega_0)z)^2)^{\frac{1}{4}}} \qquad (12)$$

It depends on the time constant of the Gaussian pulse and the modulation frequency as well as the cutoff frequency. For some chosen parameters the attenuation related to the propagation distance is in figure 3.

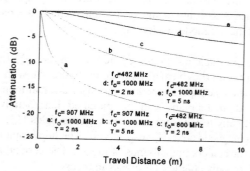

Figure 3: the attenuation caused by dispersion

It is obvious that the attenuation will be large when the cutoff frequency is close to the modulation frequency. An increase of the time constant of the Gaussian pulse will decrease the attenuation.

3. Experiment and Simulation

- Experimental facility

To give the verification of the TE mode theory, experiments have been performed by means of simulated partial discharge currents injected in GIS test facility. The test facility is shown in figure 4. It consists of busbars and junctions as well as insulated spacers. Both terminals of the facility are grounded.

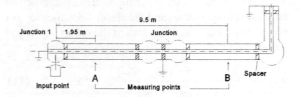

Figure 4: the test facility

The source is built in the first junction. It is a monopole probe with a diameter of 4 mm and length of 90 mm. The probe is mounted on a conical electrode and points the inner-conductor (shown in figure 5). A pulse with 100 V amplitude is generated by a reed relay. The pulse's width is about 1 ns and its repetition rate is one per second.

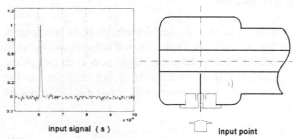

Figure 5: the injection source and the pulse current generation circuit

In the first junction of the test facility, The injecting signal causes very complicated oscillations with several modes. The oscillation frequencies are dependent on the dimension of the junction. Some modes can propagate into the busbar, and the others are non propagating modes. According to signal injecting method, we expect that the TE modes is dominant.

The propagating modes excited are measured at two points in the busbar (shown in figure 4) with the inductive sensor. The inductive sensor is a half loop (40 cm diameter) and can be mounted horizontally or vertically to measure circular magnetic fields H_φ and z axis magnetic fields H_z respectively. The horizontal loop sensor can couple all kinds of modes. The vertical loop sensor only couples TE modes in theory.

- Results

The signals from the sensors at the two points are shown in figure 6 and figure 7. Because the pulse excites not only the TE mode, but also the TEM mode, the signal from the sensors contains both mode components. The signal from the inductive sensor mounted vertically in figure 7 also contains a small TEM mode component. The velocity of the TEM mode wave is faster than the group velocity of the higher order modes. From the corresponding time delay, the source can be located. Therefore, from the measured signals , we can distinguish between TEM and TE modes approximately.

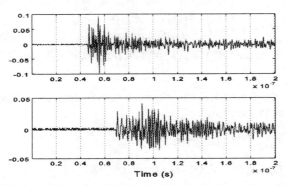

Figure 6: the signals measured by inductive sensor in horizontal at point A (above) and B (below)

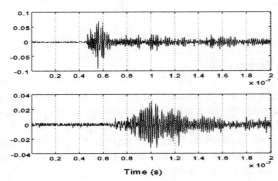

Figure 7: the signals measured by inductive sensor in vertical at point A (above) and B (below)

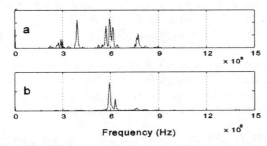

Figure 8: the frequency distribution of the signal at point B, a- horizontal position, b- vertical position

For more detailed analysis, we transform the signals in figure 6 and 7 measured at point B from time domain to frequency domain (figure 8). We see that both signals have strong components at about 590 MHz. The signal from the horizontal inductive sensor has strong components at about 750 MHz and 400 MHz. These components are probably the TEM mode. As calculated in table 1, the lowest cutoff frequency of TE mode in test

GIS facility is 482 MHz. Therefore we can identify at least the component around 590 MHz as the TE_{11} mode signal.

- Comparison with simulations

In equation 7, we choose the modulation frequency f_0 as 593 Mhz, time constant τ as 10 ns and t_0 as 5 ns according to the shape of the signal measured at first junction by the inductive sensor mounted vertically. For the TE modes propagating along the test GIS, the signal at the measured points can be calculated in figure 9. The waves are close to the measured signals in figure 7.

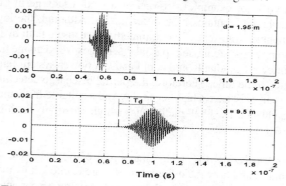

Figure 9: the simulation wave along the traveling distance

With the parameters above, we can calculate the signal attenuation as it propagates from equation 12. The relationship between the amplitude and traveling distance is shown in figure 10. From the first measured point to the second measured point, the attenuation of the signal is 3.7 dB (for 7.5 m). From the experiment, the measured attenuation is about 7.5 dB. In the test facility the attenuation is caused not only by signal dispersion but also by the insulated spacers.

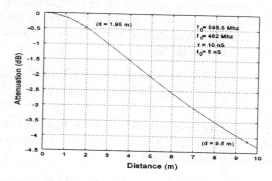

Figure 10: the TE mode attenuation along traveling distance

- Location

As mentioned above, by measuring the time delay between the TE mode signal and TEM mode signal at the same measured point, the partial discharge source can be located by:

$$l = \frac{T_d v_g v_c}{v_c - v_g} \qquad (13)$$

T_d is the time delay between the TEM and TE mode. In Table 2, we give the calculated and measured results compared with practical distance. The calculated distance is close to the realistic distance (9.5 m). The distance from the measurement result is a little larger. However, after modifying the time $t_0 = 5$ ns, the distance is very close to the realistic distance.

Table II: location of partial discharge

	cal. $(t_d - t_0)$	meas. (t'_d)	mod. $(t'_d - t_0)$
T_d (ns)	23.30	28.12	23.12
l (9.5 m)	9.52	11.49	9.45
Δ (%)	0.21	20.1	-0.53

4. Conclusion

In GIS radial discharge (partial or breakdown) will excite the TE mode apart from TEM mode.

The TE wave excited by radial discharge could contain a lot TE modes. During traveling, the wave will be split and dispersed because travel constant $\beta(\omega)$ is dependent on the angular frequency.

In GIS, apart from the attenuation factors of TEM mode, the TE or TM mode attenuation is also mainly caused by group splitting and dispersion.

Using the vertical loop sensor, the TE mode signal can be detected.

By measuring the time delay between the TEM and TE mode, It is possible to locate partial discharge.

5. Reference

1. Pearson J.S, Hampton B.F, and Sellars A.G, 1991, "A Continuous UHF Monitor for Gas-insulated Substations", IEEE Trans on EI, Vol.26, No.3, 469-478

2. Oyama M, Hanai E, et al, 1994, "Development of Detection and Diagnostic Techniques for Partial Discharge in GIS", IEEE Trans on Power Delivery, Vol.9, No.2, 811-818

3. Simon Ramo, 1984, "Fields and Waves in communication Electronics" New York, 402

4. M.C. Zhang, H. Li "High Order Mode Waves Excited in GIS", ISH99

5. Collin R.E, 1966, "Foundations for Microwave engineering", New York St.Louis

Address:

    M.C. Zhang
    Bessenvlinderstraat 113
    5641 EC Eindhoven
    The Netherlands

HIGH ORDER MODE WAVES EXCITED IN GIS

M. C. Zhang, H. Li

Xi'an Jiaotong University, Xi'an, China

Abstract

UHF signal measurement, based on the high order mode waves, is an important tool for the partial discharge detection in GIS (gas insulation substation). In this paper, the high order mode wave excitation and propagation are discussed. From experimental results, some properties are derived.

-the velocity of the high order mode waves (TE and TM modes) is less than the TEM mode velocity

-the high order mode wave in GIS is dependent on the GIS geometry and on the source of excitation

-the resonant cavity is important for the high order mode signal detection

Keywords: GIS, high order mode, excitation, discharge

1. Introduction

GIS is a coaxial system. Its properties concerning signal propagation are frequency dependent. At power frequency, it behaves like an electrical network. For fast transient signals, it can be regarded as a transmission line. For microwaves, it can also be regarded as a coaxial waveguide. Because in GIS there are insulating spacers and junctions between different sections, the impedance of GIS is not uniform. This discontinuity of the impedance causes the fast transient signal to reflect partly. For microwaves, every busbar section and every junction can be considered as a resonant cavity. Therefore, the signal propagation in GIS is complex, especially for very high frequencies[1].

In GIS fast transients can be generated by a disconnector operation or a grounding fault, and also by a partial discharge. The former generates a transient with very high amplitude, whereas the partial discharge generates a low amplitude transient. The high amplitude fast transient consists of a very strong TEM wave component. It could directly damage equipment in the power system and cause a system fault, apart from disturbing electronic devices. Hence, people pay more attention to the TEM mode,

although it could contain the high order mode waves as well. The low amplitude transient generated by a partial discharge in GIS, for example a corona or a void discharge in the insulation, contains high order modes. Although the low amplitude transient will not immediately cause a fault by in the power system itself, partial discharges will damage the dielectric gradually and could cause a power system fault after some time. Therefore, the partial discharge detecting system must have a high sensitivity and a suitable frequency bandwidth. The UHF method has a high sensitivity for high order modes [2].

For analyzing the UHF signal correctly, we have to understand how the different modes are excited, how they propagate and how they attenuate. A suitable sensor and a detection method are required to study these waves.

2. Analysis of High Order Mode Waves

The electromagnetic field in GIS can be expressed by Maxwell's equations:

$$\nabla \times \vec{H} = (\sigma + j\omega\varepsilon)\vec{E} \qquad \nabla \times \vec{E} = -j\omega\mu$$
$$\nabla \cdot \vec{E} = \frac{\varrho}{\varepsilon} \qquad\qquad \nabla \cdot \vec{H} = 0 \tag{1}$$

According to the definition of the TE and TM modes, $H_z \equiv 0$ for TM mode and $E_z \equiv 0$ for TE mode, the solutions of the TE and TM modes can be expressed as follows, by means of separation of variables [3].

$$E_z = (A J_n(kr) + B N_n(kr))(C \sin(n\varphi) + D \cos(n\varphi))$$
$$H_z = (E J_n(kr) + F N_n(kr))(G \sin(n\varphi) + H \cos(n\varphi)) \tag{2}$$

$J_n(kr)$ and $N_n(kr)$ are Bessel function of the first and second kind , respectively. For a coaxial structure, the eigenfunctions of equation 3 are:

$$TM: \frac{N_n(kr_i)}{J_n(kr_i)} = \frac{N_n(kr_o)}{J_n(kr_o)}, \quad TE: \frac{N_n'(kr_i)}{J_n'(kr_i)} = \frac{N_n'(kr_o)}{J_n'(kr_o)} \tag{3}$$

Where r_i and r_o are the radii of the inner-conductor and the

High Voltage Engineering Symposium, 22–27 August 1999
Conference Publication No. 467, © IEE, 1999

outer-conductor of the coaxial system and $k^2=-\beta^2+\mu\varepsilon\omega^2$. The cut-off frequency for the different high order modes can be calculated from the roots (k_{nm}) of equation 4:

$$\omega_c = 2\pi f_c = \frac{k_{nm}}{\sqrt{\mu\varepsilon}} \qquad (4)$$

Only if the frequency of the exciting source is higher than the cut-off frequency, the high order mode waves can propagate.

According to the dimension of the test GIS (r_i=0.045 m, r_o=0.155 m) the cut-off frequencies of the first few roots of the TM and TE mode are listed in table I. From this table, we find that the high order mode with the lowest cut-off frequency is TE_{11}, about 480 MHz.

The group velocity of the TM or TE mode can be expressed as:

$$v_g = \frac{1}{\sqrt{\mu\varepsilon}}\sqrt{1-(\frac{\omega_c}{\omega})^2} \qquad (5)$$

The group velocity of TM or TE modes is less than the TEM velocity, especially for frequencies close to the cut-off frequency.

Table I

TM mode

f_c GHz	$m=1$	$m=2$
$n=0$	1.34	2.71
$n=1$	1.43	2.77
$n=2$	1.68	2.94
$n=3$	1.99	

TE mode

f_c GHz	$m=1$	$m=2$	$m=3$
$n=0$	1.42	2.74	
$n=1$	0.48	1.56	2.82

3. High Order Mode Measurement

In the analysis, we have calculated the cut-off frequency of the different high order modes. It shows that in GIS systems there are several channels for wave propagation. However, which of the high order mode waves are actually present in GIS depends on source and way of excitation. Here we will give some experimental results obtained for different methods of excitation.

- Excitation by Fast Transients

Due to the limited bandwidth of the measuring system and the fast decay of the high order modes, these modes are difficult to find in the measured signal. We have used a differentiating sensor, a loop sensor, with a large bandwidth, to detect the high order mode waves. The test setup for generating fast transients is shown in figure 1.

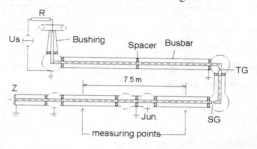

Figure 1. Test setup for fast transient generation, TG: trigger gap, SG: sharpening gap, and Jun.: junction

By firing the trigger gap (TG), we excite a fast transient voltage. For a closed sharpening gap (SG), the signal measured by the loop sensor is shown in figure 2a representing the differentiated fast transient voltage. In this signal, high frequency components corresponding to the high order modes do not appear. When the sharpening gap is opened (about 1 mm), the measured signal, shown in figure 2b, contains a large amount of the high frequency components. After propagating about 7.5 meter, this signal is attenuated strongly (figure 2c) and is delayed as compared to the TEM mode signal.

Placing the loop sensor in vertical position, we measure mainly the high order mode wave. The signal from the vertical loop sensor is shown in figure 2d. As compared to the signal of figure 2b, only the high frequency components are seen.

By analyzing the signals in frequency domain, we find that the high frequency component is about 2 GHz, which is higher than the cut-off frequency of the TM_{x1} and TE_{x1} modes calculated in table I. Further, because the sharpening gap is symmetrical along the propagation direction, only the TE_{0m} and TM_{0m} can be excited by the sharpening gap. Therefore, the high frequency components are be ascribed to the TE_{01} and TM_{01} modes.

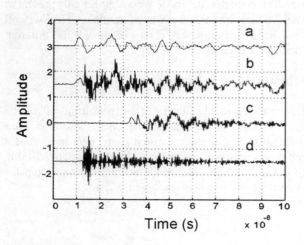

Figure 2. *Transients generated for different situation of the sharpening gap(see text)*

- Resonances in the Junction

The practical GIS consists of different components which make the impedance of the system non-uniform. At the discontinuities of the impedance, a high order mode can be excited by the fast transient voltage. Here, we mainly measure the high order modes excited in the junction. The test circuit is drawn in figure 3. A very short pulse of about 1 ns in width is injected into the circuit through a conical adapter. We measure the signal at three points, which are shown in figure 3, using a capacitive sensor with a large bandwidth /4/.

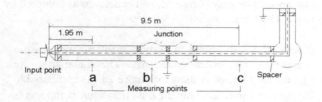

Figure 3. *The test setup for measuring the resonation in the junction*

The signals measured at the three points are shown in figure 4. The signal in figure 4b measured at point b in the junction is completely different from the signals in figure 4a and

figure 4c measured at points a and c in the busbars. Its main frequency component is about 800 MHz.

The fact that this oscillating signal was not measured at points a and c means that it could not propagate into the busbars. It is excited by the injected signal and it only oscillates in the junction. Based on the dimension of the junction, the cut-off frequencies of the TE_{01} and TM_{01} modes are about 780 MHz. The frequency of the signal measured at the junction is higher than the cut-off frequency of the TE_{01} and TM_{01} modes. This signal can not propagate into the busbars, because the cut-off frequency of the TE_{01} and TM_{01} modes are too high. The junction acts like a resonant cavity.

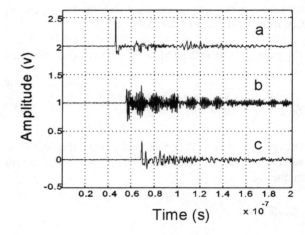

Figure 4. *The signals measured in the busbars (a,c) and at the junction (b)*

- Excitation by a Monopole Antenna

Most partial discharges take place in radial direction. This unsymmetrical discharge current can excite the TE_{nm} and TM_{nm} (n>0) high order modes besides the TEM, TE_{0m}, and TM_{0m} modes. For simulating a radial partial discharge experimentally, we have used a monopole antenna for excitation. The arrangement of the test setup is shown in figure 5. A pulse of about 1 ns width is injected into the first junction through a monopole antenna with a diameter of 4 mm and a length of 90 mm.

In the junction, the injected signal will cause a complex oscillation whose frequency is dependent on the dimensions of the junction. Some modes can propagate into the busbar, and others can not. The TE mode is expected to be dominant because of the method of excitation applied in the setup.

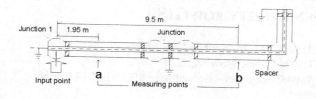

Figure 5. *The test setup for high order mode excited by monopole antenna*

The propagating modes in the busbar are measured at two points (shown in figure 5) by a capacitive sensor with a 36 mm diameter electrode. The measured signals are shown in figure 6. From them we find that the measured signals contain both TEM and high order modes. The main oscillation frequency is about 590 MHz which is higher than the lowest cut-off frequency of the TE mode only (about 480 MHz, table I). Comparing the signals measured at points a and b, we find that during the propagation the signal is distorted by dispersion. Because the group velocity of the higher order modes depends on the frequency and is slower than those of the TEM mode, the maximum of the oscillating signal component is delayed, as compared to the first peak for the TEM mode. Therefore, we conclude that the component around 590 MHz is the TE_{11} mode.

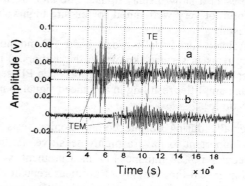

Figure 6. *The signal measured in the busbar under the monopole antenna excitation*

4. Discussion and Conclusion

The analysis and the measurements above show that the UHF signal propagation in GIS is complicated. It depends not only on the dimension of every component in GIS, but also on the exciting source and exciting method. In other words, the dimension of the GIS only provides the channels for UHF signal propagation. Which of the high order modes will actually be excited is determined by the source and the method of excitation. For the longitudinal excitation as was the case for the first two experiments above, the symmetrical high order modes of TE_{0m} and TM_{0m} are excited if the injected pulse is sufficiently fast. Because of the dimensions of the junction, a high order mode can be excited easily. The radial excitation, like a partial discharge or a system grounding fault, can excite the high order mode with the lowest cut-off frequency.

As mentioned above, the group velocity of the high order modes is dependent on the exciting frequency and the cut-off frequency. It can be used to distinguish between the TEM and high order mode waves. Combining the frequency and the mode of the measured signal, we could locate the position of the source.

The junction is the best point to measure the UHF signal for partial discharge detection, because its cut-off frequencies are lower than those of the busbar, and it can be used as a resonant cavity to achieve a higher sensitivity.

5. Reference

1. Kurrer R., Feser K. and Herbst I., 'Calculation of Resonant Frequencies in GIS for UHF Partial Discharge Detection', 7th international Symposium on Gaseous Dielectrics, in Knoxville, U.S.A.

2. Person J.S., Hampton B.F. and Sellars A.G., 'A continuous UHF Monitor for Gas insulated Substation', IEEE Trans on EI Vol.26, No.3, June 1991, p469-478

3. Simon Ramo, 'Field and wave in communication Electronics', New York, 1984

4. Zhang M.C., Qiu Y., 'Wide bandwidth Sensor for PD Measurement in GIS', 3rd international Conference of ECAAA, in Xi'an, China.

M. Zhang
Bessenvlinderstraat 113
5641 EC Eindhoven
The Netherlands

COMBINED BROAD AND NARROW BAND MULTICHANNEL PD MEASUREMENT SYSTEM WITH HIGH SENSITIVITY FOR GIS

S. M. Neuhold, M. L. Schmatz[♦], M. Hässig , M. M. Spühler[♦], G. Storf

FKH Zurich, Voltastrasse 9, CH-8044 Zurich, Switzerland

[♦]ETH Zurich Lab. for Electromagn. Fields and Microwave Electronics, Gloriastrasse 35, CH-8092 Zurich Switzerland;

email: neuhold@fkh.ch

Abstract

A computer controlled GIS (**G**as **I**nsulated **S**witch-gear) PD (**P**artial **D**ischarge) measurement system is presented which combines UHF broad and narrow band **PD** measurements with high sensitivity, multi-channel performance and near real-time response.

Each measurement channel consists of a low noise broadband sensor amplifier with automated high voltage transient protection, a full band power/envelope detector, a narrow band spectrum analyzer interface and a computerized display unit. With 0.1 to 2.0 GHz frequency range, 3 dB noise figure, 40 dB of sensor amplifier gain, full flashover and switching transient protection, twelve measurement channels, less than 0.1 second response time and a user friendly computer interface, the presented system is optimally suited for on-site commissioning tests, for laboratory testing during development and for long term monitoring of complex GIS.

This system allows the detection, localization, identification and monitoring of fault sources using a fully electrical method.

Introduction

In order to obtain the required PD measurement sensitivity [1], manufacturers of GIS are placing more and more PD sensors per GIS length [2]. Some GIS systems even allow to use some construction parts (shielding, windows, ...) as additional PD-sensors [2, 3]. Hence, about 7 to 10 sensors should preferably be monitored simultaneously during high voltage test with a typical test-section of about 10 nF.

Actually, two different PD measurement methods in the UHF frequency range are used: UHF narrow band and UHF broad band PD measurement [4, 5, 10].

UHF narrow band PD-measurement method can be characterized as follows:

+ Very high sensitivity (resonance's with high Q-factor)
+ Very insensible against external noise (allows measurement of signals near the thermal noise limit)
+ Rough localization of signal source by interpretation of PD signal spectrum is possible (the closer to the signal source, the higher the measurable frequency components of the PD signal).
+ Electrical localization with an oscilloscope is possible (full frequency range is available for measurement at the input of the spectrum analyzer).
- Only one channel can be observed simultaneously (normally only one spectrum analyzer available)
- Automation for PD monitoring is very difficult

UHF broad band PD-measurement can be characterized as follows:

+ High sensitivity, if no strong external RF noise is present.
+ Relatively simple system design (amplifier, detector, display unit).
+ Easy to implement in a PD monitoring system.
+ Parallel display of several sensors is possible (not too expensive)
- Localization with interpretation of frequency spectrum is not feasible.
- Localization by signal delay comparison method is not possible (In this circuits, the RF signal is normally not available for measurements)
- Existing UHF broad band monitoring systems present a slow display update rate and are therefore not very well suited for on site high voltage testing.
- Actually used UHF broad band PD-monitoring systems in service (known to the authors...) have cable lengths of 10 m to 30 m between PD sensor and amplifier resulting in a reduced signal to noise ratio [6].

Especially during on site commissioning tests it is very important to have a highly sensitive, flashover protected, noise immune PD measurement system with near real-time display of all measurement channels in order to localize and identify PD-sources as fast as possible [10].

These requirements imply the following specifications for a possibly 'universal' measurement system:

- Very high sensitivity (< 1 pC)
- High bandwidth (100 MHz – 2 GHz)
- Signal amplifier directly connected to the PD sensor
- Near real-time parallel display of 12 channels (<0.5 sec.)
- Full bandwidth parallel RF outputs for localization.
- For identification, it should be possible to connect the system to classical PD-measurement systems.

Measurement System

A 12 channel measurement system was realized (Fig. 1), consisting of flashover protected amplifiers connected directly to the PD sensors using a well shielded, low loss coaxial cable, a twin RF path with one output to a fast oscilloscope (for signal delay comparison method) and a multiplexed output to a spectrum analyzer for narrow band UHF PD measurements (total RF gain: 40 dB). Each channel is equipped with a low noise power detector with outputs to a 12 channel, 0.1 second response time display unit for instant overview of all signal levels.

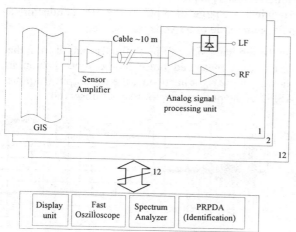

Fig. 1: Principle of realized measurement system

Sensor Amplifier

Characterization of the sensor amplifier (Fig. 2):
- Multistage circuit for transient protection (flash-over, switching) at the input
- Auto-turn-off if input power > - 10 dBm
- Flat (+/- 1 dB) frequency response (0.1-2.0 GHz)
- 50Ω noise figure better than 3 dB at 1 GHz
- Gain of more than 30 dB
- Remote powered (Bias Tee and supply regulation included)

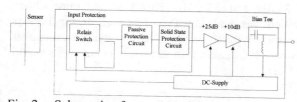

Fig. 2: Schematic of sensor amplifier

Tests have shown, that switching transients caused by disconnector switching cause more problems to the design of a protection circuit than (even higher amplitude) flashover transients during high voltage tests. An auto-turn-off circuit with additional passive protection was optimized with respect to high frequency characteristics. It is possible to use this new amplifier directly connected to a sensor in a monitoring system due to it's integrated protection circuits.

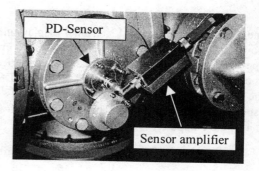

Fig. 3: Sensor amplifier connected to a PD sensor

Broad Band Power/Envelope Detector

The RF signal output of the remote powered sensor amplifier is split into 2 signal paths: A RF signal path which is amplified (10 dB) and split for multiplexed spectrum analyzer measurements and oscilloscope measurements (for PD localization), and a broad band power/envelope detector signal path (Fig. D). The broadband power/envelope detector consists of a buffer stage (40 dB), two detector diodes and a 500 kHz low pass filter-amplifier (Fig. 4).

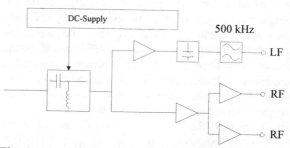

Fig. 4: Analog signal processing unit (ASPU)

Fig. 5 shows the sensitivity of the power detector as a function of frequency. The output voltage of the detector was measured with a 100 % AM modulated signal at 100 MHz, 500 MHz, 1 GHz, 1.5 GHz and 2 GHz respectively. A very good sensitivity down to − 85 dBm CW input power may be noticed over the entire frequency range [8,9].

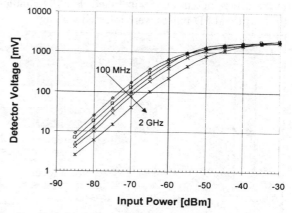

Fig. 5: Sensitivity of power detector

Display Unit

One single channel of the 12 channel display unit is shown in fig. 6. The output of the broad band envelope detector is buffered, digitized and numerically processed with a **D**igital **S**ignal **P**rocessor unit (DSP). The results of the 12 signal processor units are displayed with a Windows NT based display software on a laptop computer.

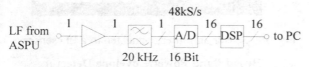

Fig. 6: Schematic of one single channel of the display unit

The display software on the laptop computer allows to set color coded reference levels individually for each channel. Three different levels are defined: Background noise level, reference level and a measurement/alarm level. In combination with the proposed sensivity check of PD sensors of GIS [1], the setting of the reference level (measured signal amplitude of an artificial PD pulse of 5 pC injected on the next sensor available) may be used (during HV testing and monitoring) as an indication of the magnitude of a real PD source (remark: UHF measurements cannot be calibrated [7]). Fig. 8 shows a part of a screenprint (3 channels) of the Windows NT based signal display. The rise and decay time of the signal display can be set, as well as different modes of maximum-hold functions. The parallel display of up to 12 sensors allows a rough but very quick overview of the PD scenery and helps to localize the PD source. Because of its standardized, computer based platform, this unit can easily be remote controlled, expanded and used as a monitoring system.

Measurements

The described system was tested with a GIS in service. The 170 kV GIS consists of an energized double busbar with 8 cable feeders and a buscoupler located in the center of Zurich, Switzerland. For PD simulation, artificial PD-pulses were injected via one

of the PD sensors. The used pulse generator has a rise time (10% - 90%) of < 70 psec into a 50 Ω load at a DC operating voltage of 300 V.

For PD measurements, two types of sensors were used: Type I is a PD-sensor, Type II is an auxiliary sensor, originally not designed as PD sensor. They have both a measured capacity C_1 to center conductor of 0.15 pF. Both sensor-types have a diameter of ≈ 100 mm and are installed in earthing switch feedtrough flanges [11].

To inject a charge of 5 pC to the HV conductor, a pulse-amplitude of 33 volts was needed.

On the left hand side of Fig. 9, the RF output from the ASPU of a measured pulse in the time domain is shown, on the right hand side, the LF output from the ASPU of the power detector may be observed.

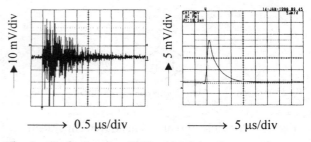

Fig. 9: Left: Output of RF path, right: Output of power detector with low pass filter

The table in Fig. 10 shows the geometrical characteristics of the sensor arrangement and the measurement results of narrow and broad band measurements. The results of the UHF-narrow band method show a very good signal to noise ratio even at very low injected charge levels. The same is valid for the UHF-broad band measurements.

The amplitude spectrum measured at sensor No. 2 has significant signal energy in the range between 1.3 GHz and 1.7 GHz (Fig. 11). The same observation was made for Type I as well as for Type II sensors. Without having a high system measurement bandwidth, these significant signal components would be lost.

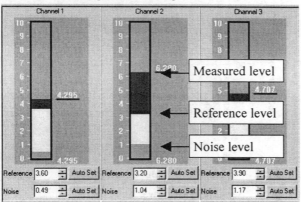

Fig. 8: Partial view of the display unit (tree of twelve channels)

	Sensor Nr											
	1	2	3	4	5	6	7	8	9	10	11	12
Sensor Type	II	I	I	I	II	II	II	II	II	II	II	II
Location (Bus Bar or Feeder)	Fd	BB	BB	BB	BB	BB	BB	BB	Fd	BB	Fd	Fd
Physical distance to Source [m]	16.3	1.1	17.5	18.6	5.4	8.5	11.0	1.3	5.2	14.6	18.8	21.3
# of T-joints between Source	6	3	7	8	2	3	4	2	2	7	7	8
# of Disconnector between Source	3	2	2	3	0	1	2	1	1	3	3	3
# of Circuit Breaker between Source	1	0	0	0	0	0	0	0	1	0	1	1
Frequency of maximum RF response @ 15 pC injected charge [GHz]	0.61	1.70	0.56	1.63	1.75	0.71	0.51	1.66	0.90	0.15	0.11	-
SNR max [dB] @ 15 pC injected charge (Bandwidth 3 MHz)	15	40	45	40	25	23	22	32	13	10	15	-
Detected envelope voltage @ 15 pC injected charge [mV]	-	149	276	3.4	37.2	18.3	17.4	28.4	1.12	15	1.2	1.2
RF – Sensitivity [pC] (Bandw. 3 MHz)	7.5	<0.8	<0.8	1.5	2.3	1.5	1.5	<0.8	9	5.3	-	-
LF – Sensitivity [pC]	>15	<0.8	<0.8	11	1.1	2.8	4.1	1.5	7.5	13.5	>15	>15

Fig. 10: Table of Measurements

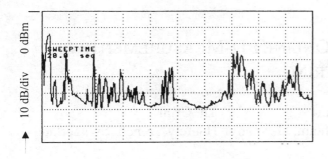

100 MHz ⟶ 1.8 GHz

Fig. 11 Spectrum measured at sensor 2

It is assumed that the high signal level at sensor No. 3 can be explained by the generation of standing waves near the end of the GIS busbar. At sensor No. 12 no signal could be measured because this part of GIS was switched off (circuit breaker open) and earthed. It was generally observed, that after passing a circuit breaker, the high frequency parts of a PD signal were highly attenuated. The sensitivity and precision of an electrical localization of a PD source was tested (Fig. 12). The artificial signal source had a distance of 1.1 m to sensor No. 2 and a distance of 17.5 m to sensor No. 3. The measurements showed a maximum sensitivity of better than 0.8 pC of injected charge at sensitive sensor localizations (S/N ratio of measured signal at 0.8 pC: 6 dB) and an average sensitivity of approximately 2 pC. The stated resolution for the localization was in the order of 30 cm with an oscilloscope with a restricted bandwidth of 600 MHz.

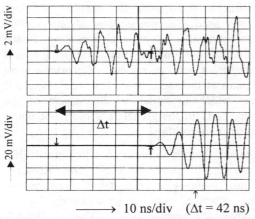

⟶ 10 ns/div   (Δt = 42 ns)

Fig. 12 Localization with signal delay comparison method

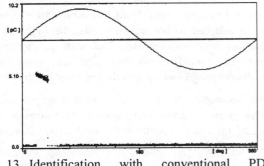

Fig. 13 Identification with conventional PD-measuring system

For identification, the down-converted output of the UHF narrow band or the output of the envelope detector can be processed with a conventional PD measuring system. In Fig. 13 the injected pulse was synchronized with the GIS power frequency (jitter due to the mechanical switching of the mercury relay).

Conclusions

A computer controlled multichannel measurement system for combined UHF broad and narrow band measurement in GIS has been realized. Due to its high bandwidth (0.1 - 2.0 GHz), 40 dB gain (RF) and 3 dB noise figure, a high sensitivity was achieved with a typical UHF sensor on site (< 0.8 pC). Due to a multistage protection circuit against flashover and switching transients, the amplifiers can directly be connected to the sensors. This solution enhances the sensitivity especially in the frequency range above 1 GHz. The combination of a high RF bandwidth and a 12 channel parallel display unit with 0.1 sec. response time makes the system ideally suited for PD source localization.

The system allows the detection, localization, identification and monitoring of PD fault sources using a pure electrical method.

References

[1] CIGRE 1998 TF 15/33.03.05 IWD 73: Sensitivity Verification for Partial Discharge Detection on GIS with the UHF and the Acoustic Method.

[2] Behrmann, G.J.; Neuhold, S., Pietsch, R.: Results of UHF measurements in a 220 kV Substation during on-site commissioning tests. 10th International Symposium on High Voltage Engineering, Montreal 1997

[3] M.D. Judd, B.F. Hampton, W.L. Brown: UHF Partial Discharge Monitoring for 132 kV GIS. 10th International Symposium on High Voltage Engineering, Montreal 1997

[4] C. Maulat, E. Fernandez, P. .Almosnino: Sensitivity of electrical Partial Discharge measuring methods on GIS with external disturbances: Experimental comparison on conventional method, UHF wide band and narrow band methods. 10th International Symposium on High Voltage Engineering, Montreal 1997

[5] A. Petit: Field Experience of Partial Discharge Monitoring with the UHF Method. Paper 5596, 9th Int. Symp. on High Voltage Engineering, Graz, 1995

[6] W. Bächthold: Lineare Elemente der Höchstfrequenztechnik, vdf verlag / Zürich 1994

[7] Sellars, A.G., MacGregor, S.J., Farish, O.: Calibrating the UHF Technique of PD Detection using a PD simulator. IEEE Trans. on Diel. and El. Ins., Vol. 2, No. 1, 1995, pp. 46-52.

[8] A.E. Bailey, Editor, 'Microwave measurements', Second edition, Institution of Electrical Engeneers, 1989, pp176 - 208

[9] R. Q. Lane, 'The determination of device noise parameters', Proc IEEE, vol. 57, pp. 1461-1462, Aug 1969

[10] W.Buesch, H..Dambach, P.Hadorn, T.Aschwanden, M.Haessig, T.Heizmann: Application of partial discharge diagnostics in GIS at on-site commissioning tests. CIGRE conference, Paris 1998

[11] Albiez, M.: Teilentladungsmessung an SF6 isolierten Schaltanlagen. ETH Zurich, Switzerland, 1992, Thesis No. 9694.

IMPROVEMENT OF SENSITIVITY IN ONLINE PD-MEASUREMENTS ON TRANSFORMERS BY DIGITAL FILTERING

K. Feser, E. Grossmann   M. Lauersdorf   Th. Grun

University of Stuttgart, Germany   ABB Mannheim, Germany   Haefely Trench, Switzerland

Abstract

Sensitive p.d. measurements can be carried out in electromagnetically shielded laboratories. Recently the p.d. measuring technique is also used for the online monitoring of apparatus in substations. In substations on-site a sufficient measuring sensitivity can only be achieved by specific filtering methods. Different digital filtering methods had been developed to improve the signal processing in order to suppress efficiently narrowband noise as the most frequent disturbance source and corona discharges which usually cannot be distinguished from the internal discharges of the p.d. specimen. Examples for a successful application of filtering principles to difficult on site conditions will be discussed as well as combinations of the new developed methods. The limits in the sensitivity of the algorithm are reported.

Introduction

Online monitoring of apparatus need a special arrangement of sensors, amplifiers and digitizing system. This input signal conditioning is necessary in order to get the full use of the abilities of the digital filtering algorithms for the reduction of disturbances. Suitable sensors for continuous online measurement of p.d. are ferrit cores mounted on the measuring tap of the bushings or in the neutral of transformers. Requirements for these devices are low inductance, which determines the voltage applied to the measuring tap in case of a lightning impulse, and a wideband, sensitive p.d. decoupling characteristic.

Followed by an amplifier and an A/D-converter , the signals can afterwards be downloaded by a personal computer, performing the digital filtering and signal processing. The developed algorithms consist of a multi-narrowband suppression and a digital bridge.

Suppression of sinusoidal Disturbances with digital Filters

Narrowband noise is a common source of disturbances in substation, caused mostly by radio stations or power electronics. The power grid and the substation expands the measuring circuit and catches conducted interferences or acts like an antenna. At first the possibilities and limits of commonly used measuring and noise reduction principles will be reviewed.

An easy to build and often used filtering technique is represented by the group of narrowband measuring devices. The filter itself is normally constructed out of analogue components and cuts out a small part (several kHz) of the frequency spectrum. Basic requirements for this procedure are at least one small frequency range, which has a satisfactory signal to noise ratio, and that the long impulse response of the system doesn't bother the evaluation process. Especially this second criteria can affect the desired ability to discriminate between close to each other following impulses.

A better way for continuing filtering can be provided by digital filters. This method is based on the computation of coefficients for a FIR-filter algorithm. Using a digital filter, the degree of suppression is limited by the number of filter coefficients used. The tremendous advantage of a FIR-filter compared to an analogue filter is the possibility of defining a multiband suppression characteristic, which can reject several narrowband disturbances at the same time. The result is a still relatively wideband signal, which consists of the remaining parts in the signal spectrum, not being damped by the algorithm.

Computation algorithms for the coefficients of FIR-filters are e.g. "Remez" or "RLS". The "Remez"-algorithmn requires a predefined function of passbands which are approximated by a Tschebbeycheff-characteristic. The "RLS"-algorithm is based on a time domain iteration suppressing spectrum bands with high amplitudes (noise) using a recursive loop to decrease the squared sum of errors. Both algorithms are the more exact adapted to the actual needs, the higher their number of coefficient are. But with an increasing number of coefficients they also take more time while running/filtering and during their calculation.

An example of an onsite measurement on a transformer is shown in figure 1. The original p.d. signal (a) and its spectrum (b) are hereby filtered by three subsequent digital filters, two with 256

High Voltage Engineering Symposium, 22–27 August 1999
Conference Publication No. 467, © IEE, 1999

coefficients and one with 128 The total spectrum after filtering is given in diagram (c). The result is a far more sensitive signal (d) in respect to partial discharges, which was in this setup improved by 18 dB (from 5500 pC to 700 pC).

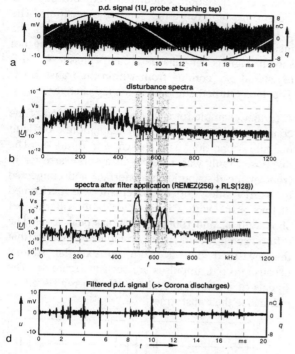

Fig. 1: Original Signal (a) and Spectrum (b), digital filtered Spectrum (c) and Signal (d)

Compensation of external Impulses by means of a "Digital Bridge"

The second common source of disturbances in substations are corona pulses and interference by impulses of power electronics. The evaluation of p.d. measurements which are carried out on transformers is restricted on the frequency range up to 1 MHz (according to IEC 60-270). Hence there is no possibility to distinguish between pulses caused by internal discharges and external corona discharges. Thus a suppression of corona discharges by using conventional filtering methods is not possible. The obvious pulses in the filtered signal of figure 1 are in general corona pulses and show the disturbance of internal p.d. by corona.

To distinguish between internal discharges and corona discharges a new detection method has to be developed. The pattern recognition procedure, which is based on the statistic long term behaviour of discharge pulses, cannot be used for the suppression of single disturbance pulses. As there is a wide variance in amplitude, frequency range and inception time in corona pulses, internal p.d. impulses are still often covered by them.

Another suppression method is based on a second channel (antenna) to blank out the corona pulses.

This takes into account that the origin of the additional evaluated electromagnetic radiation is outside of the test object. In this case the disturbance pulses in the measured p.d. signal can be directly correlated with the disturbance signal which has been received by the antenna. Thus the disturbance pulses in the p.d. signal can be detected and suppressed. However this suppression method is not applicable on large substations on-site where discharges occur outside of the measuring circuit and are interfering additionally with the antenna signal as a further disturbance source.

For the suppression of corona discharges a digital compensation method has been proved to be most efficient. This method is based on a "bridge" principle. The p.d. currents are coupled out at two independent current branches of the measuring circuit. As used in the analog bridge method, the two measuring branches have to be equalized ("balanced bridge"). For that purpose the signal of one of the two measuring branches (secondary branch) has to be filtered to adapt its transfer characteristics to the transfer function of the other connected measuring branch (specimen branch). After the application of this filtering process which is also called the "measuring circuit compensation" all pulse shaped conductor guided disturbances which occur outside of the specimen can be suppressed or recognised by using two different processing principles.

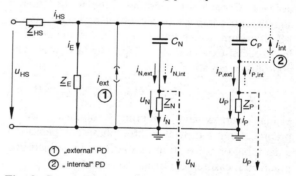

Fig. 2: Compensation of the Digital 'Bridge' Circuit, Polarity for external (1) / internal (2) p.d.

The first principle calculates the difference of the two signals of the measuring branches. Thus a common mode rejection of all external p.d. pulses can be achieved. Using the second processing principle, the product of the two measured signals is calculated. In the case of external p.d. the result of this calculation consists of a positive signal; for internal p.d. the result is a negative signal. The compensation itself is based on the following equations. Equation (1) is used for the determination of the filter $\underline{F}_K$ and its coefficients. The used signal is either a calibration impulse or a corona pulse, if the origin of the corona equals the input location for the calibration process.

External P.D.	Internal P.D.
$\underline{U}_N \cdot \underline{F}_K - \underline{U}_P = 0$ $\quad^{(1)}$	$\underline{U}_N \cdot \underline{F}_K - \underline{U}_P \neq 0$ $\quad^{(3)}$
$\underline{U}_N \cdot \underline{F}_K \cdot \underline{U}_P > 0$ $\quad^{(2)}$	$\underline{U}_N \cdot \underline{F}_K \cdot \underline{U}_P < 0$ $\quad^{(4)}$

The filter-function for the application of the measuring circuit compensation can be calculated by using different mathematical methods, as e.g. the "LS" algorithm or a frequency based windowing method. The least squares (LS) routine approximates the signal, which has to be filtered, to the reference signal by the calculation of squared differential errors. These errors are used to improve the exactness of the coefficients. Another possibility is the computation of the transfer-function $\underline{F}_K = \underline{U}_P / \underline{U}_N$ out of the frequency spectra and its back transformation. Windowing this vector the coefficients of the digital filter are obtained.

In the case of on site applications of the compensation method on three phase power trans-formers each conductor represents a separate measuring circuit. If the three conductor currents cannot be coupled out directly, then the current signal of the earthed neutral point, as a sum of all conductor currents, can be used for the compensation of the three independent measuring circuits. Thus using a common current signal of all three phases, a common compensation can be defined. Capable of handling three to be filtered signals and compensation coefficient vectors is the LS algorithm, which has to be used for this problem in a special extended (four input signals) way. The appropriate equation is:

$$\underline{U}_U \cdot \underline{F}_U + \underline{U}_V \cdot \underline{F}_V + \underline{U}_W \cdot \underline{F}_W - \underline{U}_N = 0$$

Using narrowband filtering routines as a preprocessing tool to this three phase compensation a damping of noise of up to 30 dB was obtained in a practical measurement onsite and online.

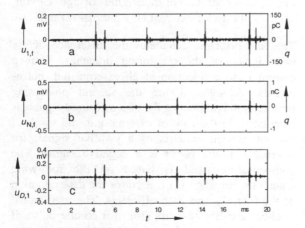

Fig. 3: Onsite Diagnosis of a 25 MVA Transformer, filtered Signals of Phase U (a) and Neutral (b) and compensated Difference Signal (c)

An example of an onsite measurement is presented in the figure 3, where in a first step the narrowband disturbances in the original signal were suppressed and afterwards the compensation method was applied. The diagram shows the difference signal (c) of the compensation, calculated from the phase U measuring tap (bushing) and the neutral. The difference hereby indicates clearly, that the measured p.d. is from an internal source. The not shown signals of phase V and phase W also registered p.d. coming from within the transformer. The defect turned out to be in the tap changer.

A further example for a successful application of the compensation principle and an adaptation to difficult on site conditions is shown in figure 4. The diagnosed fault was a defect in a surge arrester in front of the distribution transformer and connected to its HV-side. The measurement was performed offline onsite with a noise level, after filtering, of about 300 pC and 10000 pC impulse peak level. The difference shown in figure 4c indicates clearly, that the measured p.d. is from an external source, because no p.d. impulses are seen in figure 4c. The signals of phase V and phase W also registered no p.d. inside the transformer.

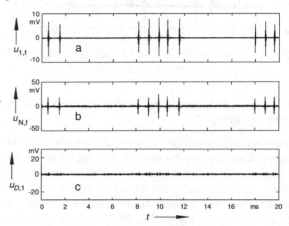

Fig. 4: Diagnosis of a 200 MVA Transformer, filtered Signals of Phase U (a) and Neutral (b) and compensated Difference Signal (c)

Influences to the Compensation Algorithm for practical Issues

The described compensation can handle an object on which the branches always have the same impulse decoupling characteristic, as for example a current or voltage transformers. However a distribution transformer contains a tap changer to balance the voltage between the coupled power grids. This adds in the circuit (Fig. 2) a variable impedance in the area between the capacitor C_P, which represents the specimen, and the coupling network Z_P. The formerly measured current ratio flowing across the couplers for external and internal p.d. is thereby changed. This can be solved by a renewal of the coefficients, thus by running the algorithm again using a calibration set which was recorded with the correct tap changer position.

A source which cannot be handled by the digital bridge is the electromagnetic radiation coupling into the measuring circuit. Under these circumstances the radiated field induces currents of opposite directions in the branches (Fig. 2), thus pretending the polarity and shape of an internal p.d. impuls. The suppression by the compensation method is therefore not very sensitive on large substations on-site where EM-pulses occur outside of the measuring circuit and are strongly interfering by their electromagnetic radiation.

Problematic in the case of on site applications are furthermore capacitive coupling between the conductors of the three phase system. Using the three phase based compensation, the corona impulses forming on the HV-line close to the transformer can be suppressed. Pulses with distant origin as well as pulses passing through capacitve highly coupled devices (e.g. GIS-substations) must be handled as three phase excitation. The result is a far smaller suppression, because the matrix compu-tation basis of the compensation are single pulses on one of the lines. This behaviour shows up especially for rainy weather, when the disturbance consists of corona of high amplitudes and high repetition frequency originating mostly on the power lines.

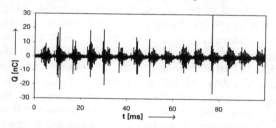

Fig. 5: Difference Signal at rainy Weather with uncompensated Corona Impulses

Unfavourable grounding conditions concerning the neutral as well as the parasitic capacitance of conductors are also influencing the compensation. For a better understanding of this behaviour, the equivalent circuit replacing the coils with a network can be looked at, as presented in figure 6. A p.d. impulse source between the conductor and the grounded tank of a transformer has, compared to an outside p.d. (corona), the same current path and is because of that reason suppressed by the compensation. Another current path of internal p.d. is build up by two conductors of the coil. For this p.d. location two extremes can be analysed, which are a p.d. close to the neutral and one close to the HV-bushing at the upper end of the coil. Close to the neutral the compensation works as described, but as closer the defect is to the bushing, the more the current flows over the parasitic capacitances of the

windings to the tank. In this latter case the algorithm "considers" the p.d. to be more and more of an origin like corona, thus suppressing it.

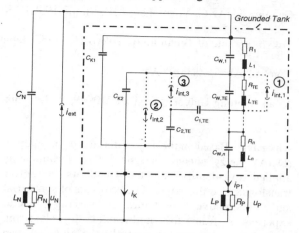

Fig. 6: Equivalent Circuit of a Transformer Coil in the Tank (including parasitic capacitances)

Conclusions

Measurement of p.d. online using digital filters provides a sophisticated way for the suppression of disturbances. The up to now applied masking of corona impulses can be replaced by the real online working analytical compensation algorithm, which prevents the blanking out of internal p.d. The combination of multi-narrowband filters and compensation allows furthermore online monitoring. This includes a noise reduction of narrowband disturbances of about 30 dB and the ability to distinguish between most external and internal partial discharges. In addition to a software implementation, not working in real time, these methods are also suitable for DSP (digital signal processor), running the filtering in real time.

References

[CIGRE, 1971] "Measurement of partial discharges in transformers", CIGRE-Publication, Electra No. 19, Nov. 1971, pp 9 - 65.

[IEC, 1981] "Partial Discharge Measurements", IEC Publication 270, 2nd edition, 1981.

[Köpf, 1994] Köpf, U., "Kontinuierliche Unter-drückung von schmalbandigen, periodischen und breitbandigen, impulsförmigen Störern bei der Teilentladungsmessung", Dissertation, Universität Stuttgart 1994.

[Lauersdorf, 1998] Lauersdorf, M., "Verfahren zur Unterdrückung von Koronastörern bei Teil-entladungsmessungen an Transformatoren vor Ort", Dissertation, Universität Stuttgart, 1998

[Stearns, 1994] Stearns, S. D.;. Hush, D. R., "Digitale Verarbeitung analoger Signale", München, Oldenbourg-Verlag, 1994

APPLICATION OF FUZZY DATA PROCESSING FOR FAULT DIAGNOSIS OF POWER TRANSFORMERS

Guanjun ZHANG, Yuan-shing LIU, Shinji IBUKA, Koichi YASUOKA, Shozo ISHII

Tokyo Institute of Technology, Tokyo 152-8552, Japan

Yong SHANG, Zhang YAN

Xi'an Jiaotong University, Xi'an 710049, P. R. China

Abstract Based on the data of dissolved gas analysis (DGA), fuzzy cluster analysis (FCA) technique is applied to identify the fault patterns of power transformers in this paper. FCA consists of some trial and instructive strategies absorbing useful experiences from the mid-results. Its clustering centers are dynamic, as a result, the approach can classify and recombine different samples successfully. Compared with the conven-tional methods, it reveals the practical advantages of unsupervised systems, including the ability to produce categories without supervision.

1. Introduction

Monitoring and on-site diagnostics of electrical equipment, in particular transformers, has attracted considerable attention for many years. Large power transformers are vital to power system with their operating state directly determining the safety of the system to a great extent. It is of great importance for the electric units to find the incipient faults of these transformers as early as possible so as to run them safely and improve the reliability of power supply. It has been proved that the technique of DGA is efficient for the recognition of incipient faults in large transformers and for the identification of the extent to which the faults have developed. As a result, DGA test has been assigned as the first test item of power transformers in the Chinese preventive test code for electric equipment issued in 1996 and put into effect in 1997, respectively.

The traditional expert system (ES) has some intrinsic shortcomings, such as the bottle-neck of acquiring knowledge and difficulty of maintaining knowledge base (KB), etc. Furthermore, more and more researches have shown that it is difficult to describe the relations between some faults and the cor-responding reasons quantitatively and it seems impossible to find a definite function between them.

Based on the traditional methods of fault diagnosis, many new effective information processing techniques such as the Artificial Neural Network (ANN) and fuzzy mathematics etc. have been gradually introduced into the field of fault diagnosis of power transformers and some satisfactory results have been obtained[1,2]. The traditional accurate mathematics theories do not perform well in the field because of the inevitable dealing of bulk fuzzy information[3]. By adjusting the boundary of diagnostic rules fuzzily and with the help of fuzzy relationship equation, the fault diagnosis of power transformers has been carried out successfully[4,5], but the shortcomings can not be avoided that the statistics of large amount of data should be completed first and the objectivity of the results is doubted because too much intervention by man is introduced during the diagnostic process. These reasons prevent fuzzy mathematics from applying to the fault diagnosis of power transformers.

The method of fuzzy cluster is feasible for the recognition of fuzzy systems and the Iterative Self-Organizing Data Analysis Technique Algorithm (ISODATA) has been successfully applied to many other fields[6]. In this paper this method is introduced to the DGA of power transformers. According to the clustering results, the new method identifies the faults of transformers, e.g., normal, arc discharge and overheating, etc. correctly and enhances the reliability of decisions greatly. In contrast, some of the DGA data will be mis-diagnosed if only the IEC approach is used. With the new method, the diagnostic conclusions of power transformer will be more reasonable than before. In addition, it shows that the preprocessing of original data is important and further researches will be done on this.

2. Principle of ISODATA

ISODATA is a kind of dynamic cluster methods. Its basic principle is that, first, a certain number of samples are selected as the clustering centers, then the other samples are classified towards their own centers respectively in the light of some certain clustering rules (e.g., the rule of minimum distance between a sample and its center), so as to result in the initial classification; secondly, the initial classifi-cation is evaluated to decide whether it is reasonable, if not the process presented in the first step should be repeated until an acceptable classification is obtained[7].

Suppose that a sample set containing n samples, marked as $X=\{u_1, u_2, ..., u_n\}$, where sample u_i is represented as $u_i=\{x_{i1}, x_{i2}, ..., x_{im}\}$. The set X can be

High Voltage Engineering Symposium, 22–27 August 1999
Conference Publication No. 467, © IEE, 1999

classified into C classes ($2 \leq C \leq n$), with the result being denoted by a Boolean value matrix $R_{C \times n}$, where r_{ij} takes different membership values according to the extent that sample x_j belongs to class i. The common ISODATA method is known as where $r_{ij}=1$ if x_j belongs to class i or $r_{ij}=0$ if not. In contrast, fuzzy ISODATA method is where the elements of matrix R are fuzzy values ranging from 0 to 1 instead of the simple Boolean values, i.e., 0 and 1.

Clustering Center

Now the concept of the clustering center vector V is introduced. While labeling the sample set X into C different classes, the vector $V=[v_1, v_2, ..., v_e]^T$ is defined as the following

$v_i=(\sum_{k=1}^{n} r_{ik}^Q X_k)/ \sum_{k=1}^{n} r_{ik}^Q$, where the index Q is a real number (≥ 1) and v_i is the clustering center of class i.

In order to acquire a reasonable and reliable clustering center, a general function $J(R, V) = \sum_{k=1}^{n}\sum_{i=1}^{c} r_{ik}^Q \|X_k - X_i\|^2$ is constructed, and here $d=\|X_k - X_i\|$ is the distance between X_k and X_i. With the acquisition of a reasonable classifying matrix R and a feasible clustering center vector V, the minimum value of $J(R, V)$ can be calculated, i.e., the satisfactory clustering result can be reached.

Calculation Procedure

(1) Determine C ($2 \leq C \leq n$), initial fuzzy matrix $R^{(0)}$, iteration times L (initially set to 0), permissible error E (>0) and index Q (≥ 1). While determining $R^{(0)}$ attention should be paid that for all possible i and k, the relationship $r_{ik} \in [0, 1]$ and $\sum_{k=1}^{n} r_{ik} =1$ should be satisfied.

(2) Calculate the clustering center

$$V=\{ v_i^{(L)} \}, v_i^{(L)} = (\sum_{k=1}^{n} (r_{ik}^{(L)})^Q x_k)/ \sum_{k=1}^{n} (r_{ik}^{(L)})^Q \quad (1)$$

(3) Make an alteration to $R^{(L)}$

$$\forall i, \forall k, r_{ik}^{(L+1)} = 1/ \sum_{j=1}^{C} (\frac{\|x_k - x_i^{(L)}\|}{\|x_k - x_j^{(L)}\|})^{\frac{1}{Q-1}} \quad (2)$$

(4) Compare $R^{(L+1)}$ with $R^{(L)}$. When $\|R^{(L+1)} - R^{(L)}\| \leq E$, stop the above iteration, otherwise increase L by 1, i.e., L=L+1 and return to step (2).

Indexes to Evaluate the Performance of ISODATA

(1) Classification Coefficient

$$F_C(R) = \frac{1}{n} \sum_{i=1}^{C} r_{ik}^2 \quad (3)$$

(2) Mean Fuzzy Entropy

$$H_C(R) = -\frac{1}{n} \sum_{k=1}^{n}\sum_{i=1}^{C} (r_{ik} \ln(r_{ik})) \quad (4)$$

According to the algorithm, the more $F_c(R)$ is close to 1 and the more $H_c(R)$ is close to 0, the better clustering results can be expected.

Usual Algorithm to Calculate Distance d

(1) Euclidean distance (D=2)

$$d=\sqrt{\sum_{j=1}^{m}(\| x_{kj} - v_{ij}\|)^2} \quad (5)$$

(2) Minkowsky distance (D=4)

$$d=(\sum_{j=1}^{m}\| x_{kj} - v_{ij}\|)^{1/P} \quad (6)$$

where P has a positive value.

3. Application of Fuzzy ISODATA for DGA

Test Samples

On the basis of the data derived from the DGA of power transformers[8], data samples are constructed. For example, a set of 15 samples is denoted as $\{u_i, i=1, 2, ..., 15\}$, where $u_i=\{x_{i1}, x_{i2}, ..., x_{i5}\}$ with x_{i1} to x_{i5} representing the concentration ($\times 10^6$) of H_2, CH_4, C_2H_6, C_2H_4, C_2H_2 dissolved in the transformer oil, respectively. These 15 groups of DGA data are shown in Table 1.

Table 1   15 Groups of DGA Sample Data ($\times 10^6$)

No.	H_2	CH_4	C_2H_6	C_2H_4	C_2H_2	IEC code	RFT
1	14.7	3.8	10.5	2.7	0.2	000	Norm
2	980	73	58	12	0	010	L.E.D
3	181	262	41	28	0	020	L.M.O
4	173	334	172	813	37.7	022	H.O
5	127	107	11	154	224	102	H.E.D
6	200	48	14	117	131	102	H.E.D
7	6.7	10	11	71	3.9	022*	Norm
8	220	340	42	480	14	022	H.O
9	170	320	53	520	3.2	022	H.O
10	27	90	42	63	0.2	021	L.M.O
11	565	93	34	47	0	011**	L.E.D
12	32.4	5.5	1.4	12.6	13.2	102	H.E.D
13	56	286	96	928	7	022	H.O
14	160	130	33	96	0	001	L.M.O
15	650	53	34	20	0	010	L.E.D

Note:
(1) Here RFT represents the real fault type and Norm, L.M.O, H.O, L.E.D, H.E.D represent normal state, low or mediate temperature overheating, high temperature overheating, low energy discharge and high energy discharge, respectively.
(2) The asterisk means false judgment by IEC code and double asterisks represent there is not corresponding fault type to the concrete IEC code.

Clustering Procedure

Set the number of classes C to 5 (the corresponding classes are C1-normal, C2-low energy discharge, C3-low or mediate temperature overheating, C4-high temperature overheating and C5-high energy discharge) and initialize the fuzzy matrix $R^{(0)}$ as Table 2 (each column embodies a separate class).

Table 2   The Initial Clustering Matrix

No.	C1	C2	C3	C5	C5
1	0.90	0.00	0.10	0.00	0.00
2	0.00	0.70	0.00	0.00	0.30
3	0.20	0.00	0.60	0.20	0.00
4	0.00	0.10	0.00	0.80	0.10
5	0.00	0.00	0.00	0.00	1.00
6	0.00	0.10	0.00	0.00	0.90
7	0.80	0.10	0.10	0.00	0.00
8	0.00	0.10	0.00	0.80	0.10
9	0.00	0.00	0.10	0.90	0.00
10	0.40	0.00	0.60	0.00	0.00
11	0.20	0.70	0.10	0.00	0.00
12	0.00	0.10	0.10	0.00	0.80
13	0.00	0.00	0.10	0.70	0.20
14	0.10	0.00	0.70	0.20	0.00
15	0.20	0.60	0.10	0.10	0.00

Then, according to the foregoing algorithm, and with E=0.00001, Q=1.04, D=4 and P=12, calculation has been fulfilled. After 12 times iteration the demand of the precision is satisfied. The result fuzzy matrix R and cluster centering vector V are shown in Table 3 and 4, respectively.

Table 3   The Final Clustering Matrix

No.	C1	C2	C3	C4	C5
1	1.00	0.00	0.00	0.00	0.00
2	0.00	1.00	0.00	0.00	0.00
3	0.00	0.00	0.99	0.00	0.01
4	0.00	0.00	0.00	1.00	0.00
5	0.00	0.00	0.00	0.00	1.00
6	0.00	0.00	0.00	0.00	1.00
7	1.00	0.00	0.00	0.00	0.00
8	0.00	0.00	0.00	1.00	0.00
9	0.00	0.00	0.00	1.00	0.00
10	1.00	0.00	0.00	0.00	0.00
11	0.00	1.00	0.00	0.00	0.00
12	0.00	0.00	0.00	0.00	1.00
13	0.00	0.00	0.00	1.00	0.00
14	0.00	0.00	0.00	0.00	1.00
15	0.00	1.00	0.00	0.00	0.00

Table 4   The Clustering Center Matrix

C1	C2	C3	C4	C5
16.13	731.64	181.01	154.73	171.77
25.37	402.33	221.50	237.38	127.67
23.97	282.22	161.33	188.58	91.13
29.37	218.26	128.00	312.71	98.55
23.78	174.60	102.40	253.26	103.10

Clustering Results and Evaluation

According to Table 3 and 4, the result of the classification is reached, with the first class being {1, 7, 10}, the second {2, 11, 15}, the third {3}, the fourth {4, 8, 9, 13} and the fifth {5, 6, 12, 14}. It can be seen that for all the 15 samples, the identification of No. 1, 2, 3, 4, 5, 6, 7, 8, 9, 11, 13, 15 is correct excluding No. 10 and 14. This means the algorithm may not be sensitive enough to the low or mediate temperature overheating. The two evaluating indexes $F_c(R)$ and $H_c(R)$, are 0.998134 and 0.005060, respectively. From the point of mathematics theories, the clustering result is encouraging. By means of ISODATA, the corresponding correct recognition rate is 86.7%, higher than the 80% that the IEC method can reach, as generally believed. When another two new samples x_1 = {33, 26, 6, 5.3, 0.2} ($\times 10^{-6}$, normal) and x_2 = {335, 67, 18, 143, 170} ($\times 10^{-6}$, arc discharge) are input, both are correctly classified into class 1 and class 5, respectively.

4. Discussion

(1) During the process of program debugging, we found that, compared with other cluster methods, the ISODATA does not demand the format preprocessing of sample data strictly and the initially assigned $R^{(0)}$ has little effect on the final result, and this is just one of the advantages of the ISODATA. In addition, Better classification result will be obtained if more bi-polarized values are assigned to the elements of $R^{(0)}$.

(2) The value of E is for precision demanded and does not have notable influence on the results. Considering the reliability of iteration, E is often assigned a value ranging from 10^{-6} to 10^{-4}.

(3) The value of Q has some remarkable influence on the whole clustering process and the final result. Appearing as an index in J(R,V), it should not be too large, or fidelity can not be guaranteed. So, the Q should be as little as possible with the premise that Q>1. Besides, it can be concluded from formula (2) that, for (Q-1) appears as the denominator of the index, Q should not be too close to 1, otherwise may cause the overflow of calculation. Caution should be taken when determining the value of Q: first, the less value of Q means the fewer iteration times L requested, the higher classifying speed and the more bi-polarized elements of matrix R (i.e., r_{ij} is closer to 0 or to 1); secondly, the feasible value of Q is often selected between 1.04 and 1.2.

(4) The selection of D determines the type of distance and judging from the debugging, it does not affect the final results remarkably. However, different values of D still lead to different clustering results. For instance, when D=4, the best result can be acquired and when D=2, the second best result can be expected but the

other values of D cause worse results.

(5) P is a parameter used in the distance calculation when D=4. It is obvious that P should be as large as possible, but concluding from Equation (6), its being too large will bring about the overflow of computer. Based on the analysis of debugging, it is somewhat ideal when P is assigned to a value between 11 and 13.

5. Conclusion

The application of fuzzy ISODATA to process the DGA data of power transformers is presented in this paper with an encouraging result. Compared with the traditional ES, the new method isn't based on inferring rules, but abstracting fault information from the original data. Thus, the disadvantages of the former may be overcome and the conclusions of diagnosis will be more reasonable.

(1) Compared with the other cluster methods, fuzzy ISODATA does not demand strict format preprocessing of the sample data and the initially set $R^{(0)}$ has little influence on the final result, as is advantageous to satisfy the need of practical engineering.

(2) The ISODATA needs some parameters and the optimization of them is of great importance. Generally, the values are determined through repetitive debugging, partially based on personal experience. The typical values presented in this paper can be looked on as a reference.

(3) In further study, it is fit to improve the ISODATA more so as to acquire a better clustering performance.

6. References

[1] N. Gao, W. S. Gao and Z. Yan, "Progress in the Application of ANN for Insulation Fault Diagnosis of Power Equipment", High Voltage Apparatus, Vol. 33, No. 6, pp. 31-35, 1997

[2] C. H. Cai, "Incipient Fault Diagnosis for Transformer by means of Artificial Neural Network", Transformer, Vol. 34, No.1, pp. 34-37, 1997

[3] L. A. Zadel, "New Progress of Fuzzy Logic ", Fuzzy System and Mathematics, Vol. 28, No. 1, pp. 1-2, 1996

[4] C. E. Lin, J. M. Ling, C. L. Huang, "An Expert System for Transformer Fault Diagnosis Using Dissolved Gas Analysis", IEEE Trans. on Power Delivery, Vol. 8, No. 1, 1993

[5] T. Y. Li and H. G. Chen, "Fuzzy Relationship Equation and its Application in Fault Diagnosis of Power Equipment", High Voltage Engineering, Vol. 19, No.1, pp. 23-28, 1993

[6] Y. Zhang, S. P. Zhou and F. Su, Fuzzy Mathematics and its Applications, Coal Industry Press, Beijing, China, 1992, Chap3, pp. 102-118

[7] Q. Shen, L. Tang, Introduction to Pattern Identification, Press of National Defense Science University, Changsha, China, 1991, Chap4, pp. 112-117

[8] N. Gao, Study on Diagnostic Techniques for Insulation of Substation Equipment Based on Fuzzy Mathematics and Neural Network, Dissertation of Doctor Degree, Xi'an Jiaotong University, Xi'an, China, 1997, Chap4, pp. 37-39

Author

Guanjun Zhang
Ishii-Yasuoka Laboratory
Department of Electrical and Electronic Engineering
Tokyo Institute of Technology
2-12-1 O-Okayama, Meguro-ku, Tokyo 152-8552, Japan
Email: zhang@iyl.ee.titech.ac.jp
Tel & Fax: +81-3-5734-2851

Acoustic emissions caused by the corona discharge in an oil - tank

T. Sakoda*,   T. Arita*,   H. Nieda*,   M. Otsubo**,   C. Honda**,   Kiyomi Ando***

* Department of Electrical Engineering, Kumamoto Institute of Technology, Japan
** Department of Electrical and Electronic Engineering, Miyazaki University, Japan
*** Center for Cooperative Research, Kyushu Institute of Technology, Japan

Abstract

Measurements of sonic and ultrasonic waves brought by corona discharges were performed based on an acoustic emission (AE) technique, to diagnose the deterioration of insulation and its location in an oil - filled pole transformer.   The detected signals in these measurements were analyzed by means of a fast Fourier transform (FFT) and their properties were discussed. It was found that the AE signals due to the Lamb waves caused in a thin steel tank can be easily recognized by the FFT analyses of initial stages of the detected AE signals.   This indicates that the accuracy on location of the AE sources in the transformer can be improved. Also, it was shown that a low pass filter was effective to improve the accuracy of location.

1. Introduction

Any deterioration of insulation in a power system should not be permitted, considering the latest development of information engineering in society. In addition, it is demanded that the diagnostic method of insulation deterioration can be applied even to the system in operation.

On oil – filled pole transformers for the power distribution systems, measurements of increased leakage current or detection of produced gases has been performed, as the diagnostic techniques of the deterioration of insulation.   However, these examinations are almost not performed in operating conditions.   Therefore, it has been attempted to diagnose the insulation deterioration by using an acoustic emission (AE) technique [1-2].   The AE consists of sound and ultrasound waves and can be observed when a stress or energy is released in a material.   The AE technique is to analyze the AE signals detected by the AE sensor set on the surface of a testing material.   In general, the first AE wave in case of transformer is radiated in oil with a partial discharge or a corona discharge, and the induced AE waves are detected on the outside wall of the earthed transformer. From the analysis of these detected AE signals, it is possible to find the occurrence of discharges in operation conditions.   Thus, the AE technique is possible to apply as an effective method to diagnose the insulation deterioration of an apparatus in the system at the real time without stopping the apparatus.

The AE technique, as known, would be used to locate the position of discharge in the apparatus by setting some AE sensors in the three dimensions. However, an evaluation of the correct location of the discharge in the pole transformer is difficult to accomplish, because acoustic emissions due to the discharge proceed to all directions in the constitution of a tank with repeating reflection and transmission, and interference.   In this case, the existent primary wave (P wave) and secondary wave (S wave) in a cylinder type of thin steel tank mutually are coupled and generate the Lamb wave [3].   As velocities of them are different, the error in the location of the discharge position will be brought.   By this reason, it is necessary to clarify in detail characteristics of the detected AE signals for correct location.   In this paper, we analyzed frequency spectra of the AE signals by means of a fast Fourier transform (FFT) in order to understand properties of the AE waves and distinguish the kinds of detectable AE waves.   Furthermore, we discussed how to improve the accuracy of location.

2. Experiments

Figure 1 shows a schematic diagram of the AE measurement system.   An oil – filled model tank of 6 kV transformer was made of 1.6 mm iron steel.   Its inner diameter and height were 304 mm and 640 mm. An electrode system of positive point to plane for generating corona discharges as the AE source was placed along the center axis of the tank.   The needle electrode had a diameter of 1.5 mm, and the plane electrode had a diameter of 5 mm.   The gap length between the tips of the needle and plane was 50 μm long.   Applying a DC voltage of 1.4 kV to the needle electrode, repetitive pulse corona discharges were generated at a pulse interval of 1 s.   The discharge current had a peak value of about 210 mA and a full

High Voltage Engineering Symposium, 22–27 August 1999
Conference Publication No. 467, © IEE, 1999

width at half maximum of 20 ns. The AE sources were arranged at the heights (DH) of 250, 275, 300, 400, 450 mm from the bottom of the tank. The AE waves were detected by using a piezoelectric AE sensor. The AE sensor was set on the outside wall of the steel tank and placed at the height of 250 mm from the bottom of the tank. The detected AE signals by the AE sensor were amplified by pre – amplifiers and a local processor. The local processor can also equip with various filters, that is, low pass (LPF), high pass (HPF) or band pass filters (BPF). The amplified AE signals were sent to a Digital oscilloscope which has a sampling frequency of 100 MHz, and to a personal computer for further analyses including the FFT.

Figure 2 shows a characteristic of the frequency sensitivity of the AE sensor. This characteristic was obtained by using two sensors connected in series, in which one side of the sensor worked as oscillation source and the other side of the sensor measured those intensity. As shown in this figure, the AE detection system used with no filter represents relatively intensive spectral sensitivity in a wide of frequency range till 1000 kHz.

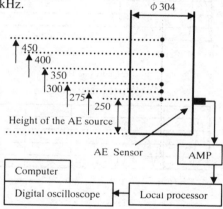

Fig.1. Schematic arrangement of the experimental apparatus.

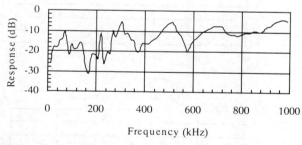

Fig.2. Sensitivity characteristic of the AE detection system.

3. Results and discussions

Figure 3 shows waveforms of the AE signal measured without filter for the case of DH = 250 mm. In this case, the heights of the AE source and the AE sensors are the same. From Fig. 3 (a), it is found that the AE signal continues for about 15 ms, in which onset times of signals obtained by the measurement number of 20 times accord well within an accuracy of $\pm 2 \%$ (2 μs). Figure 3 (b) shows the initial stage of the AE signal. Here, the early detectable AE signal will be of elastic P wave caused in the thin iron steel tank by fact that the P wave due to the corona discharge in oil reached the tank. As the velocity of the P wave in iron steel is commonly 5950 m/s (V_{Piron}), the transverse time of the elastic P wave, required to traverse thin steel tank wall till the position of the AE sensor, is about 0.27 μs. This time is much shorter as compared with the requirement time (T_o) of early detected AE signals to arrive at the AE sensor after beginning of discharge, $T_o \cong 108.44$ μs. Therefore, neglecting the required transverse time of elastic wave, it was calculated that the velocity of the P wave in oil was about 1400 m/s (V_{poil}).

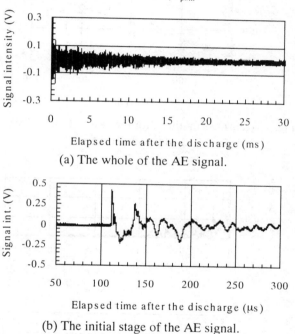

(a) The whole of the AE signal.

(b) The initial stage of the AE signal.

Fig. 3. The AE signal by the AE sensor located even with the AE sensor at DH = 250 mm.

Figure 4 shows transmission times of the AE signals obtained by setting the AE source at DH = 250, 275, 300, 350, 400 and 450 mm of different heights to the AE sensor of 250 mm. In Fig. 4, Tq means the time required of P wave in oil to directly propagate the shortest distance between the AE source and sensor. To do not correspond to Tq for the cases of DH = 300, 350, 400 and 450 mm. According to Snell's law [3], the P wave in oil directly reached the iron steel tank can be reflected from the inner surface of steel tank , when the setting position of the AE source is higher than DH = 287 mm. However, in the setting points of DH >

287 mm, as known from Fig. 4, the AE waves earlier than ones due to arrivals of the P waves in oil were observed.   These observed waves clearly are ones brought by early elastic waves occurred in 250 < DH < 287 mm and can be regarded as ones of the Lamb wave which has a much higher velocity than that of P wave in oil.   The Lamb wave is a kind of oscillation modes in a sheet and is composed of the P and S waves in the iron sheet.

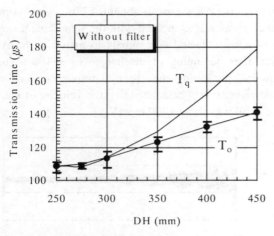

Fig. 4. Transmission times for the cases of DH = 250, 275, 300, 350, 400 and 450 mm.

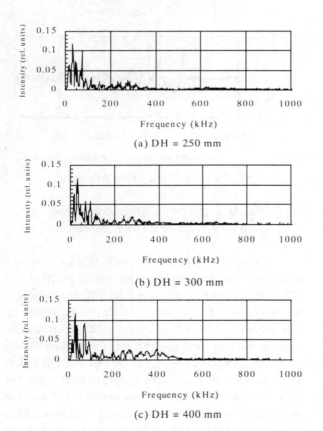

(a) DH = 250 mm

(b) DH = 300 mm

(c) DH = 400 mm

Fig. 5. Profiles of frequency spectra, in which a duration for analyses is 350 μs.

In order to investigate properties of waves, we analyzed frequency spectra by using the FFT.   Figure 5 shows profiles of frequency spectra for the cases of DH = 250, 300 and 400 mm.   The duration for the FFT analyses was 350 μs.   As seen from these figures, frequency spectra have peaks around 25 kHz.   They were very complicated with time according to transmission, reflection and interference of waves, although the detected AE signals were brought at first by the P wave in oil.   Also, there was not a large difference in their characteristics, but only spectral intensities of 200 – 600 kHz at DH = 400 mm were larger compared with those at DH = 250 and 300 mm. And so, we analyzed frequency spectra in the shorter duration of 40 μs covering a period of 25 kHz. Figure 6 shows profiles of frequency spectra for the cases of 250, 275, 300, 400 mm.   Obviously, the spectra of 400 mm are different from ones of 250 and 275 mm.   However, it is still difficult to distinguish ones of 300 mm from those of 250 and 275 mm.

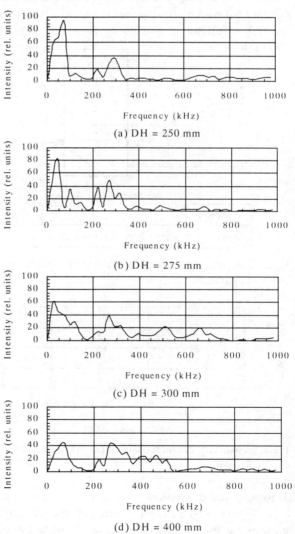

(a) DH = 250 mm

(b) DH = 275 mm

(c) DH = 300 mm

(d) DH = 400 mm

Fig. 6. Profiles of frequency spectra, in which a duration for analyses is 40 μs.

The frequency spectra for the cases of 275 and 300 mm are shown in Fig. 7, which were measured dealing with input signals of a local processor by the BPF of 200 – 1000 kHz. In this case, the large spectra of 400 – 800 kHz are observed for the case of 300 mm. Thus, the initial AE signals which were detected by setting the AE source at DH ≥ 300 mm are different from those of 250 and 275 mm, and it is considered that they are mainly composed of the Lamb wave.

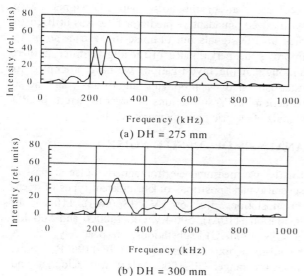

(a) DH = 275 mm

(b) DH = 300 mm

Fig. 7. Frequency spectra obtained dealing with by BPF of 200 - 1000 kHz at 275 mm and 300 mm.

In order to easily accomplish the correct location in the practical case, the information of the AE signal which are mainly composed of the Lamb wave should be rejected. Also, the measurement without catching the Lamb wave is useful as because frequency spectra of the Lamb wave have the component of 200 – 800 kHz from the FFT analyses, such as we investigated.

Figure 8 shows the AE signal obtained dealing with by the LPF of 200 kHz for the case of DH = 450 mm. The requirement time T_o of detected AE signals to arrival after discharge was 179.5 μs. This is about 38 μs shorter than that obtained without filter, and this time corresponds to Tq. In this case, the first observed signal will be of the elastic wave brought by a stress which is caused by that the P wave in oil reached the inner surface of the tank.

The transmission times obtained dealing with by the LPF of 200 kHz at DH = 250, 275, 300, 350, 400 and 450 mm are shown in Fig. 9. In Fig. 9, it is found that To correspond to Tq. Thus, the measurement of the AE signal by using LPF will be effective method to improve the accuracy of locations.

However, the presented facts in this paper are obtained under the ideal condition, in which there is no constituents of the transformer. The study about the influence of constituents is now in progress.

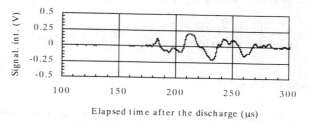

Fig. 8. The AE signal obtained dealing with by LPF of 200 kHz at DH = 450 mm.

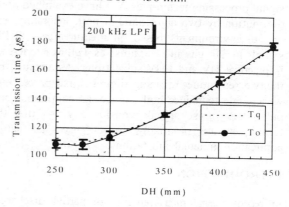

Fig. 9. The transmission time obtained dealing with by LPF of 200 kHz.

4. Conclusion

The AE waves due to the corona pulse discharge, which was generated as phenomena based on the insulation deterioration, were detected by using the AE sensor set on the outside wall of the steel tank. In order to investigate properties of the AE signals, the detected AE signals were analyzed by the FFT. From these analyses, it was found that frequency spectra were different by whether the Lamb wave is included in the AE signal. Thus, the analysis of the initial stage (about several ten micro - seconds) of the AE signal by the FFT is a very useful and powerful technique to evaluate the location of the discharge in the transformer, because the kinds of waves can be easily distinguished. Furthermore, we investigated the effectiveness of the measurement rejected components of the Lamb wave by the LPF of 200 kHz.

References

[1] R. T. Harrold: IEEE Transactions on Electrical Insulation **EI-20** (1985) P3.

[2] R. T. Harrold: IEEE Transactions on Electrical Insulation **EI-21** (1986) P781

[3] B. T. Phung, R. E. James, T. R. Blackbum and Q. Su: 7 th International Symposium on High Voltage Engineering (1991) P131

REDUCTION OF CONTINUOUS AND PULSE-SHAPE NOISE BY DIGITAL SIGNAL PROCESSING FOR RELIABLE PD DETECTION IN HV SUBSTATIONS

M KOZAKO, Z H TIAN, N SHIBATA* and M HIKITA

Kyushu Institute of Technology, Japan
Kyushu Electric Power Corporation, Japan*

ABSTRACT

This paper presents new developed noise reduction methods for partial discharge detection and measurements of electric power apparatus using digital signal processing techniques. We have examined noise diminution by two algorithms in both laboratory and on-site environments based on measurements of PD signals in SF_6 gas and real noise as well as interference in practical HV and EHV substations. As a result, by the two techniques utilizing digital signal processing, we are able to reduce both sinusoidal continuous and pulse-shape noise signals considerably, so that the techniques are promising to be used for future application in actual substations.

INTRODUCTION

In recent years, measurements of partial discharges (PD) in high-voltage electrical apparatus, especially SF_6 GIS, have been extensively made for developing reliable insulation monitoring and diagnostic techniques [1]. In order to obtain higher PD measuring sensitivity and reliable measurements, techniques for noise reduction are especially important. From this viewpoint, more sophisticated signal filtering and processing methods are imperative to meet the requirement of detecting PD signals correctly among various sorts of noise and interference like continuous and/or pulse-shape noise and interference. For this purpose, some methods of noise diminution using digital signal processing or digital filter have been investigated [2, 3]. Although a few suitable noise reduction methods have appeared for PD detection with digital signal processing under severe noise conditions of practical HV substations, it is still very difficult to distinguish PD pulse signals from various noise and interference easily and reliably. Another serious issue is to discriminate PD current pulses from pulse-shape noise so as to enhance the reliability of PD monitoring and insulating diagnosis and to avoid wrong prediction and further actions.

Therefore, this study aims to establish a digital filter algorithm which can be applied to PD measurements under actual noise conditions. We have so far examined the simulated noise reduction [4, 5]. In this paper, we have developed new noise reduction techniques toward the actual noise and interference. Electromagnetic (EM) noise in HV & EHV substations and electromagnetic waves (EMW) emitted from PD in SF_6 gas in the laboratory were measured separately. After the data of the noise and PD signals were synthesized, an adaptive predictor filter was utilized to pick up PD signals out of noise interference so as to improve the performance of the digital filter further. Throughout the investigation, we have attempted to establish new appropriate algorithms for noise reduction in HV substations in order to detect the PD signals of electric power apparatus reliably.

ANALYSIS OF ON-SITE NOISE

Firstly, we measured electromagnetic noise signals at several typical positions of four different types of open-air substations, i.e., 66 kV GIS (66G), 110 kV air insulation (110AI), 500 kV GIS (500G) and 500 kV all-GIS (500AG) substations, respectively [5]. A biconi-logperiod antenna (HP-11966P type, frequency range: 30~1000 MHz) was adopted and placed vertically or horizontally at a few meters away from the high-voltage electrical apparatus in the substations such as GIS, GCB, transformers and so on. The noise waveforms in time-domain and their frequency spectra were also measured with a digitizing oscilloscope (sampling frequency: 2.5 or 5.0 GHz) and a spectrum analyzer (Anritsu MS2661A, 9 kHz~3 GHz), respectively.

Figures 1 and 2 show typical examples of noise waveforms and the corresponding frequency spectra using a Fast Fourier Transform (FFT) measured at the 66G substation and 500G substations, respectively. From Fig. 1, it is clearly seen that frequency spectrum of the noise might consist of periodical noise arising from local radio broadcasting and mobile telephone as well as telecommunication networks etc. It was found that the 66G's noise and interference resembled the 110AI's one, because both the 66G and 110AI substations were located in residential areas. On the other hand, it is obvious from Fig. 2 that a big pulse-shape noise appears with its dominant frequency components below 300 MHz which may be due to corona on surrounding overhead HV transmission lines. Several kinds of pulse-shape noise exist in the 500G. However, such a kind of noise was found to emerge frequently in the 500G substation, but just a little in the 500 AG substation.

High Voltage Engineering Symposium, 22–27 August 1999
Conference Publication No. 467, © IEE, 1999

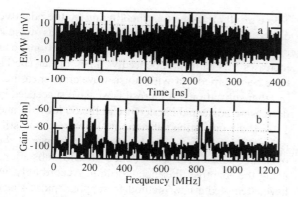

Fig. 1   Waveform and its frequency spectrum of EM
         noise measured at the 66G substation

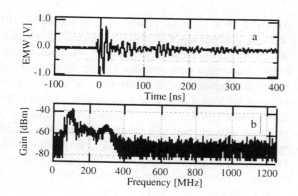

Fig. 2   Waveform and its frequency spectrum of
         EM noise measured at 500G substation

EMW MEASUREMENTS OF PD IN SF₆ GIS

Figure 3 schematically depicts the experimental
setup for measuring PD signals in a simulated 66
kV SF₆ GIS model. A needle-plane electrode with
needle tip radius 500 μm was assembled in the GIS
and the gap length between the needle and plate
electrodes was 10 mm. Pure SF₆ gas was filled into
the installed GIS at the pressure of 0.15 [MPa] in our
HV laboratory. Power frequency ac high voltage (60
Hz) was applied between the needle-plane electrode
system to produce PD inside the GIS.

Single PD current pulse waveforms were measured
using an impedance matching circuit placed below the
plane electrode. Meantime, the electromagnetic waves
emitted from PD were also measured utilizing the
biconi-logperiod antenna placed at 3 meters away from
the HV bushing of the GIS model. The detected PD
signals were measured with an ultra wideband (1.5
GHz) measuring system and digitized by a
digitizing oscilloscope (Sony / Tektronix —
TDS684B, 5 GS/s, 1.5 GHz), and the digitized
data were then sent to a personal computer through
GP-IB interface, so that the off-line signal processing
for noise reduction can be performed.

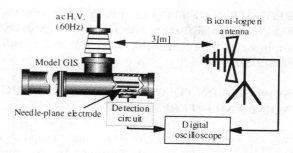

Fig. 3   Experimental setup for measuring PD
         in the SF₆ GIS model in the HV laboratory

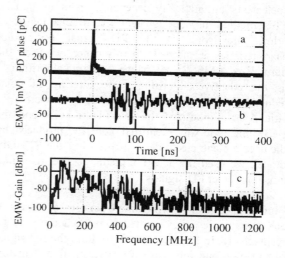

Fig. 4   Positive PD current pulse, corresponding EMW
         waveform and its frequency spectrum emitted
         from PD in the SF₆ GIS model

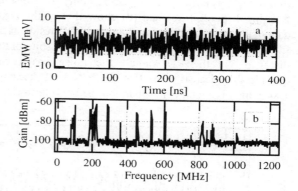

Fig. 5   Waveform and frequency spectrum of EM
         noise measured in the HV laboratory

Figures 4 (a), (b), and (c) display typical examples of a
PD current pulse waveform, and EMW signal emitted
from PD in SF₆ gas of the GIS model as well as its
corresponding frequency spectrum obtained by FFT,
respectively. Figures 5 (a) and (b) demonstrate a
typical example of a background noise waveform and
its frequency spectrum measured in the HV laboratory.
From Fig. 5, we can see a lot of periodical noise and
interference (narrow frequency band signals). When

compared Fig. 4 with Fig. 5, it is apparent that the PD signal has more and larger frequency components than that of the background noise and interference.

SUPPRESSION OF CONTINUOUS PERIODIC NOISE AND INTERFERENCE

We utilized an adaptive predictor filter incorporated with an adaptive digital filtering for reducing continuous periodic noise and interference [5]. The advantage of the adaptive filter is that it can adjust its coefficients automatically depending on input signals. This method has an apparent feature that neither a priori knowledge of the PD signal nor noise characteristics are needed. In this paper, we used the normalized LMS (Least Mean Square) algorithm [6] as the adaptive algorithm. The filtering of noise reduction needs electromagnetic noise waveforms in situ. After synthesized the data of the noise and the electromagnetic waves emitted from PD produced by the needle-plane electrode system within the SF_6 GIS model, we used the adaptive predictor filter to reduce noise or detect PD signals.

Figures 6 (a) and (b) illustrate examples of measured electromagnetic waves emitted from PD and the noise, respectively. In Figs. 6 (a) and (b), both the PD signal of 100 [pC] and the noise were measured at 3 meters away from the HV bushing of the GIS model in the laboratory and from a 110 kV GCB in a 110 kV air insulation substation, respectively. In this case, signal to noise (S/N) ratio (peak-to-peak) is only 0.28 before filtering. Figures 6 (c) and (d) show the synthesized data of the two waveforms and a typical PD result after filtering, respectively. It can be seen that, in spite of the distortion of the PD signal, the noise has been considerably decreased so that the S/N ratio became 1.84 which is much better than that before filtering.

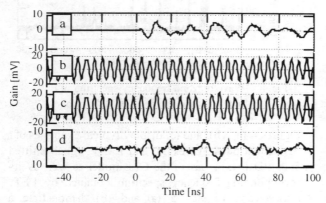

Fig. 6   Filtering results by the adaptive digital filter
(a) Measured PD signal (q = 100 [pC], d = 3 [m])
(b) Noise (110kV air insulation substation)
(c) Before filtering ((a) plus (b))
(d) After filtering by the filter (M=60, α =0.05)

These results indicate that the filtering by the adaptive digital filter technique can reduce continuous noise / interference greatly under the noisy conditions and increases S/N ratio pronouncedly. The similar checks have been carried out with noise waveforms measured in other substations. As a result, we have succeeded in decreasing noise and interference in the 66 kV GIS and 110 kV air insulation substations. However, it is still rather difficult to reduce pulse-shape interference in the open-air 500 kV GIS and 500 kV all-GIS substations using the adaptive filtering algorithm. Accordingly, we have attempted a new method to suppress pulse-shape noise.

REDUCTION OF PULSE-SHAPE NOISE

Figure 7 shows the flowchart of a so-called 2-BPF method, the new algorithm we developed, by which we are able to reduce influence of pulse-shape noise/interference with stochastic appearance. The new algorithm consists of the following processes: input pulse-shape signal is sent to two bandpass filters (BPF) which have different center frequencies fc_1 and fc_2 in order to obtain the amplitudes of both designated frequency components A1 and A2, respectively. As BPF, a 200-order (M=200) FIR (finite impulse response) windowed filter with Blackman-Harris window function was utilized. After having examined the significant difference between PD signals measured near the GIS model in the laboratory and noise measured in actual HV substations carefully, we set the center frequencies at fc_1=145 MHz and fc_2=655 MHz and their frequency bandwidths at 50 MHz, respectively, to avoid the influence of periodical noise.

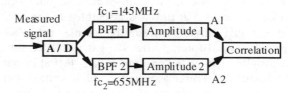

Fig. 7   Flowchart of the 2-BPF method

Figures 8 and 9 demonstrate typical examples of the processed results with actual noise waveforms measured at the 500G substation and PD signal measured in the laboratory using the 2-BPF method, respectively. When compared Fig. 9 with Fig. 8, it is obvious that the PD signal has more frequency components around 655 MHz. Also, Fig. 10 shows the relation between A1 and A2 values of the noise and PD signal, from which we can see that A2 is bigger in the PD signal group and A1 is small in comparison with the noise group. We thought that the noise group has big dispersion depending on the radiation source of noise and its transfer route as well as its attenuation.

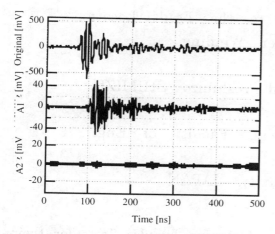

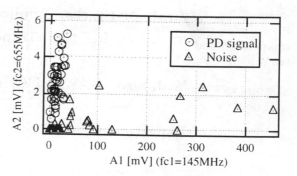

Fig. 10   Comparison of the amplitudes between two frequency bands of the pulse-shape signals

Fig. 8   Actual noise measured at the 500G substation and the processed results by the 2-BPF method

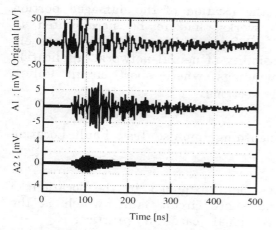

Fig. 9   Original PD signal measured in the laboratory and the processed results by the 2-BPF method

Then we calculated A2/A1 ratio and its averages in the noise and the PD signal groups. As a result, A2/A1 ratios of the noise and the PD signals were estimated at 0.6% and 13.5%, respectively. In other words, the A2/A1 ratio of the PD signal was about 20 times of that of the noise. Thus, we have successfully distinguished the pulse-shape noise and PD signal by utilizing the new 2-BPF method which is firstly proposed in this paper.

CONCLUSION

We have successfully investigated an adaptive predictor filter with LMS algorithm and a new digital 2-BPF method for PD detection in severe noise environments of practical HV & EHV substations. The effectiveness and efficiency of these techniques have been proved. Although the processed PD waveform was slightly distorted, measured continuous noise and interference have been considerably reduced.

Consequently, the adaptive predictor filtering technique enables us to effectively eliminate continuous noise without a priori knowledge of characteristics of PD signals and noise. Moreover, we proposed the new simple method to cancel the influence of pulse-shape noise and interference using two digital bandpass filters. Preliminary results confirmed that it is very useful for reducing such noise and interference in situ. More efforts will be made in these respects in the future so as to eliminate various noise and interference and to get reliable PD signals in practical HV substation environments.

REFERENCES

1.   A. Zargari et al., 1997, "Application of Adaptive Filters for the Estimation of Partial Discharge Signals in Noisy Environments", Proc. 5th ICPADM, pp. 212-215, Korea.

2.   H.N. Bidhendi et al., 1997, "Partial Discharge Location on Power Cables Using Linear Prediction", Proc. 5th ICPADM, pp. 406-409, Korea.

3.   M. Hikita et al., 1998, "Electromagnetic Noise Spectrum Caused by Partial Discharge in Air at High Voltage Substations", IEEE Transactions on Power Delivery, Vol. 13, No. 2, pp. 434-439.

4.   M. Kozako et al., 1997, "Application of Digital Signal Processing to Noise Reduction of Partial Discharge Detection in SF_6 Gas", Proc. '97 J-K Symposium on ED & HVE, pp. 63-66, Japan.

5.   M. Kozako et al., 1998, "Noise reduction and PD measurements using digital filter and signal processing technique in HV substations", Proc. '98 International Symp. on Elect. Insul Material, Toyohashi, Japan.

6.   N. Mikami: Introduction to programming of digital signal processing, CQ Publishing Co., Japan, 1993.

Neural Networks Application to the Fault Location in the Transformer Winding

Yoshihiro KAWAGUCHI and Toshihiro SHIMIZU

Department of Electrical Eng. Faculty of Eng.

KOKUSHIKAN University, JAPAN

Abstract

Transformers are usually subjected to the Lightning Impulse Tests for their withstandability, after assembled in the manufacturing plants.

Unexpectedly, their windings are sometimes broken down by excessive voltage stress due to internal oscillation. Naturally, the location of the damaged portion which might bring a long term laborious works, has to be necessitated to repair work. If location are easily done, subsequent jobs will be effectively and quickly handled. This paper shows a possibility of neural networks application to the fault location in the transformer windings, where short circuit failure is assumed to occur between some coil sections.

Keywords: Neural Networks, Transformer, Lightning Impulse Test, Fault Location

1. Introduction

In usual lightning impulse withstand test with one terminal of transformer winding grounded, reduced and full impulse voltage are applied, and the neutral current are recorded in both cases. Those two current wave forms are compared for their identity, taking into account of their magnitudes.

However, sometimes in this procedure, faulted position might not be easily found by visual inspection.

Current wave form flowing at the neutral point of the transformer varies, depending on the fault point in the transformer winding.

On the other side, even simple perceptron type neural networks(NN), which are composed by three layers are well known to be suitable to logical calculation(Boule Algebra).

Those NN has been already reported for the location ability of the grounding fault from a certain cross-over point, in JIEE Meeting in Jan. 1999.[1]

It's here extended to the location of the short circuit fault, which is the occasional case in practice.

2. Experimental Procedure.

The model for the experiments is a 1/5 model of actual 275kV-300MVA transformer with 14 coil sections.

Artificial faults are simulated by floating transistor switch for short circuit between neighboring 14 cross-over's (CO's). Transistor switch to simulate the artificial fault is triggered by impulse voltage applied on the transformer winding terminal and delayed by 4-10 μS which is adjusted by time counter setting.

The neutral currents are measured through the detecting resistance of 10 Ω, which is relatively higher than in practice, to elevate the measuring sensitivity, by Instrumental System (LeCroy CRT), with 25 MHz sampling frequency and averaging (Sweep 100) to

High Voltage Engineering Symposium, 22–27 August 1999
Conference Publication No. 467, © IEE, 1999

reduce noise.

The above-mentioned neutral currents are transferred to Personal Computer via GPIB Cable, digitized by LabView Ver.3.0.1 and the difference currents from the normal neutral current are taken.

Then, the wave up to $6\mu S$ after the fault occurrence is picked up and subjected to Fast Fourier Transformation to obtain frequency spectrum.

The representative frequency spectrum in case of short circuit faults between CO.2-3, 5-6, 8-9 and 11-12, are shown in Fig.1(a) to 1(d) respectively, for 4, 8, 10 μS delayed fault after impulse application at winding terminal.

Observed here is that the frequency spectra at any case are very similar to each other except their amplitude, independent on the fault occurring time.

It's quite understandable due to the physical nature of the system.

On the other side, it's clearly dependent on the short circuit position. This fact suggests a favorable situation to the fault location.

3. Neural Networks[2]

Perceptron type Neural Networks used here consists of three layers, which are provided 14, 28 and 4 neurons for input, middle and output layer respectively.

14 frequency components from 2.66 to 7.0 MHz every 0.333 MHz interval are selected as No. of input layer neurons.

No. of middle layer neurons are not definitely defined theoretically, but suitable 28 neurons are determined after the investigation of the effect of No. of middle layer neurons on the least learning repetition to reach the error of 10^{-4}.

Each output layer neuron should have charge so as to inform the short circuit position that 4 output layer neurons are assigned to indicate the short circuit failure between CO.2-3, 5-6, 8-9 and 11-12 respectively.

The connection between layers are made by so-called Logistic Function.

To facilitate the threshold value, a additional neuron (settled to the input of -1) is provided into input and middle layers. Now, learning is done for the short circuit faults between CO.2-3, 5-6, 8-9 and 11-12.

Before learning whole frequency components are normalized.(-1 to +1)

Then, teaching is done by Back Propagation repetitively until the learning error reaches 0.0001 and less, or until 10000 repetition.

The results, not shown here, give the satisfactory learning for all short circuit position except the one between CO.10-11, which may affect the ambiguous judgment by NN on the lower position, as mentioned bellows.

4. Execution

In view of practice, one might hardly know at which time the fault occurs.

So, the established NN should be investigated on the ability whether the NN learnt by the fault occurring at a certain time, is applicable to the fault at the unknown time.

Fortunately, the frequency spectrum of the detected neutral current is almost independent on the fault occurring time, as explained above. Therefore, the NN learnt for short circuit fault at 8 μS delayed time, may be utilized for the cases at 4 or $10\mu S$ delayed faults.

The execution results in Table 1(a) and 1(b), with the good agreement for almost the upper fault positions.

However, a few lower fault positions are not given agreeable judgment from NN, due to the original shortage of learning above-mentioned.

5. Conclusion

Irrespective of fault occurring time of grounding[1] or short circuit during impulse test of transformers, the fault location seems to be possible by the NN application.

Acknowledgement
The authors wish to thank for the management of TOSHIBA who supplied

the model transformer and for Mr. K. Takahashi who collaborated experiments and data processing.

Reference:
(1) Y. Kawaguchi et al: Neural Networks Application to the Fault Location in Transformer winding. (HV99-6 IWHV, JIEE Meeting Jan. '99)
(2) John Hertz et al: Introduction to the Theory of Neural Computation (Addison-Wesley Publishing Co.)

Address of the Authors:
D'ept of Electrical Eng.
Faculty of Engineering.
Kokushikan University.
Setagaya 4-28-1, Setagaya-ku, Tokyo, Japan.
Tel/Fax: +81-3-5481-3342
e-mail: ykawa@kokushikan.ac.jp

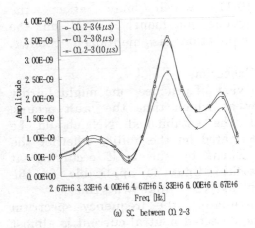

(a) SC. between CO. 2-3

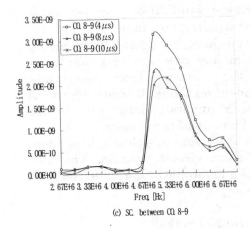

(c) SC. between CO. 8-9

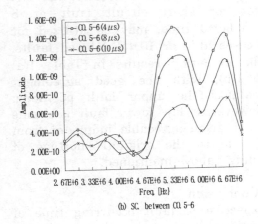

(b) SC. between CO. 5-6

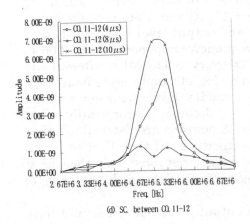

(d) SC. between CO. 11-12

Fig 1. Frequency Spectrum in case of Short Circuit between different CO.'s

At 4, 8 and 10 μS delayed fault

Table 1.   Results from NN learnt by Frequency Spectrum in Short Circuit between CO. 2-3, 5-6, 8-9, 11-12 at 8μs delayed

(a) Execution for 4μs Fault

	Output Neuron				Eva.
	1	2	3	4	
SC. 1-2	.99994	.00005	.00010	.00000	G
SC. 2-3	.98858	.00219	.00875	.00000	G
SC. 3-4	.01880	.88578	.02568	.00224	T
SC. 4-5	.01180	.99695	.00314	.00256	G
SC. 5-6	.01459	.98005	.01616	.00149	G
SC. 6-7	.01103	.25704	.55148	.00183	T
SC. 7-8	.00795	.32529	.62327	.00215	G
SC. 8-9	.12133	.00097	.96524	.00017	G
SC. 9-10	.96268	.00002	.49003	.00003	T
SC. 10-11	.99817	.00000	.05758	.00001	W
SC. 11-12	.83721	.00000	.17256	.01754	W
SC. 12-13	.01451	.00021	.08358	.73465	G
SC. 13-14	.01621	.00013	.00104	.99609	G

Shaded column corresponds to the learning point
SC. 1-2 : Short Circuit bet. CO. 1-2, Eva. : Evaluation
G : Good, T : Tolerable, W : Wrong

(b) Execution for 10μs fault

	Output Neuron				Eva.
	1	2	3	4	
SC. 1-2	.99905	.00010	.00779	.00000	G
SC. 2-3	.48852	.05436	.05547	.00005	G
SC. 3-4	.14639	.65744	.01732	.00013	T
SC. 4-5	.00519	.99815	.00091	.00826	G
SC. 5-6	.00218	.99017	.00111	.09471	G
SC. 6-7	.00245	.95838	.01180	.06577	G
SC. 7-8	.00203	.85177	.02325	.13952	T
SC. 8-9	.00677	.00906	.97947	.00370	G
SC. 9-10	.99253	.00000	.65769	.00003	T
SC. 10-11	.99996	.00000	.02736	.00000	W
SC. 11-12	.99983	.00000	.05463	.00001	W
SC. 12-13	.00228	.00197	.05778	.89609	G
SC. 13-14	.00082	.26091	.00081	.96585	G

Shaded column corresponds to the learning point
SC. 1-2 : Short Circuit bet. CO. 1-2, Eva. : Evaluation
G : Good, T : Tolerable, W : Wrong

Measurement of Flash Discharge in Explosion-proof Motors

Yu Ming, Xu Yang, Cao Xiaolong,
Qiu Changrong

People's Republic of China

ABSTRACT

The aim to detect flash discharge is to find flash in explosion-proof motors caused by high voltage and heavy current, when they were operated . Because flash in explosion-proof motors working in inflammable areas always leads to detonation, flash detection has become an important measurement method to prevent danger .This paper reports the results of an optical and electrical system used to study flash in motors . And a comparison of discharge energy and optical signal of initial spark is presented .Minimum discharge energy which can lead to detonation is found in certain circumstances .And the relationship of measuring distance and optical signal is also studied .The measurement shows that the system is acuter than the traditional method .And it can resist electrical-magnetic interference .The default can also be located.

1.INTRODUCTON

The safety run of electrical device is an essential prerequisite of industry, especially in some inflammable and explosive areas[1]. In most circumstances, fire and detonation caused by electrical devices are regarded as the most serious accidents. Therefore ,multiform explosion-proof devices like motors are much more important now . This paper mainly deals with the measurement of inner flash in explosion-proof

High Voltage Engineering Symposium, 22–27 August 1999
Conference Publication No. 467, © IEE, 1999

motors. Through the measurement of flash energy ,the judgment of the devices security can be made and so the purpose to prevent explosion can be reached.

The explosive circumstances were usually divided into three grades due to various degree. Explosion-proof motors can also be divided into many grades ,such as essential security etc .These are the main structures of explosion-proof motors we studied.

The flash discharge in motors is a kind of partial discharge brought about by deficiency in equipment , inadequate installation , or heavy current in running. At present , there are many methods to detect flash discharge , but uniform standard has not been made. It seems that three methods have been used for detection. One is the off-line method , and another is the on-line method . Though they have some advantages, they also have disadvantages. They can not provide accurate transfer process of charge, and so accurate discharge energy can not be got. It is also difficult to locate the site. In the Pre-Soviet Union , practical detection method was used .This method seems dangerous and uneconomic for large motors.

2.THE MECHANISM AND FEATURE OF FLASH DISCHARGE

Partial discharge in motors has many different forms. Most discharge has the ignition capacity.

There are many places prone to appear flash discharge ,such as gas discharge in insulation ,discharge between winding and stator slot ,discharge in rotor .Because inner discharge can not light up inflammable gas ,we mainly study the visible discharge on end plate.

Ordinarily ,when all the outsider electrical parameters are the same ,the discharge may not consistently light up the explosive gas .Sometimes light and big spark can not light it up, while small and invisible one can do it .Whether it can light up the gas or not depends on the energy and times of the discharge .Generally, the parameter used to describe the ignition capacity is ignition rate(P=M/N) .It is a ratio of times of the gas being ignited and total times of the experimental .

The light radiated by different discharge has different wavelength .The wavelength of corona light is relatively short, below 400nm, which belongs to the scope of ultraviolet rays wavelength .The wavelength scope of flash discharge ranges from 400nm to exceeding 700nm. Most of them fall into visible wavelength.

3.TEST SETUP

Figure 1 shows the structure of the whole system. From Figure 1, we can see that the main parts of

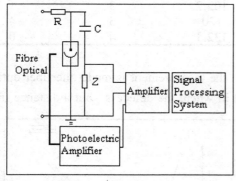

Figure 1:Flash discharge setup,with implemented PD defect

the system are the shiner, transmit medium ,photoelectric device ,amplifier ,and signal processing system .

The shiner is the spark in motor to be

detected .The transmit medium used is highly developed fiber optical .The fiber optical was selected , because it has many advantages ,such as electromagnet interference protection, flexibility ,low loss ,wide waveband , especially suitable to be used in explosive-proof motor.

Because of its high gaining for small signal, photoelectric amplifier tube is used .It also has suitable spectrum and same sensitiveness at cathode.

Due to electrical signal is very weak after the photoelectric transmission ,the signal must be magnified and filtered .Considering the low frequency interference ,when we designed the filter ,we choose a appropriate filter ,which waveband ranges from 20kHz to 300kHz .The filter is used also thinking of the abundant-contain signal spectrum in the waveband scope[2] .

In order to study the relationship between the light impulse amplitude and the discharge energy, the optical measurement should be done at the same time of the measurement of the discharge energy. With regard to the discharge pattern ,two methods

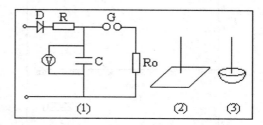

Figure 2: Stimulating discharge producers

are used .One method is to measure the impulse current as we usually did .The discharge energy can be got by the charge and discharge voltage . Regarding to sphere gap, the second method calculates the discharge energy by charging energy of capacitance .

In order to study the relationship between flash in explosion-proof motor and its optical energy, three simulating discharge producer are made, corresponding to pin-half sphere discharge ,pin-plate discharge ,and sphere gap discharge ,which circuit is showed in Figure 2.

When the voltage on the capacitance is high enough to break the gas gap, the discharge takes place .The discharge energy is considered as the same stored on the capacitor .It can be calculated using formula (1).

$$E=1/2CU^2 \qquad (1)$$

When all the parts of the system is determined , the amplitude of optical impulse L is in proportion to light intensity I_p .

$$L=k\times I_p(t) \qquad (2)$$

The energy generated by the discharge is in proportion to amplitude L.

$$E_0=A/B\times L \qquad (3)$$

A, B are the coefficients correlative to filter optical, photoelectric amplifier tube, measuring circuit ,and measuring distance .

4.ANALYSIS OF EXPERIMENT RESULTS

It follows that the discharge energy can be got by

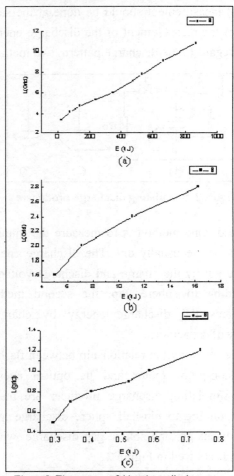

Figure 3: The energy of (a) sphere discharge ,(b)pin-plate discharge ,(c) half-sphere discharge and optical signal amptitude

the amplitude of optical impulse .The Figure 3 shows the relationship of amplitude of optical impulse and discharge energy.

From the graph ,we can see that with the increasing of

discharge energy the amplitude of optical signal also increases correspondingly .We can also see that when the discharge energy is great ,the E seems linear to L .

Why E is not linear to L when the energy is small needs second thoughts. Generally ,we can use L to indicate E ,because they increase at the same time. We also study the minimum energy leading to explosion. Sealing the sphere gap in a container filled with a mixer of air and hydrogen at fixed ratio

21% , the minimum energy can be tested at certain circumstances .Repeating the experiment more than once , we can find the Table 1.

From the Table 1,we can select 120 uJ as minimum energy at this environment .We hope the result will be helpful to decide the standard of spark in explosive-proof motor .

Table 1- Times of explosiveness and discharge energy(uJ) (Only the energy near the minimum energy is selected)

Discharge energy(uJ)	Number of times explosiveness	Number of times not explosiveness
103.7	3	1
120	3	4
132.3	4	0

When the measurement is made ,the fiber optical is not in where the defect is , but a distance from

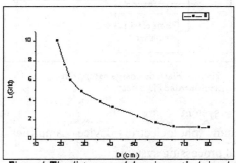

Figure 4: The distance and damping optical signal

it .The signal may damp through the distance .Calibrating measurement must be made .Figure 4 shows we can find a scope in which optical signal changes less .This is very important for practical usage.

4.CONLUSION

Research shows that there are corresponding relationships between optical signal and discharge energy in explosive-proof motors .Through it ,we can use amplitude of optical signal to express the discharge energy .

Elementary studies were made to find the minimum discharge energy which can cause an explosion and serve for constituting corresponding standard of explosive-proof motor in the future.

A computer-aided testing system to measure the inner flash discharge is made, and it has been successfully used in practice.

5.REFERENCES

[1]M.Massudo,Y.EHARA, . July,1991 " Relationships between Partial Discharge Magnitude and Discharge lumination Insulation" , In Proceeding of the 3th Inter. Confer. On Properties and Application of Dielectric Material, Tokyo Japan:P234~238.

[2]P.S.Excell, 1998 , "Measurement of Minimum Ignition Energy Using a 9.4Ghz Pulsed Radar Source " , In Fourth International Conference on Electrical Safety in Hazardous Areas, IEEE Conference Publication,

TIFDES – AN EXPERT SYSTEM TOOL FOR TRANSFORMER IMPULSE FAULT DIAGNOSIS

P. Purkait
Department of Instrumentation Engg.
Haldia Institute of Technology
West Bengal, India

S. Chakravorti   K. Bhattacharya
Department of Electrical Engg.
Jadavpur University
Calcutta, India

Abstract: The objective of this paper is to present TIFDES - an expert system tool for impulse fault diagnosis in transformers. These important and complex decision tasks are, in general, performed by highly experienced testing personnel, whose knowledge can be expressed as a set of production rules. The knowledge base of this expert system is derived from IEEE and IEC standards, literatures and expert human knowledge to include as many diagnosis rules as possible. To identify and locate a fault, an inference engine is developed to perform deductive reasoning based on the rules in the knowledge base and statistical techniques. The basic idea of TIFDES is to provide a non-expert with the necessary information and interaction in order to make a fault diagnosis in a very friendly environment by a windowed display system.

Keywords: Expert System, Fault Detection Rule, Impulse Fault Diagnosis, Knowledge-Based System, Transformer Fault.

1. INTRODUCTION

Impulse tests are performed on power transformers to assess its insulation integrity, during which it is subjected to a special sequence of voltages as per standards (IEC 722) and the resulting current oscillograms are recorded. Fault detection using these current records is performed on the basis of the presence of any deviation amongst the records.

In the earlier days, visual examinations of the oscillographic traces were performed for detecting any incumbent insulation fault [1-4].

Later on, with the availability of transient digitizers these records were acquired and stored enabling its post-processing employing methods like the Transfer Function Approach [5].

Satish et al [6] mentioned a Multiresolution Signal Decomposition Method for partial discharge measurement in power transformer during impulse tests.

Very recently Wang et al [9] presented a combined ANN and Expert System Tool for transformer fault diagnosis using Dissolved Gas-in Oil analysis.

In reality, however, it is difficult to grasp manually, the fault conditions and identify the fault location in a short time because sophisticated know-how and plenty of experiences are required to judge fault conditions precisely. This kind of sound knowledge base may not be available in all the testing laboratories, particularly in the developing countries. Moreover, unless the heuristics of fault detection are well documented in a rule-base, the controversies in the judgement procedure due to human factor will continue to persist.

In electric power systems the use of expert systems has already been presented by several researchers [7-8].

In this paper the proposed expert system tool, TIFDES, aims at assisting the testing personnel to obtain unambiguous and quick decision in impulse-failure diagnosis of transformers. To free the engineers from the strenuous job of manual fault-diagnosis, a set of heuristic rules are compiled through consultation with the standards, literatures and the experts to ensure that the expert system detects a fault with high confidence. These rules are coded into computer programs and are stored in the knowledge base of the expert system. The inference engine, which combines the heuristic rules of the rule-base along with mathematical and statistical methods of the expert system, emulates the behavior of well-trained testing personnel.

The expert system, which is implemented on a personal computer, has been extensively tested on an analog model of a 33/11KV, 3 MVA transformer. It is found from the test results that the expert system is capable of identifying and locating a variety of winding faults in a very efficient manner.

The expert system TIFDES has been developed in the High Voltage Laboratory of Jadavpur University, Calcutta, India and it is being used for both testing and educational purposes.

High Voltage Engineering Symposium, 22–27 August 1999
Conference Publication No. 467, © IEE, 1999

2. DESIGN OF THE EXPERT SYSTEM

The structure of the designed expert system designed is shown in Fig.1.

It consists of the functional modules as described below.

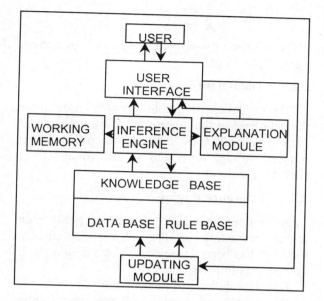

Fig.1. TIFDES Configuration

1.1 User Interface

An effective expert system should provide a user-friendly communication environment with the end-user. An interactive user-interface (UI) using simple menus as shown in Fig.2. has been implemented with the help of which the user can ask the expert system to identify the fault, locate the fault as well as update the data and rules in the knowledge-base .

2.2 Knowledge-Base

The specialised knowledge required for the development of the expert system has been acquired from:

i) Expert engineers working for years at the High Voltage Testing Laboratory, Jadavpur University, Calcutta, India.

ii) Data accumulated over the years from the testing of different types of transformers.

iii) IEEE, IEC and Indian Standards along with the information collected through literature survey.

The developed knowledge-base consists of a rule-base and a data-base.

2.2.1 Rule-Base

A rule-base has been identified by the knowledge acquired as described above. The heuristic expert knowledge is encoded in sets of production rules with IF-THEN structure in a tree-like form as are discussed below.

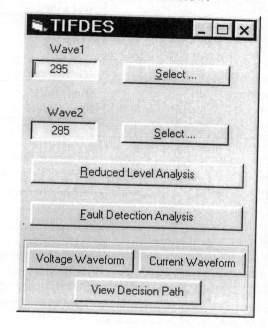

Fig.2. Main Window

i) **If** (Reduction in voltage wave Tail-Time)
   Then (Shunt fault near line-end)
ii) **If** (Oscillations in the current wave under observation are less than those in the unfaulty current wave)
   Then (Shunt fault in the middle section)
iii) **If** (The two current waves, taken in p.u basis, deviate too much , provided any possible deviation in the input voltage have been taken care of)
   Then (There is a series winding fault)
iv) **If** (Deviation in the two current wave magnitudes all throughout the time span)
   Then (Series fault near middle)
v) **If** (There is good amount of oscillations near the front of the current wave only)
   Then (Series Fault near line end)
   Else (Series Fault near earth end)

2.2.2.1 Fault Detection algorithm

The algorithm of fault-detection, identification and its localisation, as discussed in the above rule-base is described by the flow-chart of Fig.3.

RT50   = Ratio of Tail-Time of voltage waves

ROSC = Ratio of oscillation magnitude of the current waves after 30 µs

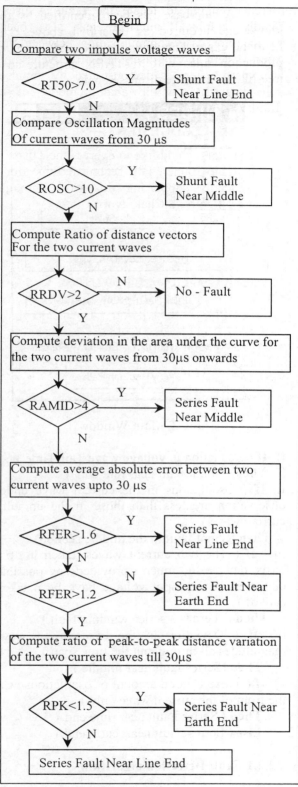

Fig.3. Knowledge Tree for Fault Diagnosis

RRDV = Ratio of Distance Vectors
Distance Vector is defined as :

$$DV = \frac{1}{N} \sqrt{\sum (C1 - C2)^2}$$

where C1 and C2 are ordinates for same abscissa for the waveforms under comparison.
RAMID =Area-ratio under the two current waves
RFER = Ratio of average absolute error for the front of the two current waves,
where Average Absolute Error is defined as:

$$AAE = \frac{1}{N} \sum |(C1 - C2)|$$

RPK = Ratio of oscillation magnitude of current waves upto 30 µs

2.2.2 Data-Base

The digital oscillographic records required for the data-base in the present work are voltage and current records (by tank-current method) at reduced voltage level and at full voltage level (BIL) respectively .

2.3 Inference Engine

An inference engine capable of performing logical reasoning on the rules and statistical methods has been designed for fault diagnosis. It proceeds as follows:
i) Use the rule-base to identify fault
ii) Designate the section of the transformer winding where the fault has occurred
iii) Provide the user with the logic behind the fault diagnosis through the **explanation module** .

2. IMPLEMENTATION OF TIFDES

The expert system has been implemented on a PC. To demonstrate the effectiveness of the designed expert system, impulse fault diagnosis has been performed on the HV winding of the analog model of a 11/33 kV , 3 MVA, 3-phase, DY-11, 50 HZ transformer. In the present diagnosis, voltage sensitive disc-to-disc short circuit faults as well as disc-to-earth shunt faults have been simulated. Faults have been simulated in discs over the entire length of the winding to properly represent the winding sections, namely the line-end, mid-winding and earth-end sections. Performance of these coils are studied with the whole winding being subjected to impulses from a Recurrent Surge Generator (RSG). The winding response, captured by the tank-current method is acquired by the PC through a Tektronix [TDK 320] digital oscilloscope.

Fig.4 shows typical oscillographic records of impulse currents at reduced as well as full-impulse level. The current waveform at full impulse level correspond to a voltage sensitive short-circuit of 1% of the winding length at a distance of 10% of the winding length from the earthed end. For this case, the expert system immediately flashes a fault notification on the monitor to the user.

Then if the user wants to find out the logic behind the inference drawn, he may do so through the explanation module. Through the UI the user then receives back the decision-making path used by the inference engine as shown in Fig.5.

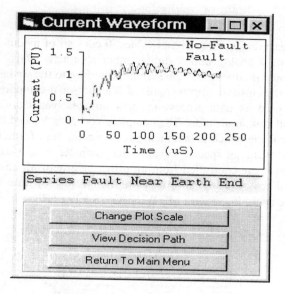

Fig.4. Waveform Display Window

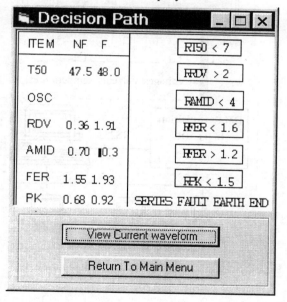

Fig.5. Explanation Module

3. CONCLUSION

In this paper, an expert system has been developed to serve as an operational aid for the detection and location of impulse faults in a transformer. The heuristic rules gained by experience over the years have been embedded in

the rule base and these rules form the basis for the deductive reasoning performed by the inference engine to locate the fault. A user-friendly graphical interface has also been designed for easy communication between the expert system and the user. The effectiveness of the expert system is demonstrated by impulse fault diagnosis on the analog model of a 11/33kV, 3 MVA, 3-phase transformer.

4. ACKNOWLEDGEMENT

The authors would like to express their thanks to 'All India Council For Technical Education' for financially supporting this work through project grant to Dr.S.Chakravorti.

5. REFERENCES

[1] L.C.Aicher, "Experience With Transformer Impulse Failure Detection Methods", AIEE Trans. Vol. 63, 1944, pp. 999~1005.

[2] F.Beldi, "The Impulse Testing of Transformers", The Brown Boveri Review, Vol. 37, 1950, pp. 179 ~ 193.

[3] J.H.Hagenguth, J.R.Meador, "Impulse Testing of Power Transformers", AIEE Trans., Vol.71, 1952, pp. 697~704.

[4] C.K.Roy,J.R.Biswas, "Studies on Impulse Behavior of a Transformer Winding With Simulated Faults by Analog Modelling", IEE Proceedings-C ,Vol. 141, 1994, pp.401~412.

[5] R.Malewski,B.Poulin, "Impulse Testing of Power Transformers Using the Transfer Function Method",IEEE Trans. on Power Delivery, Vol.3, 1988, pp.476~490.

[6] S.K.Pandey,L.Satish, "Multiresolution Signal Decomposition: A New Tool for Fault Detection in Power Transformers During Impulse Tests",IEEE Trans. on Power Delivery, Vol.13, No.4, 1998.

[7] Y.Hsu.,F.C.Lu, Y.Chien, J.P.Liu, J.T.Lin, H.S.Yu, R.T.Kuo, "An Expert System For Locating Distribution System faults", IEEE Trans. on Power Delivery, Vol.6, 1991, pp.366 ~ 371.

[8] M.V.Ernesto, L.Oscar, M.Chacon, J.Hector, F.Altuve, "An On-Line Expert System For Fault Section Diagnosis in Power Systems", IEEE Trans. on Power Systems,Vol.12, 1997, pp.357 ~ 362.

[9] J.Wang, Y.Liu, P.J.Griffin, "A Combined ANN and Expert System Tool For Transformer Fault Diagnosis", IEEE Trans. on Power Delivery, Vol.13, 1998, pp. 1224~1229.

Online Monitoring of Power Transformers - System Technology and Data Evaluation

T. Leibfried

Transformatorenwerk Nürnberg

Siemens AG

Germany

Abstract: During the last few years, the electricity market has changed in many countries worldwide. The more competitive situation leads to a higher cost consciousness on the part of utilities. One aspect for saving costs is reducing effort required for maintenance of power network equipment. Power transformers are among the most expensive single elements of the high-voltage transmission system. Monitoring systems can help to decrease the transformer life cycle costs and to increase the high level of availability and reliability. In 1997 two large power transformers were equipped with the new Siemens monitoring system. The structure of the system as well as a first data evaluation are presented in this paper.

I. INTRODUCTION

In many countries the electricity market was opened by a political legislation. The consumer has the choice from which supplier he wants to purchase energy. Certainly, the utility which offers the most favourable conditions has the best chance to survive on the market. This results in a more competitive situation and finally in a higher cost consciousness on the part of utilities. An important field of cost saving is investment and maintenance of power network equipment, especially of power transformers. Online monitoring systems can help to reduce life cycle and maintenance costs of power transformers. The more accurate lifetime estimation leads to an improvement in investment strategies. Furthermore, a higher loadability of power transformers expected in the future by some experts can only be realized with minimum risk by permanent condition measurement. This contribution reports about the evaluation of data obtained by the Siemens monitoring system which was mounted on two transformers (200 MVA, 220 kV and 300 MVA, 400 kV) in 1997.

II. STRUCTURE OF THE MONITORING SYSTEM

A. System concept

Fig. 1 shows the hardware structure of the monitoring system installed on the 200 MVA generator transformer. The system hardware consists of two main parts: the data aquisition unit and the personal computer [2]. The data aquisition unit converts the analog

measurements into digital values. It consists of modules with 4 analog and 4 digital input channels. Thus, a modular structure of the system according to the wishes of the customer can be realized. The personal computer is used for data processing, data storage, visualization and communication. The complete system is located in a cubicle which is mounted directly on the transformer. By means of special software the complete system can be remote-controlled from an office or a network control center. The platform of the monitoring software is the operating system MS WINDOWS 95 or NT.

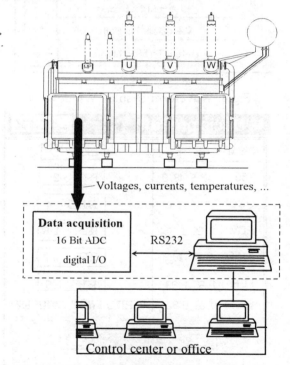

Voltages, currents, temperatures, ...

Data acquisition
16 Bit ADC
digital I/O

RS232

Control center or office

Fig. 1 System concept of the monitoring system

B. Sensor equipment

Table 1 shows the measuring quantities which are picked up by the monitoring system installed on the 200 MVA generator transformer. Voltages are measured at all three high-voltage bushings. A special sensor allows measurement from rated frequency (50 Hz) up to the MHz range. Currents into the hv windings are measured by current transformers mounted in the bushing domes. Together with the tap changer position, the load of the transformer can be calculated.

High Voltage Engineering Symposium, 22–27 August 1999
Conference Publication No. 467, © IEE, 1999

Measuring quantity	No. of sensors
Voltage (HV)	3
Current (HV)	3
Tap changer position	1
Temperatures	7
Failure gas-in-oil (trend)	1
Moisture of the oil	1
Humidity of air in the conservator	1
Circuit state of pumps and fans	4 + 4
Velocity of oil	2
Oil level in the compensator and the tap changer tank	1+1

Table 1 Measuring quantities picked up by the Siemens power transformer monitoring system installed on the 200 MVA transformer

Temperature measurement is one of the most important elements of a transformer monitoring system since the thermal condition inside has major impact on the ageing of the oil/paper insulation system. The 200 MVA transformer is water-cooled with a forced and directed oil flow (ODWF). Fig. 2 shows a principle diagram of transformer and heat exchanger which consists of 4 cooling units. Oil temperatures are measured at input (ϑ_1, ϑ_3) and output (ϑ_2, ϑ_4) of the heat exchanger. Furthermore, the input ($\vartheta_{W,in}$) and output temperature ($\vartheta_{W,out}$) of the cooling medium water as well as the temperature at the top of the tank are picked up. The oil temperatures measured at the input to the heat exchanger (ϑ_1, ϑ_3) are with good accuracy equivalent to the oil temperature at the top of the windings. The maximum temperature inside the transformer (hot spot temperature) determines the ageing behaviour of the oil/paper insulation system. According to IEC 354 (Loading Guide) [1] the hot spot temperature is calculated by

$$\vartheta_h = \vartheta_{Oil,top} + k \cdot \left(\frac{I}{I_N}\right)^y \cdot \Delta\vartheta_{wo}\Big|_{I=I_N} \quad (1)$$

Here, $\Delta\vartheta_{wo}$ denotes the winding-oil temperature rise, $\vartheta_{Oil,top}$ is the oil temperature at the top, I_N is the rated current, I is the load current, k is the hot spot factor ($k = 1.3$ for power transformers) and y denotes the winding exponent with $y = 2.0$ for OD-cooled transformers. According to a basic model for thermal ageing, well-known as the Montsinger equation [1], a thermal ageing rate can be defined and calculated from the hot spot temperature according to

$$V = 2^{\frac{\vartheta_h - 98^\circ C}{6^\circ C}} \quad (2)$$

The thermal ageing rate doubles with every 6 K temperature rise of the hot spot temperature above 98 °C. Equation (2) allows a first rough estimation of the lifetime loss. However, there is meaningful impact on the transformer condition by other stresses like short circuit forces and overvoltages in the power network.

The well-known Hydran sensor is used for measurement of fault gases dissolved in the oil. This provides in combination with moisture measurement information about the condition of the oil and insulation material. Both kinds of sensors make up a reliable early warning system. Furthermore, the dessicant can be checked by comparing humidity of the air in the conservator and the moisture content in the oil. The circuit state of pumps and fans is used for calculating the operating time of this equipment. Knowledge of the velocity of the oil through the cooling system in combination with load and temperature measurement allows determining of the mass and heat transfer to the heat exchanger. Measurement of oil level in the compensator and the tap changer tank is used for leakage detection - a first step towards a condition-based maintenance strategy.

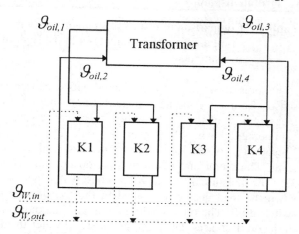

Fig. 2 Principle diagram of transformer and heat exchanger with the measured oil ($\vartheta_{oil,i}$) and water temperatures ($\vartheta_{W,i}$)

III. Data Evaluation

Fig. 3 shows measurements recorded on the 200 MVA generator transformer during one year from December 1997 to December 1998. The load is almost constant at about 70 % ($\approx$ 150 MVA) with only a few exceptions. Thus, the temperatures are also very constant (Fig. 3c). The maximum value of the hot spot temperature is about 80 °C which corresponds with an ageing rate of 0.125 according to (2). Thus, thermal ageing is negligible in this operation mode for this transformer.

The content of several fault gases dissolved in the oil is measured by the Hydran sensor. This sensor is sensitive not only to hydrogen but also to about 18 % to carbon monoxide (CO) and to about 8 % to acetylene (C_2H_2). The accuracy of the sensor is $\pm$ 10 % and $\pm$ 25 ppm for hydrogen. In order to obtain a better understanding of the sensor and the interpretation of measured results, oil samples taken from the transformer in the mid-January 1998 and at the end of April 1998 were analysed in the laboratory. The results are shown in Table 2. There is a change only of the CO

content from January to April 1998. The H₂ content
and the content of all other fault gases were constant
and very low during the entire time range. This means
the increase of the fault gas content measured by the
Hydran (Fig. 3d) is caused by the increasing CO content. However, such behaviour is not unusual for a new
transformer and does not indicate a transformer failure.
This example shows that a correct interpretation of
measured results requires additional information about
the transformer like load, operating conditions and age.

Gas/Date	January 1, '98	April 29, '98
H₂ [ppm]	5	7
CO [ppm]	216	386
CₓHᵧ [ppm]	4	9
Hydran display	71	125

Table 2 Comparison of results from laboratory gas-
in-oil analysis of oil samples and the Hydran
measurement

The moisture of the oil is measured with a capacitive
sensor from an American manufacturer and for the
measurement of the air humidity in the conservator a
commercially available sensor is used. There is no
"breathing" of the transformer when the load is constant. Thus, moisture of oil and air do not vary (Fig.
3e). The moisture of the air in the conservator increases
when the load goes down or when the transformer is
switched off. The oil is cooling down and the moisture
in the oil decreases. Diffusion processes of water inside
the transformer lead to a change of the equilibrium of
water in the oil/air system. When the temperature of the
oil decreases, the water diffuses from the oil into the
solid insulation material (pressboard). With warming
up of the transformer the water goes from the winding
insulation into the oil. However, a moisture content of a
few ppm measured with the above-mentioned sensor is
excellent for a large power transformer.

The measured oil flow (Fig. 3f) is constant and almost
independent of the oil temperature. Due to this behaviour it is evident that oil velocity measurement does not
provide much information concerning condition assessment. The velocity of the oil flow is measured in a
pipeline with a diameter of D=200 mm. Under the sim-
plifying assumption of a constant oil velocity in the
cross-section of the pipeline, the oil flow rate can be
calculated using the measurement according to Fig. 3f
by

$$V_{oil} = v_{oil}\frac{\pi}{4}D^2 \approx 1.8\frac{m}{s}\cdot\frac{\pi}{4}(0.2m)^2 = 204\frac{m^3}{h} \quad . \quad (3)$$

The oil flow rate through each of the 4 cooling units is
about 116 m³/h according to information from the heat
exchanger manufacturer. The oil flow through K1/K2
and K3/K4 is twice this value (232 m³/h) which is in
good accordance with the result of equation (3).

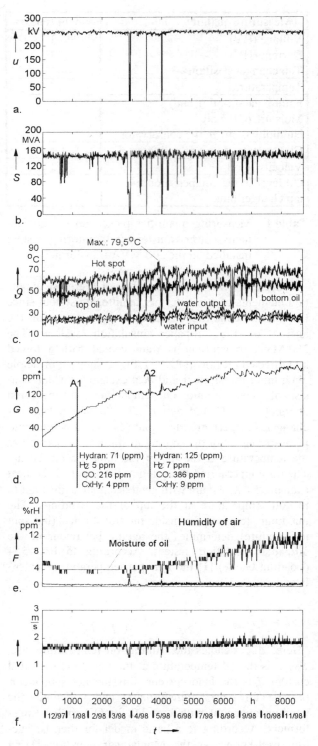

Fig. 3 Measurements obtained from a 200 MVA generator transformer
a. Voltages at all three HV phases
b. Load
c. Temperatures
d. Gas-in-oil, trend (*: Hydran measurement)
e. Moisture content of the oil (**: Harley measurement, in ppm) and humidity of the air in the conservator (in %rH)
f. Velocity of the oil in the pipeline to the heat exchanger

The stress on the tap changer depends besides other factors mainly on the number of switching operations and the current through the tap changer contacts at the moment of switching. By investigating those data the relationship between lifetime of and stresses on a tap changer can be determined. As an example, Fig. 4 shows the number of switching operations into all tap changer positions of the 200 MVA transformer during one year.

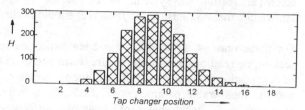

Fig. 4 Frequency distribution H of tap changer positions obtained during one year in the example of the 200 MVA generator transformer

Delayed purchasing of power network equipment is one method of cost reduction. Some experts expect a higher loadability of existing power network equipment in future due to such a new purchasing strategy. This would lead to a higher failure risk without any kind of permanent condition monitoring. For power transformers one possibility for reducing this risk is using an early-warning system consisting of an online gas-in-oil monitor like Hydran. The other aspect is avoiding thermal overload. Due to time constants of the oil in the range of some 100 minutes depending on the size of transformer, methods for a predictive estimation of the top oil temperature are required when a certain load is assumed. Therefore, a thermal model is required. The following well-known equation describes the relationship between load and steady-state top oil temperature:

$$\vartheta_{O,E} = \vartheta_{O,N} \left(\frac{1 + \left(\frac{P_k}{P_0} \right) \cdot \left(\frac{S}{S_N} \right)^2}{1 + \left(\frac{P_k}{P_0} \right)} \right)^x . \quad (4)$$

Thereby, P_0 is the no-load loss and P_k denotes the load loss. $\vartheta_{O,N}$ and $\vartheta_{O,E}$ are the steady-state top oil temperatures at rated load S_N and actual load S respectively. For modeling the dynamic behaviour usually a first-order differential equation with the oil time constant T_O is used:

$$T_O \cdot \frac{d\vartheta_O}{dt} + \vartheta_O = \vartheta_{O,E} . \quad (5)$$

For computer calculation the differential equation has to be transformed into a difference equation under consideration of the sampling time T_S (1 min in our case):

$$\Delta\vartheta(k) = \frac{T_S}{T_O} \left(\vartheta_{O,E}(k) - \vartheta_O(k) \right)$$
$$\vartheta_O(k+1) = \Delta\vartheta(k) + \vartheta_O(k) \quad (6)$$

The calculation starts with an initial value $\vartheta_O(k=1)$ for the top oil temperature. Then the difference $\Delta\vartheta$ and

the next value ϑ_O are calculated in a loop. The unknown parameters $\vartheta_{O,N}$, x, T_O and $\vartheta_O(k=1)$ have to be determined from onsite measurements of load and top oil temperature. Fig. 5 shows the verification of the thermal model. The difference between calculation and measurement shows that the model can be used for predicting the top oil temperature on the basis of a load assumption with sufficient accuracy. However, there is an influence of the cooling medium (water) temperature which has not been modeled up to now.

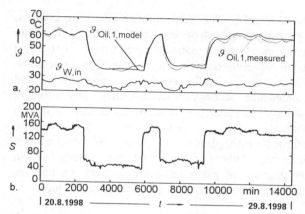

Fig. 5 Verification of the thermal model for the 200 MVA generator transformer
a. Top oil and water temperature
b. Load

IV. CONCLUSIONS

On-line monitoring of power transformers helps to obtain more detailed information about a transformer's condition. The probability of an unexpected outage can be minimized. Furthermore, monitoring systems open up the possibility for extending the operating time of power transformers, reduce the risk of expensive failures and provide potential for changing the maintenance strategy. Future developments of monitoring equipment have to focus on three points: Onsite measurement evaluation procedures in order to realize condition assessment, sensor technology and standardizing hard- and software interfaces of high flexibility

V. REFERENCES

[1] Loading Guide for oil-immersed transformers (International Electrotechnical Commission, IEC, IEC 354, 1972)

[2] Leibfried, T; Knorr, W.; Viereck, K.; Dohnal, D.; Kosmata, A.; Sundermann, U.; Breitenbauch, B.: On-line Monitoring of Power transformers - trends, new developments and first experiences. CIGRE 1998, paper 12-211

Address of the Author:

T. Leibfried, Transformatorenwerk Nürnberg, Siemens AG, Katzwanger Str. 150, 90461 Nürnberg, Germany

DETECTION AND CHARACTERISATION OF PARTIAL DISCHARGE ACTIVITY ON OUTDOOR HIGH VOLTAGE INSULATING STRUCTURES BY RF ANTENNA MEASUREMENT TECHNIQUES

B G Stewart, D M Hepburn, I J Kemp, A Nesbitt and J Watson[#]

Glasgow Caledonian University, UK and [#]First Engineering Limited, UK

ABSTRACT

Flashover of outdoor high voltage insulating systems, e.g. string or post-type insulators, bushings, etc., can result in electrical outages with severe social and economic consequences. It is therefore imperative, within an overall condition monitoring and maintenance strategy, that effective and efficient means are developed to monitor the condition of such systems for degradation to permit refurbishment/replacement prior to catastrophic flashover. This paper reports the results of an investigation of partial discharge activity on outdoor insulation systems utilising RF antenna techniques.

INTRODUCTION

The measurement of partial discharge activity has proved an effective method for assessing the integrity of both the surface and bulk properties of solid insulating systems of many high voltage plant systems. The same is true of the outdoor structures mentioned above. Degradation of the electrical properties of such structures is generally associated with the build-up of pollution, be it industrial dust, salinity, etc. Associated changes in electrical stress profile result in increased leakage currents and, if left unchecked, potentially, flashover. Partial discharge activity will be prevalent at the surface of these insulating structures as degradation proceeds and it is reasonable to presume that, as with other forms of degradation, the measurable parameters of partial discharge activity will provide data from which an assessment can be made of the nature, form and extent of degradation present.

Unfortunately, unlike most solid insulating systems associated with high voltage plant, these outdoor structures present specific problems to the measurement of partial discharge activity. In general, these structures are relatively inaccessible. To make connections to such structures to allow measurements of the electrical pulses associated with partial discharge activity (the most common method in use), would take an unacceptable length of time. In addition, unlike other items of high voltage plant where there may be only a few of critical importance, there are potentially hundreds of these structures to be monitored in a given electrical section or system, to be secure in the knowledge that no catastrophic flashover should occur.

For these reasons, little effort has been expended in seeking correlations among the nature, form and extent of degradation prevalent on polluted outdoor structures and the measurable parameters of discharge activity. However, a technique of partial discharge measurement and characterisation which did not include the time-consuming process of making direct connections and which could be carried out quickly towards an assessment of structural integrity would be of great advantage.

This paper reports the results of an initial investigation into the measurement and potential characterisation of partial discharge activity through the use of antennae to measure their associated electromagnetic spectra. The approach is based on the premissive that if electromagnetic spectra of partial discharges could provide sufficient data to make characterisation of the activity possible, then this could be correlated with prevalent degradation and provide a possible means for monitoring the condition of outdoor insulating structures.

It has long been realised that Radio Noise (RN) and Radio Frequency Interference (RFI) due to partial discharge activity on high voltage lines occurs over a wide range of frequencies from below 100 kHz up to 1 GHz [1]. However, most references in the literature to RN measurements have reported on discharge activity occurring on transmission line rods operating in the 100-1200 kV regions. For such measurements the frequency activity appears to occur mainly below 25 MHz, with typically a significant RN power in the lower MHz regions [2-5]. There therefore appears to be little past or recent investigation into possible correlations between RFI and the nature or identifications of specific classes of partial discharge. The purpose of this paper is to report on some initial and preliminary measurements of higher frequency RN from different types of insulators undergoing discharge activity at the lower, 30 kV, region. A simple goal is to identify possible variations and characteristics of discharge phenomena which may provide mechanisms for future exploration of RN techniques.

High Voltage Engineering Symposium, 22–27 August 1999
Conference Publication No. 467, © IEE, 1999

The following section outlines briefly the system utilised for RN measurement. The next section presents some measured results for aged ceramic and polymer insulators undergoing discharge activity. The last section is a brief conclusions section which also outlines some proposals for future investigative work.

THE ANTENNA MEASUREMENT SYSTEM

The system comprised a simple balanced multi-element UHF antenna mounted securely at a safe distance from the discharge samples within the high voltage test cage. The antenna output signal was fed out of the cage through coax and connected to a Hewlett Packard 4195A network/spectrum analyser for frequency analysis and data storage.

Though the earthed high voltage test cage provides significant screening of external radio interference, it is still possible to measure the frequency spectrum of external RF signals which gain access to the cage. With the spectrum analyser filter bandwidth set to 30 kHz the measured background RN spectra for the range 0 – 60 MHz is shown in Figure 1, the frequency scale is linear. It is possible to see all external active communication channels in operation. These are characterised by the narrow frequency spikes which reflect the carrier frequencies of broadcasts and/or narrow band modulation information. It should be noted that the bandwidth of the communications channels present are much less than the spectrum analyser filter bandwidth, thus no variations of spectral information should be apparent during communication transmissions.

The background spectrum information in Figure 1 is critical for the assessment of RN measurements as it provides the reference spectrum which makes it possible to determine any RN activity generated by the samples under discharge (see later).

INITIAL MEASUREMENTS

Two types of aged and polluted insulator samples which had been taken out of service were chosen for partial discharge measurement:
(i) an 11 (6 large, 5 small) shed polymer insulator of 36 cm length, 17 cm and 11 cm shed diameters, and 5 cm core diameter; (ii) an 8 shed ceramic insulator of 40 cm length, 16.5 cm shed diameter, and 7 cm core diameter. Diagrams for these insulators are depicted in Figures 2 and 3.

Operating with dry insulators, the voltage to the HV terminal was incremented to 30 kV ac at which there

was no indication of discharge activity for either sample.

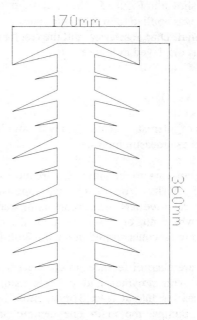

Figure 2 : Sketch of polymer insulator

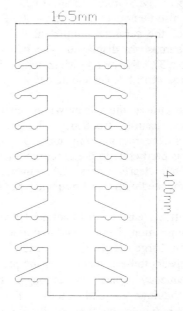

Figure 3 : Sketch of ceramic insulator

The samples were then soaked in a sea water, to simulate coastal pollution, and the experiment repeated. On application of 30 kV, audible noise was present in addition to visible scintillation on the surface of the samples. The frequency spectrum over the 0 – 60 MHz range associated with the resulting discharge activity for the polymer sample is detailed in Figure 4. To extract the frequency spectrum information associated purely with the discharge activity, a helpful and powerful technique is to subtract the background spectrum of Figure 1 from the measured spectrum of Figure 4. The resulting spectrum for the polymer sample is shown in Figure 5. The same subtraction technique was applied to the ceramic sample undergoing discharge activity, and the resultant spectrum is displayed in Figure 6.

OBSERVATIONS

A number of interesting features may be highlighted in regard to these measurements.

(a) There appears to be little RN frequency activity below 25 MHz. This may be contrasted with previously reported RN measurements of transmission line rods where all, or if not most, of the activity is considered to be concentrated in the 0 – 25 MHz range.

(b) There are clear differences in the measured spectra associated with polymer and ceramic samples. The polymer sample has at least 3 discernible broadband frequency humps, the main one centred around 30 MHz, with 2 smaller bulges at about 21 MHz and 36 MHz respectively. Other small higher frequency humps may also be noticed in the 55 - 60 MHz region. In respect of the ceramic sample, again discernable broadband humps are displayed. The two main ones occur at about 33 MHz and 48 MHz, with smaller ones perceived around 16.8 MHz and 36 MHz.

(c) The spectra exhibit bandwidth characteristics around the centre frequency bulges, ranging somewhere in the region from about 3 – 6 MHz in width. This is partly due to the spectrum analyser filter bandwidth, but clearly infers that a measurable RN power is associated with these frequency ranges.

(d) It is relatively easy to differentiate between active radio communication channels and that associated with apparent discharge activity. A radio channel's signature appears to be evidenced by the presence of a narrow frequency spike in amongst broadband discharge behaviour. A good example of this may be seen on the polymer spectrum of Figure 5 where a strong radio communication previously not existing in the background RN measurements has been picked up by the antenna during measurement.

CONCLUSIONS

A positive result from this work is that measured partial discharge activity associated with a polymer and a ceramic insulator under test appear to exhibit RN frequency spectra characteristics in the 0 - 60 MHz region. Encouragingly, distinctive and different broadband frequency components may be observed

across the spectra of both samples suggesting that a possible relationship may exist between insulator discharge phenomena and specific RN discharge activity. A key outcome would also appear to relate to RFI. Depending on the relative strengths of signals, the frequency content in these spectra would clearly cause interference with many HF/VHF mobile and fixed communications channels allocated to these frequencies.

Whilst these preliminary results are extremely encouraging, a little note of caution is warranted. There may as yet be an unappreciated influence of the high voltage system itself generating forms of damped oscillations which may be reflected in the antenna measurements. However from knowledge of circuit characteristics, this may be expected to exhibit a lower frequency phenomenon in comparison to the frequencies measured in the experiments above.

Future work will involve:
(i) extending the frequency range for discharge measurements; (ii) investigating possible influences on the RN measurements associated with the high voltage system itself; (iii) attempting to correlate any relationship between RN frequency components associated with discharge activity and pollution deterioration of samples; (iv) making comparisons with other current condition monitoring techniques currently in operation.

ACKNOWLEDGEMENTS

We would like to thank Mr. A. Wilson and the staff at First Engineering Limited, Glasgow for the use of their high voltage facilities to conduct the experiments presented in this paper.

REFERENCES

[1] Pakala W E, Chartier V L, "Radio Noise Measurements on Overhead Power Lines from 2.4 to 800 kV",
IEEE Trans. Power Apparatus and Systems, PAS-90, 1155-1165

[2] Halasa G H, "Radio Noise Meter for the Detection of Incipient Faults on High Voltage Transmission Lines",
4th European Conf. On Electronics, Stuttgart, Germany, 1980, 484-488

[3] Azernikova T I, Perelman L S, Rokhinson P Z, Timashova L V, Yemel'yanov N P, "Corona-Generated Radio Interference on EHV and UHV Lines",
Int. Conf. on Large High Voltage Electric Systems, 2, 36.03.1-36.03.8

[4] Sarmadi M and Tudor J R, "The use of corona radio noise in detecting and locating incipient faults in high-voltage power systems", Proc. American Power Conf., 51, 598-603

[5] Sarmadi M, "Slow changing parameters influencing radio noise level from high-voltage transmission lines", Proc. American Power Conf., 59-1, 460-465

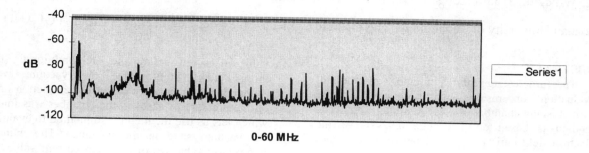

Figure 1 : Signal received by antenna - no volts applied

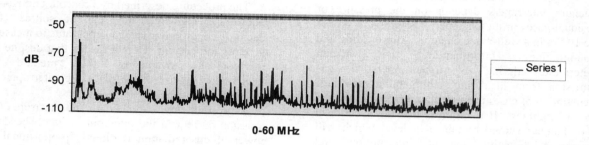

Figure 4 : Signal received by antenna - 30kV ac applied, wet polymer

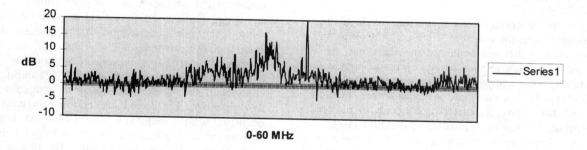

Figure 5 : Signal received by antenna - 30kV ac applied, wet polymer, subtraction of Fig. 1 from Fig.4

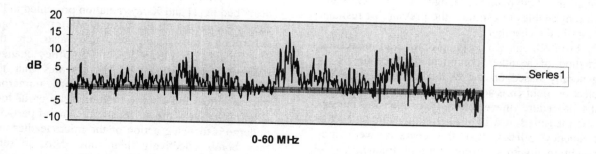

Figure 6 : Signal received by antenna - 30kV ac applied, wet ceramic, subtraction

EVALUATION OF INTERNAL PRESSURE OF VACUUM INTERRUPTERS BASED ON DYNAMICS CHANGES OF ELECTRON FIELD EMISSION CURRENT AND X-RADIATION

K. Walczak, J. Janiszewski, H. Mościcka-Grzesiak

Poznan University of Technology, The Institute of Electrical Power Engineering, Poznan, POLAND

ABSTRACT

The article presents a method of indirect evaluation of residual gas pressure in the extinguishing chamber of a vacuum interrupter. The method is based on a change analysis of the electron field emission current between electrodes and a dose power of X-radiation.

1. INTRODUCTION

Proper work of the extinguishing chamber of the vacuum interrupers depends on the pressure of residual gases in the chamber. The pressure should be 10^{-6}Pa in a well-made chamber, after heating and conditioning. Unfortunately, the vacuum deteriorates during interrupter operation, gas emission from insulating and metallic materials, penetration of gases through chamber walls, and the loss of tightness. It is required that the vacuum in the chamber should remain at a level which will guarantee effective current interruption and preserving insulation properties of the intercontact gap for at least 20 years. This means that the residual gas pressure should not exceed the value of 10^{-2}Pa [1].

After all conditioning operations made by the manufacturer, the chamber is cut off the pumps and the vacuum measurement system. Therefore, there is no possibility to measure gas pressure inside it, and due to operational reasons, it is recommended to check the vacuum state in the chamber. In such a situation, there are different concepts of indirect methods of pressure evaluation of residual gases.

There is a certain group of methods of indirect vacuum evaluation in an extinguishing chamber using measurements of electron field emission current between open electrodes. The analysis of dynamics changes of the emission current enables to estimate the pressure of residual gases in the chamber.

Frontzek and König [1] proposed an interesting method of a rather complicated procedure, it is generally based on the evaluation of changes of electron field emission current. In the first stage of the procedure, the electron field emission current was measured at a high alternating voltage of the frequency 50Hz. Next the contacts were arc-conditioned with a current of a high frequency (1 – 50kHz). This was followed by an immediate return to the voltage of 50Hz of the value fixed before and it was found that the electron field emission current increased manifold. This current decreased to the

value measured before the arc-conditioning the more quickly the worse the vacuum was. A functional dependence of current extinction time on the pressure of gases in the chamber was found which lets us use the mentioned method to evaluate the vacuum state in the chamber. The authors interpret the changes of emission current values as an effect of gas desorption from the contact surface during arc conditioning and next, readsorption and related change of electron work function.

The procedure described by Frontzek and König requires a high-frequency current source (for electrode conditioning) and an apparatus to measure the emission current. Thus it results that despite its metrological advantages, this method is inconvenient to use at the place the interrupter is installed.

X-ray radiation is related to the electron field emission current in the intercontact gap. The dose power of this radiation is directly proportional to the electron field emission current and to the square voltage between electrodes. Thus the measurement of the emission current can be substituted by the measurement of the X-ray radiation dose, whose advantage is that the investigation can be made without the cable connection of the measurement setup with the chamber.

Fink [2] proposes vacuum evaluation in the extinguishing chamber on the basis of an amplitude analysis of the X-ray radiation dose emitted from the chamber. He found its certain fluctuations occurring with a frequency from 3 to 13Hz, which corresponded to pressure changes in the chambers from 10^{-4} to 10^{-2}Pa, respectively. He relates the described fluctuations to gas desorption and adsorption on electrode surfaces.

The effect of the emission current changes described in [1] and X-ray radiation presented in [2] can be explained on the basis of the effect of gas desorption and adsorption, and also on the mechanism of electrode microprofile changes proposed by Slivkov [3]. He assumes that the electrode surface is rough and has numerous microneedles, and electric field intensity on its tops is amplified very much. Ions and charged particles undergo a focusing action of the microneedles and they bomb selectively their tops. This, in turn, causes blunting the microneedles, decrease of electric field intensity on its tops and, as a result, decrease of the electron field emission current. Therefore, the more charge carriers in the interelectrode space, the faster extinction of the electron field emission current. Certainly, depending on the method of electrode conditioning, the effect of gas sorption or the effect of electrode microprofile changes can be predominant, or they both occur together.

2. AIM AND RANGE OF THE PAPER

The aim of the paper is to evaluate residual gas pressure inside an interrupting extinguishing chamber using indirect methods based on an analysis of the electron field emission current and X-ray radiation.

It was assumed that the method of vacuum state evaluation in the chamber must be simplified in order to be able to apply it at the place where the interrupter is installed. It was proposed to apply the simplest way (relatively) of electrode conditioning and to substitute the analysis of the electron field emission current by an analysis of X-ray radiation.

In this paper, the following research range is planned:

- taking measurements in extinguishing chambers, commercially manufactured,
- electrode conditioning with a possibly simplest method, i.e., successive flashovers at high alternating voltage 50Hz,
- measurement of the electron field emission current for each electrode (in the positive and negative voltage half-cycle) using a bridge circuit,
- measurement of a selected quantity characterizing the X-ray radiation taken simultaneously with the measurement of the field emission current,
- tracing changes of the field emission current and a dose of X-ray radiation in a determined time range (of the order of ten minutes),
- determining the vacuum in chambers with the magnetron method for verification.

3. EXPERIMENT

3.1. RESEARCH OBJECTS AND MEASURE-MENT CIRCUITS

The research objects were extinguishing chambers of vacuum interrupter, commercially manufactured of the rated voltage 12kV and a rated electrode gap equal to 10mm. All the chambers underwent series of different laboratory tests and they were stored for a few years, therefore it was expected that their vacuum deteriorated in comparison to the vacuum of new chambers.

The measurement of the electron field emission current was done in a high-voltage unbalanced bridge [4] enabling to separate the electron field emission current from the capacitive current between the electrodes. Each of the electrodes emits the electron current only in one voltage semiwave. The oscilloscope applied in the bridge circuit enabled a measurement of the field emission

current emitted by both the electrodes, which is essential because as a rule values of these currents are different.

The measurement dose of the X-ray radiation emitted from the chamber was done using a dosimetric probe and a universal transistor radiometer [5]. The setup is presented in Fig. 1.

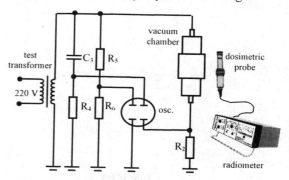

Fig. 1. Bridge circuit to measure the electron emission current and equipment to measure the X-radiation dose

In order to verify the results from the mentioned above measurements, the magnetron method was applied, which enables to determine residual gas pressure in chambers without interference into their internal structure [6].

3.2. RESEARCH METHODS

As a result of numerous reconnaissance measurements, such research parameters were determined so that all chambers, even differing substantially, could be tested in the same experimental conditions.

The change analysis of the emission current always requires properly conditioned electrodes. Electrodes are conditioned in order to eliminate microdischarges in a limited time range. Microdischarges appear, especially at larger electrode gaps (of the order of 1cm), at a considerably lower voltage than the emission current and they lead to a flashover very often, before the emission current occurs. Elimination of microdischarges using a selected technique of electrode conditioning is a necessary condition to measure the emission current.

Electrode conditioning was performed using the method of successive flashovers. The electrode distance was 10mm. The flashovers occurred at a voltage in the range from about 10 to 90kV, which depended on the chamber quality, particularly on electrode smoothness and purity. Usually the effect of full conditioning was obtained after about twenty flashovers.

Next the electrode distance was reduced to 2mm and the voltage was increased to 60kV. The emission current of the upper electrode (I_{e1}), lower electrode (I_{e2}), and a dose (X) of X-ray radiation were measured. The measurements were taken for about 20 minutes.

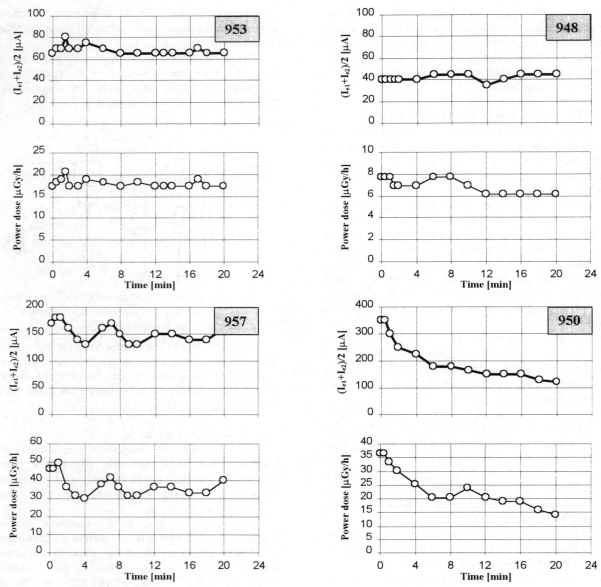

Fig. 2. Changes of the electron field emission current and power dose of X-ray radiation as a function of time

4. RESEARCH RESULTS AND THEIR INTERPRETATION

Fig. 2 shows some selected characteristics of the emission current and X-ray radiation dose as a function of time. The current is an arithmetic mean of field emission currents for each of the electrodes.

It was found that there is a trend to decrease the value of the electron field emission current as the measurement lasts on the basis of an analysis of the collected experimental data, carried out according to the presented before theoretical assumptions. However, the dynamics of the occuring phenomena is varied for each of the chambers. Chamber 953 shows relatively slowest changes of the current and X radiation. This lets us suppose that it is characterized by the lowest value of residual gas pressure. Assuming increase of dynamics changes

of the field emission current and X radiation with pressure increase, chamber 950 is the one of the worst vacuum level, according to the hypothesis presented before. At this approach, chambers 957 and 948 can be assigned mid features. In their case, the curves of the current and X-ray radiation are additionally characterized by clearly exposed fluctuations.

The validity of the presented above suggestions was verified by determining the pressure in the chambers using the magnetron method [6]. The research was done in a measurement setup, which enabled modeling and measuring the vacuum state in a pattern extinguishing chamber, situated in axial magnetic field of an adjustable value of magnetic field induction. The graduation characteristic of the pattern chamber was done at voltage of 5kV and magnetic field induction of 10mT. Results of the vacuum measurements in the interrupting chambers were plotted onto the graduation characteristic presented in Fig. 3.

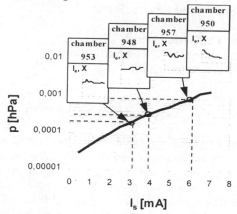

Fig. 3. Graduation characteristic of the magnetron circuit with plotted results of pressure measurements in the investigated extinguishing chambers

Generally, we can conclude that all the chambers were characterized by a low vacuum. Chamber 953 was distinguished by the best vacuum of $2 \cdot 10^{-2}$Pa. This is confirmed by conclusions resulting from studies of stability of the electron emission current and X-ray radiation intensity.

Chamber 948 shows a greater change dynamics of these quantities in time and their certain recurrence. Chamber 957 behaved in a similar way, but the frequency of the recurrent changes was greater. The pressure in those chambers was of $3 \cdot 10^{-2}$ and $9 \cdot 10^{-2}$Pa respectively. The conclusion concerning the cyclical changes of the emission current whose frequency is greater when the vacuum in the chamber is worse, is in agreement with Fink's conclusions [2], but in his research, the fluctuation frequency was considerably greater. Probably, there is an effect of a large change frequency superposing onto slow current fluctuations with the time of a single period of the order of 1 minute. In these studies, the slow fluctuations were measured. The explanation of this effect can be based on the mechanism of microneedle blunting on the contact surface. The effect of the focusing action of the microneedles and their blunting by bombing particles "migrates" on the electrode surface from one microneedle to another, which is accompanied by fluctuations of the emission current and X-ray radiation.

Whereas in the case of chamber 950, it turned that determining the vacuum was not possible in such conditions as for the remaining chambers, i.e., when supplied by direct voltage of 5kV, which confirms its preliminary unsatisfactory grade. The vacuum in this chamber was the worst ($>10^{-1}$Pa), and the characteristic of the emission current as a function of time shows the greatest steepness.

5. CONCLUSIONS

As the research shows, the change dynamics of the electron field emission current can be an evaluation criterion of the vacuum in extinguishing chambers. The more stable the course of this current in time is, the better vacuum is inside the chamber. The recurrence of its changes is the first signal of internal pressure increase, and emission current decrease in a large range of values means that the process of chamber degradation is advanced.

The frequency of cyclical changes of the electron field emission current and intensity of X-radiation increase with the deteriorating vacuum.

In order to evaluate the pressure of residual gases in the chambers, it is possible to use the measurement of the dose power of X-ray radiation, which is related by functional dependence with the electron emission current. The measurement itself can be done remotely, without electric connection with high-voltage circuits.

It is expected that the found regularities will be conceived quantitatively in the nearest future on the basis of extended laboratory research.

REFERENCES

[1] Frontzek F. R., König D., „Measurement of Emission Currents Immediately After Arc Polishing of Contacts", IEEE Transactions on Electrical Insulation, August 1993, pp. 700-705

[2] Fink H., „Measures of ionizing radiation for diagnostics of residual gas pressure in Vacuum interrupters", Elektryka № 38, Poznan University of Technology, 1989, pp. 241-245

[3] Сливков И. Н., „Электроизоляция и разряд в вакууме", Атомиздат, Москва 1972, p. 50

[4] Mościcka-Grzesiak H., „O zagadnieniu elektrycznej wytrzymałości powierzchniowej materiałów i układów wysokonapięciowych w próżni", Poznan University of Technology, 1973, p. 12

[5] Górczewski W., „Badanie promieniowanie rentgenowskiego emitowanego z próżniowej komory gaszeniowej", Elektryka № 33, Poznan University of Technology, 1986, pp. 24-35

[6] Kutzner J., Janiszewski J., „Pomiar ciśnienia gazu w próżniowych komorach gasze-niowych", Elektryka № 35, Poznan University of Technology, 1987, pp. 6-31

ACKNOWLEDGMENT

The investigations described in the paper were sponsored by Polish Committee of Science Researches, grant № 101/T10/97/13

EXPERIENCES FROM ON-SITE PD MEASUREMENTS USING OSCILLATING WAVE TEST SYSTEM

Frank J. Wester, Edward Gulski *, Johan J. Smit * and Paul N. Seitz**

Energie Noord West N.V., Amsterdam, The Netherlands
*Delft University of Technology, High Voltage Laboratory, Delft, The Netherlands
**Seitz Instruments A.G., Niederrohrdorf, Switzerland

Abstract

In this paper experiences of partial discharge measurements on medium voltage power cables in the field are discussed. Several examples are used to indicate different aspects of diagnosing service aged power cables. It is shown that the PD activity in power cables usually originates from cable accessories, but sometimes from the cable insulation too. Furthermore, it is shown that measured PD activity may not always originates from a clear location in the cable sections.

1. Introduction

In recent years utilities have growing interest in analysing partial discharges (PD) in existing medium voltage cable networks. The main goal of such analysis is the detection, location and recognition of possible insulation failures in cable and cable accessories at an early stage. As a result maintenance actions can be precisely planned to prevent unexpected discontinuities during operation of the cable network.

Detection and location of PD at power frequencies and voltage levels similar to normal operating conditions is desirable. One of the methods for testing cable system on PD activity is the Oscillating Wave Test System (OWTS) [1,2]. In this paper, field experiences with OWTS are described.

2. Oscillating Wave Test System (OWTS)

With this method, the cable sample is charged with DC power supply over a period of just a few seconds to the usual service voltage (figures 1 & 2). At this moment, a specially designed solid state switch connects an air-cored inductor to the cable sample in a closure time of less than 1 μs. Now series of voltage cycles starts oscillating with the resonant frequency of the circuit:

$$f = 1\big/(2\pi\sqrt{LC}),$$

where L represents the fixed inductance of the air-core and C represents the capacitance of the cable sample.

The air-core inductor has a low loss factor and design, so that the resonant frequency lies close to the range of the power frequency of the service voltage (within the range from 50 to 1000 Hz).

Due to the facts that the insulation of the power cables usually has a relative low dissipation factor, the Q of the resonant circuit remains high depending upon cable (30 to more than 100) [1]. As a result, a slowly decaying oscillating waveform (decay time 0.3 to 1 second) of the test voltage is applied to energise the cable sample. During tens of power frequency cycles the PD signals are initiated in a way similar to 50 (60) Hz inception conditions. All of these PD pulses are measured using a fast digitiser. Further description of the Oscillating Wave Test System is given in [1,2].

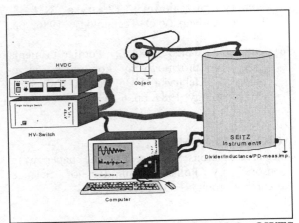

Figure 1: Schematic representation of the OWTS measuring method.

Figure 2: OWTS on-site testing of a 2700 m 10 kV power cable section: a) 50 kV HVDC power supply, solid state switch, processing unit; b) air/core inductance, HV and PD divider (in circle PD calibrator).

High Voltage Engineering Symposium, 22–27 August 1999
Conference Publication No. 467, © IEE, 1999

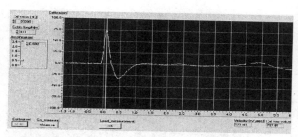

Figure 3: Example of PD calibration of a cable sample. Three parameters are determined: the [pC] reading of the PD detector, the propagation velocity of PD pulses and the pulse attenuation.

3. Field experiences

Using OWTS, the past 12 months systematic field measurements have been performed. Based on results obtained, several aspects of interpreting measuring results have become relevant. On the basis of three examples, the aspects will be discussed here:

① 3-core 10 kV paper/oil cable section with a length of 840m;
② 3-core 10 kV paper/oil cable section with a length of 775m;
③ 3x1-core 50 kV mass cable section with a length of 3235 m.

In all cases the cables were calibrated using an IEC 60270 PD calibrator (figure 2 & 3). As a result three PD measuring circuit parameters are estimated:

❑ the [pC] reading of the PD detector for the particular cable circuit, ① 2000 ② 2000 ③ 2000;
❑ the velocity in [m/µs] of PD pulses in a particular cable circuit, ① 155 ②159 ③ 138;
❑ the pulse attenuation in [mV/m], ① 1.2 ② 1.2 ③ 2.9.

The PD measurements with OWTS are performed one by one phase at several voltage levels: 4, 6, 8, 10, 12, 14 and 16 kV_{top}. In this way, the inception voltage of the PD signals can be determined. During testing of a particular phase, the other phases were grounded.

3.1 Example ①

Measurements are performed on a 3-core paper/oil cable section with power voltage $U_0 = 6kV_{rms}$. The section was fitted in 1982 with a total length of 840 m (240 mm² Al), in which six cable joints and two cable terminations are located.

Table 1: Measured PD at various voltages obtained from 10 kV power cable with OWTS.

Utop	phase U	phase V	phase W
6 kV	1400 pC	1400 pC	1600 pC
9 kV	2900 pC	3000 pC	2900 pC
16 kV	3000 pC	3100 pC	3400 pC

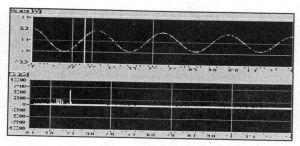

Figure 4: PD pattern measured after a 6 kV oscillating voltage charging a 840 m long 10 kV paper/oil power cable section.

Table 1 shows the measured PD amplitudes at three different measuring voltages. The table shows that PD pulses are measured at test voltages lower than the operating voltage. The first PD signals, as shown in figure 3, are measured at an oscillating voltage of 6 kV_{top}, which is approximately 0.7 U_0.

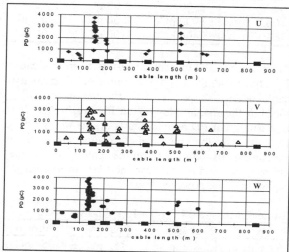

Figure 5: PD location diagram obtained from all three phases of a 840 m long 10 kV paper/oil cable section, using OWTS with voltages up to 2U_0. The cable accessories are reflected by ■.

Figure 5 shows the evaluation of the PD measurements for each of the single phases of the cable section. The evaluation was obtained from several PD measurements at oscillating voltages up to 16 kV (~ 2 U_0). All three phases clearly show a concentration of PD activity at 150 metres, with amplitudes up to 4000 pC. This PD location corresponds to the location of one of the cable accessories (the black squares in the graph). The cable accessories at 200m and 500 m, also gives some PD activity at all three phases. Furthermore, the cable joints at 265 and 370 m only give concentrations of PD in phase U, with amplitudes up to 3000 pC.

The evaluation of the measurements of this cable section show, that the PD > 1 nC occur at the negative half of the voltage period. The PD locations in each phase correspond the locations of the cable accessories in the section, according to the network documentation. To determine the seriousness of these discharges for this particular type of cable accessories, more field experiences and laboratory investigations necessary on service aged cable objects.

3.2 Example ②

This example handles the measurements on a second 10 kV 3-core paper/oil cable section. This cable section of 775 m length was fitted in 1963 (95 mm2 Cu), with a renewed cut of 200 m fitted in 1972 (150 mm^2 Al). In the cable section, five cable joints and two terminations are located.

Table 2: Measured partial discharges at various voltages obtained from 10 kV power cable.

Utop	phase U	phase V	phase W
6 kV	2500 pC	2500 pC	2500 pC
10 kV	2900 pC	3000 pC	2900 pC
16 kV	3400 pC	3100 pC	3500 pC

Table 2 gives the measuring results from this cable. Like the cable discussed in example 1, this particular cable section shows PD activity already at 6 kV$_{top}$ too, but the magnitudes of the PD are higher. It is not usual, that PD activity starts at this low voltage. In table 3 the measuring results from other measurements, performed on the same type of cable, are reflected and in that case the PD start at voltages higher than the power voltage.

Table 3: Measured PD on a 10 kV power cable.

Utop	phase U	phase V	phase W
6 kV	<50 pC	<50 pC	<50 pC
10 kV	< 50 pC	<50 pC	<50 pC
12 kV	2500 pC	2200 pC	2500 pC

In figure 6, the measured PD signals at different test voltages are reflected for phase U of the cable section. In this figure it is shown, that the PD pattern changes when the test voltage is increased. For this particular cable section, the amplitude of the pulses changes slightly. On the other hand it is clear, that the number of pulses increase as the voltage increases. At higher voltages, PD pulses also occur at the second and third period of the oscillating wave. The same is true for the other phases.

Figure7 shows the evaluation of the PD location in this cable section. It follows from the figure, that for all three phases there is a concentration of PD activity at 175 m. According to the utility network documentation, no cable joint is present at this location. Moreover, the comparison of PD magnitudes in the phases provides the conclusion that from all six possible discharge sites (3 x phase to ground and 3 x phase to phase) , all phases have the same PD levels and patterns.

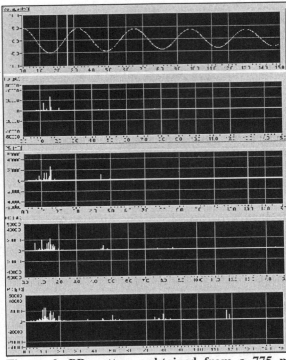

Figure 6: PD patterns obtained from a 775 m long 10 kV paper/oil cable section charged with several test voltages. From top to bottem:
- Oscillating voltage with frequency of 322 Hz;
- PD pattern of 2500 pC$_{top}$ measured at 6 kV$_{top}$;
- PD pattern of 2900 pC$_{top}$ measured at 10 kV$_{top}$;
- PD pattern of 3100 pC$_{top}$ measured at 12 kV$_{top}$;
- PD pattern of 3400 pC$_{top}$ measured at 16 kV$_{top}$.

As a result mechanical damages at this place cause local degradation of the cable insulation.

3.3 Example ③

This example handles the measurements performed on a 3x1-core 120 mm^2 50 kV mass power cable, with a length of 3235 m. The measuring result from OWTS of this cable section are reflected in table 4. The cable was measured with OWTS up to 22 kV$_{top}$. The table shows that a slight growth of the PD pulses as the voltage increases.

Simultaneously, PD measurements are performed on this cable section using a 55kV/50 Hz series resonant on-site test system and conventional PD detection. The PD inception voltage as well as PD magnitude obtained during 50 Hz AC charging of these cables are the same as obtained with OWTS.

Table 4: Measuring results from a 50 kV power cable from 1953.

U$_{top}$	phase U	phase V	phase W
5 kV	150 pC	200 pC	150 pC
8 kV	220 pC	350 pC	250 pC
12 kV	380 pC	400 pC	400 pC
16 kV	500 pC	500 pC	550 pC
22 kV	800 pC	800 pC	850 pC

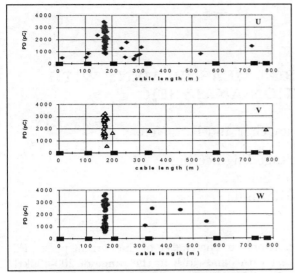

Figure 7: PD location diagram obtained from the three phases of a 775 m long, 240 mm², 10 kV paper/oil cable section, using OWTS with voltages up to $2U_0$. The cable joints and terminations are reflected by ■.

Figure 8 shows the measured PD patterns from these power cables. The PD signals follow the shape of the sine, so at the top of the sine, the PD signals are also at their top.

A notable fact with the performed measurement is, that no particular location of the measured PD activities, as shown in examples 1 & 2, could be determined. Based on this fact and that PD magnitudes are in the low range the PD activity originates from the total length of the cable, caused by degradation of the insulation material during the 45 years of exploitation.

4. Conclusions and Suggestions

In this study, field experiences on PD detection and location in power cables using OWTS are described. Based on these results, the following conclusions can be made:

1. Using Oscillating wave test system it is possible to energise on-site medium voltage cable sections and to detect, to measure in [pC] and to locate sensitively partial discharges. Moreover, this method for on-site testing of medium voltage power cables provides the same results as the combination of 50 Hz AC continuous energising and conventional PD detection.
2. Discharges in a cable section can be related to different conditions of a cable system:
   a) degradation of a particular cable accessories (junction or termination);
   b) local degradation of the cable insulation;
   c) overall degradation of the cable insulation without a partial discharge site.

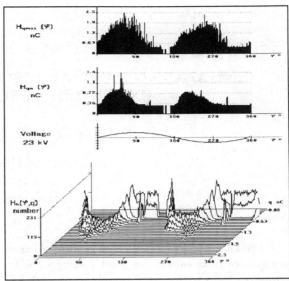

Figure 8: The PD pattern (2D and 3D) obtained from a 50 kV mass power cable originating from 1953.

3. On the base of measured PD magnitudes in [pC] only, it is difficult to judge the condition of a cable section. Therefore, depending on the cable type and comparative measurements in [pC] on similar cable section, the presence of discharge at voltages below operation voltage can be used as an indicator of the insulation degradation.
4. Based on the present experiences with the OWTS, it should be realistic to develop a diagnosis support system for maintenance purposes of medium voltage power cable system. For this purpose further systematic field investigation are necessary supported by laboratory experiments on service aged cables and accessories.

5. Acknowledgements

The authors would like to acknowledge NKF Kabel B.V. in Delft, The Netherlands for experimental support during on-site testing and interesting discussions about on-site cable testing.

6. References

[1] *PD measurements On-site using Oscillating Wave Test System.* E.Gulski, J.J.Smit, P.N.Seitz, J.C.Smit, IEEE International Symposium on EI, Washington DC, USA, 1998.
[2] *On-site PD diagnostics of power cables using Oscillating Wave Test System.* E.Gulski, J.J.Smit, P.N.Seitz, J.C.Smit, 11th ISH, London, UK, 1999.

ENHANCING PARTIAL DISCHARGE MEASUREMENT WITH EXTENDED RESOLUTION ANALYSER

S. Senthil Kumar, M.N. Narayanachar and R.S. Nema

Department of High Voltage Engineering
Indian Institute of Science
Bangalore 560 012, India

Abstract: Partial discharge fingerprints are generally obtained for model samples with known defects with computer-aided partial discharge analysers. These fingerprints help in generating a data base for classification of PD defects. Due to differences in PD detector characteristics, the measurements can be different with different PD detectors. The fingerprints obtained then, are dependent on the characteristics of PD detectors and the fingerprints obtained may lead to wrong classification and interpretation. The present paper proposes a solution of parallel detector scheme which will help obtain standard fingerprints independent of the PD detector characteristics.

1 Introduction

Computer-aided partial discharge measurements are commonly in use and partial discharge quantities /1/ are used as quality index which forms an important tool for diagnosis of insulation defects. PD is conventionally measured as a voltage drop caused by the movement of apparent charge across a series impedance (known as measuring impedance or quadrapole) in the PD circuit. The voltage across the measuring impedance is suitably amplified with a wide band or a narrow band amplifier and measured as a pulse height on a CRO or a PD analyser. The measuring impedance along with the amplifier does quasi-integration on the PD pulse whereby the pulse height is proportional to the apparent charge.

1.1 PD Detection System

Partial discharge current pulses have pulse widths of the order of 10-100ns. The frequency spectra of such pulses exceed tens of mega Hertz. The PD detector performs quasi-integration in frequency domain with the help of a filter which responds to the frequencies within the flat region of the PD pulse spectrum. Integration in frequency domain results in the response of the detector proportional to the charge of the current pulse. Filters with bandwidth in the range of 200-400kHz are called wide band systems. The upper limit on the bandwidth is to minimise the influence of the external high frequency interference as well as to ensure that the bandwidth of the filter remains in the flat region of the PD pulse spectra. If the bandwidth of the detector exceeds the bandwidth of the PD pulse, integration errors occur /2/. Hence, the upper limit is restricted to around 400kHz. The lower cutoff frequency is generally greater than 10kHz to avoid interference of power frequency signal and its harmonics into the detection system. Narrow band system has a bandwidth of about 10kHz with the center frequency tuned to a frequency in the flat region of the PD frequency spectrum where the influence of the external interference is minimum.

Superposition error will result in the measurement of pulse height if pulses occur within the resolution time of the detector /3/. Typically, wide band detectors have pulse resolution time of the order of $10\mu s$ and narrow band detectors have resolution time of the order of $100\mu s$.

1.2 Computer-aided Measurement of PDs

Computer-based PD analysers are used in conjunction with a conventional PD detector. These PD analysers give records of the pulse height, the phase of occurrence of the pulse with respect to the zero crossing of the applied voltage and the pulse counts familiarly known as $\phi - q - n$ distribution. This distribution is taken to characterise the PD phenomenon and to obtain fingerprints for diagnosis purposes /4/. It is to be noted that the analysers have an associated dead time which is the time duration required to process the pulse. The dead time is mainly due to the ADC conversion time and the time to transfer the ADC data to some temporary storage (buffer) or to the computer. Analysers with dead time of $10\mu s$ or less are easily available today.

High Voltage Engineering Symposium, 22–27 August 1999
Conference Publication No. 467, © IEE, 1999

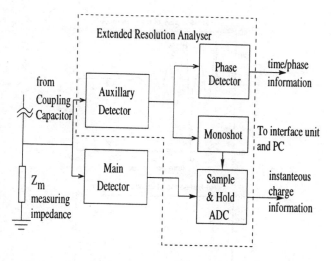

Figure 1: Block diagram of the proposed PD detection system.

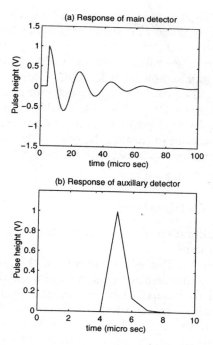

Figure 2: Response of (a) the main detector ($80\mu s$ resolution) and (b) the auxillary detector ($4\mu s$ resolution).

2 Errors in the Measurement of PD

It is important to know the detector's pulse response characteristics and the analyser's mode of data acquisition as mismatch in the detector resolution time and the analyser dead time can result in erroneous measurement. The detector resolution time can either exceed or be within the dead time of the analyser.

Dead time of the analyser smaller than the resolution of the detector may result in multiple measurements for the same pulse /5//6/. This error can be avoided by setting the dead time of the analyser higher than the resolution time of the detector. Another method is to shape the PD pulses to unipolar pulses by envelop detection.

Given that the dead time is higher than the resolution time, errors in the measurement can still appear due to overlapping pulses within the resolution time /3/.

Thus it can be seen that the PD pulse measurement is dependent on the measuring instrument performance and the fingerprints obtained therefore may lead to wrong classification and interpretation of the results.

3 PD Measurement System with Extended Resolution Analyser

The present paper gives a novel scheme by which the resolution of the detector can be extended to match the dead time of the analyser. This is made possible by evolving a computer-based system which integrates the PD detector and the analyser.

The resolution problem is mainly due to the limitations of the detector. The lower the bandwidth poorer is the detector resolution /7/. Bandwidth of the PD system is mainly decided by the measuring impedance and the amplifier. The frequency characteristic gets defined for the system which permits to quantify the apparent charge as proportional to the pulse height. This filter characteristic also defines the time domain response of the detector. The fixed time domain characteristic of the detector is used to advantage in the present scheme of measurement.

The block diagram of the Extended Resolution Analyser (ERA) is shown in Figure 1. The ERA includes an auxillary detector, phase detector, monoshot and a sample & hold ADC. The main detector in Figure 1 is the conventional PD detector which can have a narrow band or a wide band characteristic. The measuring impedance has a high bandwidth and gives a pulse with settling time less than the dead time of the analyser.

Pulses can occur within the resolution time of the main detector resulting in superposition errors. The ERA provides for a pulse correction algorithm which helps in correcting superposition errors in the measurement. A normalised lookup table is used to convert the amplitude error due to superposition depending on the time of occurrence of the pulse.

The auxillary detector has a wider bandwidth so as to give a settling time less than the dead time of the analyser. The auxillary detector along with the phase detector detects the instant of occurrence of the pulse with a time resolution of $1\mu s$ or less. The auxillary detector can be a logic comparator with an appropriate threshold for noise immunity. On the occurrence of a pulse, the monoshot generates a programmable delay trigger to the sample & hold ADC. This will help capture the peak of the main detector response.

The preparation of a lookup table required to correct for superposition errors and the tuning of the monoshot to capture the peak of the pulse are two important steps required for the ERA calibration.

3.1 Calibration

The calibration of the ERA system is performed in two steps.

Step 1. The pulse response of the main detector is digitized with the help of a digitizer. It should be ensured that the detector response is not modified by the envelop detector which is generally in the narrow band system. It is suggested that the sampling time of the digitizer should be such that the difference in the magnitude between two successive samples is less than 5%. Normalisation of the response is performed by dividing each sampled value by the peak amplitude of the pulse so that the whole response lies between -1 and 1. A lookup table can be created to give the normalised value at different sample times. Table 1 gives one such normalised lookup table for the example considered (Figure 2(a)) in the utility demonstration (section 3.2).

Table 1: Normalised lookup table

time (μs)	Digitized response				
1-5	1.000	0.905	0.732	0.506	0.253
6-10	0.000	-0.229	-0.414	-0.542	-0.606
11-15	-0.607	-0.549	-0.444	-0.307	-0.153
16-20	-0.000	0.139	0.251	0.329	0.368
21-25	0.368	0.333	0.269	0.186	0.093
26-30	0.000	-0.084	-0.152	-0.200	-0.223
31-35	-0.223	-0.202	-0.163	-0.113	-0.056
36-40	-0.000	0.051	0.092	0.121	0.135
41-45	0.135	0.122	0.099	0.068	0.034
46-50	0.000	-0.031	-0.056	-0.073	-0.082
51-55	-0.082	-0.074	-0.060	-0.042	-0.021
56-60	0.000	0.019	0.034	0.045	0.050
61-65	0.050	0.045	0.036	0.025	0.013
66-70	0.000	-0.011	-0.021	-0.027	-0.030
71-75	-0.030	-0.027	-0.022	-0.015	-0.008
76-80	-0.000	0.007	0.013	0.016	0.018
81-85	0.018	0.017	0.013	0.009	0.005
86-90	-0.000	-0.004	-0.008	-0.010	-0.011
91-95	-0.011	-0.010	-0.008	-0.006	-0.003

Step 2. The time required for the pulse response to peak is estimated from the lookup table prepared in calibration step 1. The monoshot is tuned to this time so that the peak of the main detector response is captured on detection of the pulse.

These two steps completes the calibration required for the ERA.

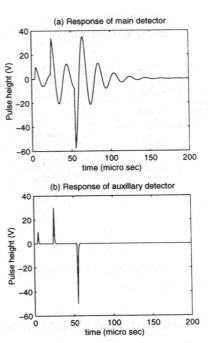

Figure 3: Response of (a)the main detector and (b) the auxillary detector with three pulses appearing within the resolution time of the main detector.

3.2 Utility Demonstration

The utility of the ERA is demonstrated with results of a simulation study carried out for the case when superposition of pulses occurs. Figure 2 shows the response of the main detector and the auxillary detector. The settling time of the main detector is taken as $80\mu s$ and that of the auxillary detector is taken as $4\mu s$. Figure 3 shows the case when three pulses appear within the resolution time of the main detector. The time difference between the first pulse and second pulse is $20\mu s$ and the time difference between the second pulse and the third pulse is $30\mu s$. The three pulse considered have the charge value of 10pC, 30pC and -50pC. The third pulse is considered negative to demonstrate the effectiveness of the system. Figure 3(a) shows the response of the main detector for case of three pulses considered. It can be seen that the three pulses cannot be resolved and have superimposed on the response of the earlier pulses. However, the auxillary detector (Figure 3(b)) can resolve the pulses and the times of occurrence of the pulses are measured by the phase detector. The charge magnitude is obtained with the help of the pulse correction algorithm as explained in section 3.3.

3.3 Pulse Correction Algorithm

When pulses appear well beyond the resolution time of the main detector, the correct peak value of the response is captured and there is no need for any correction. However, corrections are required when pulses appear within the resolution time of the main detector depending on the instant of occurrence of the following pulses. As a general rule, whenever a pulse is detected, it should be checked whether the previous pulses have appeared within the resolution time of the main detector. Figure 3 shows two pulses appearing before the response of the first pulse has settled. Therefore, the second and the third pulse require amplitude corrections. The second pulse will have magnitude correction due to the effect of the first pulse and the third pulse will have magnitude correction due to the effect of both the first and the second pulse. Table 2 demonstrates the procedure used to calculate the exact pulse peak magnitudes in the example considered.

The first pulse peak is measured to have a magnitude of 10pC. The time difference ($20\mu s$) between the first pulse and the second pulse is measured. As the time difference is less than $80\mu s$, superposition error occurs. Hence, the value (33.67) measured by the sample & hold ADC will give the algebraic sum of the first pulse component and the second pulse peak. The normalised value of the first pulse component (0.368, at $20\mu s$) can be determined from the lookup table. The normalised value is multiplied by the peak magnitude of the first pulse to give the magnitude of the first pulse component (3.68) at that instant. This value when subtracted from the measured value, gives the peak magnitude of the second pulse (29.99). Similarly, the peak magnitudes for the subsequent pulses are also corrected. The algorithm worked out is given in Table 2.

Table 2: Pulse correction procedure demonstrated for the example considered

Pulse number	1	2	3
Charge measured q_{meas} (pC)	10	33.68	-57.51
Measured time of occurrence (μs)	4	24	54
Time difference between the nth and $(n-1)^{th}$ pulse (μs)	-	20	30
Is correction required? (resolution $80\mu s$)	-	yes	yes
Normalised value q_{norm} (from lookup table)	-	0.368	-0.223
Correction required Δq $q_{prev} \times q_{norm}$ (pC)	-	3.68 (0.368×10)	-6.69 (-0.223×30)
Corrected value q_{meas}-Δq (pC)	-	29.99	-50.82
Time difference between n^{th} and $(n-2)^{nd}$ pulse (μs)	-	-	50
Is correction required? (resolution $80\mu s$)	-	-	yes
Normalised value (from lookup table)	-	-	-0.082
Correction required Δq $q_{prev} \times q_{norm}$ (pC)	-	-	-0.82 (-0.082×10)
Corrected value q_{meas}-Δq (pC)	-	-	-50
Final value (pC)	10	29.99	-50

4   Conclusions

To conclude,

1. the PD measuring system can measure pulses to the limit of the dead time of the analyser.

2. it is possible to standardise the measurement for a given dead time of the PD analyser. Hence, standard finger-prints of the PD phenomenon can be obtained.

3. the system is independent of the main PD detection system. The advantages of both the wide band and the narrow band detectors can be fully exploited in so far as noise suppression, the polarity information and the resolution is considered.

5   References

/1/ IEC publication 270, Partial Discharge Measurements, 1981.

/2/ P. Osva'th, E. Carminati and A. Gandelli, *A Contribution on the Traceability of Partial Discharge Measurements*, IEEE Trans. Elec. Insul., Vol. 27, pp. 130-134, 1992.

/3/ P. Osva'th, *Comment and Discussion on Digital Processing of PD Pulses*, IEEE Trans. Diel. and Elec. Insul., Vol. 2, pp. 685-699, 1995.

/4/ A. Krivda, *Automated Recognition of Partial Discharges*, IEEE Trans. Diel. and Elec. Insul., Vol. 2, pp. 796-821, 1995.

/5/ E. Gulski, *Digital Analysis of Partial Discharge*, IEEE Trans. Diel. and Elec. Insul., Vol. 2, pp. 822-837, 1995.

/6/ R. Bozzo, G. Coletti, C. Gemme and F. Guastavino, *Application of Design of Experiment Techniques to Measurement Procedures. An Example of Optimisation applied to the Digital Measurement of Partial Discharges.*, IEEE Instrumentation and Measurement Technology Conference, Ottawa, Canada, pp. 470-475, May 19-21, 1997.

/7/ W.S. Zaengl, K. Lehmann and M. Albiez, *Conventional PD Measurement Techniques used for Complex HV Apparatus*, IEEE Trans. Elec. Insul., Vol. 27, pp. 15-27, 1992.

APPLICATION OF ARTIFICIAL NEURAL NETWORK (ANN) IN SF$_6$ BREAKDOWN STUDIES IN NON-UNIFORM FIELD GAPS

Sandeep Chowdhury and M.S. Naidu

Department of High Voltage Engineering, Indian Institute of Science
Bangalore, India.

ABSTRACT

In SF$_6$-filled electrical equipments, the electric field distribution is kept rather uniform. However in practice, the electric field in the gas gap is distorted by non-uniformities. For this reason, the inhomogeneous field breakdown in SF$_6$ has been extensively studied by various researchers and the breakdown characteristics of compressed SF$_6$ have been reported. Obtaining experimental data under all conditions is not possible. Therefore, an attempt has been made in the present work to apply an Artificial Neural Network (ANN) to obtain such data.

The Projection Pursuit Learning Network (PPLN) has been used as the ANN model. Breakdown data for four different voltage waveforms were used to train the network for SF$_6$ pressures of 1-5 bar and rod diameters of 1-12 mm in a rod-plane geometry. The ANN was first trained with these data so as to obtain a smooth regression surface interpolating the training data. The regression surface thus obtained, was thereafter used to generate the breakdown and corona inception voltages with in the range of gas pressures and non-uniformities studied, where no data is available.

1. Introduction

Sulphur hexafluoride (SF$_6$), a strongly electronegative gas is being increasingly used as an insulating medium in electrical power equipments. In SF6-filled electrical equipments, in practice, the electric field in the gas gap is distorted by field non-uniformities created by metallic particles present in the gas and surface protrusions on the electrodes. For this reason, the inhomogeneous field breakdown in SF$_6$ was extensively studied and in these studies the inhomogeneous field distribution has been simulated by rod-plane gaps, varying either the rod diameter or the gap spacing and the breakdown characteristics have been extensively reported[1,2,3].

In the absence of any mathematical model describing a relationship between the experimental parameters (gas pressure, rod diameter and gap spacing) and the corresponding breakdown voltages for a particular voltage waveform, an attempt has been made in the present work to apply an Artificial Neural Network (ANN) to obtain such a correlation. The Projection Pursuit Learning Network (PPLN) has been used as the ANN model. Breakdown data for four different voltage waveforms were used to train the network-DC (positive and negative polarity), AC, Standard LI (+1.2/50µs) and SI (+250/2500µs).

High Voltage Engineering Symposium, 22–27 August 1999
Conference Publication No. 467, © IEE, 1999

2. The Artificial Neural Network (ANN) model

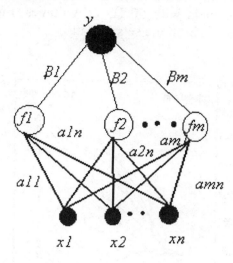

Figure 1: The PPL Network used in the present study

Fig 1. Shows the architecture of the Projection Pursuit Learning (PPL) network 1,2,3 used in the present study[4,5]. The PPLN is a two-layer feed forward learning network whose trainable parameters are:

1. The hidden layer weights denoted by $\{\alpha_{kj}, j=1,\ldots,n\}$, connecting all the inputs units to the k_{th} hidden neuron.

2. The unknown (trainable) "smooth" activation functions denoted by $\{f_k, k=1,\ldots,m\}$ of the k_{th} hidden neuron, which are hermite

$$f_k(z_k) = \sum_{r=1}^{R} c_{kr} h_r(z_k) \ldots (1)$$

functions defined as
where the c_{kr} s are the coefficients of the hermite functions h_r s, and R is called the order of the hermite function f_k, which is a constant set by the user. The hermite functions are orthonormal and defined by

$$h_r(z_k) = (r!)^{-1/2} \pi^{1/4} 2^{-(r-1)/2} H_r(z_k)\phi(z_k) \ldots (2)$$

where $H_r(z_k)$s are the hermite polynomials constructed in a recursive manner as

$$H_0(z_k) = 1 \quad \ldots (3)$$
$$H_1(z_k) = 2\,z_k \quad \ldots (4)$$
$$H_r(z_k) = 2[z_k H_{r-1}(z_k) - (r-1)H_{r-2}(z_k)] \quad \ldots (5)$$

$$r = 2,3,4,\ldots$$

and $\Phi(z_k)$ is the weighting function

$$\phi(z_k) = \frac{1}{\sqrt{2\pi}} e^{-z_k^2/2} \ldots (6)$$

3. The output layer weights are denoted by β_k connecting the k_{th} hidden

$$\hat{y}_l = \sum_{k=1}^{m} \beta_k f_k(\sum_{j=1}^{n} \alpha_{kj} x_{jl}) \ldots (7)$$

neuron to the output unit. The output of the network is mathematically expressed as
where $\hat{y}_l$ is the actual output of the network computed by the ANN for the l_{th} training pattern.

The training of all the weight parameters, β_k, $\{c_{kr}\}$ and $\{\alpha_{kj}\}$, is based on the criteria of minimizing the MSE(mean-square-error), or the so called L_2 loss function defined as

$$L_2 = \sum_{l=1}^{p} [y_l - \hat{y}_l]^2 \ldots (8)$$

$$= \sum_{l=1}^{p} [y_l - \sum_{k=1}^{m} \beta_k f_k(\alpha_k^T x_l)]^2 \ldots (9)$$

which is the sum squared error between the desired and the actual output over all the training patterns.

The characteristic features of the training process of the PPL network are as follows:

1. The network is built gradually, starting from a single hidden neuron; additional hidden units and weights are added subsequently during the training process till a minimum architecture is obtained which gives

a satisfactory solution for the given problem.

2. Instead of training the whole network after a new hidden unit is added, the PPL algorithm first trains only the new hidden unit. After the new hidden unit is trained, the parameters associated with the previously installed hidden units are updated one unit at a time. Thus the PPL network is trained in a neuron-by-neuron manner and not as a whole entity.

3. Further simplification is obtained by training a single hidden unit in a layer-by-layer manner. For this purpose, the parameters that have to be trained after adding the kth hidden unit are divided into three groups (α_k, c_k, and β_k). Each group is updated separately for that particular neuron.

3. Results and Discussion

For DC applied voltage (both polarities), the breakdown data[1] consisted of both the corona inception and breakdown voltage characteristics of rod-plane gaps. The field non-uniformity was varied by varying the rod diameter values in the range 1.0 to 6.3 mm for a fixed gap spacing of 20 mm. The gas pressure was varied in the range 1 to 5 bar. The breakdown voltage data were trained in the corona-stabilized region while the corona onset data were trained separately in the full pressure range. The ANN had two inputs, one corresponding to the pressure values, and the other to the rod diameter values. The maximum training error for both polarities was around 5% of the desired output value.

Similar breakdown data for AC applied voltage had a pressure range of 1 to 6 bar and rod diameter range of 1 to 10 mm for a fixed gap spacing of 60 mm[2]. The maximum training error was 5% of the desired output value, when this data was used for training. Using

this, new data was generated for different values of the experimental parameters with 50 Hz ac voltages. Fig 2. shows both the sets of data.

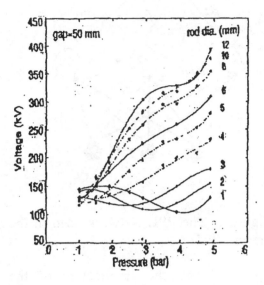

Fig 2 : SF$_6$ Breakdown Characteristics for positive LI (1.2μs/50μs) in a 50mm Rod Plane Gap. (The solid lines correspond to training breakdown voltage inputs and the broken lines to breakdown voltages computed by the ANN).

For impulse voltage waveforms, namely, standard LI (1.2/50 μs) and SI (250/2500μs)), the breakdown data used for training consisted of breakdown voltage values only. In this case, the field non-uniformity was varied by varying separately two experimental parameters - the rod diameter as well as the gap spacing. The ANN had three inputs, corresponding to pressure, rod diameter and gap spacing. The training data for types of impulse waveforms had pressure range of 1 to 5 bar, rod diameter range of 1 to 12 mm and gap spacing range of 10 to 100 mm. The maximum training error for both waveforms was around 10% of the desired output value. Using this, new data was generated for experimental parameters for which no data is available in the literature. Both the sets of data has shown in Fig. 3 (for standard LI voltage) for a gap spacing of 60 mm. Data at other gap spacings for LI and SI are not presented due to lack of space.

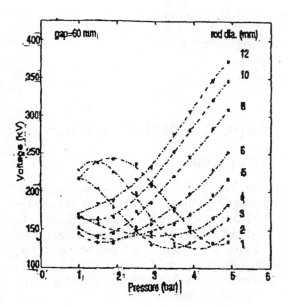

Fig 3: SF$_6$ Breakdown Characteristics for positive SI (250μs/2500μs) in a 60mm Rod Plane Gap. (The broken lines correspond to the breakdown voltages computed by the ANN).

The generalization of the network for all the applied voltage waveforms showed fairly good consistency with the general nature of the experimentally obtained breakdown voltage characteristics reported in the literature. For DC and AC voltages, the corona inception characteristics are linear while the breakdown voltage characteristics are non-linear and exhibit corona stabilization effect[1,3]. The corona stabilization effect was more pronounced with an increase in the field non-uniformity. Negative polarity DC had lower corona inception voltages but its breakdown voltage characteristics exhibited a more pronounced corona stabilization effect compared to the positive polarity voltages. The SI breakdown characteristics were similar to those of AC and DC voltages while in the case of LI breakdown characteristics,

the corona stabilization effect was seen only for higher field non-uniformities.

The breakdown characteristics obtained by application of ANN show trends similar to those obtained experimentally with fair accuracy of about 5%. It appears to be a useful tool to generate breakdown data for a variety of conditions that exist in power apparatus.

References

1. N.H. Malik and A.H. Quereshi : "The influence of voltage polarity and field non-uniformity on the breakdown behaviour of rod-plane gaps filled with SF$_6$", IEEE Trans. on Elect. Insul, vol. EI-14, pp 327-333, 1979.

2. M. Zwicky : "Breakdown phenomenon in SF$_6$ and very inhomogeneous large rod-plane gaps under 50-Hz AC and positive impulse voltages", IEEE Trans. on Elect. Insul., vol. EI-22, pp 317-324, 1987.

3. B. S. Manjunath : "Inhomogeneous electric field breakdown in pure SF$_6$ gas under AC and impulse voltages", M.Sc.(Engg.) thesis, Indian Institute of Science, May 1991.

4. J.N. Hwang, S.S. You, S.R. Lay and I.C. Jou :"The cascade correlation learning: a projection pursuit learning perspective" , IEEE Trans. on Neural Networks, vol. 7, pp 272-289, 1996.

5. T.Y. Kwok and D.Y. Yeung : "Use of bias term in projection pursuit learning improves approximation and convergence properties", IEEE Trans. on Neural Networks, vol. 7, pp 1168-1183, 1996

CHARACTERISATION OF PARTIAL DISCHARGES IN OIL IMPREGNATED PRESSBOARD INSULATION SYSTEMS

Y.P.Nerkar, M.N.Narayanachar and R.S.Nema

Department of High Voltage Engineering
Indian Institute of Science
Bangalore 560 012, India

Abstract:

Partial discharge (PD) measurement and interpretation is one of the most useful diagnostic tools for quality assurance testing during the design, manufacture and life assessment of electrical equipments. Pressboard-oil insulation continues to be a major component in the insulation design of EHV transformers. As tests on practical systems are not always feasible due to cost and time factors, model systems tests are gaining importance. In the present work a model insulation system consisting of oil impregnated pressboard with uniform field electrode arrangement has been tested for its behaviour under PDs. It is shown that by defining a modified scale parameter for the Weibull distribution it is possible to take into account the effect of both smaller and larger magnitude PD pulses. The modified scale parameter serves as a good index for characterising PD behaviour in oil impregnated pressboard insulation system studied.

1 Introduction

Pressboard-oil insulation continues to be a major component in the insulation design of EHV transformers. Partial discharges are considered to be hazardous for the satisfactory performance of the insulation system. The gradual deterioration due to PDs can lead to premature failure of the system. Two or mixed parameter Weibull functions are conveniently used to fit the experimental data /1/. Model insulation systems are found to be convenient for the study of PD behaviour /2,3,4,5/. In the present work, oil impregnated pressbaord samples have been tested for their PD behaviour at predetermined stress levels. Records of PD pulse distribution patterns are acquired with the help of the multichannel analyser. The data has been analysed using Weibull statistics. A modified scale parameter of the mixed Weibull distribution is found quite useful for monitoring aging behaviour.

High Voltage Engineering Symposium, 22–27 August 1999
Conference Publication No. 467, © IEE, 1999

2 Experimental Procedure

2.1 Test Cell

A pair of $7\pi/12$ Rogowski brass electrodes of 30 mm overall diameter was used for the study of PD characteristics in a uniform field. A PMMA (Poly Methyl Methaacrylate) cell of volume 130 mm x 130 mm x 150 mm was used for conducting experiments. The pressboard samples were placed between electrodes in the PMMA cell filled with transformer oil.

2.2 Preparation of samples

Pressboard sheets with nominal thickness of 2.0 mm, mat finished on both sides were used for making samples of 50 mm diameter. Samples were vacuum dried at a temperature of 100 degree centigrade. Filtered and vacuum treated transformer oil was used for impregnation. The impregnated samples were stored in vacuum tight dissicator containing moisture absorbent.

2.3 Test Procedure

The voltage across the samples was raised gradually at an average rate of 2 kV/s till inception of PD. The inception voltage was noted. PD distributions were acquired using MCA (*EG&G ORTEC*), Model 921. Further the voltage was raised to a predetermined stress level as explained in the following section.

2.4 Criterion for selecting the stress levels

Initially three samples were tested at a stress level of 12.0 kV/mm. The stress level of 12.0 kV/mm was close to the inception level (11.0 kV/mm average). PD pulse distributions were recorded continuously. It was found that failure of the sample took place only in one case after 230 minutes of application of overvoltage stress. The

Table 1
Results of PD tests at different
stress levels

Sample number	Inception stress level kV/mm	Applied stress level kV/mm	Time to failure min.
1	11.60	13.0	92
2	11.00	13.0	120
3	11.10	13.0	67
4	11.00	13.0	113
5	11.10	13.0	80
6	10.40	13.0	92
7	11.00	13.5	65
8	11.80	13.5	78
9	11.10	13.5	35
10	12.50	13.5	52
11	11.00	13.5	64
12	11.10	13.5	206
13	10.50	14.0	40
14	10.50	14.0	21
15	12.50	14.0	60
16	11.00	14.0	24
17	11.10	14.0	60
18	11.80	14.0	52

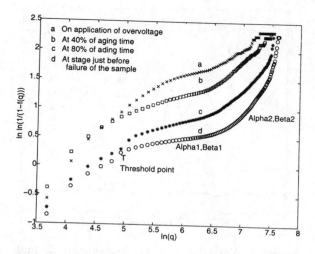

Figure 1: Typical Weibull plots for the sample (no.8) at 13.5 kV/mm

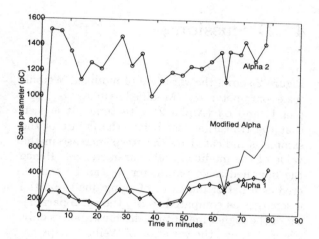

Figure 2: Variation of scale parameters α_0, α_1 and α_2 with aging time for a typical sample (no.8) at 13.5 kV/mm

other two tested samples did not fail even after aging for 24 hours at which stage the experiments were terminated. In order to accelerate the aging process due to PDs an overvoltage stress of 15.0 kV/mm was tried for two samples. The failure times were 30 minutes and 41 minutes which were found too short for observing the PD behaviour. Hence an intermediate overvoltage stress levels of 13.0 kV/mm, 13.5 kV/mm and 14.0 kV/mm were selected for further experimentation. PD pulse distribution patterns were acquired continuously with each record containing pulses for 10 seconds duration.

3 Analysis of results

Weibull distribution was used for analysis of PD pulse distribution patterns. The Weibull distribution is given by

$$F(q) = 1 - e^{\left(\frac{-q}{\alpha}\right)^{\beta}}$$

where F(q) is the cumulative Weibull distribution function. The α, known as scale parameter, defines 63.2% probability for the group. The β, known as shape parameter. q is the discharge magnitude.

A typical Weibull distribution plots are shown in figure 1. The distribution has been considered in two sections representing α_1, β_1 and α_2, β_2 parameters. A threshold discharge magnitude was decided and initial portions from Weibull plots were excluded. In order to explain the PD behaviour a modified scale parameter called α_0 has been defined as

$\alpha_0 = P_1\alpha_1 + P_2\alpha_2$ P_1 is the probability of the first section of the distribution

P_2 is the probability of second section of the distribution.

$P_1 and P_2$ are calculated as

$$P_1 = \frac{N_1}{(N_1+N_2)} \ and P_2 = \frac{N_2}{(N_1+N_2)}$$

N_1 = total number of pulses in section 1 and
N_2 = total number of pulses in section 2.

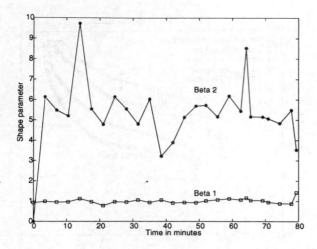

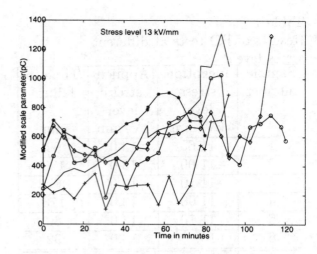

Figure 3: Variation of shape parameters β_1 and β_2 with aging time for a typical sample (no.8) at 13.5 kV/mm

Figure 4: Variation of modified scale parameter α_0 with aging time at 13.0 kV/mm for all the samples

4 Discussion

Figure 2 shows the variation of modified Weibull scale parameter α_0 (Modified Alpha), α_1 (Alpha 1) and α_2 (Alpha 2) with time. It is seen that the transition just before the failure of the sample is indicated by the steep increase in the value of the modified scale parameter α_0. Hence the modified scale parameter α_0 can be considered as a better indicator of aging due to partial discharges as compared to α_2, the scale parameter of the higher magnitude discharge pulses only. Figure 3 shows the variation of Weibull shape parameters β_1 and β_2 with time. A modified shape parameter β_0 has not been defined in view of the fact that the large value of β_2 (in the range from 3 to 6) of the larger magnitude pulses is a better indicator of the aging rate than the value of β_1 (in the range from 1 to 2) given for the smaller magnitude discharge pulses.

The variation in the modified scale parameter α_0 and shape parameter β_2 for all the system tested at different stress levels are shown in figures 4 to 9. It may be seen from figures 4, 6 and 8 that initial α_0 is significantly higher ($\gg 500$) for the samples at 14 kV/mm. But at the end of aging period, just before the failure of the samples, the values of α_0 are in the range (1000 to 1200 pC) for all the stress levels. As shown in figures 5, 7 and 9, the range of the β_2 values are 3.2- 5.5 at 13.0 kV/mm, 3.5- 6.5 at 13.5 kV/mm and 4.6- 6.5 at 14.0 kV/mm indicating a gradual increase in the rate of deterioration of the samples with increase in stress levels. This fact is also reflected in the time to failure (Table 1).

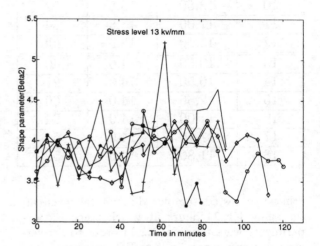

Figure 5: Variation of shape parameter β_2 with aging time at 13.0 kV/mm for all the samples

5 Conclusions

It is shown that by defining a modified Weibull scale parameter the PD behaviour of pressboard insulation system can be studied more effectively than by just considering two modes of the distributions.

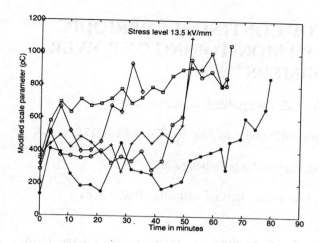

Figure 6: Variation of modified scale parameter α_0 with aging time at 13.5 kV/mm for all the samples

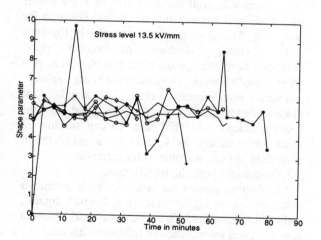

Figure 7: Variation of shape parameter β_2 with aging time at 13.5 kV/mm for all the samples

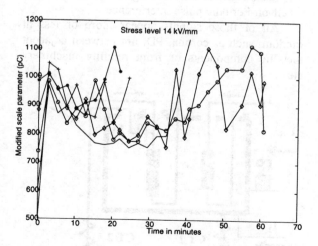

Figure 8: Variation of modified scale parameter α_0 with aging time at 14.0 kV/mm for all the samples

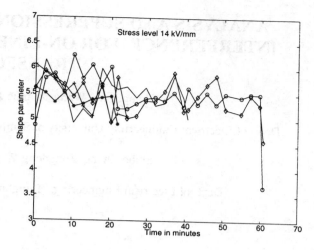

Figure 9: Variation of shape parameter β_2 with aging time at 14.0 kV/mm for all the samples

6 References

/1/ A.Contin,E.Gulski,M.Cacciari and G.C.Montanari "A Weibull approach to the investigation of partial discharges in aged insulation' system", Conference record of the IEEE International Symposium on Electrical Insulation, Montreal, Quebec, Canada, June 16-19,1996,pp. 416-419.

/2/ Palmer S and Sharpely W.A. " Electric strength of transformer insulation", IEE Proc., Vol.116, No.12, December 1969, pp.2029-2038.

/3/ Mohsin M.M., M.N.Narayanachar and R.S.Nema "A study of partial discharge characteristics in oil impregnated pressboard insulation",Conference record of the IEEE International Symposium on Electrical Insulation, Montreal, Quebec, Canada, June 16-19,1996, pp.79-82.

/4/ K.Raja, M.N.Narayanachar and R.S.Nema "A study of phase angle distribution of partial discharges in oil pressboard insulation system", Conference record of the IEEE International Symposium on Electrical Insulation, Montreal, Quebec, Canada, June 16-19,1996, pp.75-78.

/5/ Karlernst Giese "Electrical strength of pressboard and components for transformer insulations" IEEE Electrical Insulation Magazine, Vol 12,No.1, January-February,1996, pp.29-33.

ANALYSIS AND SUPPRESSION OF CONTINUOUS PERIODIC INTERFERENCE FOR ON-LINE PD MONITORING OF POWER TRANSFORMERS

Changchang Wang, Xianhe Jin, T.C. Cheng, IEEE Fellow

Dept. of Electrical Engineering, University of Southern California, Los Angeles, CA 90089-0271, U.S.A

Shibo Zhang, Zongdong Wang, Xuzhu Dong, Deheng Zhu

Dept. of Electrical Engineering, Tsinghua University, Beijing - 100084, P. R. China

Abstract

The key to on-line partial discharge (PD) monitoring of power transformers is how to suppress interference effectively. This paper reports an interference qualitative and quantitative analysis on site using a Fast Fourier Transform (FFT) technique. Various periodic interferences have been found, which emanated from the carrier communication, thyristor switching, and high frequency protection signals of power system as well as radio broadcasting signals.

LMS adaptive digital filter and FFT threshold digital filter can be used to eliminate continuous periodic noises for on-line PD monitoring. However, they suffer from their own intrinsic weaknesses. Therefore, a multi-band-pass digital filter based on FFT has been developed. It is an much better method for the purpose of suppressing continuous periodic interferences.

Introduction

Electric-Magnetic interferences will reduce the on-line PD monitoring sensitivity seriously. Sometimes, PD signal may be total merged in interferences when we use pulse current method to detect PD in a trans- former of substation. Therefore, how to suppress the interferences is one of the critical tasks for building an on-line PD monitoring system of transformers.

A lot of techniques can be used for interference rejection on site. But all of the methods have their own intrinsic advantages and disadvantages, no one can suppress all of the interferences on site with a same effect. Therefore we should use different methods to suppress different interferences according to their characteristics.

Sources of Interference

The interferences sources and coupling routes are shown as Fig.1. I_1 is from high-voltage transmission line, whose sources are radio broadcast, carrier communication, corona of transmission line or adjacent equipment, PD in others equipment etc. I_2 comes from distribution line or 220V power line, such as the power line of electric fan used for coo- ling. The sources of I_2 are mostly thyristor swit- ching, random interferences etc. I_3 corresponds to the grounding system which is a complicated network. The sources of I_3 are various random interferences etc. I_4 stands for electromagnetic interferences of air space, such as radio broadcast etc. Therefore we can summarize those interferences as follows in a subs- tation according to their waveform characteristic:

1.Continuous periodic interference

All of these signals are sinusoid, such as carrier communication, high frequency protection signals, radio broadcast etc. We can also call them a narrow- band interference in frequency domain .

2.Pulse interference

(1)Periodic pulse interference

Such as thyristor switching etc.

(2)Non-Periodic pulse interference

All of those signals appear more or less at random, such as corona, PD, arc between adjacent metallic components or from rotating machinery etc.

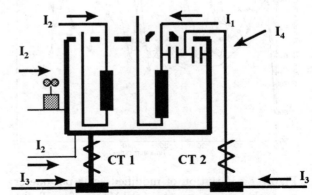

Fig.1 Interferences sources and their coupling routes

High Voltage Engineering Symposium, 22–27 August 1999
Conference Publication No. 467, © IEE, 1999

Methods of Interference Eliminated

We can summarize different methods of interference eliminated as follows according to hardware or software.

1.Hardware methods:[1,2]

(1)Filter: It can be used to suppress continuous periodic interferences, such as: band-pass or band-rejection filter, combination filter etc.

(2)Differential balance system: It can suppress common-mode interferences including continuous periodic or pulse interferences.

(3)Pulse polarity discrimination system: It can suppress common-mode pulse interference.

2.Software methods:

(1)Averaging technique.

(2)Digital filter: It can suppress continuous periodic interferences. [3,4]

(3)Pulse recognition technique: It can suppress pulse interference.

We developed several hardware to suppress interferences and discussed them in the previous paper.[1,2]. We will focus our attention on how to use the digital filter to suppress continuous periodic interferences in this paper.

Interference Analysis on Site

We have measured the interference signals by current transducer (bandwidth: 62.4-621 kHz) clamped on tank grounding wire of a 500kV transformer in a substation as shown in Fig.2. Sampling frequency of the transient recorder is 2.5MHz. From the frequency spectra of interference's signals we can see that continuous periodic signals are the main interference sources.

Table 1. Main frequency contents of Fig.2

f, kHz	PR, %	Source	f, kHz	PR, %	Source
95.2	0.5	CCS	329.4	9.5	CCS
117.2	3.4	CCS	380.9	7.5	CCS
190.4	8.9	CCS	415.0	58.5	CCS
256.3	1.5	CCS	981.4	0.2	RS
300.3	7.2	CCS	-	-	-

Note: PR-Power ratio,

$$PR=(\sum A_{2i}^2 / \sum A_{1i}^2) \times 100\%;$$

A_{2i}-Signal amplitude of some band in time domain;

A_{1i}-Amplitude of total signal by monitoring after eliminating zero shift,

f-Central frequency;

CCS-Carrier communication;

RS-Radio broadcast.

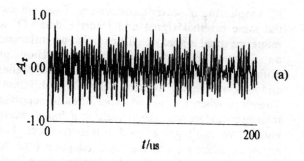

(a)

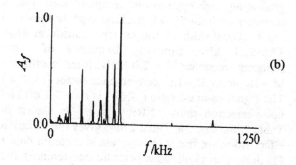

(b)

a - Waveform ; b - Frequency spectrum

Fig.2 Interference signal from tank grounding wire of a 500kV transformer

Compared with carrier communication band-width of this substation most of the narrow-band interfe- rences are from the carrier communication as shown in Table1. Table1 shows that total power ratio of periodic interference signals is 97% of total signal. We also found strong periodic pulse interference signals coupled with current transducer through 220V power line of the transformer which is from thyristor switching. Most of their frequency are under 132.8 kHz.

Digital Filtering Technique

1.LMS Adaptive Filter

The principle of LMS adaptive filter has discussed in previous paper[4]. Sometimes it can suppress narrow-band interference effectively[4]. But in practice, how to choose the key parameters properly, such as delay D, convergence factor μ and filter coefficient M which have influence on the convergence and filtration effect, is a prolem. And the choice of these parameters depends on experience. When non-narrow-band interference is quite large, such as white noise, convergence process will be deteriorated or can't achieve convergence. In addition, LMS adaptive filter need two input signals, which are original input signal (PD signal plus interference signals) and reference input signals (interference signals). But we can't get two input signals like those for on-line

PD monitoring of power transformers. Thus, we use the same original signal and insert a delay D as reference input signal. It will exercise an influence on filtration effect, because the correlation of different frequency contents between original signal and original signal inserted delay D is different. Therefore when we use adaptive filter to suppress interferences on site which include a lot of white noises, we can't get a good filtration effect. For example, when we use the LMS adaptive filter to process the signal which was detected from tank grounding wire by injecting a square wave into a transformer from 500kV bushing tape as shown in Fig.3. (Bandwidth of the current transducer B_W= 37.8-52.1 kHz, sampling frequency of the transient recorder f_s = 2.0 MHz, filter coefficient M =16, delay D =1.0, convergence factor μ = 0.2.) The signal-to-noise ratio (SNR) is only 3.7 dB and noise-rejection ratio (NRR) is 3.1 as shown in Table 2 and Fig. 3b. Table 2 also gives the filtration effect when we use it to suppress simulation data (PD pulse plus continuous periodic interference) and simulation data plus white-noise. For the former, its filtration effect is good, SNR = 11.3, NRR = 16.6, but the latter its filtration effect is unsatisfactory, SNR = 0.9, NRR = 6.1.

2.FFT Threshold Digital Filter

We also have developed a threshold digital filter[3] based on FFT to suppress continuous periodic interference. A PD pulse signal appears as a uniform spectrum and a continuous periodic interference appears as a narrow-band pulse in frequency domain. We can eliminate some pulses which are greater than pre-determined threshold in frequency domain. After eliminating, we can transform it into time domain by Inverse Fast Fourier Transform (IFFT). Actually this is a multi-band rejection digital filter. Its filtration effect is also as shown in Fig.3 and Table 2. Because the white-noise distributes in the whole detecting band, filtration effect is unsatisfactory either (Table 2). For field data, its effect is much better than LMS adaptive filter (Fig. 3 c and Table 2). Its NRR is 5.4 dB. All of the test parameters are same as the above descriptions.

The disadvantage of FFT threshold digital filter is the uncertainty of choice of its band rejection, because its filtration effect will change with the threshold. Besides the practical PD signal detected by a current transducer is often a attenuation oscillation wave. It is not a uniform spectrum in frequency domain and distributes at some band-width. In this case the PD signal will lose more frequency content after filtering. Sometimes FFT threshold digital filter is not satisfied for on-line PD monitoring of power transformers (Fig.3c and Table 2).

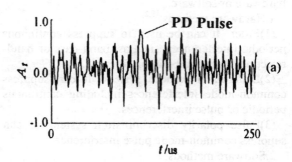

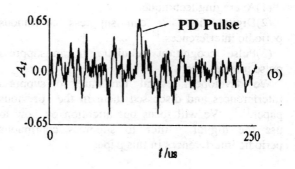

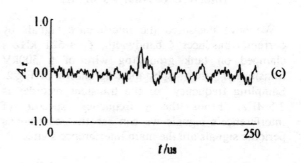

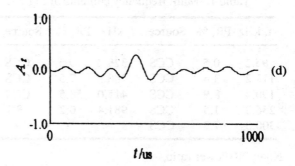

a-Input waveform;
b-After filtering by a LMS adaptive digital filter;
c-After filtering by a FFT threshold digital filter;
d-After filtering by a multi-band-pass digital filter

Fig.3 Filtration effect of different digital filters for field data

Table 2 Filtration effects for different digital filters

		Data 1	Data 2	Data 3
LMS adaptive	SNR, dB	11.3	0.9	3.7
filter	NRR, dB	16.6	6.1	3.1
	t, s	4.0	4.0	8.0
FFT threshold	SNR, dB	12.4	4.1	6.0
filter	NRR, dB	17.7	9.3	5.4
	t, s	1.0	1.0	1.0
Multi-band-	SNR, dB	17.5	5.0	12.0
pass filter	NRR, dB	22.8	10.2	11.4
	t, s	1.0	1.0	1.0

Note:Data1-Simulation data, attenuation oscillation
signal plus narrow interference signal;
Data 2-Data 1 plus white-noises;
Data 3-Field data from tank grounding wire
of a 500 kV transformer;
t-Processing time

3.Multi-Band-Pass Digital Filter

It is also based on FFT like the threshold digital filter. We can remain (or reject) one or several bands and eliminate (or remain) the others in frequency domain. It is like an ideal multi-band-pass (or multi-band-rejection) digital filter.

A multi-band-pass digital filter has some advantages:

(1)It is convenient to select one or multi-band-pass or band-rejection used for monitoring bands. We can find out the best monitoring bands and make their SNR maximum. It can offer a scientific basis for the choice of monitoring band. When we select multi-band-rejection, its filtration effect is similar to a FFT threshold filter. But its band-rejection is definite and filtration effect is stable too, different from FFT threshold filter. Because the practical signal is often a attenuation oscillation signal, its energy concentrates in a quite narrow band, if select suitable band and avoid narrow interference, we can get the best filtration effect for field data like Fig.3d. Its SNR is 12 dB and NRR is 11.4 dB. The monitoring band of this multi-band-pass filter is 7.8-37.1 kHz. All of the test parameters are same as the above descriptions.

(2)PD signal having a very small power related to the whole signal (including severe interference) can be picked up after filtering. For example, in Fig.3d, the power ratio of the PD related to the whole signal is only 0.9%.

However, if the central frequency of PD attenuation oscillation signal is equal to or very close to some narrow band interference or non-narrow band interference or white noises is very strong, this kind of filter can't get satisfactory filtration effect either.

Conclusions

1.The continuous periodic signals (or narrow-band signal) and periodic pulse signals are the

main interferences in a substation for on-line PD monitoring of power transformers. The main content of narrow-band signal is carrier communication. Source of the periodic pulse signal is thyristor switching.

2.Digital filter may be used to suppress narrow-band interference. But some of them still have some disadvantages and can't always get good filtration effect on site. Multi-band-pass filter is much better than the others according to the filtration effect on site.

3.Multi-band-pass filter not only can be used for a filter, but also a method. It can offer a scientific basis for the choice of monitoring band and the design of a hardware band-pass filter.

References

[1]Changchang Wang, Zhongdong Wang, Fuqi Li, Xianhe Jin, Anti-interference Techniques Used for On-Line PD Monitoring, Proceedings of the 4th ICPADM, July 3-8, 1994, Brisbane, Australia.

[2]Changchang Wang, Zhongdong Wang, Wei Tao, Deheng Zhu, A Pulse Polarity Discrimination Sys- tem Used for On-Line PD Monitoring, Proceedings of the 3th China-Japan Conference on Electric Insu- lation Diagnosis, Osaka, Japan. September, 1994.

[3]Changchang Wang, Zhongdong Wang, Xianhe Jin, Deheng Zhu, A Digital Filter Technique Used for On-Line PD Monitoring, Proceedings of the 2nd China-Japan Conference on Electric Insulation Diagnosis, Shanghai, China, October, 1992.

[4]Sher Zaman, Deheng Zhu, Xianhe Jin, Kexing Tan, A new adaptive Technique for On-Line PD Monitoring, IEEE Trans. On EI-2(4): 700-707, 1995.

INFLUENCE OF THE DEGREE OF PARALLEL PARTIAL DISCHARGES ON THE DEGRADATION OF THE INSULATION MATERIAL

K Temmen née Engel

University of Dortmund, Germany

ABSTRACT

In order to assess the aging state of complex insulation systems it is necessary to understand the intricate aging processes occuring in the insulation. Test objects out of epoxy resin with different degrees of parallel discharges are produced in order to determine the relation between the degree of parallel discharges and the degradation of insulation material. The damaging effect of the pd is proved by scanning electron microscopy (sem) and roughness measurements. A model based on simulation of the relevant gas discharge processes is used to interpret the results.

INTRODUCTION

Partial discharge (pd) measurements play an important part in the field of quality assurance of power supply devices. The goal of a continuous registration of the pd-activity can be reached by means of an automated measuring system (Gulski (1)).

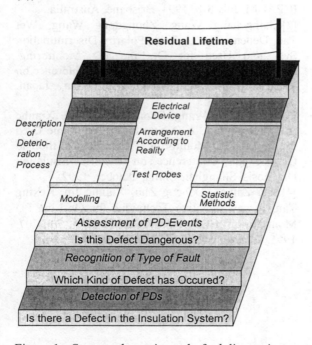

Figure 1: Steps to the main goal of pd diagnosis

The thereby obtained data, which supply beside information about the apparent charge also the phase angle and the instantaneous value of the test voltage as well as the interval between two succeeding pulses and the polarity of the pd, require further

interpretation in order to be used for pd-diagnosis (Schnettler and Tryba (2)). The main goal of all investigation dealt with aging of insulating material is the determination of residual lifetime. In order to reach this last step investigations have to answer several questions (Figure 1). One of the most important questions contains the assessment of pd events. Especially in flat cavities the evaluation of the measured data is complicated by the occurence of parallel discharges, which cannot be differentiated from single pulses of the same apparent charge due to the restricted bandwidth of commercial measuring systems (Engel and Peier (3)). Figure 2 illustrates the occurence of parallel discharges in a flat cavity.

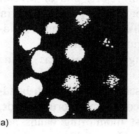

a)   b)

Figure 2: Light emission of parallel discharges in a flat cavity on the 'anode' a) and 'cathode' b) of the insulation material (Heuser (4))

In Engel (5) the determination of the degree of parallel discharges with the help of statistical data obtained by a commercial measuring system is presented. The next step presented in this paper, which is located on the ‚test-probe-stage‘ of the staircase in Fig. 1, contains the determination of the connection between the degradation of the insulation material and the degree of parallel discharges.

1   EXPERIMENTAL WORK
1.1   Measuring Arrangement

Experimental examinations are performed for test objects out of epoxy resin with a defined air-filled flat cavity. The occuring pd-current pulses are registered by a measuring system for pd-monitoring with included synchronous pd current measurement, while the test objects are aged with an ac test voltage. With the help of pd current measurement pd pulse trains, which are predominatly due to parallel pulses can be distinguished from pd pattern which are mainly the effect of single discharges (Engel and

High Voltage Engineering Symposium, 22–27 August 1999
Conference Publication No. 467, © IEE, 1999

Peier (3)). Thus, the degree of parallel discharges P_p of pd measurements can be determined. To analyse the influence of the degree of parallel pd on the degradation of the insulation material test objects out of the same material have to be constructed which are characterized by different degrees of parallel discharges. Figure 3 illustrates the test objects.

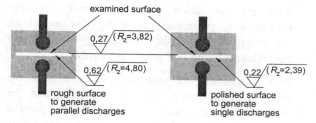

Figure 3: Test objects: Epoxy resin with embedded electrodes and a flat cavity

The left test object differs from the right one in a very rough surface inside the cavity. This leads to a high degree of parallel discharges, because of electron emission at micro pits. By way of contrast, the right test object contains a polished surface. This leads to the occurence of of high single discharges caused by delay time. The upper surface of both cavities posses of the same roughness. This surface will be examined after pd aging by scanning electron microscopy (sem) and roughness measurements to determine the material degradation due to pd behaviour.

1.2 Measuring Results

Figure 4 shows the pd pattern caused by different surface roughnesses in flat cavities in epoxy resin. Although the pd patterns are characterized by the same maximum apparent charge, a significant difference in the degree of parallel discharges can be established. Whereas in test objects with polished surface almost single discharges occur (degree of parallel discharge $P_p=5\%$), in the test object with rough surface parallel discharges are also likely to occur (degree of parallel discharge $P_p=90\%$).

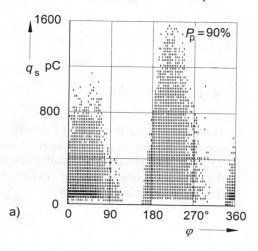

a)

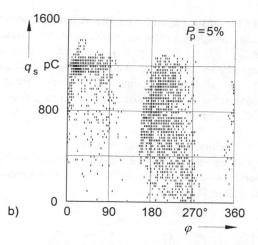

b)

Figure 4: Phase resolved pd pattern and degree of parallel pd P_p produced by flat cavities in epoxy resin due to a) rough and b) polished surface

2 DEGRADATION OF THE INSULATION MATERIAL

After an aging time of 10 hours the upper surfaces are proved by scanning electron microscopy and roughness measurements.

2.1 Scanning Electron Microscopy

Figure 5 shows the differrence in the structure of the surfaces after being aged by parallel and single discharges.

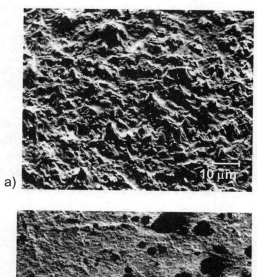

a)

b)

Figure 5: Scanning microscope photographs due to pd aging a) by parallel and b) by single discharges

While the surface aged by parallel discharges shows a higher roughness over the whole area, the surface aged by single discharges of the same apparent charge shows deep holes. These craters will be the start of electrical treeing. Thus, pd pattern predominated by single discharges lead to a shorter residual lifetime

2.2 Roughness measurements

The roughness of a surface can be described by two parameters (DIN ISO 1302). Figure 6 illustrates these values R_a and R_z.

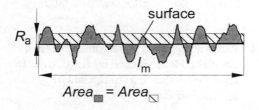

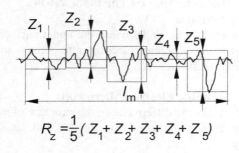

$$R_z = \frac{1}{5}(Z_1 + Z_2 + Z_3 + Z_4 + Z_5)$$

Figure 6: Graphical explanation of the parameter of surface roughness according to DIN ISO 1302

While R_a represents the interated difference to the mean value of the surface contour, in R_z the maximum and minimum values are considered.

The results of roughness measurements are presented in Figure 7. The significant rise in the roughness of the surface damaged by single discharges can be seen in contrast to the small rise in the case of surfaces aged by parallel discharges.

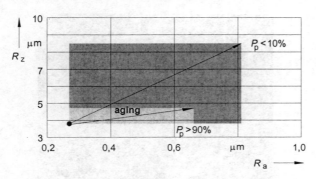

Figure 7: Connection between the degree of parallel discharges P_p and roughness of the aged surface expressed in the parameters R_a and R_z according to DIN ISO 1302

Thus, scanning electron microscopy and roughness measurement prove the dangerous insulation degradation due to single discharges caused by low electron emission.

3 INTERPRETATION AND DISCUSSION

Partial discharges can be simulated with regard to gas- and surface-discharge processes. In Engel and Peier (6) a model based on physical understanding used for simulation of the relevant discharge processes is presented. By using this model an interpretation of the different degradation mechanisms found in Figure 5 is possible.

In Figure 8a and 9a measured pd currents of the same apparent charge are shown. In Figure 8 the measured charge consists of two smaller parallel discharges, whereas Figure 9 presents a single discharge. With the help of the model energy densities can be determined.

Because of the amplitude of the pd current in Figure 8 the field strength can be determined as the minimum inception field strenght. The time-integrated surface energy densities caused by electrons spread out over the dielectric surface at the 'anode' and UV-radiation in the cathode fall are shown in Figure 8b.

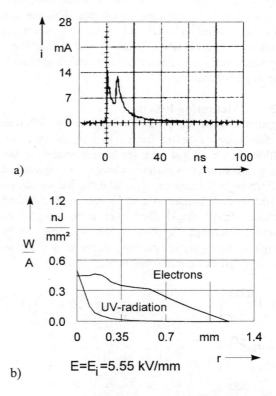

Figure 8: a) Measured pd current and b) simulated energy densities of UV radiation at the 'cathode' and electron bombardement at the 'anode' surface of the cavity at minimum inception conditions

The comparison of figure 8b to figure 9b shows the influence of field strength on the pd current and the energy densities. Caused by delay time the field strength in the cavity and the amplitude of the pd current rises. The delay time can be traced back to the polished surface.

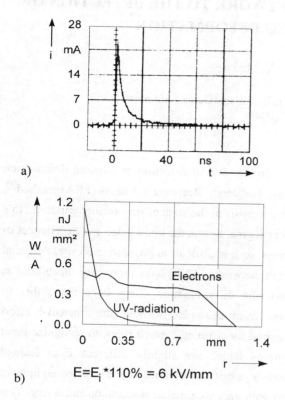

a)

b) $E = E_i * 110\% = 6 \text{ kV/mm}$

Figure 9: a) Measured pd current and b) simulated energy densities of UV radiation at the 'cathode' and electron bombardement at the 'anode' surface of the cavity 10% above minimum inception conditions

These simulation results are good tools to interpret the measured value 'apparent charge', because a comparison of the energy densities shows, that inspite of the same value of the apparent charge of the measured pd current, the UV-radiation energy in the 'cathode' centre is two times higher for single discharges.

Thus, the deep holes and craters on the dielectric surface aged by single discharges (figure 5b) can be traced back to the high energy density of UV-radiation in the 'cathode' centre of the discharge due to time delay.

CONCLUSION

The electrical insulation is the responsible element for the limited lifetime of high voltage apparatus. As pd in the dielectric are the prime cause of insulation degradation pd measurements are of great importance with regard to quality assurance. Therefore the evaluation of measured pd pattern is

an important but also difficult task. Especially the occurance of parallel discharges in flat cavities complicates the interpretation. To determine the relation between the degree of parallel discharges and the degradation of insulation material test objects out of epoxy resin with different degrees of parallel discharges are produced. The damaging effect of the pd is proved by scanning electron microscopy (sem) and roughness measurements. The results are interpreted with the help of a model based on simulation of the relevant gas discharge processes. The different structures of the aged surfaces are interpreted by means of calculated energy densities on the dielectric surface as a function of the inception field strength.

REFERENCES

1. Gulski, E., 1995, "Application of Modern PD Detection Techniques to Fault Recognition in the Insulation of High Voltage Equipment", 9th International Symposium on High Voltage Engineering, paper No. 5642

2. Schnettler, A., Tryba, V., 1993, "Artificial Self-Organizing Neural Network for Partial Discharge Source Recognition", Archiv für Elektrotechnik, 76, 149- 154

3. Engel, K., Peier, D., 1997, "Physically Based Interpretation of Partial Discharges in Flat Cavities", 10th International Symposium on High Voltage Engineering, 189-192

4. Heuser, C., 1984, "Zur Ozonerzeugung in elektrischen Gasentladungen", PhD Thesis, RWTH Aachen

5. Engel, K., 1998, "Bewertung von Teilentladungen in spaltförmigen Isolierstoffdefekten", PhD Thesis, University of Dortmund, Shaker Verlag, Aachen, Germany

6. Engel, K., Peier, D., 1997, "The Influence of Dielectric Material on Partial Discharges in Flat Cavities, International Conference on Dielectrics and Insulation, 229-232

AUTHOR

Dr.-Ing. Katrin Temmen née Engel
Institute of High Voltage Engineering
University of Dortmund
D-44221 Dortmund Germany
Katrin.temmen@ha1.e-technik.uni-dortmund.de

APPLICATION OF ARTIFICIAL NEURAL NETWORK TO THE DETECTION OF THE TRANSFORMER WINDING DEFORMATION[*]

D.K. Xu, C.Z. Fu, Y.M. Li

High Voltage & Insulation Dept., School of E.E., Xi'an Jiaotong University
Xi'an, P.R.China. 710049

Abstract: The application of the artificial neural network technique to the Frequency Response Analysis method (FRA) for the detection of transformer winding deformation is presented. A set of simulating experiments is performed in order to obtain the information of deteriorated winding. The fingerprints of the state of transformer windings obtained from simulating tests of deformation with different types and positions are learned by a multilayer feedforward neural network using back-propagation algorithm. Results show that after being trained, the neural network could well discriminate the state of transformer windings.

1 INTRODUCTION

In power systems, transformer is one of the essential elements and failures of large transformer can cause serious problems in electric utility operation. The rate of transformer failures caused directly or indirectly by the winding deformation remains high. According to the latest information about the reliability analysis of 220kV and above transformers in China in 1997, the unplanned offstream time due to the winding takes 79.49% of the whole in 220kV gradation, and 72.31% in 330kV, 98.92% in 500kV. Therefore the study on methods of detecting the transformer winding deformations is of practical value to avoid unexpected accidents.

In the field of detection of winding deformation using Frequency Response Analysis (FRA) method[1], the diagnosis of the state of transformer windings, to a great degree, lies on the knowledge and experiences of expert, so it is difficult to popularize on site. Artificial neural networks (ANN) have promising capabilities as automatic discriminators. They have the ability to learn from examples without the intensive effort required for defining explicit rules, to recognize input patterns which are slightly different from learned ones, to adapt to differentiating pieces of equipment and with only minimal customization necessary. This paper deals with behavior of a multilayer feed-forward neural network using back-propagation algorithm for the automatic discrimination of the state of winding.

2 STRUCTURE OF ANN

The structure of the ANN adopted in this research is a simple three-layer system with feedforward connections, which has been found to be one of the typical structures of ANN[2]. Fig.1 illustrates an associative ANN consisting of input, hidden and output layers. The relationship between input values x_i and output vales y_k could be represented by the

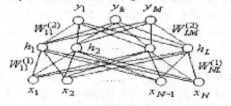

Fig.1 Structure of ANN

[*] Supported by NSFC Combined with North-East Electric Power Company of China（No．59637200）

following equations

$$h_j = f\left(\sum_{i=1}^{N+1} w_{ij}^{(1)} x_i\right) \qquad j=1,2,\dots,L \quad (1)$$

$$y_k = f\left(\sum_{j=1}^{L+1} w_{jk}^{(2)} h_j\right) \qquad k=1,2,\dots,M \quad (2)$$

$$f(x) = \frac{1}{1+exp(-x)} \qquad (3)$$

Here, $w_{ij}^{(1)}$ (i=1,2,…,N ; j=1,2,…,L) is the connection strength between the neuron i in the input layer and the neuron j in the hidden layer. $w_{jk}^{(2)}$ (j=1,2,…,L; k=1,2,…,M) is the connection strength between the neuron j in the hidden layer and the neuron k in the output layer. The value h_j is the output value of neurons in the hidden. The function $f(x)$ is the sigmoid function.

The ANN learns deformation patterns by itself, using the back-propagation learning method. The back-propagation algorithm performs the input to output ' mapping by making weight connection adjustments following the discrepancy (error) between the computed output value and the desired output value (teacher signal).

3 SIMULATING TEST

The transformer winding can be regarded as a passive, linear two-port network composed of resistance, inductance and capacitance. After the deformation of windings the parameters, mainly the inductance and the capacitance, have a change and so does the performance of the network. FRA method is chosen to be the detecting method and the frequency of input signal is changing from 10kHz up to 1MHz. The changes of transfer function can be adopted as the criterion of the winding deformation. The schematic diagram of implemented instrumentation system is given in Fig.2.

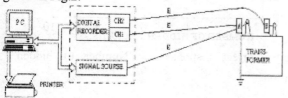

Fig.2 Schematic diagram of FRA system

The FRA method is utilized in lab condition on a 35kV/10kV three-phase model transformer which has 54 tapping points in the primary winding of phase B. Some capacitors are connected between tap points or between tap points and earth point to simulate deformations of different types and on different locations.

4 RECOGNITION OF STATE OF WINDING

The neural network, when adequately designed and trained, can synthesize a useful nonlinear mapping between input and output patterns. This is a key property for winding deformation detection. Fig.3 outlines the general procedure for detection of winding deformation.. Extracted features from the signature signal of the experiment data are used as inputs to train a standard feedforward layered perceptron artificial neural network.

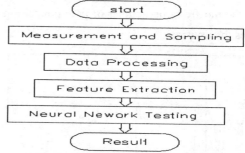

Fig.3 Winding deformation detection procedure

4.1 FINGERPRINTS OF WINDING STATE IN FRA

Obviously, the key of correctness of recognition is the fingerprint picked-up from the test data, which should represent the state of windings roundly. In this paper, correlation coefficient

$$\rho_{xy} = \frac{\sum_{i=1}^{N} x_i y_i}{\sqrt{\sum_{i=1}^{N} x_i^2 \sum_{i=1}^{N} y_i^2}} \qquad (4)$$

and standard deviation

$$E_{xy} = \sqrt{\frac{\sum_{i=1}^{N}(x_i - y_i)^2}{N-1}} \qquad (5)$$

$\{x_n\}, \{y_n\}_{n=\overline{1,N}}$: two sets of test data are adopted as the fingerprints to describe difference between two sets of test data.

Generally, the deformation only causes changes of partial pole points which relate to the variance in certain section of response curve, and the ρ_{xy} and E_{xy} in the whole frequency range may not change so much, even for some seriously destroyed windings. Therefore, the measured frequency range is divided into three sections: high frequency section, middle frequency section and low frequency section. The correlation coefficient ρ_{xy} and the standard deviation in not only the whole range but also every one of three frequency sections are chosen as the eigen parameters. Because of the diversity of structure of windings, the distribution of pole points of transfer function is different. And in order to reflect the state of winding by these eight parameters, division of the frequency range must be reasonable. So, each section is ordered to have the mostly same number of pole points. Fig.4 illustrates the division.

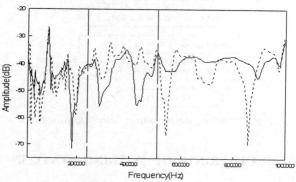

Fig 4. The frequency range is divided into three section

Due to low value of winding resistance of transformer, the movements of pole points only take distinct effect to nearby range. According to this dividing method, the changes of different sections are mainly caused by the changes of pole point positions.

4.2 PARAMETERS OF ANN

According to the fingerprints mentioned above, eight eigen parameters are chosen as the input patterns of the ANN, so the input vector is an eight dimensions one $\square$ $x_1, x_2, ..., x_8$ $\square$, thereinto the standard deviation is normalized, and the number of neurons in the input layer N=8 .

The choice of output in this particular problem is very straightforward. A binary output is sufficient to indicate whether the winding has deformation or not. In this paper, a value of 0 indicates 'NORMAL' and 1 indicates 'ALARM' respectively, and the number of neurons in the output layer M=1.

After the inputs and outputs are defined, the next task is to incorporate hidden layers in the network. The selection of hidden layer is a matter of trial and error. However, it has been observed that in most applications, one hidden layer is sufficient [2]. In order to obtain the appropriate number of neurons in the hidden layer, networks with different number of hidden neurons are trained by 24 test stylebooks and 24 desired output values. The allowed error in training is 0.5%. Fig.5 shows the learning counts needed for networks with different hidden neurons. After being trained, the networks are tested by 10 stylebooks randomly obtained from the remainders. Fig.6 shows the percentage of correct response for different networks. Taking account of these two aspects, the number of neurons in the hidden layer is chosen L =5.

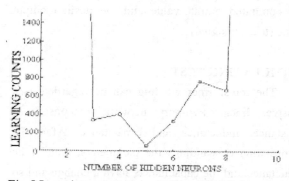

Fig.5 Learning counts vs. number of hidden neurons

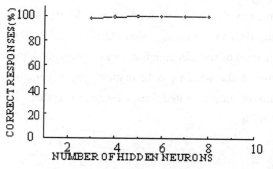

Fig.6 Correct response vs. Number of hidden neurons

In order to select the optimal learning step η and inertia coefficient α of the network, the numbers of neurons in three layers are fixed on, the relations between η, α and learning counts are studied respectively. The allowed error is 0.5%, too. From Fig.7 and 8, it can be seen that η =1.0, α =0.8 are the best.

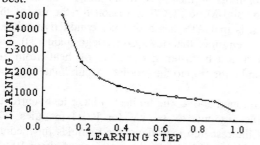

Fig.7 Learning counts vs. learning step η (α =0.8)

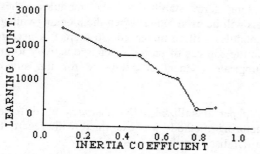

Fig.8 Learning counts vs. inertia coefficient α
(η =1.0)

4.3 RESULTS AND DISCUSSION

The proposed ANN should satisfy the following requirement: it must possess a high level of ability to correctly discriminate input patterns with deformation from those without deformation.

While recognizing the all remained stylebooks, the percentage of correct responses reaches 100%, since the decentralization of simulating test data are small and training package includes the deformation types in test package.

In further researches it is found that, for new input patterns obtained from simulating data on model transformer with different types from those old ones, the percentage of correct responses falls down quickly. And for new input patterns obtained from engineering field, the percentage of correct responses also falls down quickly and the decentralization of recognition results remains high, because the types,

positions and extent of winding deformation differ from the training ones so much.

While the network is trained by the old input patterns and the new ones together, the percentage of correct responses rises up to 95% again. It means that the proposed network is able to detect winding deformation with high degree of accuracy, and it is important to collect kinds of input patterns to train the network continually, thus the network can possess more and more experiences to improve the ability of recognition. The test results are listed in the Table 1.

Table 1. Test Results

ANN Structure	
Input Neurons	8
Hidden Neurons	5
Output Neurons	1
Patterns	
Training Patterns	24
Testing Patterns	20
Test Results	
Correct Response	>95%
Allowed Error	0.5%

5 CONCLUSION

An ANN based algorithm for deformation detection of transformer winding is studied using simulating test data of deformation with different types and positions. The results show that the algorithm is capable of distinguishing between normal and failed state quite satisfactorily and thus successfully establish the efficacy of the proposed method. Therefore, for modern transformer windings with hidden deformation, the ANN aided FRA method would be more effective. The method is simple but robust and easy to implement.

REFERENCE

1 E.P.Dick, C.C.Erven, "Transformer Diagnostic Testing by Frequency Response Analysis", IEEE *Trans on PAS-97,* 1978(6): 2144-2152
2 H.Suzuki and T.Endoh, "Pattern Recognition of Partial Discharge in XLPE Cables Using a Neural Network", IEEE *Trans. Tran. on Electrical. Insulation*, vol. 27, no. 3, 1992: 543-549.

INFLUENCE OF MANUFACTURING PARAMETERS ON THE PD-BEHAVIOUR OF AlN-SUBSTRATES

T. Ebke, D. Peier, K. Temmen née Engel

University of Dortmund, Germany

ABSTRACT

Metallized aluminiumnitride - substrates (AlN) for power semiconductor modules are examined in regard to their partial discharge behaviour. The samples used differ in the size of the metallization and in the duration of the structuring etching process, which has an effect on the vertical geometry of the edges. Other examined samples were metallized by sputtering a thin layer of copper onto the surface of a blank AlN-plate instead of the usual bond technology. These samples differ in size and shape of the metallization.

It is found out that the duration of the etching process has a decreasing influence on the appearance of partial discharges of high apparent charge. This is because a longer etching process makes the edges of the metallization becoming steeper and some thin copper on the surface of the ceramic along the edges disappear. As in these examinations sets of samples with different sizes of the metallization are regarded the desired effect can be seen in each set of samples. When comparing the sets to each other, an influence of the size of the metallization on the pd-behaviour with bigger samples showing more pd can be recognized. This is proved by pd-measurements of the samples with sputtered metallization. In regard to the shapes of the metallization the influence of the angles of a corner is examined. It can be seen that smaller angles cause more partial discharge because of higher field strength at the tip.

INTRODUCTION

In recent years the development of silicon semiconductors reached voltage levels of several kilovolts. For example IGBT (Insulated Gate Bipolar Transistor) chips are available for a blocking voltage of 3.3 kV (eupec (1)). Since the modules are designed for higher currents as well, power losses in the silicon chips increase strongly. To avoid overheating and therefore destruction of the chips the insulating substrate-material of high power semiconductor modules was changed

from aluminiumoxide (Al_2O_3), which is still commonly used for most applications, to aluminiumnitride (AlN) (Lefranc and Mitic (2)). The reason for the use of this material is that AlN has a thermal conductivity that is about seven times that of Al_2O_3 (Kriegesmann (3)). This allows a much faster dissipation of heat from the semiconductor chip to the (cooled) groundplate.

As mentioned above, the highest voltage level currently available amounts to 3.3 kV for IGBT-modules. The aim of the research is to develope modules for medium voltage range applications, i.e. voltages of about 10 kV. In commercially available IGBT-modules even at current voltage levels partial discharges cause problems in the long term stability of the modules. These problems will be even bigger when the blocking voltage of the modules will be increased. So there is a need to examine the sources of partial discharge activity to get an appropriate connection geometry for the silicon chips.

In this paper metallized AlN-substrates of different manufacturing methods are examined with regard to their partial discharge behaviour.

TEST SAMPLES AND VARIED MANUFACTURING PARAMETERS

In the examination a total of 19 samples is used. An overview of all samples is given in table 1. There are three sets of four samples, differing in size of the metallization and duration of the etching process (table 1, samples 1-4, 5-8, 9-12 , three with different sizes of the metallization (13-15), metallized in a different way and four samples with different shapes of the metallization (16-19). Samples 1-12 are metallized using the Direct Copper Bonding-Technology (DCB), the metallization of samples 13-19 is made by sputtering a thin copper layer of about 200 nm onto the surface of a blank ceramic-plate. Samples 1-12 differ in the duration of the etching process. A single pass of a standardized etching process takes away about 35 μm of

TABLE 1 – Overview of the examined samples

No.	d_{AlN} / μm	d_{Cu} / μm	Bonding Technology	Shape of metallization	Lateral length of metallization / mm	No. of etching processes
1-4	635	300	DCB	square	5	10, 12, 14, 16
5-8	635	300	DCB	square	15	10, 12, 14, 16
9-12	635	300	DCB	square	47	10, 12, 14, 16
13-15	635	0.2	sputtering t.	square	5, 10, 15	1
16-19	635	0.2	sputtering t.	different shapes (figure 4)		1

High Voltage Engineering Symposium, 22–27 August 1999
Conference Publication No. 467, © IEE, 1999

copper. Since the metallization is about 300 μm thick, it takes approximately nine passes until the blank ceramic surface appears. The following passes only affect the vertical geometry of the edges of the protected square that remains. The four samples of each set pass the etching process 10, 12, 14 or 16 times, differing therefore in the vertical geometry of the edges. These samples are metallized in the Direct Copper Bonding method (DCB) described by Kluge-Weiss and Gobrecht (4). This method was originally developed for the metallization of Al_2O_3 ceramics. The bonding is made by a very thin layer of copperoxide. To apply this method to AlN, the surface has to be oxidized first to get a layer of Al_2O_3. After this the technique corresponds to the metallization of pure Al_2O_3. The oxidation of the surface must be carried out very carefully, so that the oxide layer is thick enough without Al_2O_3 remaining underneath the copper after bonding. If the oxide layer is too thin elementary nitrogenium is set free in a chemical reaction between copper and AlN. This would cause voids in the bonding layer that are sources of partial discharge. Samples 13-19 are made of a blank AlN-ceramic with a layer of copper (200 nm) sputtered on both sides. Using these test objects the influence of size and shape of the metallization without any disturbance of other pd-sources is examined. Because of the production technology no cavities occur underneath the sputtered metallization.

MEASUREMENTS AND RESULTS

All measurements are made using phase-resolved partial discharge measurement. Because of the similarity of the phase-angle-distributions in the examination of the influence of size and shape of the metallization the results are shown by comparing some characteristic values. These values are the maximum apparent charge, the number of discharges in a measuring interval or the summarized apparent charge of all measured discharges.

Influence of the etching process

The duration of the etching process has an influence on the geometry of the edge of the 300 μm copper metallization. When the etching time is as short as possible (10 passes of the standardized process) a thin copper layer remains on the AlN-surface along the edge of the metallization. This copper disappears with further passes of the etching process. Figure 1 shows photomicrographs of (a) a corner of a sample that was etched 10 times and (b) a corner of another sample etched 16 times. The broken line in figure 1a indicates the edge of the metallization as desired. As it is seen in the photograph only the areas above and on the left of the broken line are in focus because the copper layer is very thin in this area. The 300 μm copper layer is out of focus. The thin foothills are critical in two ways: first there is an increased field strength at the tip. As the copper is very thin this field strength could lead to partial discharges. Even more critical is the fact that the tip lies on the surface of the insulating material, the aluminiumnitride ceramic. Therefore the increased field strength is tangential to a material interface. These interfaces are known to have a low breakdown strength. For this reason partial discharges along the surface are very likely to occur. Measurements are carried out in an insulating liquid to get nearly realistic circumstances. High power semiconductor modules are filled with silicone gel to prevent partial discharge on the surface so measurements in air would lead to distorted results. Partial discharges along material interfaces appear in the phase-angle-distribution between zero and maximum voltage (Blasius and Weck (5)) and have an increasing apparent charge at higher voltages because of the rather unlimited space they have to develop.

Figure 2 a and b show the phase-angle-distributions (PAD) corresponding to the photomicrographs in figure 1 a and b. The measurements are carried out at a peak voltage of 9.0 kV for one minute. The phase-angle-distributions are modified to get a closer view of the very rare discharges of high apparent charge.

a)

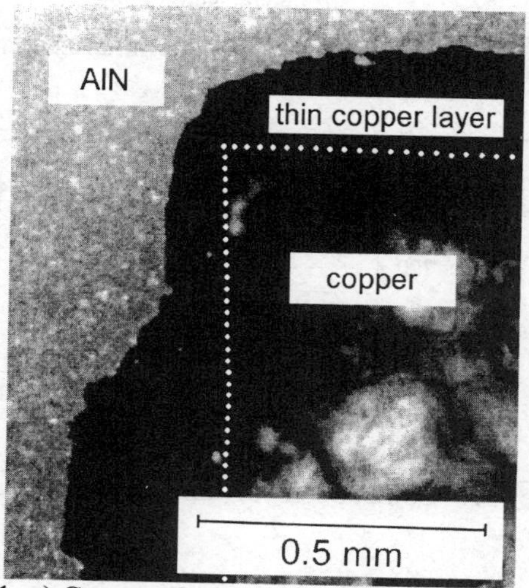

b)

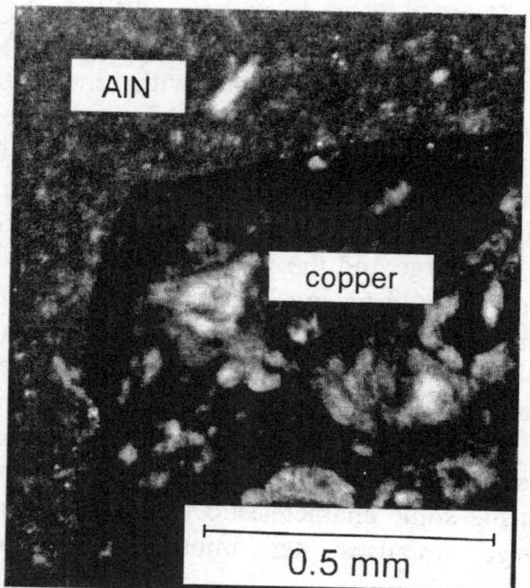

Figure 1: a) Corner of a sample etched 10 times b) Corner of a sample etched 16 times

a)

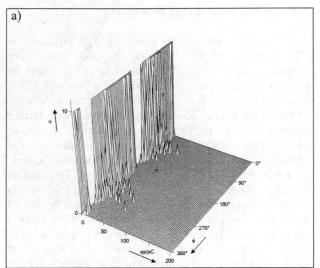

b)

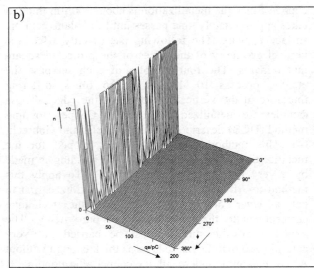

Figure 2: a) PAD of the sample etched 10 times   b) PAD of the sample etched 16 times

The number of discharges is cut at a value of 10 to adjust the scale of the n-axis and the whole image is turned around so that the interesting peaks are visible. The phase-angle-distributions show that the number of impulses with high apparent charge is reduced. This leads to a decrease in the maximum value of the apparent charge from around 60 pC to nearly 10 pC. This is very important because international standards about partial discharge only refer to this value. In most cases standards give a limiting value and a single discharge of a higher apparent charge causes the test device failing the test. The large number of pd of small apparent charge is not effected by the etching process. To improve this means of reducing the electric field strength well known in high voltage engineering have to be applied to semiconductor modules regarding the possibility of realization.

Influence of the size of the metallization

Comparing the examinations of samples 1-12 bigger samples seem to have more partial discharges than smaller ones. The partial discharge inception voltage of the samples with a copper square of 47 mm x 47 mm on both sides amounts to just about half of the inception voltage of the samples with a metallization of 5 mm x 5 mm. To support this theory samples 13-15 are tested. The sputtering technique is chosen to avoid any gas inclusions between the aluminiumnitride and the copper. These gaps would be another source of partial discharge and the impulses would superimpose those that should be investigated. The impulses in all measurements of all sizes appear between zero voltage and the following maximum voltage as it is expected for partial discharges along surfaces (Blasius and Weck (5)). For this reason the phase-angle-distributions look quite similar and the influence can be seen better by comparing some characteristic values. The number of discharge impulses per minute at 10 kV, the

summarized apparent charge of all impulses and the maximum apparent charge are shown in figure 3 a-c.

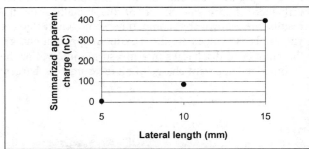

Figure 3a: Comparison of the numbers of pd

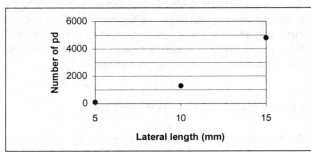

Figure 3b: Comparison of the summarized apparent charges

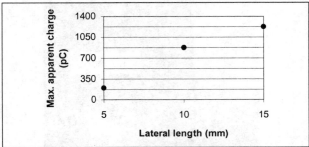

Figure 3c: Comparison of the maximum apparent charges

Samples 13-15 show the same pd-behaviour as samples 1-12 regarding the size of the metallization. A relation between the extends of the metallization and the level of partial discharge is therefore detected. This is caused by the bigger area surrounding the copper. The influence of the surface charges of a single partial discharge decreases because of their constant extends.

Influence of the shape of the metallization

The examination of the influence of the shape of the metallization is done using samples 15-18. Because of the sputtering technique voids between ceramic and copper are suppressed. The shape of the phase-angle disributions is similar to those in the previous examinations. Therefore again the influence of the shape of the metallization is shown by comparison of characteristic values. The shapes examined are shown in figure 4.

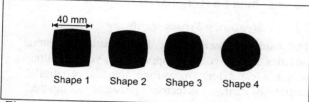

Figure 4: Shapes of the metallization

The values compared here are the number of partial discharge impulses per minute and the summarized apparent charge per minute. The measurements are carried out at a peak voltage of 7.0 kV. Figure 5 shows the comparisons of the characteristic values.

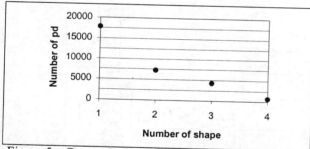

Figure 5a: Comparison of the numbers of pd

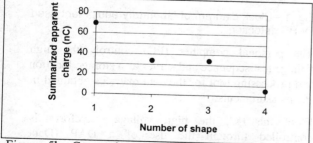

Figure 5b: Comparison of the summarized apparent charges

Obviously the samples show more partial discharge the sharper the corners of the metallization (small shape

numbers) are. This relation can be understood by analyzing the electric field strength. There is an increase in field strength near small radii and therefore the corners of the square-like shapes cause high intensities of the electric field. By increasing the angles of the corners the electric field is homogenized and partial discharges reduced.

CONCLUSION

In this paper the influence of four different manufacturing parameters on the partial discharge behaviour of metallized aluminiumnitride-substrates for high power semiconductor modules is examined. It is found out that the duration of the structuring etching process influences the appearance of pd of high apparent charge while the large number of pd of low apparent charge seems to be not effected at all. This shows the necessity of introducing means of field strength reduction. Small geometries without sharp corners seem to be more suitable for the design of semiconductor modules than larger ones.

REFERENCES

1. eupec GmbH +Co. KG, 1998, catalogue, Warstein, Germany

2. Lefranc, G., Mitic, G., 1997, „Untersuchungen von AlN-Substraten für zukünftige Leistungshalbleiter-module", VTE 9, 5, 246-252

3. Kriegesmann, J. (Publ.), 1989, DKG, Technische Keramische Werkstoffe, Dt. Wirtschaftsdienst, Köln

4. Kluge-Weiss, P.and Gobrecht, J., 1985, „Directly Bonded Copper Metallization of AlN Substrates for Power Hybrids", Mat. Res. Soc. Symp. Proc., 40, 399-404

5. Blasius, P. and Weck, K.-H., 1993, „Bewertung von Teilentladungen", in: König, D. and Rao, Y.N. (Publ.), Teilentladungen in Betriebsmitteln der Energietechnik, vde-verlag, Berlin and Offenbach, Germany, 105-122

ACKNOWLEDGEMENT

This work is funded by the Deutsche Forschungsge-meinschaft (DFG, German Research Society) and the Maschinenfabrik Reinhausen GmbH, Regensburg, Germany. It is supported by the Siemens AG, Munich, Germany by preparation and provision of samples.

AUTHOR

Dipl.-Ing. Thomas Ebke
Institute of High Voltage Engineering
University of Dortmund
D-44221 Dortmund, Germany

MEASURING SYSTEM FOR PHASE-RESOLVED PARTIAL DISCHARGE DETECTION AT LOW FREQUENCIES

Juleigh Giddens*, Hans Edin*, Uno Gäfvert**

*Kungl. Tekniska Högskolan, Stockholm, Sweden; **ABB Corporate Research, Västerås, Sweden

Abstract

This paper addresses the measurement of partial discharges (PD) for a range of low frequencies of applied voltage and explores the frequency dependence of PD patterns for a few objects. The aim is to determine the existence of frequency-dependent trends in PD patterns and how these trends can be expressed through use of statistical parameters calculated from the PD patterns.

An experimental setup was developed to measure phase-resolved PD patterns for a range of low frequencies, from 1 mHz to 100 Hz. Tests were performed on a few PD-infested objects, including a corona gap, a needle-film object, and a generator stator bar, to demonstrate the measuring system and analysis methods. Statistical operators were used to quantify the PD patterns, and the resulting parameters were then considered as functions of frequency.

1. Introduction

The detection of partial discharges (PD) is useful in testing high-voltage insulation to determine its quality and age and to recognize defects in the insulation. PD measurements made at frequencies lower than power system frequencies require less power during testing than 50-60 Hz tests do, especially for highly capacitive objects, and could provide an insight into physical mechanisms of partial discharges [1].

In this study, the frequency dependence of the PD behavior has been investigated through use of statistical analysis of probability density functions obtained from the measured PD patterns.

This paper presents the variable-frequency PD measuring system, based in part on a previously developed variable-frequency measuring system [2], and some sample results of PD tests on a few test objects for demonstration of the system.

2. The Measuring System

The variable-frequency measuring system used in this study is based on a PD Systems Digital Partial Discharge Acquisition System, by Power Diagnostix Systems, GmbH. This ICM PD detector was designed for use at frequencies between 32 Hz and 400 Hz, so in order to study frequency dependence of phase-resolved PD patterns, additional software and hardware were devised to allow measurements in the range from 1 mHz to 100 Hz.

2.1 Measuring System Hardware

The hardware used in the PD measuring system includes a high-voltage supply, high-voltage filter, coupling capacitors, measuring impedance, digital partial discharge detector, personal computer, calibrator, and other accessories. These devices are linked together as indicated in the block diagram of Fig. 1.

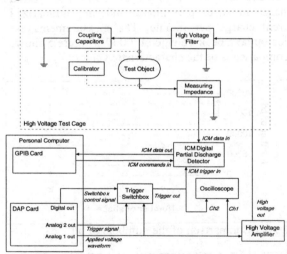

Fig. 1—Block diagram of laboratory equipment used for PD detection

The personal computer (PC) controls the high voltage test sequence and serves as a storage location and processing tool for the data matrices obtained in PD measurements.

From the PC, the high voltage waveform is controlled through the use of a DAP (Data Acquistion Processor) card, which generates a voltage waveform of a specified amplitude and frequency for a given number of cycles. This waveform is then amplified by the high voltage supply, a TREK 20/20 A amplifier.

High Voltage Engineering Symposium, 22–27 August 1999
Conference Publication No. 467, © IEE, 1999

The PC is also used to activate and control the digital partial discharge detector, to which it is connected through a GPIB (general purpose interface bus). Prewritten commands for the GPIB card and ICM PD detector for use in a C++ program were provided by the manufacturer, PD Systems. After a PD test, the PD matrix is downloaded from the PD detector memory to the PC.

Once the test results are sent to the PC, they are stored in the standard ICM data format, which can be analyzed using appropriate software on any PC.

2.2   Measuring System Software

The software that was used to manage a sequence of low-frequency PD tests was written in Borland C++, running under Windows 95. This software reads a sequence of test specifications from a settings text file, generates a sinus waveform of a given amplitude and frequency for a given number of periods, stores the measured PD data, and repeats until all of the tests listed in the text file are complete. The C++ program also reads a list of hardware settings for the PD detector from the settings text file, allowing the user to control gain, triggering, calibration, and other detector settings.

As mentioned earlier, the digital partial discharge detector that was used for this study was designed for use in 32 Hz-400 Hz tests, so new software had to be written to allow accurate data collection at frequencies under 32 Hz. In the end, three frequency ranges were defined, with a different method of generating the waveform or triggering the ICM implemented in each range. The frequency ranges and the methods used in each are described next.

<u>Range 1: 100 Hz ≥ frequency > 0.05 Hz</u> In the highest frequency range, the data acquistion processor card functions within its designed limits, but the PD detector is unable to trigger properly on sinusoidal frequencies less than 32 Hz. To trigger the PD detector consistently, a simultaneous squarewave of equal frequency is generated by the DAP card along with the voltage-supply input sinewave.

<u>Range 2: 0.05 Hz ≥ frequency ≥ 2.5 mHz</u> In the middle frequency range, the partial discharge detector shifts the data by a frequency-dependent number of phase channels. The phase shift was experimentally determined to be a linear function of frequency, so a compensating phase shift is applied by the C++ program for tests in this frequency range.

<u>Range 3: 2.5 mHz > frequency ≥ 1 mHz</u> In this range, the C++ program must send commands to the DAP once per cycle instructing it to generate a single waveform and replicate each of its points several times in succession to obtain the correct period. At these low frequencies, the PD detector is software-triggered by a command sent from the C++ program at the beginning of each cycle.

2.3   Verification of the Measuring System

The measuring system is accurate to within one phase channel (out of 256 total phase channels) at frequencies higher than 0.05 Hz, and is accurate to within three phase channels at lower frequencies. To test the measuring system, a pulse was generated at 180 degrees and at 360 degrees for several frequencies. The resulting plots are shown in Fig. 2.

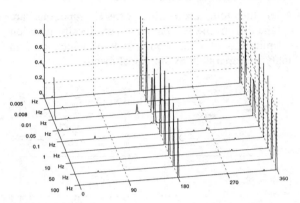

Fig. 2—Pulse test results demonstrating measuring system at different frequencies

3. Quantifying the Density Functions

Work by Gulski, Kreuger, and others has shown that statistical analysis can provide a means of quantifying probability density functions obtained from partial discharge tests [3].

A brief description of the process used in this study for statistical analysis of the measured PD test patterns is given next.

3.1   The Probability Density Functions

The data collected by the PD detector is arranged in the form of a 256 x 256 matrix whose row numbers represent the height of each PD impulse in pC, and whose column numbers represent the phase "channel" of the applied voltage during which the discharge occurred. The width of one phase channel is 360 degrees divided by 256.

Each element in the PD matrix is an integer count of the number of discharges that occurred in the phase channel determined by the element's column number and had the charge magnitude determined by the element's row number.

From the PD pattern, four probability density functions are obtained and arranged as functions of phase angle, and then used in statistical analysis of partial discharges [4]. They are:

- Pulse count density function $h_n(\varphi)$, defined as the number of discharges in each phase channel.

- Discharge amount density function $h_{qs}(\varphi)$, defined as the sum of the discharge magnitudes in each phase channel.

- Mean pulse height density function $h_{qn}(\varphi)$, defined as the average discharge magnitude in each phase channel. $h_{qn}(\varphi)$ is equal to the discharge amount density function $h_{qs}(\varphi)$ divided by the pulse count density function $h_n(\varphi)$.

- Maximum pulse height density function $h_{qm}(\varphi)$, defined as the maximum single discharge observed in each phase channel.

3.2   Phase vs. Shape Method

After the four probability density functions are obtained from the PD pattern, each density function is split into two functions, each of which is then analyzed separately by statistical means.

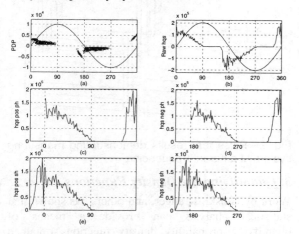

Fig. 3—(a) PD pattern; (b) resulting h_{qs}; (c) phase-broken positive h_{qs}; (d) phase-broken negative h_{qs} (e) shape-broken positive h_{qs}; (f) shape-broken negative h_{qs}

One method often used for splitting these functions is strictly according to the phase of the applied voltage. In this "phase" method, the so-called positive half of the density function is the portion from 0° to 180°, and the so-called negative half is the portion from 180° to 360°.

However, in many cases, the phase method causes extreme sensitivity in the statistical operators to small changes in the functions' behavior near applied-voltage zerocrossings. Therefore, in this study a "shape" method for breaking the functions is used, in which the functions are broken according to their positive and negative shapes, regardless of phase. Fig. 3 shows a probability density function broken by phase and the same function broken by shape.

3.3   Statistical Analysis

To quantify the halved density functions, the familiar low-order statistical moments mean and variance are calculated for each half function. Since higher order moments can provide more detail about the density functions, adjusted forms of the skewness and kurtosis are calculated also, as in Gulski's work [3]. Skewness (Sk) is a measure of asymmetry about the mean, and kurtosis (Ku) is a measure of the flattening of a curve near its mean.

For skewness, the 3rd central moment divided by the cube of the standard deviation is used. For kurtosis, the 4th central moment is divided by the fourth power of the standard deviation, and three is subtracted from the quotient to set the kurtosis of a Gaussian density function to be equal to zero.

4. Partial Discharge Test Results

When considering the frequency-dependent behavior of the PD test results, it is important to bear in mind that voltage dependencies in PD patterns often express themselves in similar ways to frequency dependencies. Varying voltage, like varying frequency, can alter the number of partial discharges, the phase location of the discharges, and the shape of the PD patterns. Therefore, when looking for frequency dependence in PD tests, it is advisable to compare only tests that have been performed at the same voltage and other conditions in order to minimize possible causes of discrepancy among the test results.

In some cases, a frequency dependence of the PD patterns is observable only above or below a certain voltage level. Therefore, the PD tests in this study were performed on each object at a few different voltages to allow observation of possible joint voltage-frequency dependence for the object.

The results of the tests and parts of the subsequent statistical analyses are presented next.

4.1   Corona Gap Test Results

A point-hemisphere corona gap is tested here to demonstrate the variable frequency PD test results for a familiar object. The h_{qs} appears in Fig. 4.

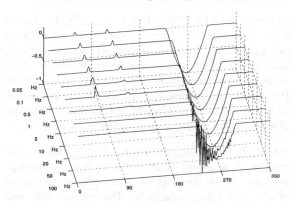

Fig. 4—h_{qs} for corona discharges (scaled to -1)

The corona gap shows little frequency dependence, especially in the negative half-cycle. At lower frequencies, when each phase channel lasts a longer time, more discharges appear per cycle.

The positive half-density-function is empty for all frequencies above 10 Hz. At and below 10 Hz, two distinct peaks appear in the positive half-density-function. The peaks represent the voltage ranges in which positive streamers occur. Between the peaks, there is steady "glow" discharge behavior, which is not recorded by the PD detector. At higher test voltages, the two positive peaks begin to appear at higher frequencies.

In the interests of space, no plots of statistical parameters are included for corona, since the mean, variance, skewness, and kurtosis are almost constant for the negative corona half-density-function.

4.2   Needle-Film Object Test Results

The needle-film object tested is a narrow steel needle suspended 1 mm above a Kapton film. Fig. 5 shows some frequency dependence, particularly in the relative size of peaks into which the positive h_{qs} half-density-function resolves itself. This is reflected by a decreasing Ku with frequency until about 1 Hz, when Ku becomes constant, shown in Fig. 6.

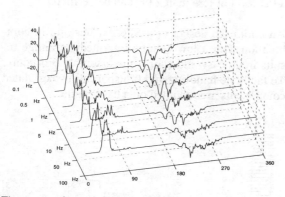

Fig. 5—h_{qs} for needle-film object

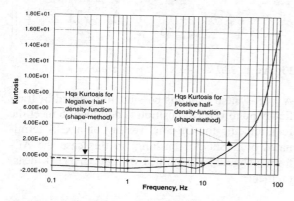

Fig. 6—Kurtosis of h_{qs} for needle-film object

4.3   Stator Bar Test Results

The stator bar tested is a used, asphalt-mica-insulated stator bar from a generator. The insulation contains delaminations that result in PD activity.

The stator bar exhibits frequency dependence in the number of discharges per cycle, the charge magnitudes, the phase position, and the shape of the PD patterns. Fig. 7 shows the $h_{qs}(\varphi)$ for the stator bar, indicating that the number and/or magnitude of the discharges decreases with decreased frequency. Fig. 8 plots the number of positive and negative discharges per cycle versus frequency. Mean and variance also decrease with frequency, but skewness and kurtosis show no consistent trend.

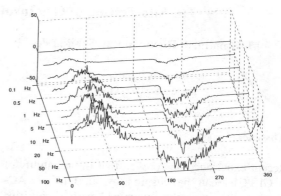

Fig. 7—h_{qs} for the stator bar

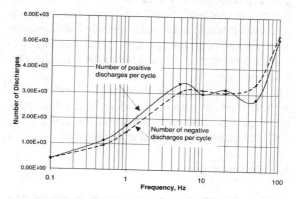

Fig. 8—Number of discharges per cycle, stator bar

5. Conclusions

A measuring system has been developed for measuring PD in the frequency range 1 mHz to 100 Hz. Frequency dependence does exist to varying extents for different types of PD-infested objects. In this study, the stator bar showed the strongest frequency-dependent trend.

The frequency-dependent characteristics of the PD patterns are indeed reflected by changes in the statistical parameters calculated from the PD patterns, although specific information regarding the density function shapes is lost if only statistical parameters are considered and no visual inspection of the PD patterns is performed.

References

[1] Holbøll, J.T., and H. Edin. "PD Detection vs. Loss Measurements at High Voltages with Variable Frequencies." 10th International Symposium on High Voltage Engineering. *Conference Proceedings*, Vol. 4, 1997.

[2] Staiblin, Jasmin. *Realization of a Measuring System for Partial Discharge Measurements at Low Frequency.* Thesis paper at Kungl. Tekniska Högskolan, August 1996.

[3] Kreuger, F.H., E. Gulski, A. Krivda. "Classification of Partial Discharges." *IEEE Transactions on Electrical Insulation*, Vol 28, No 6. December 1993

[4] Gulski, E. *Computer-Aided Recognition of Partial Discharges Using Statistical Tools.* The Hague, Netherlands: Delft University Press, 1991.

On-Site Testing of Gas Insulated Substation with AC voltage

M. Gamlin, J. Rickmann, P. Schikarski

Haefely Trench
Basel, Switzerland

R. Feger, K. Feser

University of Stuttgart
Germany

Abstract

Recent experience has proven that the reliability of a GIS/GIL can be ensured by an On-Site AC-Testing after assembly combined with a sensitive partial discharge measurement.

Either conventional 50/60 Hz transformer systems or resonant test systems with variable inductance as well as resonant test systems with variable frequency have been used to perform the GIS/GIL On-Site Testing.

The paper describes the different HV test systems with their advantages and limitations and the applicable partial discharge measuring methods. The UHF PD measuring method is emphasised.

1.   Introduction

GIS/GIL testing has been performed either by AC-voltage, lightning impulse voltage or switching impulse voltage or a combination of them.
The experience showed that no single test method detects all fault causes with the same sensitivity.
A GIS/GIL AC-testing with an increased voltage level combined with a sensitive partial discharge measurement has met best the requirements for detecting the different fault causes [1].

Present AC test procedures of several GIS/GIL suppliers consists of a prestressing at a voltage level of about $(1.1 \cdot \sqrt{3}) U_m / \sqrt{3}$ for some seconds up to some minutes followed by the 1 minute test voltage in the range of $(1.1-1.5) U_m$ and finished by a PD measurement at a voltage level of $1.1 \cdot U_m / \sqrt{3}$.

The AC test voltage for On-Site Testing can either be generated by a SF6 insulated test transformer system, a conventional series resonant test system at 50/60 Hz or a series resonant test system with variable frequency (30 .. 300 Hz). The series resonant test systems are mainly applicable for to a higher test capacitance and longer test duration.

2.   On Site AC Test Systems

2.1   On Site Testing with SF6-insulated test transformer systems of the type TES

SF6 insulated test transformer systems are designed for a short time duty. The max. output voltage and 15 min output power is up to 1000 kV and 375 kVA

at a weight to power ratio of approx. 10 kg/kVA. The TES-system can either be directly flanged to a GIS or used in combination with a SF6/Air-bushing.
If directly flanged to the GIS via a multiflange housing an encapsulated coupling capacitor can be integrated. With such an arrangement and the application of the conventional PD measuring method a PD level of less than 2 pC can be verified.

An optimised design of a SF6/Air bushing in combination with a low volume intermediate flange results in an easy and short time assembly procedure. To facilitate handling and transport all encapsulated components are mounted on a base frame.

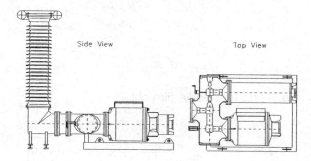

Figure 1: *230 kV, 45 kVA test system*

A TES-system which is possible to be tilted hydraulically up to vertical position is designed to meet the requirements for On-Site Testing of a GIS terminated by a SF6/Air bushing. In this case the AC-testing should be combined with an UHF PD measurement.

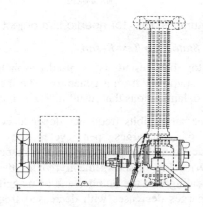

Figure 2: *510 kV, 105 kVA test system*

High Voltage Engineering Symposium, 22–27 August 1999
Conference Publication No. 467, © IEE, 1999

2.2 On-Site Testing with conventional series resonant test systems of the insulating cylinder type RZ

Conventional oil insulated series resonant reactors of the insulating cylinder type are designed for a longer test duty and higher power rating. Typical performance data of a two module tower is a output voltage up to 800 kV and a 1h output power of up to 3200 kVA at a weight to power ratio of approx. 5 kg/kVA. The test voltage is applied to the test object via an Air/SF6 bushing. Using this design several modules can be cascaded to increase the test voltage but a fully encapsulated On-Site Test arrangement is not possible to achieve. Therefore the UHF PD measuring method is preferred when using this test arrangement.

The auxiliary equipment such as exciter transformer, electrodes, base frames and the measuring capacitor is stored in a 20" container. Furthermore a control room is integrated in the container as well. The reactor itself is transported in a crate.

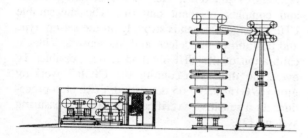

Figure 3: *800 kV, 4 A, 50/60 Hz series resonant test system*

2.3 On-Site Testing with SF6-insulated resonant test systems with variable frequency of the type RSS

With a SF6 insulated series resonant reactor with fixed inductance an encapsulated test set up with a higher output power rating compared to a SF6 insulated test transformer can be achieved. The weight to power rating ratio can be optimised to approx. 0.6 kg/kVA.
The test circuit is tuned to the resonant condition by means of adjusting the frequency of a high clock rate frequency converter (1). The frequency range of the test system can be calculated by the following formula:

$$\frac{f_{max}}{f_{min}} = \sqrt{\frac{C_{max}}{C_{min}}}$$

Similar to SF6 test transformer systems the resonant reactor system (3) can either be connected directly

to a GIS/GIL by using an intermediate flange or by means of a SF6/Air bushing.

Disturbances generated by the semiconductor switching in the range of some nC require a filtering and shielding (2) of the power supply in order to allow a sensitive PD measurement during the AC testing.
Using a fully encapsulated arrangement a PD level of less than 1 pC can be verified.
A signal windowing of the PD detector is not necessary the full phase angle range can be recorded.

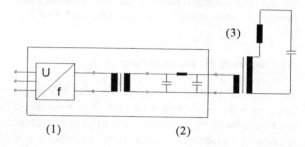

Figure 4: *Circuit diagram of a resonant test set with variable frequency*

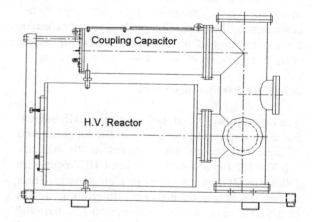

Figure 5: *400 kV, 3 A test system with 0.5 nF coupling capacitor*

If the resonant system is designed for resonant frequencies above 80 Hz/100 Hz the GIS can be tested with installed voltage transformers.
This allows testing of the fully assembled GIS without any temporary disassembly of voltage transformers which are designed for a 50/60 Hz system.

3. On-Site partial discharge measurements on GIS

3.1 Conventional PD measurements

For on-site PD measurements on GIS several different measuring techniques are applicable. The conventional PD measuring method according to IEC 60270 is the most known since it is in use already for many years in various fields of PD measurements. PD signals are measured using a coupling capacitor which can be fully enclosed or externally connected. The PD signal is not significantly damped in the GIS so that this method can be calibrated. Since PD signals are measured in the kHz range excessive electromagnetic interference of much higher amplitude, such as corona discharges of incoming overhead lines, makes detection of PD in GIS difficult. To obtain an acceptable background noise level the use of costly filter and noise suppression methods is required. For this reason an on-site PD measurement with the conventional method is only suitable in fully enclosed test set-ups.

3.2 Acoustic PD measurements

The acoustic method is another sensitive PD measuring technique applicable on GIS. It is immune to electromagnetic noise in the substation and because it cannot be calibrated, a sensitivity verification was developed by the CIGRE working group WG15/33.03.05 [2]. Since the measuring range is limited to one gas compartment acoustic PD measurements on a complete GIS are very time-consuming. Therefore, this method is rather useful for the location of an eventually occurring defect than for an on-site PD test of a GIS arrangement.

3.3 UHF PD measurements

The measuring of PD in GIS using the UHF method has been discussed intensively in the literature and is meanwhile commonly accepted in the industry [3]. The PD signals are detected at UHF frequencies (300MHz- 3GHz) with built-in plate sensors. Since corona discharges do not create interfering signals in the UHF range on-site measurements on partially open test set-ups become possible. The sensitivity of the UHF method is comparable to the conventional method. Measurements can be performed either in the frequency-domain with a bandwidth of a few MHz, which is commonly referred to as the narrow-band method, or in the time-domain over a broad frequency range up to 2 GHz, which is usually called the broad-band method (fig. 6). The PD detection using the narrow-band method is done with a commercial spectrum analyzer and low noise / high gain UHF preamplifiers and results in the frequency spectrum of the PD signal. For the broad-band method the PD signals are detected from an UHF peak detector. Then, a suitable PD measuring device displays a PRPD pattern similar to those obtained with the conventional method. For noise suppression and protection of the measuring equipment the use of a high pass filter with a cut-off frequency of about 250-300MHz in series is necessary.

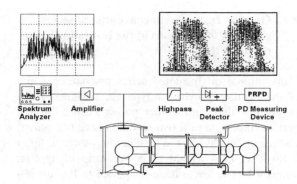

Figure 6: *Principle of the narrow-band (left) and broad-band (right) UHF detection method with typical obtained signals of a free moving particle*

Partial discharges in the GIS excite electromagnetic waves propagating along the coaxial waveguide constituted by the inner and outer conductor. Due to the very fast rise time of the discharge currents waves are not only excited in the TEM mode but also in several higher order TE and TM modes. Reflections at various discontinuities of the waveguide (spacers, T-junctions, etc.) cause standing waves and complex resonant patterns. The measurable UHF PD signal depends strongly on the set-up, type and position of the defect, and the sensors. Thus, a calibration of the UHF method is not possible. To overcome this short-coming the CIGRE working group WG15/ 33.03.05 developed a two step procedure for a sensitivity verification of UHF measuring systems [2].

Investigations have proven that pulses of a certain characteristic injected via a sensor into a GIS excite a UHF spectrum similar to that of a free moving particle. The rise time of the pulse has to be less than one nanosecond, the pulse repetition rate less than 100 kHz, and the time to half-value needs to be more than a few ten nanoseconds, depending on the pulse shape. In the first step, the magnitude of the artificial pulse has to be determined by placing a particle close to a sensor in a compartment of the laboratory set-up (fig. 7). When the PD signal of the moving particle meets an apparent charge of 5 pC the UHF spectrum is measured at sensors located in other compartments of the set-up. Injecting artificial pulses with varying magnitudes to the sensor near to the defect the obtained UHF spectra are compared to the one of the particle until the spectra are equal with an accepted tolerance of 20% (fig. 8).

For the second step, the sensitivity verification on-site, the same type of test equipment has to be used as during the laboratory test. Artificial pulses of the same shape and magnitude are injected into a sensor of the GIS on-site. If the signal can be measured at a neighboring sensor, the sensitivity verification is successful for the GIS section between both sensors.

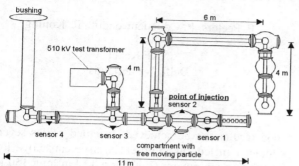

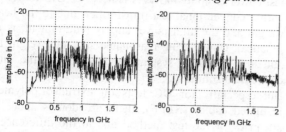

Figure 7: *Experimental set-up of the 420 kV GIS with location of sensors and free moving particle*

Figure 8: *UHF signals measured at sensor 3*
left: *free moving particle, 5pC, l=4mm,*
$\hat{U}=380kV$
right: *injected pulse at sensor 2, $\hat{U}=7V$*

With the new sensitivity verification of UHF measuring systems developed by the CIGRE working group WG 15/33.03.05 its acceptance not only for monitoring of GIS but also for on-site testing will certainly grow further. In the near future a replacement of the lightning impulse test on GIS through a sensitive on-site UHF PD measurement seems to be possible. Generally, both UHF methods are suitable for on-site PD tests on GIS. The broadband method is certainly less cost intensive while the narrowband method is slightly more sensitive and offers the possibility to suppress high frequency disturbances like TV transmitters or mobile phones.

4. Summary

AC testing with an increased voltage level combined with a sensitive partial discharge measurement is now a common test method used for GIS/GIL On-Site Testing after assembly [4-6].
Depending on the GIS/GIL design these tests can be performed by a fully encapsulated test arrangement or by means of an Air/SF6 bushing.

A fully enclosed On-Site Test arrangement can be achieved either by using a conventional SF6 insulated test transformer system of the type TES or by using a SF6 insulated series resonant system with variable frequency of the type RSS.

For „open" arrangements with a SF6/Air bushing each of the introduced On-Site Test systems can be used depending on the output power and test duration.

Fully encapsulated test arrangements provide the lowest background noise level in order to perform a PD measurement either by conventional or with UHF method with the necessary sensitivity.

The UHF method is preferred for „open" On-Site Test arrangement with interfering external signals.

References

[1] Dieter König, Y. Narayana Rao
Partial Discharges in Electrical Power Apparatus, VDE Verlag 1993

[2] Joint CIGRE TF 15/33.03.05, 1998: Sensitivity Verification for Partial Discharge Detection on GIS with the UHF and the Acoustic Method, (15/33.03.05 IWD 73

[3] R.Kurrer, K. Feser: The Application of Ultra-High-Frequency Partial Discharge Measurements to Gas Insulated Substations, IEEE Trans. on Power Delivery, Vol.13, No. 3, July 1998

[4] C. Neumann: GIS-Vorortprüfung aus Sicht des Betreibers, Highvolt Kolloquium '97

[5] W. Buesch, et al: Application of partial discharge diagnostics in GIS at On-Site Commissioning Tests, Cigre Session 1998, Paper 15-104

[6] G.J. Behrmann, S. Neuhold, R. Pietsch: Results of UHF measurements in a 220 kV GIS Substation during on-site commissioning tests, ISH 1997, Montreal

TWO YEARS OF EXPERIENCE WITH A MOBILE RESONANT TEST SYSTEM FOR TESTING OF INSTALLED MEDIUM- AND HIGH VOLTAGE POWER CABLES

P. Schikarski. M. Gamlin. J. Rickmann

Haefely Trench AG
Basel. Switzerland

P. Peeters. P. v.d. Nieuwendijk. R. Koning

NKF Kabel
Delft. Netherlands

Abstract

Since October 1996 one of the largest series resonant test system with variable frequency has been in operation. More than 270 installed medium- and high voltage cables have been tested using this equipment in the Netherlands. Germany. Belgium and in the United Kingdom. The tested cable lines consist of either XLPE. oil-paper or a combination of both insulating materials. The principle of testing is briefly described and the results of these tests. depending on the voltage level. the age and construction of the cable. are statistically evaluated. The experience in on-site testing of power cables is summarized and future testing trends and recommendations are discussed.

1.   Introduction

Testing on installed cable systems is gaining importance. After laying tests on new cables prove that the cable laying and the accessories installation were carried out correctly. The after laying test fills the "quality assurance gap" between the routine test of the cable drum at manufacturer site and the commissioning of the complete cable system on-site (Fig. 1). Tests on old installed cables should establish that the quality of the cable. the cable joints and cable end terminations is sufficient to allow further operation of the cable system.

For on-site tests on installed medium- and high voltage cables several test methods like withstand tests with DC. 0.1 Hz. AC or impulse voltages are known. Depending on the specific test application more or less

good results can be achieved. Installed cables several kilometer in lengths represent a large capacitive load for the test equipment. (E.g. for a 13 km long 150 kV XLPE-insulated cable. test voltage 220 kV. 50 Hz a test power of up to 40 MVA is needed). On-site testing with a conventional test transformer. due to the heavy weight and the high power supply requirement. is not practical. Considering reasonable size and weight DC and 0.1 Hz test systems seems to be the most convenient solution.

But especially for DC tests on extruded polymeric insulated cables (XLPE). there are doubts about the adequacy of the test method due to the difference in stress distribution of DC and AC voltages. DC tests can even be dangerous for the cable under test because of weak points inside the cable insulation caused by electrostatic space charges. This effect could lead to breakdowns during the tests or could decrease the lifetime of the cable system dramatically [1].

The very low frequency test system (0.1 Hz) is normally used for medium voltage cables. For high voltage cable testing this equipment is not available on the market as a commercial product. Furthermore the sensitivity to detect failures with test voltage is only half as compared to the AC voltage test systems [2].

To have. on one hand. the advantage of testing with frequencies close to power frequency and on the other hand a test system which can be handled easily on-site. resonant test systems with variable frequency were developed. This paper describes the more than two years of experience made by a well-known cable manufacturer using this kind of on-site test system.

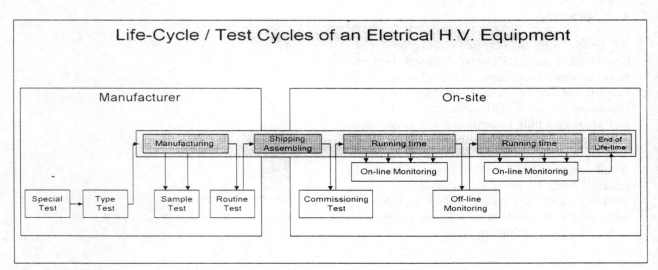

Figure 1:   *Life-cycle test cycles of a cable system*

High Voltage Engineering Symposium, 22–27 August 1999
Conference Publication No. 467, © IEE, 1999

2. Design of the on-site AC test system

The resonant test system with variable frequency mainly consists of the frequency converter. the exciting transformer. the coupling capacitors and two high voltage reactors with fixed inductance (Fig. 2). To be independent from a stationary power supply a motor-generator set is used. The frequency converter generates a variable voltage and frequency output which is applied to the exciter transformer. The exciter transformer excites the series resonant circuit consisting of the reactor's inductance L and the cable capacitance C. Even though the frequency converter creates harmonics the output voltage shows a pure sine wave [3]. The resonance is adjusted by tuning the frequency of the frequency converter according the formula:

$$f = \frac{1}{2\pi \cdot \sqrt{(L \cdot C)}}$$

The tuning range of the test system is determined by the converter's frequency range:

$$\frac{C_{max}}{C_{min}} = \left(\frac{f_{max}}{f_{min}}\right)^2$$

Depending on the voltage level and the capacitance of the cable the test system can be extended by a second reactor. For longer power cables up to 150 kV and more than 10 km. the second reactor is connected in parallel. For test voltages up to 440 kV a series connection of the two reactors has to be realized (Fig. 3).

The quality factor "Q" of the resonant system determines the relation between testing power and the required power supply [4]. For the XLPE cable testing a system quality factor in the range of 100 - 150 and for paper-insulated cables 50 - 90 were achieved. The high quality factor of the high voltage reactors leads to a compact and lightweight system design.

The ratio between weight and testing power of this resonant test system is only a quarter of a comparable conventional test transformer.

The power supply has only to deliver 1 to 3 % of the required testing power. The technical data and the test capabilities of the resonant test system are shown in Table 1.

	1 reactor (1S1P)	2 reactors parallel (1S2P)	2 reactors in series (2S1P)
Nominal voltage	220 kV	220 kV	440 kV
Nominal power	14.5 MVA	29 MVA	29 MVA
Max. test load	1.6 µF	3.2 µF	0.8 µF
Frequency range	30 Hz - 200 Hz		
Tuning - range	1 : 44.44		
Load duty cycle	1 h ON / 2 h OFF, 6 x in 48 h		
Weight including trailer	40 t		
Length	13.6 m		
Width	2.5 m		
Height	4 m		

Table 1: *Technical data of described resonant test system*

Figure 3: *Test set-up to reactors in series (2S1P)*

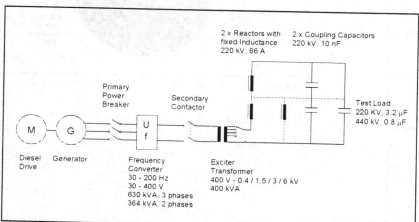

Figure 2: *Electrical diagram of the described resonant test system*

3. Operating experience

Since October 1996 the 440 kV. 29 MVA resonant test system has been in operation. More than 270 installed medium- and high voltage power cables with a total length of approx. 1'600 km has been tested. The demand for on-site testing on cables is increasing rapidly. E.g. between 1997 and 1998 the growth rate was 70 % (Fig. 4)!

To operate an on-site test system economically a short set up time has to be considered. Due to the easy operation and the fast connection of the test system to the cable under test the typical installation time is approx. an hour.

Most of the tested circuits consist of polymeric (XLPE) cables. Also fluid filled (FF) and fluid filled mixed with polymeric (XLPE&FF) has been tested (Fig. 5).

The performed on-site tests have been made on new laid cables (38 %) as well as on old installed cables (68 %) (Fig. 6).

Depending on the voltage level. type and age of the cable system different test procedures have been used. Often the customer defines his own special test procedure. Most tests have been performed according the Dutch Standard NEN 3630 [5]:

New XLPE-insulated cable systems

Cable Class [kV]	Test Voltage	Test duration [min]
50	2.5 U_0	10
110	2.5 U_0	10
150	2.5 U_0	10

Old XLPE-insulated cable systems

Cable Class [kV]	Test Voltage	Test duration [min]
50	2.5 U_0 x 80 %	10
110	2.5 U_0 x 80 %	10
150	2.5 U_0 x 80 %	10

Old paper-insulated cable systems

Cable Class [kV]	Test Voltage	Test duration [min]
50	2.5 U_0 x 80 %	15
110	2.5 U_0 x 80 %	15
150	2.5 U_0 x 80 %	15
380	2.5 U_0 x 80 %	15

mixed cable systems

Cable Class [kV]	Test Voltage	Test duration [min]
50	2.5 U_0 x 80 %	10
110	2.5 U_0 x 80 %	10
150	2.5 U_0 x 80 %	10

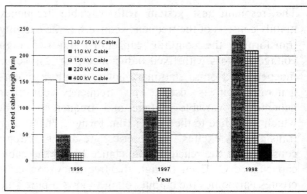

Figure 4: *Tested cable length by NKF Kabel from 4.10.96 till 14.12.98 with the 440 kV resonant test system*

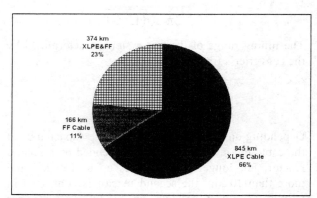

Figure 5: *Statistic of tested cables depending on the cable type*

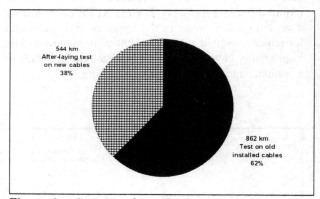

Figure 6: *Statistic of tested cables depending on the cable age*

In some cases the circuit capacitance had to be increased by additional load capacitors to fulfill the customers specification of a test frequency e.g. less than 100 Hz.

Some test arrangements due to limited space on-site or because of cable terminations installed within a building required an additional 200 m feeding cable connecting the test system to the cable under test. This feeding cable has also been used as an additional load capacitance for testing short cable length to lower the test frequency.

Of 230 cable systems tested, two failed in the cable insulation and four in the accessories. Three installed accessories broke down after 5 to 20 min of applied test voltage. Only one accessory and two cable insulation failed during the voltage rise of the test voltage.

Five of the six failed cable circuits were in operation several years. The only new cable circuit which failed during the after-laying test was due to an accessory.

From the tested cable circuits which fulfilled the test requirements no failure has occurred during service until now.

4. Conclusion

More than 270 cable circuits has been tested with the variable frequency series resonant test system. Due to the easy handling, low specific weight and high reliability of resonant systems with variable frequency and the achieved test experiences the AC high voltage on-site testing is a useful tool to prove the quality of the commissioning work of a new laid cable system and to verify the reliability of an old installed cable system. Therefore a further increase of AC high voltage on-site tests on high voltage cables could be expected.

When on-site testing becomes a subject of standardization different test procedures and test levels for after-laying tests on new cable system and tests on old installed cable circuits should be taken into consideration. The testing experience documents shows that the CIGRE recommendations „After laying tests on high voltage extruded insulation cable systems" [1] have not been applied. The tests showed that the breakdown voltage of the failed accessories were in any cases higher than the test levels given in the CIGRE recommendations. E.g. for the 150 kV equipment the breakdown voltage of the accessories were ≥ 2 U_0 (CIGRE recommendation 1.7 U_0) and for the 220 kV equipment ≥ 1.7 U_0 (CIGRE recommendation 1.4 U_0).

If a sensitive partial discharge locating system is used a reduction of the above described test levels could be expected. For operating of this PD measuring system pairs of sensors have to be installed in the joints of the laid cable. Based on the time differences of the PD signal measured by each sensors the PD in cable joints can be located and verified [6].

To reduce the test level for installed old cable systems the quality of the cable insulation as well as the quality of the accessories has to be tested by a sensitive partial discharge measurement. Therefore a further improvement of the conventional PD measurement has to be done. This might be realized by using digital filters to suppress interfering noise [7].

Reference

[1] CIGRE WG 21.09.
After laying tests on high voltage extruded insulation cable systems
ELECTRA, No. 173, August 1997

[2] Dipl.-Ing. Gundolf Schiller
Breakdown behavior of cross linked polyethylene depending on voltage wave form and pre-stress
Dissertation Universität Hannover
February 1996

[3] P. Mohaupt, M. Gamlin, R. Gleyvod, J. Kraus, G. Voigt
High voltage testing using series resonance with variable frequency
10th ISH Montreal, 1997

[4] A. Jenni, M. Pasquier, R. Gleyvod, P. Thommen
Testing of high voltage power cables with series resonant systems and water terminations
7th ISH Dresden, 1991

[5] Dutch Standards NEN 3630
Cables with insulation of cross-linked polyethylene for voltages of 50 kV up to and including 220 kV
June 1996

[6] A. Miyazaki, M. Yagi, S. Kobayashi, C. Min, K. Hirotsu, H. Nishima
Development of partial discharge automated locating system for power cable
CEIDP Annual Report 1998, Volume II

[7] K. Feser, E. Grossmann, M. Lauersdorf, T. Grun
Improvement of sensitivity in online PD-measurement on transformers by digital filtering
Submitted as paper for 11th ISH London, 1999

EQUIPMENT FOR ON-SITE TESTING OF HV INSULATION

T. Grun   M. Loppacher   J. Rickmann   R. Malewski

Haefely-Trench   Instytut Elektrotechniki

Basel, Switzerland   Warsaw, Poland

Abstract

On-site testing has often been requested by the utilities following installation of new HV cables, and also to check the condition of HV power transformers suspected of a damaged insulation.

Specialized equipment for on-site tests has to withstand off-road transportation, to pass under road-bridges, be equipped with its own power source, and what is most difficult, to make measurements at HV substations which are notorious for intense electromagnetic interference.

At present, a HV resonant test set-up with the partial discharge (PD) measuring instrument, and the C-tgδ bridge with built-in test voltage source are the most frequently used mobile test equipment.

This paper presents a mobile HV resonant test set-up for installed cable testing, and measuring instruments designed for on-site transformer insulation testing.

1. INTRODUCTION

1.1 Resonant test set-up

Considering a large capacitance of the HV cable line, and a high test voltage level, the resonant test set-up is the only practical test voltage source. It can provide the required high reactive power to the examined cable, and it draws only a small active power from the supply source.

Different design options of such resonant test set-up have been proposed for on-site testing of HV power cables. Initially, the capacitive reactance of the examined cable was compensated at power frequency, by an inductive reactance of the HV reactor with a mechanically regulated gap in the magnetic core. This design is still employed at HV laboratories for testing short cable sections. However, the weight and size of a controlled gap reactor preclude its use as a mobile test set-up for on-site testing of long cable sections.

The first resonant test set-up composed of a fixed HV reactor, but a variable frequency voltage supply was developed twenty years ago [1], for commissioning tests of gas insulated substations (GIS). A relatively small capacitance of the test object has required only a

modest output power. The first variable frequency mobile resonant test set-up was composed of four reactors rated each at 200 kV, 1.2 MVAr (10 min. load), and such rating was sufficient to test GIS.

An order of magnitude higher output power is available from the modern, variable frequency, mobile resonant test set-up [2]. This is a major improvement, since for the first time the resonant test set-up provides a sufficient output power to test a few kilometer HV cable line, such as encountered in power transmission systems. An original design of the test set-up composed of two reactors rated each at 220 kV, 14.5 MVAr (60 min. load) is described in the following paragraphs.

1.2. C-tgδ bridge for on-site measurements.

Reliability of HV power transmission systems depends largely on the reliable operation of power transformers. Experience has shown a surprisingly long life of some HV power transformers in service. A CIGRE study has revealed a significant number of such transformers operating satisfactorily for more than 40 years. On the other hand, an impressive service record of paper-oil insulation resembles a "time-bomb" situation, where the transformer integrity depends on an aged paper and contaminated oil.

Utilities would like to keep older transformers in service, but check the condition of their HV insulation. The most common procedure consists in periodic checks of the dielectric loss factor - tgδ, and capacitance - C of the transformer paper-oil insulation.

Although the C-tgδ bridge was invented by Schering more than half a century ago, its operation in HV substations is hampered by the strong electromagnetic interference. The original version of this instrument has been modified to enable on-site periodic checks of the transformer insulation. The conventional decade resistors and capacitors were replaced by the current comparator circuit which reduces the effect of such interference, and the bridge was equipped with a portable test voltage source regulated up to 12 kV.

High Voltage Engineering Symposium, 22–27 August 1999
Conference Publication No. 467, © IEE, 1999

1.3 HV insulation recovery voltage measurements

Besides the C-tgδ bridge, another instrument may be employed to assess an overall degradation of the transformer paper-oil insulation, using the dielectric polarization effect.

The polarization recovery voltage method (RVM) allows for an assessment of the water content in the paper-oil transformer insulation, and also helps to reveal an aged insulation. However, the recorded recovery voltage waveforms have to be evaluated by a skilled operator. In particular, paper-oil insulation composed of a multilayered structure may produce results, which do not correspond to the reference curves obtained on samples of dielectric material.

2. ON-SITE TESTS OF XLPE HV CABLES

2.1 Standards and established practice

After-laying tests of XLPE cables rated at 220 kV and 400 kV have not yet been covered by international standards [3], but a test specification was proposed by KEMA, the leading Dutch HV research and test institution.

According to KEMA, a new 400 kV XLPE cable after installation has to be stressed with $2\times400/\sqrt{3}=462$ kV test voltage applied for 30 minutes. This test may also be performed by applying 420 kV for 120 minutes, according to the proposed IEC 840 Standard. A mobile test set-up was executed with a rated output power of 29 MVAr, and this allows for an application of 440 kV test voltage for 60 minutes.

2.2 Design features of the mobile, variable test frequency resonant test set-up

Several important factors have to be taken in consideration while designing a resonant test set up for on site testing to ensure its reliable operation under all circumstances occurring during transportation and testing.

Figure 1 shows the circuit diagram for a resonant test set with variable frequency. The frequency converter (1) is supplied with three phase, power frequency voltage, and generates a single phase rectangular pulse of a variable repetition rate, and a controlled pulse width. The repetition rate is adjusted to the resonance frequency of the HV reactor (3) connected to the examined cable. Impedance of this circuit is very low at resonance frequency, but large at higher harmonics. In consequence, the test voltage waveform at the cable terminal is close to sinusoidal [4].

The frequency converter generates intense disturbances, which may reach some nC, during the semicoductor switching. A complex filtering system (2) has been developed to ensure a disturbance free voltage supply. Using such a filter, partial discharges can be measured on the cable under test without gating and/or windowing of the recording instrument. This solution is

preferred with respect to other design based on blocking the PD detector at the time of the frequency converterswitching.

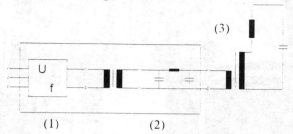

Figure 1 *Circuit diagram of a resonant test set with variable frequency*

A detailed analysis of the electric field distribution inside the oil tank was carried out with the aid of the Weidmann reference curve method [5], as illustrated in Fig. 2. The safety margin derived from these calculations was increased to account for the lower dielectric strength of oil, reduced by shaking during the road transportation. Nonetheless, a certain rest period is required after the trailer installation on the site, before an application of the test voltage.

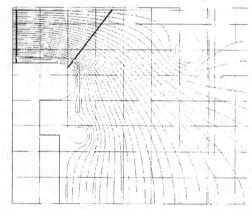

Figure 2 *Electric field distribution at the reactor top electrode. A safety margin of paper-oil insulation was checked along the trajectories marked on this field plot, using the Weidmann reference curve method.*

Thermal capacity of the winding has been dictated by the resonant test set-up duty cycle: 1 hour-"on", 2 hours-"off", up to 6 cycles in 48 hours. Oil forced, air forced cooling system was adopted to keep the winding and oil temperature rise within the 60°K and 65°K limit, respectively. These temperatures are very conservative for a test set-up, since they are normally specified for power transformers with 30 year life expectancy.

A ratio of the reactive power S oscillating in this resonance circuit, and the active power P supplied to cover the losses, determines the circuit factor Q=S/P. Owing to the high Q factor, usually exceeding 100 for XLPE cables, a limited power drawn from the mobile motor-generator set is sufficient to generate the required high reactive power in the cable under test.

Figure 3 *After-laying test of 420 kV cable with the aid of the mobile test set-up. One HV reactor and coupling capacitor standing on the trailer and the second reactor on a platform isolated for 220 kV.*

The complete resonant test set-up can be moved on a platform to the cable installation site, as it is shown in Fig. 3, and perform the required test in a relatively short time. The cable manufacturer who acquired this mobile test system has already performed more than 100 tests.

3. ON-SITE TESTING OF THE OF HV POWER TRANSFORMER INSULATION

3.1. C-tgδ bridge for periodic checks of HV insulation of transformers in substation.

An invention of the capacitance and the dielectric loss factor (C-tgδ) measuring bridge by Schering in 1920-ies. enabled an early detection of incipient faults in paper-oil insulation of HV transformers. However. the Schering bridge low voltage part composed of decade capacitors and resistors. as well as the vibration galvanometer operate at the milivolt level. Even a modest stray capacitance between them and a near by HV bus bar results in stray currents induced in the sensitive bridge circuits. Such interference effectively precludes the Schering bridge operation in HV switchyards.

An introduction of the current comparator. invented by Kusters and Petersons in the 1960-ies. reduced the effect of stray capacitance between the bridge low and high voltage circuits. The current comparator is composed of the primary and secondary windings wound on a toroidal magnetic core in such a way. that the ampere-turns of the primary tend to cancel the ampere-turns of the secondary winding. A tertiary winding is wound on an inner core made of a high permeability material. and installed inside the main core. The tertiary winding detects an imbalance magnetic flux in the comparator core. but is shielded against the stray flux of the primary. as well as the secondary winding.

At the balanced condition. there is no magnetic flux in the comparator core. and the primary and secondary windings have very low input impedance. determined by the resistance of copper wire. This low input resistance makes the current comparator naturally immune to electromagnetic interferences.

The current comparator with an additional system of power amplifiers has been implemented in a portable C-tgδ bridge [6] designed for operation in substations. This bridge comprises a test voltage source regulated up to 12 kV. which supplies the examined HV transformer insulation. A current flowing through the insulation is brought to the first winding of the current comparator. shown in Fig. 3 on the left side of the core.

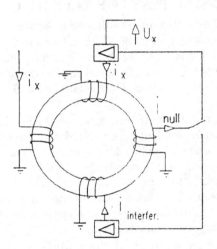

Figure 4 *The current comparator at input stage of the C-tgδ bridge effectively reduces the electromagnetic interference from HV circuits.*

At first. the bridge is operated with the test voltage switched off. and the input current i_x is induced only by the electromagnetic interference. The imbalance detecting winding (shown in Fig. 4 on the right side) provides the current i_{null} to the lower amplifier. The amplifier generates a current $-i_{interfer}$, which cancels the interference current i_x. This interference canceling current is then "frozen"; i.e. is maintained during the whole measuring session. Subsequently. the test voltage is applied to the examined insulation. and the input current i_x is then composed not only of the interference. but also of the sought current from the test object. The i_{null} signal is then switched to the upper amplifier. which produces a current equal to $i_x - i_{inter}$, being the replica of the sought current.

Effectively. the current comparator separates the sought current of the examined insulation from the interference current [6]. The current comparator also provides an output voltage Ux proportional to the sought current.

3.2. Transformer paper-oil insulation ageing assessment based on the recovery voltage method.

An ageing of cellulose and increased water content have an influence on the insulation dielectric polarization properties. An assessment of the cellulose insulation can be derived from the polarization frequency characteristic. however it is quite difficult to perform such measurements on a transformer in substation.

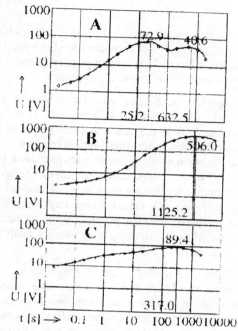

Figure 5 *The polarization recovery voltage as a function of charging time. The first graph A reveals a maximum which precedes the peak corresponding to the normal ageing. The first maximum indicates a premature ageing of paper in a particularly exposed area, whereas the whole bulk of paper insulation is much less affected. The two other curves B and C show only one hump, which indicates a normal ageing process.*

The time (rather than frequency) domain method has been proposed by Hungarian research team [7]. and implemented in an instrument for measurement of the polarization recovery voltage. The instrument applies a direct voltage to the examined insulation for a sufficiently long time to polarize the dielectric. and then this voltage is suddenly removed. A relaxation of dielectric material produces a recovery voltage (or current in an external circuit) which is initially increasing at a certain rate. and then decreasing to zero.

As shown in Fig. 5. a maximum recovery voltage occurs at a certain time. which is an indicative of the insulation ageing. and of the water content. A quantitative assessment of the insulation condition should be based on a comparison of the recovery

voltage waveform recorded on a new transformer and after a certain time of service.

An interpretation of readings obtained on a HV transformer in substation requires a certain knowledge of the examined insulation geometry. and in particular the relative thickness of paper and oil space in the stressed paper-oil insulation.

6. CONCLUSIONS

- Owing to the increased capacity of mobile resonant test set-up. after laying tests can be performed on the highest voltage XLPE cable lines. The HV reactor constitutes the critical part of such set-up. and recent improvement of the reactor design enabled an on-site testing of 400 kV. 4 km long XLPE cable line.

- Periodic checks of the transformers in substation include a measurement of C-tgδ of the paper-oil insulation. A special bridge was developed for on-site measurements of C-tgδ with an interference suppression circuit based on the current comparator principle. A portable source of 12 kV test voltage is integrated with this bridge.

- An instrument was developed for an on-site assessment of the transformer insulation water content and ageing. based on the dielectric polarization effect.

7. REFERENCES

1. Bernasconi. F. . Zaengl. W.. Vonwiller. K.. "*A New Series Resonant Circuit for Dielectric Tests*". 3-rd International Symposium on High Voltage Engineering (ISH). Milano. 1979. paper 43.02.
2. Mohaupt. P.. Gamlin. M.. Gleyvod. R.. Kraus. J.. Voigt. G.. "*High-Voltage Testing Using Series Resonance with Variable Frequency*". 10-th ISH. Montreal. 1997.
3. "*After laying tests on HV extruded insulation cable systems*". ELECTRA. Nr. 173. 1997.
4. Schufft. W.. Hauschild. W.. et al. "*Powerful Frequency Tuned Resonant Test System for After-Laying Tests of 110 kV XLPE Cables*". 9-th ISH. Graz. 1995. paper 4486.
5. Derler. F.. Kirch. H.J.. Krause. Ch.. Schneider. E.. "*Development of A Method for Insulating Structures Exposed To Electric Stress in Long Oil Gaps and Along Oil Transformerboard Interfaces*". 7-th ISH. Dresden. 1991. paper 21.16.
6. "*Type 2818/5283 automatic test system for dielectric measurements (C-tgδ) on HV apparatus*". Application pamphlet #515e. Tettex Instruments. Dietikon. Zurich. Switzerland.
7. Bognar. A.. Kalocsai. L.. Csepes. G.. Nemeth. E.. "*Diagnostic test of HV oil-paper insulating systems (transformer insulation) using DC dielectrometrics*". CIGRE. 1990. paper 15/33-08.

SENSITIVITY OF METHODS FOR DIAGNOSTICS OF POWER TRANSFORMER WINDINGS

Tomáš Hasman, Vladislav Kvasnička and Jonko Totev

Czech Technical University in Prague, Czech Republic

Abstract

The paper deals with comparison of several methods for diagnostics of transformer windings from the viewpoint of their sensitivity. The procedure of comparison is based on the theory of statistical prediction intervals. On a chosen level of confidence it is compared how particular methods detect two types of failure in the transformer winding.

Introduction

Winding of power transformers is exposed to severe electrical and/or mechanical stresses not only during transformer tests, but also in service, when near short circuits or overvoltages in power systems can cause a failure of transformer winding. It is important for both manufacturer and purchaser of tested transformer to be sure that there are no changes in the state of winding after the test and that transformer is able to work properly in service. Similarly, operators of electrical power systems should be interested in the winding state of their power transformers, because an inter-turn short circuit or a displacement of transformer winding can be a cause of catastrophic transformer failures.

There are some methods recommended for detection of failures in winding insulation during high voltage tests [L1] and other methods for detection of movement of windings during tests of ability to withstand short circuit [L2]. Some methods have been developed for diagnostics of power transformer windings in service [L3, L4, L5, L6].

All the methods for diagnostics of transformer winding are in principle similar. They compare the response of some configuration of transformer winding to an input voltage. As the response depends on the winding configuration, any change of this configuration causes a change of the response. Methods mentioned above differ in the way of response measurement and/or in the way of response changes evaluation.

There are two main approaches to the response measurement. The first one is measurement of the winding response to a voltage impulse supplied to an winding input (this way of measurement is called measurement in time domain), the other approach is based on measurement of responses to sine voltages in a frequency interval. This type of measurement is called measurement in frequency domain.

Changes of responses can be evaluated by comparison of response shapes, in case of measurement in time domain it is more sensitive to compare responses after their transformation into frequency domain by the Fourier transformation (or by the Fast Fourier Transformation – FFT). Methods using frequency domain for evaluation of changes in winding response are called Frequency Response Analysis methods. Another possibility for detection of response changes is using the wavelet analysis [L7].

Methods intended for diagnostics of transformers in service have to meet some special requirements in comparison with method used during transformer tests. As time period between measurement of compared responses can be quite long (months or years), there should be considered situation that the voltage source used for the first measurement is not available and the source used for further measurement has a little different parameters of output voltage. The difference in output voltage parameter causes different response which can be misinterpreted as a change of the winding state. Therefore methods for diagnostics of transformer winding in service should be independent on the amplitude and shape of the input voltage. The independence is usually reached by calculation of ratio of input voltage and its response.

One of very important property of methods for diagnostics of transformers winding is their sensitivity to changes of responses. In further text the sensitivity of different methods is compared in case of detection of two types of winding failures.

Compared methods

Four methods altogether were compared, three of them are based on measurement in frequency domain, one in time domain. All of them evaluate the change of responses in frequency domain.

Frequency Characteristic Method I

The Frequency Characteristic Method I (FCH – method, [L5]) is based on measurement of voltages

High Voltage Engineering Symposium, 22–27 August 1999
Conference Publication No. 467, © IEE, 1999

V_1 and V_2 in frequency domain in the range from 1 kHz to 10 MHz (Fig. 1 – voltmeter 1 measures voltage V_1, voltmeter 2 measures voltage V_2, both voltages are rectified by probes P_1 and P_2). The result of measurement is an array of 400 values of voltage attenuation $A = 20.\log (V_1/V_2)$.

Frequency Characteristic Method II

This method differs from the method describe above in the way of measurement of V_1 and V_2. As it is shown in Fig. 2, these voltages are measured by

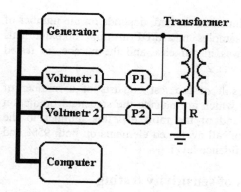

$$Z_{in} = \frac{V_1}{I_1} \; \Big| I_2 = 0 \qquad Z_{tr} = \frac{V_2}{I_1} \; \Big| I_2 = 0$$

Figure 1

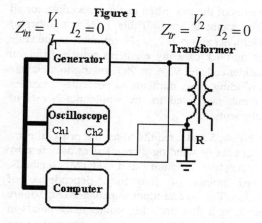

Figure 2

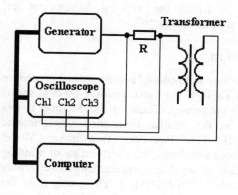

Figure 3

oscilloscope. This way of measurement enables to measure, besides the attenuation of voltages, also their phase. The result of measurement are two arrays of 400 elements, one array for attenuation, the second one for phase of voltages.

Two-Port Transformer Model Method

This method (TPTM method) is based on detection of changes of input and transmission impedance of transformer [L8]. The input impedance Z_{in} and transmission impedance Z_{tr} are defined according to following expressions:

$$Z_{in} = \frac{V_1}{I_1} \; \Big| I_2 = 0 \qquad Z_{tr} = \frac{V_2}{I_1} \; \Big| I_2 = 0$$

V_1 and I_1 are primary voltage and current, V_2 and I_2 are secondary voltage and current. Frequency dependence of input and transmission impedance can be measured according to scheme in Fig. 3. The result of measurement is four 400 element arrays with values of magnitude and phase of both input and transmission impedance, which are measured in the frequency range from 1 kHz to 10 MHz.

Transfer Function Method

This method is based on measurement of an input voltage impulse and the winding response in time domain (the scheme of measurement can be the same as for *FCHT II* - Fig. 2). After transformation by FFT the frequency dependence of transadmittance $R.V_1/V_2$ is calculated for harmonic frequencies. The result of measurement is an array of frequency dependence of this transadmittance, which is called transfer function. Number of calculated harmonics (and number of array elements) depends among others on the number of samples used for recording of both input voltage and winding response. In case of the measurement described in this paper it was used 512 samples for voltage recording and 256 number of harmonics were calculated. No special signal processing was used.

Procedure of sensitivity testing

Changes in state of a transformer winding are detected by methods described above as a change of frequency dependence of ratio V_1/V_2 in case of FCH methods, as a change of input and transmission impedance in case of TPTM Method or as a change of the transfer function. But, owing to some inaccuracy of measurement systems, results of any measurement always differ one from each other. It may happen that a difference of a measured frequency dependence can be caused only by an inaccuracy of the used method or that it can be caused both by inaccuracy of the method and by a change in winding state. A method is sensitive to a change of winding state in case, when a difference

Table 1 – Results of sensitivity test procedure

	FCHM I			FCHM II			TPTM-Z_{in}			TPTM-Z_{tr}			TFM		
	N	95%	99%	N	95%	99%	N	95%	99%	N	95%	99%	N	95%	99%
Short-circuit	400	286	267	400	201	177	400	19	13	400	12	8	256	25	22
		72%	67%		50%	44%		5%	3%		3%	2%		10%	9%
Displacement	400	372	354	400	246	208	400	226	184	400	137	100	256	33	23
		93%	89%		62%	52%		57%	46%		34%	25%		13%	9%

caused by the inaccuracy of the method is essentially smaller than the difference caused by the winding change.

The way of sensitivity testing is based on following procedure. The same state of the winding of transformer BEZ 200 kVA, 10.5 kV/400 V was ten times measured by every method described above (every element x_i of the output array of all methods was ten times measured). After that the state of winding was changed at first by a short-circuit of one winding turn, than by a displacement of one quarter of measured winding. For each of this state the winding was one times measured by all methods.

By the probability-based procedure [L9] it was tested, if points y_i of the frequency dependence measured on the short-circuited winding or points z_i measured on the winding with displacement were assigned to a statistical prediction interval, which is defined on a confidence level as:

$$x_{im} = \pm t \cdot s \sqrt{1 + \frac{1}{n}}$$

In this expression x_{im} and s are the mean value and variation of elements $x_{i,j}$, $j = 1, 2,..., n$ (in our case n equals ten) and t is a variable of Student (t) distribution with n - 1 degrees of freedom. When a tested value y_i (z_i) is outside the statistical prediction interval on the confidence level e. g. 95%, it means that there is probability 95% that the change of response is not caused by inaccuracy of the method, but it is caused by a change of the winding state. Similarly, in case the element y_i (z_i) is outside the statistical prediction interval on the confidence level 99%, there is probability 99% that the change of response is caused by a change of the winding state.

According to the described procedure all of 400 elements in case of method measuring the winding response in frequency domain were tested. The number of elements, which can be tested in case of

transfer function method, depends on the number of voltage samples measured in time domain. This number was in our case and the number of tested elements was 256.

A result of the test consists of a number of elements which are outside the statistical prediction interval and of the ratio of this number to the number of all evaluated elements on both 95% and 99% confidence level.

Results of sensitivity testing

Results of the described testing procedure for all described methods are in Table 1 (N is number of array elements x_i, i = 1, 2, ..., N, numbers in the table express number of array elements which differ on the confidence level 95% or 99% owing to a change of the winding state, numbers in percentage express these numbers related to the whole number N of array elements).

In Fig. 4, 5 and 6 results of testing procedure are shown in case of winding displacement detection by FCH, transfer function and TPTM methods. Measured points of frequency dependence of attenuation (Fig. 4) and input impedance (Fig. 6) are connected by a thin line, big points mark points in which the winding displacement was detected on the confidence level 95%. In Fig. 5 the points of transfer function, in which the winding failure was not detected, are marked by "+".

Conclusion

According to results of sensitivity test procedure shown in Table 1, the most sensitive method seems to be the Frequency Characteristic Method I, less sensitive is the similar Frequency Characteristic Method II. The Two-Port Transformer Model Method does not detect the short-circuit of winding turn, in case of winding displacement its sensitivity is comparable with FCHM II. The worst from the viewpoint of sensitivity is the transfer function method. Nevertheless, it should be emphasise that the sensitivity of the transfer function method can be

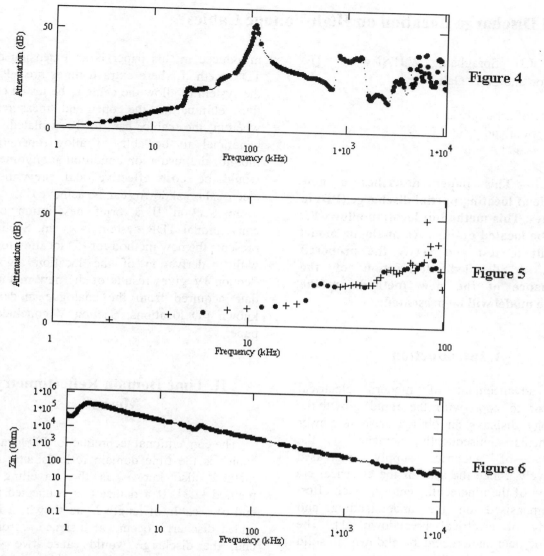

Figure 4

Figure 5

Figure 6

influenced by parameters of used digital recorder (authors used 8 bits, 200 MSa recorder) and that sensitivity of this method can be increased by using a suitable signal processing method for transformation from time to frequency domain.

References

[L1] Malewski R., Poulin B.: Impulse testing of power transformers using transfer function method, IEEE Transaction on Power Delivery, Vol. 3, No. 2, April 1988

[L2] IEC Publication 76-5, Power Transformers, Part 5: Ability to withstand short circuit

[L3] Lech W., Tyminski L.: Detecting transformer winding damage by the Low Voltage Impulse method, Electrical Review, No 21, Vol. 179, November 1966

[L4] Lapworth J. A., Noonan T. J.: Mechanical condition assessment of power transformers using Frequency Response Analysis, 62nd Annual International Conference of Doble Clients, 1995, Massachusetts, U. S. A.

[L5] Hasman T., Kvasnička V.: Method for Testing of Transformer Windings, 7-th International Simposium on High Voltage Engineering, Dresden 1991, (74.07)

[L6] Leibfried T., Feser K.: On-Line Monitoring of Transformers by Means of the Transfer Function Method, IEEE International Symposium on Electrical Insulation, Pittsburgh, 1994

[L7] Wang Y., Li Y. M., Qiu Y.: Application of Wavelet Analysis to the Detection of Transformer Winding Deformation, 10th ISH, Montreal 1997

[L8] Hasman, T.: Reflection and transmission of traveling waves on power transformers, IEEE Transactions on Power Delivery, Vol. 12, No 4, October 1997

[L9] Vardeman V. B.: Statistics for Engineering Problem Solving, IEEE Press, 1993

Partial Discharge Location on High Voltage Cables

I.Shim, J.J. Soraghan, W.H.Siew, †F. McPherson and ‡P.F. Gale

University of Strathclyde,
Glasgow, Scotland.

Abstract - **This paper describes a new technique of locating partial discharge (PD) in HV cables. This method of location allows PD sites to be located on-line. An analogue model was built to test and verify the proposed method. A detailed description of the performance of the new method on the analogue model will be presented.**

I. Introduction

The advancement of modern electrical equipment to cope with the rapid growth of technology displays an obvious rise in power requirements. Consequently, demands on the reliability of electric supply increased distinctively. Since the start of the electrical era at the end of the nineteenth century, much effort was emphasised on the understanding and predicting of electrical breakdown [1]. The subject of degradation and breakdown of solid insulating materials is of constant interest.

The development of PD mechanisms in voids is significant for the degradation of an insulation material under electrical stress. PD is caused by the cavities or voids in cable insulations within the dielectric material or on boundaries between the solid and the electrodes. Every PD event deteriorates the material by the impact of high-energy electrons.

The techniques for PD location in cables described in [4,5,6] are based on the time domain reflectometry (TDR) method. The TDR method is only capable of locating PDs at a limited length of a cable due to the difficulty of the presence of more than one PD. The method introduced in this paper is an extension of the TDR method where extra features are added to the system to allow the cable to be tested on-line thus eliminating the open-end measurements wherein the cable has to be isolated. It is beneficial to be able to allow operators to monitor the insulation condition at anytime. This would be cost effective and prevention of electrical breakdowns can be achieved.

In Section II a brief description of the conventional TDR system is given. Section III presents the new method for PD location together with a derivation of the location algorithm. Section IV gives results of the new method for data acquired from the analogue model with known PD locations. Section V concludes the paper.

II. Time Domain Reflectometry Method

The conventional technique for PD location in cables is the time domain reflectometry (TDR) method, also known as the travelling wave method [2,3]. If a detector is connected to one end of a cable of length L, as shown in Fig. 1, and a discharge occurs at distance x from one end, the discharge would cause two voltage pulses to originate from the void one travelling towards the detector as a wave with propagation velocity, v m/s and the other travelling towards the other end of the cable.

One impulse appears at the detector after x/v sec. The other impulse, having been reflected at the end of the cable, would appear at the detector after having travelled $2L-x$ m. The second impulse would be $(2L-2x)/v$ sec from the first pulse.

From this observation, x may be computed. This approach has been used successfully in [4,5,6] for single PD location.

† ScottishPower Power Systems, Glasgow, Scotland

‡ Hathaway Instruments Ltd., Hertfordshire, England

High Voltage Engineering Symposium, 22–27 August 1999
Conference Publication No. 467, © IEE, 1999

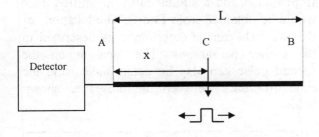

Fig. 1. PD at xm from the detector

III. New On-line PD Location Method

A block diagram of the new method is given in Fig. 2. Two measurements are made on the cable at the two selected locations. For synchronisation purposes a periodic timing signal is used at one end.

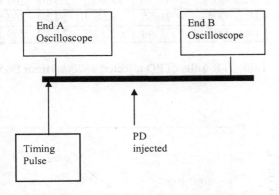

Fig.2. Block Diagram of the new method

An analogue model has been built utilising the TDR method to generate data. Two oscilloscopes were used to acquire data from a selected length of cable. The oscilloscopes were placed one at End A and the other at End B, as illustrated in Fig.2. The cable used is a standard coaxial cable. A pulse generator is used to inject a pulse into the cable, which would act as the PD pulse. The PD injected has a pulse width of $1\,\mu$ sec and a magnitude of 10V. Another pulse generator is used to transmit continuous timing pulses into the cable. The timing pulse is transmitted with a frequency of 50kHz and the pulse width of 100nsec, at 0.1V. The timing pulses act as a means to synchronise the pulses that would appear in end A and end B.

Since the PD is injected at random, there is ambiguity in determining the point where $t = 0$. Therefore the synchronising timing pulses allow a pulse appearing at one end to be verified with the pulse appearing at the other end to be the same PD. This eliminates the problem when long lengths of cables are being tested. It also allows on-line measurements to be taken.

To determine the location of the PD, when a cable is tested with a single PD, the display at End A is obtained as shown in Fig. 3 and similarly the display at End B in Fig. 4. By aligning the signals of the two ends using the PD point as the reference, i.e. making $t = 0$ be at the point of the PD, the location of the PD can be determined. As demonstrated in fig. 5, when the PDs are aligned, the difference in time between the two timing pulses, t3 is the time taken for the pulse to move from the PD point to each end.

Therefore,

$$t3 = t1 - t2 \qquad (1)$$

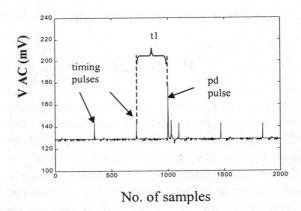

No. of samples

Fig. 3. Recording obtained from End A

The PD is located at

$$d = \frac{t3}{2}v_1 \qquad (2)$$

where d is the distance of the PD from End A and v_1 is the velocity of propagation for the cable.

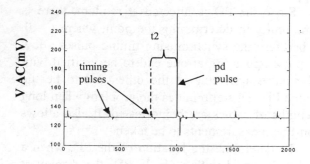

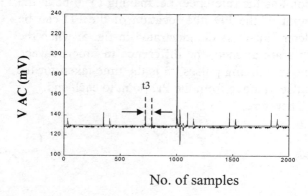

Fig. 4. Recording obtained from End B

Fig. 5. Aligning both ends at the PD pulse

IV. Results

Using the set-up in Fig. 2, a 300m cable has been used to verify the method described earlier. The PD is injected at 100m from End A, then at 200m from End A. The results obtained are shown in Table 6 and 7. Each experiment has been carried out 5 times in order to confirm that the readings taken are correct. The velocity of propagation for the coaxial cable has been calculated to be 192.3m/μ sec.

From Equation (2), the distance d from End A was calculated. To complicate matters further, two PDs has been injected simultaneously into the cable, one at 100m from End A and the other at 200m from End A. Fig. 8 is the plot obtained

from End A and a similar but time-shifted trace would be obtained from End B. The locations of the two PDs can be determined as described in the earlier paragraphs. The location on the second pulse can also be determined using the time difference between the pulses as shown below.

Data Name	t1	t2	t3	d(m)
Data_001	13.44	12.36	1.08	103.8
Data_002	15.52	14.48	1.04	99.99
Data_003	2.16	1.16	1.00	96.15
Data_004	13.24	12.20	1.04	99.99
Data_005	7.60	6.56	1.06	101.19

Table 6. Results of PD injected at 100m from End A

Data Name	t1	t2	t3	d(m)
Data_001	11.12	9.04	2.08	199.92
Data_002	11.48	9.36	2.12	203.84
Data_003	14.76	12.60	2.16	207.68
Data_004	10.68	8.56	2.12	203.84
Data_005	5.44	3.32	2.12	203.84

Table 7. Results of PD injected at 200m from End A

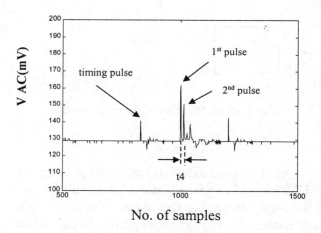

Fig. 8. Two PDs injected into the cable

Table 9 illustrates the results obtained from the experiment. $t4$ is the time taken between the first and second pulse injected at 100m and 200m from End A respectively.

Using Equation (2), distance *d* from End A is approximately 100m from End A. Utilising the first pulse as the reference, i.e. t=0, the distance of the first pulse from the second pulse is

$$t4 \times 192.3 = 103.8m$$

Therefore the distance of the second PD from End A is

$$100 + 103.8 = 203.8m$$

Data Name	t1	t2	t3	t4
Data_001	6.84	5.68	1.16	0.52
Data_002	2.96	1.88	1.08	0.56
Data_003	7.40	6.32	1.08	0.56
Data_004	12.16	11.08	1.08	0.56
Data_005	13.88	12.80	1.08	0.52

Table 9. Results of PD injected at 100m and 200m from End A

The results obtained indicated that the PD was located at the correct location.

V. Discussion and Conclusions

Complications would occur when more than one PD is injected into the cable. The results shown above showed straightforward answers as the discharges which were injected one at 100m and the other at 200m were too far apart for any of the pulses to overlap each other. Another limiting factor to this technique would be the maximum length of cable that can be tested.

The method used for detecting PD has proved to be feasible through the results obtained. The analogue model generates data of the PD that was injected at random into the cable, and the data is then processed further to obtain the location of the PD. The location of a single artificially created PD into the cable was easily calculated using the derived equation.

The results of this study lead to the necessity of inflicting more than one PD into a cable at random locations where the pulses are randomly injected with different amplitudes to create a more realistic model of a cable with PD problems. Location procedure becomes more difficult when there are more than one discharge occurring at different sites in the cable. Each discharge is observed separately by changing the trigger level of the oscilloscope. However, when discharges are unstable the problem is enhanced. This is the subject of our ongoing research.

References

[1] Petrus Henricus Franciscus Morshuis, "PD MECHANISMS, Mechanisms leading to breakdown, analysed by fast electrical and optical measurements", Delft University Press/1993.

[2] Brunt, K.L.Stricklet, J.P. Steiner, S.V.Kulkarni, "Recent Advances in Partial Discharge Measurement Capabilities at NIST", IEEE Trans. on Electrical Insulation, Vol.27, No.1, pp 114-129, Feb 1992.

[3] M. Beyer, W.Kamm, H.Borsi, K. Feser, "A New Method for Detection and Location of Distributed Partial Discharges (Cable Faults) in HV Cables Under External Interference", IEEE Trans. On Power Apparatus and Systems, Vol. 101, pp 3431-3438, 1982.

[4] Yafei Zhou, A.I.Gardner, G.A.Mathieson, Y.Qin, "New Methods of PD Measurement for the Assessment and Monitoring of Insulation in Large Machines", Industrial Research Ltd.

[5] Zhifang Du, P.K. Willet, M.S. Mashikian, "Performance Limits of PD Location Based on Time-domain Reflectometry", IEEE Trans. On Dielectric and Electrical Insulation, Vol. 4, No. 2, pp182-188, 1997.

[6] N.H. Ahmed, N.N. Srinivas, "On-line Partial Discharge Detection in Cables", IEEE Trans. On Dielectric and Electrical Insulation, Vol. 5, No. 2, pp 181-188, 1998.

ON-SITE APPLICATION OF AN ADVANCED PD DEFECT
IDENTIFICATION SYSTEM FOR GIS

A. Lapp[1], H.-G. Kranz[1], T. Hücker[2], U. Schichler[2]

[1]Bergische Universität - GH Wuppertal, Germany
[2]Siemens AG, Germany

Abstract

This contribution examines the identification reliability of automated Partial Discharge (PD) diagnosis systems for common PD defects in Gas Insulated Switchgear (GIS) under practical industrial conditions. The influence of the reference data base on diagnosis results as well as new PD evaluation tools like $\Delta u/\Delta\varphi$-Pattern based on phase resolved pulse sequence analysis (PRPSA) are investigated. Especially the abilities of the PRPSA are examined to identify PD defects with about 70-200 useful PD pulses per second near inception voltage (50Hz ac stress).

Introduction

PD monitoring is a useful tool to assess the insulation condition of high voltage equipment. Therefore most of the GIS above 400 kV put into service recently were equipped with ultra high frequency (UHF) couplers for a continuous and sensitive PD monitoring [1].

An essential point in PD monitoring is the interpretation of the recorded PD data. This PD data evaluation shall answer the questions whether there is a PD defect present and if so what kind of defect it is. These findings help to assess the failure risk of the GIS caused by partial discharges. This knowledge assists to schedule maintenance work to keep the outage times as short as possible.

The PD data evaluation can either be done by human experts using visualisation tools or by computer based automated PD diagnosis systems mostly based on methods of pattern recognition. Recent investigations have shown that automated diagnosis systems can reach a higher identification reliability than humans when the automated system is well adapted to the measuring circuit [1]. 673 PD measurements relevant for GIS were examined by human experts as well as by an automated PD diagnosis system. Humans were only able to identify 30% of the investigated PD measurements correctly. Whereas on the same data stock automated diagnosis systems, based on conventional methods of pattern recognition, are capable to identify 75% correctly [1].

In this paper an extended data stock of 1247 PD measurements (PDGIS), acquired by the Siemens UHF⁻ PD monitoring system for GIS [2], of the following PD defects were investigated:

- BOUNCING PARTICLE
- FLOATING POTENTIAL
- INTERNAL VOID

- PARTICLE ON DIELECTRIC
- TIP ON HIGH/EARTH POTENTIAL

The correlation between the PDGIS measurements and the causal defects were documented in detail (PDGISDoc). With an advanced automated PD diagnosis system these PD recordings - acquired under on-site and practical industrial conditions - were additionally evaluated to develop an improved diagnosis system.

This investigation was performed with the automated PD diagnosis system WinTED 1.0 of Wuppertal University which uses a redundant diagnosis concept to achieve a more reliable diagnosis result [6, 7]. This Redundant Diagnosis System (RDS) employs the methods of pattern recognition which are based on the key components: feature extraction, classifier and reference data base [6]. The implemented feature extraction techniques and classification algorithms of WinTED have achieved excellent results when laboratory measurements were investigated [5]. Therefore as a first step it was examined whether a set-up of a well structured reference data base for the PDGIS measurements provides more reliable diagnosis results.

Influence of the Reference Data Base Set-up

Fig. 1 displays a dendrogram where for a clearer diagram only 50 PDGIS measurements of the PD defects TIP ON HIGH POTENTIAL, FLOATING POTENTIAL and BOUNCING PARTICLE are listed on the x-axis. The y-axis shows the differential distance, which the feature vectors $\vec{m}$ of these measurements have from each other in the feature space [6]. These distances are indicated by the horizontal lines i.e. the lower such a line is the higher is the similarity of the features.

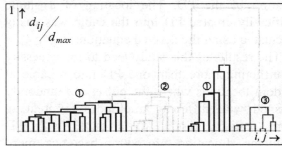

Fig. 1: Dendrogram of 50 PD measurements acquired at GIS of the PD defect types: ① TIP ON HIGH POTENTIAL, ② FLOATING POTENTIAL, ③ BOUNCING PARTICLE

It is visible in Fig. 1 that the recordings of the PD defect TIP ON HIGH POTENTIAL do not cluster at a single location in the feature space. Usually this fact

High Voltage Engineering Symposium, 22–27 August 1999
Conference Publication No. 467, © IEE, 1999

can be observed for on-site investigated PD defects. This also indicates the necessity that in most applications more than one reference has to be employed to identify one defect type reliable [4, 7].

It was shown in [7] that not only the choice of the references but also the set-up of these references in a data base can influence the diagnosis result, even when using the same measurements. Therefore a program (WinCLUST) was developed which automatically sets up a reference data base out of a data stock using methods of cluster analysis.

Tab. 1 shows the diagnosis results when WinTED uses a reference data base to identify the PDGIS measurements which was set up by a human expert (manually) as well as automated by WinCLUST. The computer generated reference data base set-up not only saves the expert time but also improves the diagnosis result considerably. Using the identical PDGIS data stock the number of correctly identified PD measurements (documented PD defects) was increased to over 85% and also the number of miss-classifications could be decreased to a little more than 5%.

recognised as:	data base set-up: manual	data base set-up: automated
original defect	743 (59,6%)	1061 (85.1%)
noise	94 (7,5%)	13 (1.0%)
unknown	201 (16,1%)	110 (8.8%)
wrong defect	209 (16,8%)	63 (5.1%)

Tab. 1: Identification results of 1247 measurements using a manual or automated reference data base set-up. Diagnosis system WinTED [6]

Although these diagnosis results are conclusive one can ascertain that during the WinCLUST set-up of the reference data base 220 PD measurements out of 1247 were put into a reference class "NOISE". All these recordings contain only a small number of useful PD pulses. So of the 1061 correctly identified measurements 209 are from the class NOISE. Using the documentation (PDGISDoc) these measurements were replaced from the class NOISE into their "natural PD defect class". Nevertheless they were identified as noise measurements by the RDS again, due to the low number of PD pulses. Tab. 2 shows the diagnosis results for this case. About 18% of the measurements where not assigned to their documented PD defect type but identified as noise.

Although the number of miss-classifications were decreased in such a diagnosis set-up, a reduced potential to identify defects can be observed. This demonstrates the necessity to change the feature extraction method in such a way that PD measurements with only a few useful PD pulses can be characterised more significantly.

recognised as:	automated data base set-up with noise adjustment:
original defect	852 (68.3%)
noise	229 (18.4%)
unknown	110 (8.8%)
wrong defect	56 (4.5%)

Tab. 2: Diagnosis results of WinTED using an automated (WinCLUST) reference data base set-up, adjusted of noise measurements

Improved Feature Extraction

Recent investigations have shown that PRPSA is especially qualified to identify PD defects near inception or extinction which have a lower number of PD pulses. One relevant new diagnosis tool in PRPSA is the $\Delta u/\Delta\varphi$-Pattern for feature extraction.

A PD current can be considered as a sequence of PD pulses $q(\varphi_i)$ with their apparent charge q or with the magnitude in mV for UHF-measurements. The angel φ_i determines the phase of occurrence within an ac cycle of the applied test voltage (Fig. 2). Since the PD pulse occurrence is related to the applied test voltage, digital PD acquisition systems should also record the magnitude of the test voltage u. So a PD activity can be acquired with all the information considered relevant as a sequence of PD pulses $q(\varphi_i,u(\varphi_i))$ over an arbitrary number of cycles. Such a PD recording with the essential information about the development of the PD activity is called a phase resolved pulse sequence measurement. A $\Delta u/\Delta\varphi$-Pattern can be calculated from such a measurement. Equation (1) describes the slope m_i between two consecutive PD pulses $q(\varphi_{i-1})$ and $q(\varphi_i)$ (Fig. 2) where φ_i is the phase position and $u(\varphi_i)$ the instantaneous voltage value when a PD pulse i occurs.

$$m_i = \left.\frac{\Delta u}{\Delta\varphi}\right|_i = \frac{u(\varphi_i) - u(\varphi_{i-1})}{\varphi_i - \varphi_{i-1}} \qquad (1)$$

The slope m_i approximates the voltage gradient which is necessary to excite a consecutive PD pulse i. When the frequency of the slopes of consecutive PD pulses m_i and m_{i+1} are put into a co-occurrence matrix where m_i is on the x-axis and m_{i+1} on the

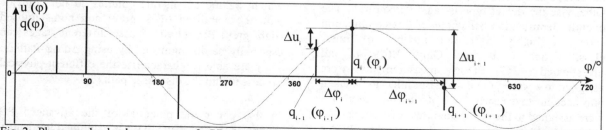

Fig. 2: Phase resolved pulse sequence of a PD signal: $q(\varphi_i)$ where q = apparent charge in pC or the magnitude in mV of UHF-measurements, φ_i = phase position and • = $u(\varphi_i)$ instantaneous voltage at phase position φ_i in kV

y-axis and the frequency N of a (m_i, m_{i+1}) relation is on the z-axis, a histogram $N(m_i, m_{i+1})$ is created (Fig. 3). The number of repetitions N is indicated by the colour respectively by the grey value of the pattern. The $\Delta u/\Delta\varphi$-Pattern as well as the common Δu-Pattern [3] approximate a conditional probability $p(m_{i+1} | m_i)$, that a given slope value m_i is followed by a slope value m_{i+1}.

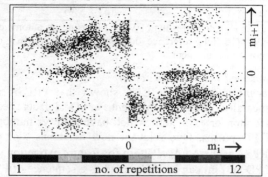

Fig. 3: $\Delta u/\Delta\varphi$-Pattern H (m_i | m_{i+1}) of an INTERNAL VOID

The feature vector extracted from this pattern is based on a discrete two dimensional sine and cosine transform [8]. The characteristics from $\Delta u/\Delta\varphi$-Patterns (fingerprints) are used in combination with an Euclidean distance classifier (L2) [6]. With this new pattern more than 85% of the PDGIS measurements can be assigned to the correct PD defect (Tab. 3). In comparison to the results of the RDS in Tab. 2 an improvement of the diagnosis potential is perceptible. Especially the identification abilities of measurements with few useful PD pulses were improved. But an unacceptable increment of the number of miss-classifications can be observed because there is no ability to classify a measurement as unknown.

recognised as:	L2 with PRPSA
original defect	1061 (85,1%)
noise	51 (4,1%)
wrong defect	135 (10,8%)

Tab. 3: Diagnosis results of PRPSA based features and a Euclidean distance (L2) classifier

Modified Redundant Diagnosis System

Therefore for an advanced diagnosis the RDS was combined with the $\Delta u/\Delta\varphi$-Pattern based L2-classifier. The identification results of these two systems where checked against each other. A measurement is only assigned to a PD defect when the results of the two diagnosis systems correspond. Otherwise the defect type is rated "unknown". The overall identification behaviour of this combination shows a highly decreased number of miss-classifications (Tab. 4). Only 1.6% of all investigated 1247 PDGIS measurements were assigned to a wrong PD defect. But it has to be taken into account that nearly 30% off the measurements were assigned to the group unknown PD source.

When looking more closely at the PDGIS measurements which were assigned to the wrong PD defect one discovers that several recordings of the PD defect INTERNAL VOID are identified as a PARTICLE ON DIELECTRIC. By comparing the $\Delta u/\Delta\varphi$-Pattern of such a measurement (Fig. 4e) with an example of a PARTICLE ON DIELECTRIC (Fig. 4a) an expert can comprehend the automated diagnosis result. Therefore a plausibility check of this PD data is necessary.

recognised as:	RDS + L2 with PRPSA
original defect	828 (66.4%)
noise	36 (2.9%)
unknown	363 (29.1%)
wrong defect	20 (1.6%)

Tab. 4: Diagnosis results of WinTED combining RDS and PRPSA results

Fig. 4 displays measurements of different discharge paths of the PD defect types PARTICLE ON DIELECTRIC and INTERNAL VOID. The measurements of one defect type are in one column.

The entire PD defect INTERNAL VOID was investigated more profound in a laboratory set-up which enables the acquisition of locally resolved optical PD information and the electrical PD signal simultaneously. In combination with an additional microscopic analysis of the dielectric surface in the void the electrical PD data can be correlated to the deterioration stage of this defect [8]. This allows to point from an evaluation of the electrical PD signal to the deterioration stage and the failure risk of a PD affected insulation. Fig. 4b shows a "stabilised PD activity" which can result in an initial damage inside the material due to the concentration of the PD activity to a restricted area. This leads to an accelerated breakdown of the insulation. Whereas the $\Delta u/\Delta\varphi$-Patterns of Fig. 4c-d represent a "stochastic non-stable" PD activity. Although measuring the same defect an acute danger for the insulation material is not to be expected because this PD activity takes a significantly longer time till a final breakdown.

Examining the industrial PDGIS measurements these different PD activities can be observed too. Fig. 4e-f point to a stabilised PD activity whereas Fig. 4g represents a non-stable discharge activity in the same void. Therefore it is not possible to consider these measurements of the subjectively documented PD defect INTERNAL VOID as a consistent defect type. They represent different deterioration stages as indicated in Fig. 4 by differently marked frames. It is visible that PD measurements of the same PD defect which are located in one column do not coincide with the stabilised (dark grey) and non-stabilised (light grey) PD activity. These different stages can especially be distinguished by using $\Delta u/\Delta\varphi$-Pattern. They are able to characterise the different physical process within the discharge path [8].

Conclusion

In this paper the potential of the advanced PD diagnosis system WinTED was examined to identify

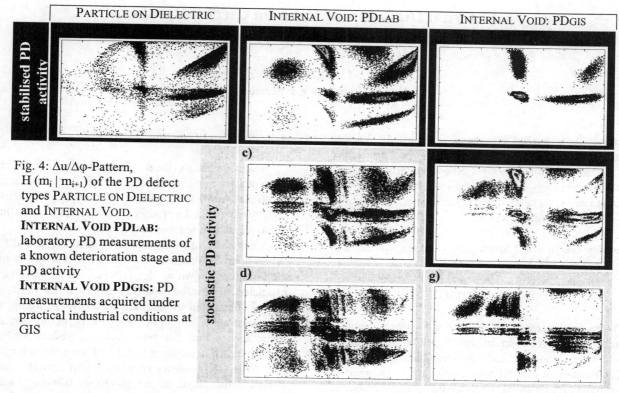

Fig. 4: Δu/Δφ-Pattern,
H (m$_i$ | m$_{i+1}$) of the PD defect
types PARTICLE ON DIELECTRIC
and INTERNAL VOID.
INTERNAL VOID PDLAB:
laboratory PD measurements of
a known deterioration stage and
PD activity
INTERNAL VOID PDGIS: PD
measurements acquired under
practical industrial conditions at
GIS

1247 PD measurements acquired at GIS under industrial conditions.

It was demonstrated that the diagnosis results of any redundant diagnosis system can be distinctively improved when a well structured reference data base is used. Such a data base with clear cut reference classes can be set up automated without expert knowledge by the program WinCLUST which was developed at Wuppertal University.

This paper also shows that the development of additional evaluation methods based on phase resolved pulse sequence analysis (e.g. Δu/Δφ-Pattern) have an increased identification potential, especially for PD measurements with only a few PD pulses (70–200 per second). These measurements near PD inception can also be identified more reliable by using pattern with a physical interpretation ability.

The combination of the redundant diagnosis system WinTED with a classifier, using Δu/Δφ-Pattern based features, achieves defect identifications where miss-classifications can be avoided almost completely. The rates for correct identified defects have to be improved in further investigations.

It was also shown that the application of the Δu/Δφ-Pattern enables a distinction of different deterioration stages of the same PD defect. These stages are represented by different PD activities which are caused by distinct physical processes. Therefore the reference data base should not only be set up with respect to the common PD defect types but also accordingly to the physical process of the discharge path.

References

[1] Beierl, Hücker, Katschinski, Kirchesch, Neumann, Ostermeier, Rudolph, „Intelligent Monitoring and Control Systems for modern AIS and GIS Substations", CIGRÉ 34-113, 1998

[2] Hering R., Schichler U., Gorablenkow J., „On-line Partial Discharge Monitoring System for 420/550kV GIS", Labein, VI Jornadas Internationales de Aislamiento Eléctrico, October 1997

[3] Hoof M., Patsch R., „Detection of Multiple Discharge Sites Using Pulse/Pulse Correlation", I.C.D.I, pp. 213-216, Budapest, Hungary, 1997

[4] Hücker T., Gorablenkow J. „UHF Partial Discharge Monitoring and Expertsystem Diagnosis for GIS" IEEE Transactions on Power Delivery 1998

[5] Hücker T., Kranz H.-G., Lapp A., „A Partial Discharge Defect Identification System with Increased Diagnosis Reliability", 9th ISH, pp. 5612, Graz, Austria, 1995

[6] Lapp A., Kranz H.-G., „Neuro-Fuzzy Diagnosis System with a Rated Diagnosis Reliability and Visual Data Analysis", IDA-97, London, UK, Lecture Notes in Computer Science 1280, Springer Verlag, 1997

[7] Lapp A., Kranz H.-G., „The Influence of the Quality of the Reference Data Base on PD Pattern Recognition results", 10th ISH, Vol. 6, pp. 317-321, Montréal, Canada, 1997

[8] Lapp A., Kranz H.-G., Guzmán, H., „Quantitative Deterioration Assessment of Polymer Surfaces", ISEIM, pp. 83-86, Toyohashi, Japan, 1998

Address of Authors
Prof. Dr.-Ing. H.-G. Kranz, Dipl.-Ing. A. Lapp
Bergische Universität GH Wuppertal / FB13
Fuhlrottstr. 10, 42119 Wuppertal, Germany
E-Mail: kranz@uni-wuppertal.de, lapp@uni-wuppertal.de

Dr.-Ing. T. Hücker, Dr.-Ing. U. Schichler
Siemens AG, EV HG 81, 13623 Berlin, Germany

External Diagnosis of Vacuum Circuit Breaker
Using X-ray Image Processing

S.Nakano, T.Tsubaki and Y.Yoneda

Hitachi Engineering & Services, Ltd.

Abstracts: As a method of external diagnosis for Vacuum Circuit Breaker, X-ray projection technology has been applied. New method of high accuracy projection, 3 dimensional image processing technology by application of X-ray CT, method of detecting more accurate condition of the interior, by application of X-ray and has been developed.

Keywords: Gas Insulated Switchgear, Vacuum Circuit Breaker, X-ray, Imaging Plate, X-ray Tomography

Ⅰ.INTRODUCTION

With the increase in the power system installed capacity and interconnection, concentration of demand in urban area, sophisticated information infrastructure, further balance in reliability and economy is required for power transmission and distribution equipment is required. This requirement makes the specification for the equipment much more stringent.

For example, 500 kV class GIS increasing interrupting capacity, reduction in the number of interrupters and further capacitate current interrupting capability of disconnecting switch, inductive current interrupting capacity of grounding device etc. makes the equipment much more complicated.

In order to accomplish well balanced economy and reliability, external diagnostic technology applying state of the art technology is regarded as very effective tool. Previous diagnostics is limited mainly to circuit breakers, however nowadays all other contacts are also involved, such as those of disconnecting switch etc. and points required for diagnostics has been increased several times of the previous.

On the other hand with the increase in dependence on electric power effects of faults on social life is increasing, preventive maintenance applying external diagnostic system is becoming more important.

In general, external diagnostics for power transmission and distribution equipment has been made by individual sensors internal projection by X-ray. This paper introduces improved internal diagnosis by higher precision portable X-ray , and earlier detectors of faults, and also applications by 3 dimensional image processing by application of X-ray CT.

Ⅱ.DEVELOPMENT OF FLUOROSCOPE TECHNOLOGY

A. Application of Imaging Plate(IP).

Imaging Plate device is a Radiography, as a detector of radio active rays, applying Imaging Plate consisting of photo stimulated fluorescence. IP is coated 2 dimensional high density thin layer using *special fluorescent having photo stimulated luminescence..

Mechanism of radiation is shown in Fig. 1 and Fig. 2. When X-ray is irradiated on IP, energy is stored in trap of the crystal as shown in Fig. 1. When He-Ne laser beam is irradiated to IP, X-ray energy stored in the trap of the crystal is radiated as luminescence (light) through pump and generator. This phenomenon is photo stimulated luminescence and the time series is shown in Fig. 2. Sensitivity, time required is 1/10 of the conventional type film. When fluorescence receives radio active rays such as X-rays radiate light, after stopping of stimulus, radiation attenuates. However, during the process, if light with longer wave length or infrared is irradiated, the radiation of light increases. IP device have high sensitive, wider dynamic range, higher density storage, processing image processing functions.

IP device consists of, reader device of IP, image processing device to process the image which is read, film image recording device, CRT image display, photo-disc image filing device. IP data reading, after X-ray irradiation, is performed by scanning the surface of IP surface by laser beam.(See Fig. 3). From Irradiated IP photo radiation in proportion to the quantity of X-ray is obtained. This photo signal is transformed to electronic signal through photo detector and photo electronic amplifier, and changed to digital signal through analog to digital converter.

-Dynamic range at photo transformer   max.4 digit
-Size of image element at image reading   150 μ m
-Density of image element reading   10 bit (1024)

Image transformed to digital signal will be processed through image processing device.

This is the first time IP device is applied to inspection of power transmission and distribution (T&D) equipment. X-ray film have been used for the inspection of T&D equipment. Fig. 4 shows a valve for vacuum circuit breaker using X-ray film and IP. IP, having broader dynamic range compared with X-ray film, clearly, shows, in wider range the internal construction of the vacuum valve. See Fig. 4. By using conventional film, it is

---

*associated europium barium fluorohalied
BaFX:Eu^{2+},(X=Cl, Br, I).

High Voltage Engineering Symposium, 22–27 August 1999
Conference Publication No. 467, © IEE, 1999

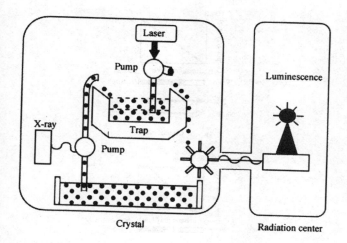

Fig. 1. Radiation mechanism of Imaging Plate

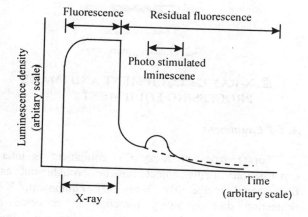

Fig.2. Photo stimulated luminescence

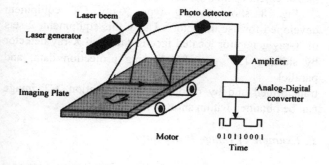

Fig. 3. Reading method of Imaging Plate

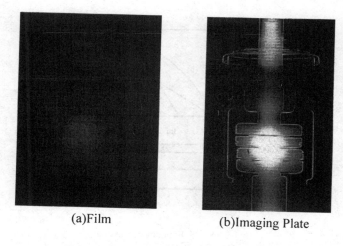

(a)Film   (b)Imaging Plate

Fig. 4. Image of valve for vacuum circuit breaker

necessary to take some shots under various conditions, further, IP, having higher.

B. Euphony Processing

As an example of image processing, euphony processing will be shown. Euphony shows grade of brightness and euphony value of IP equipment is 1024. In euphony processing, density level of image (gray level) i.e. euphony value (for IP equipment, $0 \sim 1023$) will be processed by characteristic functions. Characteristic function consists of straight lines or curves as shown in Fig. 5. Fig. 5.(a) for image whose euphony value (density level) is concentrated in a certain range. Every possible euphony value (density level) is taken and extended, the contrast is improved. Fig. 5.(b) and Fig. 5.(c) shows γ correcting curve and (b) shows to shift the image to darker side (for greater gray level) and (c) shows to shift the image to brighter side (for smaller gray level).

Interrupting part of Vacuum Circuit Breaker by X-ray film is shown in Fig. 6.(a) image after euphony processing in Fig. 6.(b) and Fig. 6.(c). Fig. 7.(a)$\sim$(c) shows euphony processing characteristics used in Fig. 6.(a)$\sim$(c).

In Fig. 6.(a) euphony processing of original image is being original image. Input euphony value and output euphony value correspond to 1:1 in the range of $0 \sim 1023$ and input euphony value and output euphonic value becomes equal. As input euphony value becomes larger, material to be inspected would be thinner, X-ray value to be detected could be greater and gray level becomes greater. In original image of Fig. 6.(a) when X-ray is irradiated directly on IP gray level becomes larger. However parts other than interrupting parts, quantity of X-ray is attenuated due to the thickness of circuit breaker containment vessels, and euphony value becomes less than 1023. Image becomes lower contrast and faint color as a whole.

Fig. 6.(b) shows the original image euphony processed. In this figure, as shown in euphony characteristics in

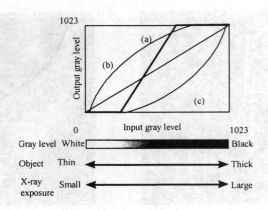

Fig. 5. Euphony processing

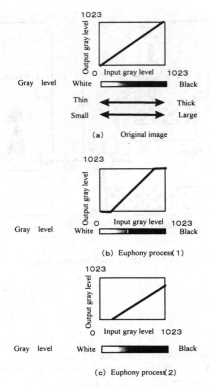

Fig. 7. Euphony processing

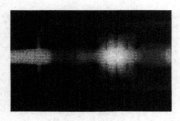

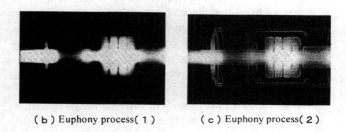

（a）Original image

（b）Euphony process（1）　　（c）Euphony process（2）

Fig. 6. Euphony processing image of vacuum circuit breaker

Fig. 7.(b) greater side of input euphony value, i.e. image to which greater amount of X-ray is irradiated, is diminished and smaller side of input euphony is diminished, and the balance of the euphonic value is extended. As a result, contrast of whole image has been improved, and image of stationary and moving contacts became clearer.

Fig. 6.(c) shows image of greater side of output euphony being cut, Fig. 7.(c) shows its euphonic characteristics. Contacts and bellows which have been dim in Fig. 6.(a) become clearer. This enables to identify the wear of the top of stationary contact, or whether bellows are exactly in position or not.

Thus, original image of thick or thin object can be selected by euphonic processing and, by emphasizing contrast, internal structure of the objects can be easily defined.

Ⅲ. X-RAY CT EQUIPMENT AND IMAGE PROCESSING EQUIPMENT

A. CT Equipment

Characteristics of X-ray CT equipment is internal construction of the object can be brought out as 3 dimensional image. By X-ray CT equipment, X-ray projection data in every direction is collected and tomography is obtained by performing image reconstruction operation such as filter connected inverse projection, and internal defects, detection of foreign material, measurement of density distribution, measurement of internal construction can be achieved.

Fig. 8 shows swing type X-ray CT equipment developed for use in factory. Using this equipment, 2 sets of X-ray generator located face to face and X-ray detector by shifting 130°, and collecting projection data, and parallel

transformed and by reconstruction operation, CT image can be obtained within short period.

B. Example of Image Processing

By image synthesizing 2 dimensional cross sectional image data by volume rendering method, and displaying data in 3 dimension, inspection of internal construction of the object becomes easier.

As an example of CT image processing, photo graphing of valve for vacuum circuit breaker using swing

Fig. 8. Swing type X-ray CT equipment

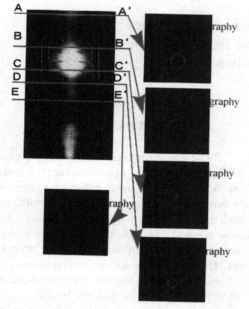

Fig. 9. X-ray CT image of valve for vacuum circuit breaker

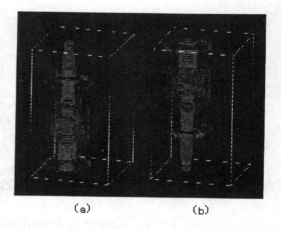

Fig. 10. Three dimensional image of valve for vacuum circuit breaker

type X-ray CT equipment is shown in Fig. 9 and Fig. 10. Fig. 9.(a) shows X-ray fluoroscope and a part of tomograpy. Fig. 10 shows each cross section of CT image being summarized into 3 dimensional image and this is the first successful attempt in applications for transmission and distribution equipment.

IV. CONCLUSION

New method of X-ray projection technology has been developed in order to accomplish more accurate external diagnosis for vacuum circuit breaker.Condition of the interior has been detected more accurately and shortly by this development.

V.REFERENCES

[1] T.Fukuoka , "Digital Radiography Using Imaging Plate", JSNDI 45(10), 1996, pp.720-724, 1996.

[2] K.takahashi, et al., "Imaging Plate", Plus E 6, pp.100-105, 199.

[3] T.Fukuoka, et al., "FCR Digital Radiography with Imaging Plate", Radiation Symposium(1) 45(10), pp.178-183, 1995. FCR technical report Fuji Photo Film Co.,Ltd.

[4] Onoe, et al., "Handbook of Image Processing", Shoukendo Co., Ltd.

[5] M.Takagi, et al., "Handbook of Image Analysis", Tokyo University.

[6] Y.Imazato, et al., "Medical Image Processing", Shoukendo Co.,Ltd.

[7] M.Fujii, "Development and Application of Industrial X-ray CT",Inspection Engineering, 3(10), PP.1-6(1998).

Address of author
Tooru Tsubaki
Hitachi Engineering & Services, Ltd.
38,Shinkou-cho,Hitachinaka-shi,
Ibaraki-Ken,312 Japan
E-mail:ttsubaki@hesco.hitachi.co.jp

DESIGN AND CONSTRUCTION OF A PC-BASED PD DETECTOR AND LOCATOR FOR HV CABLES USING FPGA

Narong Tongchim[*], Santi Yodpetch[**],
Rattapoom Vudhichamnong[*], Samruay Sangkasaad[*]

[*]Center of Excellence in Electrical Power Technology
Chulalongkorn University, Bangkok 10330, Thailand
[**]Electricity Generating Authority of Thailand

ABSTRACT

Rapid development in Field Programmable Gate Arrays (FPGA) technology opens up new opportunity to develop re-configurable real time DSP to reduce disturbance during PD measurements under on-site conditions. This paper apply this technology in the design and construction of a real-time PD detector and locator. The detector and locator consists of analog circuits and digital circuits most of which are programmed into an FPGA chip. Then, the FPGA chip, the analog circuits and digital circuits are fabricated into a PCB that can be inserted into a slot of a PC. The instrument results in a low-cost and high-performance digital PD measuring system.

1. INTRODUCTION

Partial discharge (PD) within the dielectric of HV cable cause progressive deterioration and may lead to ultimate failure. Therefore detection and location of PD is important for quality control of HV cable. The principle of PD detection and location in HV cables are well-known. A simplified test circuit which widely used method for PD detection and location, is shown in Figure 1 [1-5]

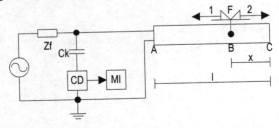

Figure 1 PD detection and location test circuit.

High Voltage Engineering Symposium, 22–27 August 1999
Conference Publication No. 467, © IEE, 1999

2. SYSTEM DESIGN

The measuring instrument was designed as a PC-based system. The advantage of using PC is to use its own graphic display system to show output and reduce pulse shape disturbance in real time processing by software using time window method. In addition, the system is able to store the data for further analysis or printing report. The design concept for analog circuits of the proposed system follows the conventional PD detection and location circuits[1-3,5]

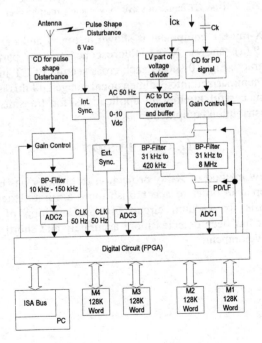

Figure 2 Block diagram of the measuring instrument.

This instrument as shown in figure 2 consists of the Intel Pentium pro processor 200 MHz PC system, coupling devices, signal cables and the PC expansion card. The working concept of the system can be described as follows :

1). PD locator circuits start from receiving the input at ADC1 to convert analog signal into digital signal at sampling rate of 40 MS/s (12 bit). Then, storing the data into the memory unit M1 before sending them to computer to locate the PD location.

2). PD detector circuits process in real time starting from the input at ADC1 to convert analog signal into digital signal at sampling rate of 10 MS/s (12 bit). The collected data are sent to digital circuits to analyze the peak value and the phase angle of PD pulse[6]. Then, storing all the data for each cycle into the memory unit M1 and M2 alternately with synchronization signal for M1 and M2 is from the CLK 50 Hz signal. before sending them to computer.

3). Disturbance suppressing circuits works like the PD detector except for receiving disturbance signal from the antenna[7,8]. The suppression of disturbance is done by comparing the phase angle of each pulse between the PD detector circuits and antenna circuits. If the result is match then that pulse is eliminated.

4). Synchronized signal circuits from either test voltage or internal voltage from power supply is used to synchronize and generate reference phase angle for PD detector circuits.

5). Test voltage measuring circuit is used to indirectly measure test voltage from low voltage terminal of coupling capacitor.

The detail of digital circuits implemented in the FPGA chip is described as shown in Figure 3. All digital circuits are shown in figure 4. The resulted expansion card is shown in figure 5. The resulted system is shown in figure 6.

3. TEST & RESULTS

The designed measuring system was tested to locate the PD of a 20 kV 50 sqmm XLPE cable with 229.54 m in length. The test sample cable was made to be a defect at 179.86 m. The test results showed that the designed measuring system could detect the PD location with accuracy of 0.1%. The detail of the test results and the histogram of PD location are shown in figure 7 and 8 respectively.

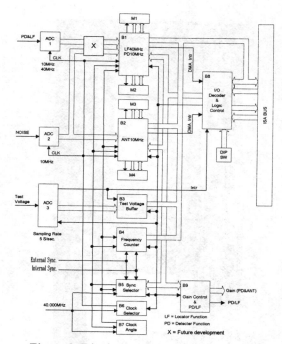

Figure 3. Block diagram of digital circuits implemented in the FPGA chip

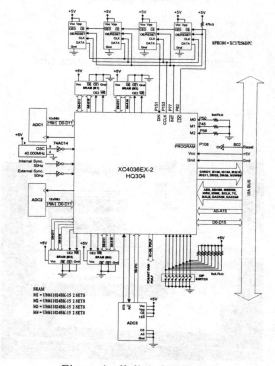

Figure 4 all digital circuits

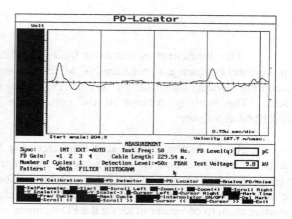

Figure 7. Test result for PD location

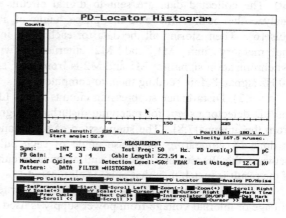

Figure 8. Histogram of PD location

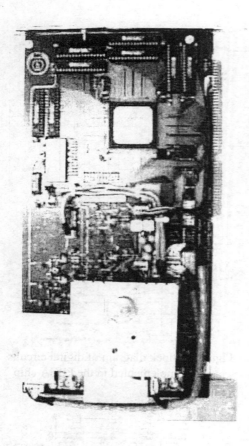

Figure 5. The resulted expansion card

Figure 6. The resulted measuring instrument(MI)

4. CONCLUSION

The design and construction of a PC-based PD detector and locator for HV cables by using FPGA technology ,the following can be concluded.

1) The high speed and complex digital circuits can be designed and fabricated.

2) The conventional PD detector and locator can be designed as an expansion card inserting into a slot of a PC results a high performance and low-cost measuring system.

3) With this expansion card, digital signal processing technology and data analysis can be applied.

4) Due to the rapid development in FPGA technology and FPGA design-aided software[9]. We expect that real-time noise reduction can be achieved efficiently by using re-configurable real time DSP such as a digital filter in "x" Block of Figure 3.

5. ACKNOWLEDGEMENTS

We would like to thank Mr.Warut Thaweesub and Mr.Krit Athikulwongse for their contribution of this paper.

6. REFERENCES

1. Kreuger, F. H. Discharge Detection in High Voltage Equipment ,pp. 7-85. Temple Press Book Ltd., London, 1964

2. Kreuger, F. H. , Wezelenburg, M.G. , Weimer, A.G. and Sonneveld, W. A. "Error in the Location of Partial Discharges in High Voltage Solid Dielectric Cable"IEEE EI-6 (1983)

3. IEC 60270(1998),Partial Discharge Measurements.

4. Gulski, E. , Samual, A. R. , Kehl, L. and Geene, H. T. F. "Digital Discharge Location in HV Cables with the Travelling Wave System" Conference Record of the 1996 IEEE International Symposium on Electrical Insulation , Montreal ,Canada , 1996

5. Weiringa , L. ,"Location of Small Discharges in Plastics Insulated High Voltage Cable", Trans, IEEE PAS-104(1985), pp.2-8.

6. Gulski, E. "Digital Analysis of Partial Discharges" IEEE Transactions on Dielectrics and Electrical insulation Vol.2 , No.5 , 1995

7. Robinson Instruments Limited, "Pulse Discrimination Partial Discharge Detector " Model 803 : Instrument Operation Manual

8. Hariri, A. et. al. "Field Location of Partial Discharge in Power Cables using an Adaptive Noise Mitigating System" Conference Record of the 1996 IEEE International Symposium on Electrical Insulation , Montreal ,Canada , 1996

9. Xilinx, Inc. DSP LogiCORE

EFFECTS OF AIR HUMIDITY AND TEMPERATURE TO THE ACTIVITIES OF EXTERNAL PARTIAL DISCHARGES OF STATOR WINDINGS

E. Binder, A. Draxler

VERBUND-Elektrizitäts-
erzeugungs-GmbH
Austria

M. Muhr, S. Pack, R. Schwarz

Institut für Hochspannungstechnik
mit Versuchsanstalt
Technische Universität Graz
Austria

H. Egger, A. Hummer

Kärntner Elektrizitäts AG
KELAG
Austria

Abstract

According to some unexpected partial discharge (PD) test results during off-line measurements of generators it must be assumed, that a relative high humidity content of the ambient air considerably influence the activities of external partial discharges. This assumption could be verified by means of PD tests performed on models of stator bars exposed to various climatic conditions. These tests have shown, that activities of external PD are strongly reduced, if the ambient air humidity exceeds 50 %. Consequently there is a risk of false diagnosis under such circumstances.

Keywords

Generator, stator winding, partial discharge, PD-pattern, humidity, ozone

1. Introduction

In case of important components of power plants as HV generators condition based maintenance can be considered as an optimal strategy for the efficient assignment of restricted budget resources. Prerequisite for this maintenance policy are reliable and precise diagnoses. As the diagnostic of generators is not yet fully satisfactory, further improvements of various measuring methods are required.

The progress of the insulation diagnostic, in particular of the PD measuring technique, is object of a research project of VERBUND. In part the tests are performed in co-operation with KELAG and the TECHNICAL UNIVERSITY GRAZ.

A partial project deals with the effect of the air humidity and temperature to surface discharges of stator bars [1]. This investigation was motivated by the supposition that the activity of external partial discharges are strongly influenced if the air humidity is high or changes during off-line tests.

2. Situation

The dominant importance of PD measuring methods for condition assessments of insulation systems results from following facts:

- Apart from effects of the voltage stress, PD activities are characteristically determined by the geometry and the nature of damages to the insulation system [2]. Since digital PD measuring techniques are available, these relations can optimally be utilised to identify insulation defects [3]. For this purpose the visualisation of the PD activities by means of phase related PD histograms (nqφ-diagrams, PD patterns) have been proven successfully.

- The ozone, produced by PD in air, can be utilised for identification of external partial discharges. Therefore, recently the O_3-analyses of the cooling air of generators have been introduced to our diagnostic system. First experience have shown, that this method is excellent suited for recognition and quantitative assessment of damages to the corona protection and to the stress grading of stator bars [4]. Hence, this method complements the conventional PD measuring technique and improves the reliability of insulation diagnoses.

- Ultrasonic or optical signals radiated by partial discharges as well as propagation time effects of PD pulses can be utilised to localise insulation defects.

Concerning the application of off-line PD test results there is, however, a less-known source of error. Occasionally the diagnosis is confused by significant changes of the PD activities occurring during off-line tests.

This phenomenon is illustrated by Figure 1 showing PD patterns of a stator winding, measured twice in the same arrangement and at the same voltage stress at a distance of some hours.

High Voltage Engineering Symposium, 22–27 August 1999
Conference Publication No. 467, © IEE, 1999

Figure 1a clearly shows characteristics of slot discharges caused by small damages to the corona protection. This diagnosis have been verified by means of a visual inspection. In contrast, no symptoms of slot discharges are visible in Figure 1b and hence a false diagnosis would result from this pattern. As only the humidity of the ambient air has changed between the two tests (from 33 % to 58 %) this alteration was assumed to be the cause of this phenomenon.

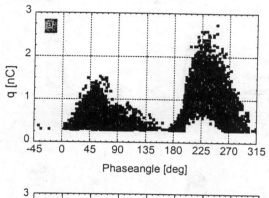

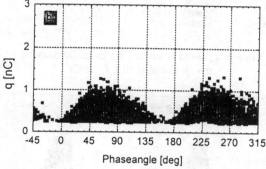

Fig.1. PD pattern of a stator winding with damaged corona protection
a. Ambient air: relative humidity 33 %
b. Ambient air: relative humidity 58 %

3. Investigations

In order to get more information in which way PD activities are influenced by various environmental conditions, systematic and reproducible investigations with models of generator bars have been started in the laboratory at the TECHNICAL UNIVERSITY GRAZ [5]. For this investigations the PD activity in correlation to the temperature and humidity as well as the ozone production are of interest.

Therefor one iron slot model and some generator bar models of a stator winding with a length of 400 mm have been made (Fig.2). All stator bar models were carried out with new outer corona protection and end potential grading to ensure similar starting conditions and a very low basic interference level. Two types of model setups were carried out. One setup with a generator bar model on a fixture (Fig.2b) directly surrounded by the climate in the test (sur-

face discharges) and the other one fixed in the iron slot model to generate partial discharges between the insulation surface and the stator iron (slot discharges).

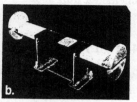

Fig.2. Slot model (a) and generator bar model (b)

For the measurements the experimental setup is placed in a temperature and humidity regulated high voltage testing room. There are the high voltage source and the measurement circuit located too. The model itself is located in the center of an additional closed volume to enable constant conditions for the climatic (temperature, humidity, ozone) and to protected it against external influences (Fig.3).

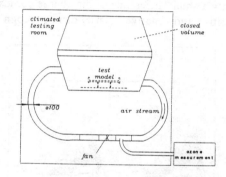

Fig.3. Schematics of the experimental setup with the test model

For the PD measurements a test circuit according to IEC270 is used (Fig.4).

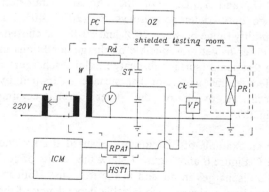

Fig.4. Measurement circuit based on IEC270
VP, RPA1, ICM ... PD measurement system
OZ, PC ... ozone measurement system
PR ... test setup

The basic test setup using a coupling capacitor of 1 nF is nearly free of partial discharges (basic interference level < 2 pC). With the digital partial discharge recording system ICM the pulse high distribution is recorded under constant and reproducible environmental conditions using a sampling time of 60 s. All measurements were carried out in a shielded testing room.

4. Test results

Results from the on-line and off-line measurements of generators during on site conditions represent the basis of the applied investigation parameters. Looking at these results the registered effects are assumed to have an influence caused by different temperature and humidity conditions.

Before starting systematic investigations a reference PD pattern including the complete measuring system and test setup was taken (Fig.5). The maximum charge of < 45 pC (10 kV) gives a clear distance to the expected values in case of small damages to the outer corona protection. It should be pointed out, that only a small difference of the charge comparing the positive and negative polarity can be recognized.

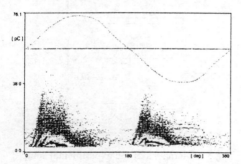

Fig.5. PD pattern of the test setup without defects in the outer corona protection (10 kV, 40°C, r.h. 13 %)

The voltage for all test series was varied between 0,2 U_N and 1,0 U_N. For the climatic parameters based on two temperature values (20°C, 40°C) a humidity range between 20 % and 80 % was chosen. In order to get reproducible discharge conditions in the test setup (surface discharges, slot discharges) a defect at the generator bar surface is applied by partial removing of the outer corona coating (Fig.2b).

As an example of PD pattern measured at 8 kV and 40°C Figure 6 and Figure 7 show the results of surface discharges in air and in the iron slot. Both results show a clearly recognizable dependency of the PD pattern in correlation to the humidity of the circulating air in the test volume.

With respect to the upper PD pattern in Figure 6 and Figure 7 for 20 % and 40 % humidity the partial discharge activity (surface discharges/slot discharges) is about 800/1000 pC for the positive and

about 4700/3500 pC for the negative voltage which documents the enormous polarity effect. For 60 % humidity a PD activity of 80/760 pC (positive) and 600/1300 pC (negative) have been measured, the polarity effect decreases in both setups. At the very high humidity of 80 % the polarity effect almost disappear. Values between 20 and 30 pC for the remaining PD activity were measured.

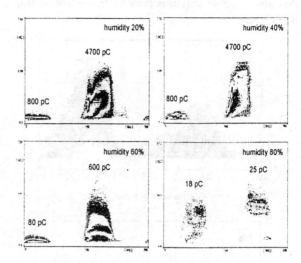

Fig.6. PD pattern of surface discharges by varying the air humidity (8 kV, 40°C)

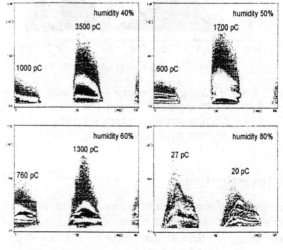

Fig.7. PD pattern of slot discharges by varying the air humidity (8 kV, 40°C)

A comparison of the recorded results of the two setups at a temperature of 40°C within a voltage range from 2 kV up to 10 kV are shown in Figure 8 and Figure 9. Therefore a large number of measurements in systematic order and under constant and reproducible conditions were carried out. In this diagram 5 cycles with a relative humidity between 20 % and 80 % were compared. The matter of fact, that the PD level is a function of the applied voltage is already well known. The significant relation between PD values and humidity, which appeared at the on-site measurements, were afterwards confirmed with measurements at the laboratory.

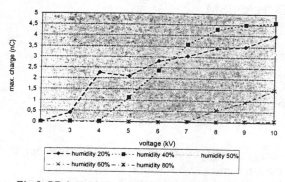

Fig.8. PD intensity of surface discharges vs. voltage at a temperature of 40°C

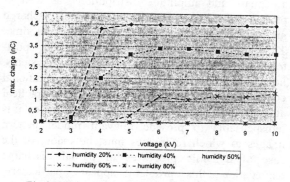

Fig.9. PD intensity of slot discharges vs. voltage at a temperature of 40°C

The effects of the ozone production during the PD activity versus time are shown in Figure 10. The measured ozone concentration strongly correlates to the time when a constant voltage (e.g. 7 kV) is applied. The ozone concentration in the circulating air decreases when the relative humidity increases. It has to be taken into consideration, that the PD activity depends on the relative humidity as shown before.

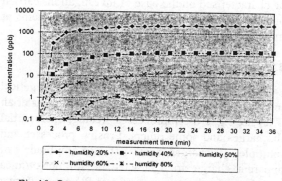

Fig.10. Ozone concentration vs. time (7 kV, 40°C)

The circumstance of a reaction between the ozone concentration and the relative humidity is known. A published analysis [6] carries out the monotonous decreasing of the ozone concentration for increasing humidity when applying constant PD activity.

5. Conclusions

Laboratory tests have shown, that the activity of external partial discharges (PD) of stator bars will diminish or even extinct when the humidity of the ambient air exceeds 50 %. The test results have verified the assumption that the correlation between PD activity and relative humidity as well as the ozone production can not be neglected during on-site off-line measurements.

If these relations are not taken under consideration at off-line tests false diagnoses can be concluded. In case of on-line tests, however, this problem is neglectable as the humidity in the generator is usually low and the temperature of the stator bar is higher than the cooling air.

References

[1] S. Pack, M. Muhr, E. Binder, A. Kogler: Verhalten von Teilentladungen an der Oberfläche von Generatorstäben, 43rd International Scientific Colloquium, Technical University of Illmenau,1998.

[2] F.H. Kreuger: Partial Discharge Detection in High-Voltage Equipment, Butterworth & Co., 1998.

[3] E. Binder, A. Kogler, K. Zikulnig: Diagnosis of Stator Winding Defects based on PD Test Results. CIGRE/EPRI Colloquium on Maintenance and Refurbishment of Utility Turbogenerators, Hydrogenerators and large Motors, Florence, 1997.

[4] E. Binder, A. Draxler, H. Egger, A. Hummer, M. Muhr, G. Praxl: Experience with On-Line and Off-Line PD Measurements of Generators, CIGRE Sess. 1998, Paris, Paper 15-106.

[5] S. Pack, R. Schwarz: Einfluß der Umgebungsparameter auf das Verhalten von Oberflächenentladungen an Generatorstäbe, Technical Report HS97191, Institute and Research Institution of High Voltage Engineering, Technical University Graz, 1998.

[6] D. Peier, M. Malejczyk, D. Klockow: Chemische Untersuchungen zur Luftzersetzung durch elektrische Teilentladungen, Elektrizitätswirtschaft, Jg.87, 1988, Heft 23.

THE CORRELATION OF PD-CHARGE AND -ENERGY PATTERNS IN THE STUDY OF PARTIAL DISCHARGE MECHANISM IN THERMOSETTING INSULATION

B. Florkowska and P. Zydron

University of Mining and Metallurgy, Electrical Power Institute, Kraków, Poland

Abstract

The objective of this paper is the interpretation of distributions and phase-resolved patterns of partial discharges apparent charge and energy. Results of the measurements performed have been obtained in long-term tests at increased electrical and thermal stresses. This issue is of vital cognitive importance for the assessment of material degradation mechanisms; it is also very important while diagnosing the high voltage insulating systems.

Test and measurements were carried out using a thermosetting insulation system in high voltage machines in which PD occurred at an increased temperature and increased working intensity of the electrical field was applied in investigations has been used as the object of tests and measurements. After these tests the decrease of the PD pulses magnitude and the increase of the number of small pulses were observed in the PD charge distributions. This means the transition of partial discharges to a high repetition form - multiply tiny discharges. In this case, it is also observed in the PD pulses energy distribution, the presence of the extremum in the range of small values of energy. This effect provides new information about the degradation process.

1. Introduction

Investigations upon the correlation between destruction mechanisms of insulating materials and the quantitative characterisation of partial discharges phenomenon are till now one of the topical questions in this domain. Results from those investigations would have both a cognitive and practical importance since they would allow for the recognition of physical mechanism of the material destruction and the determination of the material resistance against the partial discharge actions; on the base of those results it would also be possible to develop corresponding standardised numerical criteria pertaining to the material resistance.

To recognise the degradation mechanism, it is necessary to study hypotheses for the influence of energy conversions occurring within the region of partial discharge. Those hypotheses deal with various types of the impact by the energies of discharges in their sources, for example thermal energy, energy of electrons and ions movement, UV radiation energy, mechanical energy.

Presently, the most frequently applied methods to quantitatively describe the mechanism of discharges are phase and amplitude distributions, as well as phase-resolved images of charges and of discharge number obtained by the registration in real time. This procedure is usually used in the post-technological qualification of insulating systems, however, its application under operational conditions is also considered, and adequate practical attempts have already been made. In both cases, it is indispensable to provide relevant model pattern images to which references must be made. In the case of dynamic processes of partial discharge degradation, this problem appears particularly difficult. There are many papers that deal with the problem of the recognition of partial discharge patterns that are a visual representation of mechanisms and occurring degradation processes. Thus, a question appears which asks about the mutual correlation of different partial discharges representing various quantities and changing during the degradation processes. For example, this problem refers to the correlation between distributions of partial discharges, of charges' images and charges quantities, and the discharges' energy distributions and images. This problem shall be characterised on the base of investigations carried out on thermosetting machine insulation systems at increased electrical and thermal stresses.

2. Partial discharges energy

The existence of defects in areas of energetical unbalances leads to local destruction of material. They are e.g. erosion processes, treeing, micropartial discharges and some others. Erosing processes and multiple discharges with small amplitude of apparent charge are characteristic for epoxy/mica multilayer insulation as a result of PD action during the long term tests with electrical and thermal stresses [1,2].

It is necessary to investigate the partial discharge energy as a reason of structural changes of insulating material. In both theoretical and empirical investigation on the influence of such factors like partial discharge source geometry [3], properties of material surface [4], inception and extinction voltage values, voltage kind - on partial discharge energy is the aim of the recognition. The apparent charge and partial discharge energy depend on the volume and the shape of gaseous inclusion being the source of discharges. However, contrary to the energy, the apparent charge depends also on the insulation thickness, sample geometry and localisation of partial discharge source. Therefore, the energy can be a more useful parameter for the characterisation of partial discharges and their effects. In many recent investigations, the total energy in the selected time interval, e.g. ac voltage period, was an object of measurements and analysis. However, such an approach to energy is not sufficient for the description of degradation processes in insulation. It is true that the final result is the effect on single discharges cumulation, but it is sufficient to calculate the correlation of effects and energy values in such a manner only in the case of single discharge sources.

In basic studies, the total partial discharge energy within an ac-test voltage period, is as follows:

$$W = k \cdot Q \cdot U \qquad (1)$$

where:

  Q - apparent charge of a discharge,
  U - test voltage value at the moment of discharge inception,
  k - coefficient

The above indicated formula does not express a statistical character of discharges owing to many random factors. Impulses of discharges in their individual single sources show a stochastic character and differ between themselves by instant values of ac voltage at the moment when subsequent discharges are initiated; they also differ by their apparent charges. Individual sources of partial discharges generate different pulse distributions, they differ from each other by the values of average initial voltage values and pulse charges, as well as by discharges' numbers in one half-part of the ac voltage period. Pulses from all PD sources within an insulation system, and regarded within set time interval create a set the elements of which have different apparent charges and initial voltage inception values. They also create a two-dimensional distribution of random Q – and U values, with a density function $f(Q,U)$.

Data of the empirical $f(Q,U)$ function have been obtained using pulse height analysis. Thus, a shape of the energy density function $f(W)$ depends on the total density function $f(Q,U)$.

Now it is possible to measure the energy changes even in single discharges by means of real-time measurements which enable to register of consecutive partial discharge parameters and their

pulse-high and phase analysis. This problem is presented here.

In the basic equivalent scheme of insulating system (model a-b-c) [5], the energy of a single discharge is:

$$w_{ri} = q_{ri} \cdot \delta U_i \qquad (2)$$

or

$$w_{ri} = C_c \cdot \Delta U_i{}^2 \cdot \left(1 + \frac{C_a}{C_b}\right)^2 \qquad (3)$$

where:

  q_{ri} - real charge of a single discharge,
  δU_i - voltage drop in discharge source,
  ΔU_i - voltage drop on the object
  C_c - capacity of inclusion,
  C_a - capacity of solid dielectric,
  C_b - capacity of solid dielectric arranged with the discharges' source in series.

These quantities are unknown and impossible to be measured. Therefore, the determination of value of energy in discharge source is not possible.

The only mean of energetical description is the determination of energy basing on an apparent charge value measured on the object:

$$w_i = q_i \cdot U_i = q_i \cdot U_m \cdot \sin \varphi_i \qquad (4)$$

where:

  q_i - apparent charge of a single discharge,
  U_i - voltage value on object in the moment of discharge inception,
  U_m - crest value (amplitude) of ac voltage,
  φ_i - voltage phase of discharge pulse.

PD energy distributions $Dn(w)$ and apparent charge distributions $Dn(q)$, as well as phase-resolved images $D(q,\varphi,n)$ and $D(w,\varphi,n)$ can be quite different. A probabilistic description of PD energy suggests that it would be very useful to distinguish at least three pulse groups, i.e.:

- pulses with the greatest probability of occurrence,
- pulses of the greatest energy,
- pulses with energy corresponding with the local extremum in a $Dn(w)$ distribution.

3. Measuring system and test object

The measuring system used in the experiment enabled to register of the PD pulses in real time simultaneously with a corresponding phase angle of the test voltage [6]. This measuring technique allows for the following:

- registration of PD pulse and determination of its
  height and polarity,
- phase location of PD pulse corresponding to ac voltage period,
- PD pulse registration in a programmed ac

voltage period number.

Measurement results can be performed for the basic PD intensity quantities, including the energy of individual PD pulse. The two or three dimensional distributions of these quantities, with regard to the phase location of PD pulses, can be made as next. In this way the partial discharge charge and energy

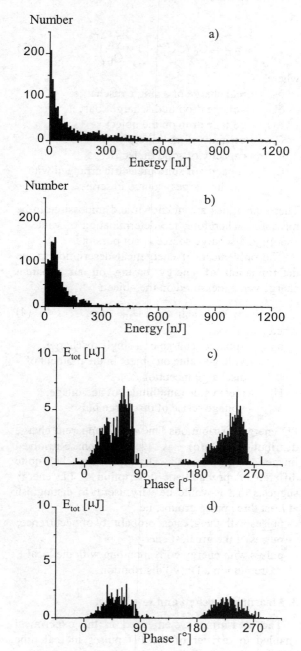

Fig.1. Two groups of energy distributions $Dn(w)$ and $Dw_{tot}(\varphi)$:
a), c) at the beginning of the test (A) at 20°C,
b), d) at the end of the test at 20°C (test D).

distributions were calculated: $Dq(\varphi)$ and $Dw(\varphi)$ - phase distributions; $Dn(q)$ and $Dn(w)$ - amplitude distributions. The following quantities were used to interpret results: N - number of discharges in ac voltage period, q_m - maximum charge, Q_{tot} - total

charge in registration time, w_m - maximum energy, W_{tot} - total energy in registration time, q_{av} - average charge of single discharge, w_{av} - average energy of single discharge. During the measurements for 50 Hz testing voltage, the resolution time was preset on 5,56 µs i.e. whole 20 ms period was divided into 3600 basic phase windows $\Delta\varphi$. The software delivers

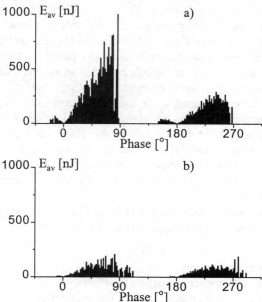

Fig.2. Phase distributions of the PD pulse average energy $Dw_{av}(\varphi)$:
a) at the beginning of the test at 20°C,
b) at the end of the test at 20°C (test D).

not only the distribution „quantity - phase" accumulated to one period, but also allows for subsequent tracking period (one by one) during the whole measuring time. In order to make the graphical presentation more clear and for mathematical analysis, the pulses registered in near phase windows $\Delta\varphi$ could be accumulated in the one phase window $\Delta\varphi' = k\Delta\varphi$ (usually k=10), then for the visualisation the purpose, period of voltage is divided into 360 windows $\Delta\varphi'$. For the phase distribution: total, accumulated values of charge or energy could be calculated or, according to pulse-phase distribution $Dn(\varphi)$ or to the number of registered ac periods, distribution of averaged values of selected quantity in each time window could be calculated.

The results presented in the paper were obtained from the tests carried out with the use of specimens from coils 6 kV motors. Prior to the long term tests, they were controlled at the initial stage (A): electrical stress up to 4 kVmm[-1], ambient temperature 20±2°C. Next, the specimens underwent a test (D) with a combined electrical (3 kVmm[-1]) and thermal (135°C) stress. The test time was 2500 hours and the measurements were taken during the tests. The analysis of results is based on the final measurements, that revealed the essential differences with the initial measurements.

4. Results and discussion

The analysis of changes of apparent charge distributions of thermosetting insulation after a long-term multistress test was studied [1,2] and some meaningful differences in their shape were stated. Here, the changes of energy distribution Dn(w) and Dw_{tot} are presented (Fig.1 and Fig. 2).

At the initial stage, the average energy of PD pulses is approximately 220 nJ (Fig.1a) and after the long term test (Fig. 1.b), it diminishes to 70 nJ, and at the Dn(w), a distribution extremum value is present at the energy of approximately 50 nJ.

The slight decrease in the total pulse number after the long time test is also characteristic (Fig. 3.)

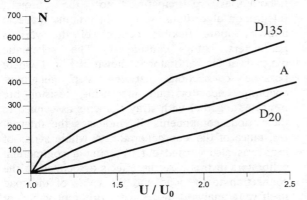

Fig. 3. PD pulse number N vs. test voltage U/U_0:

A-the initial stage,

D_{135}-after the test D at 135°C,

D_{20}-after the test D at amb. temp.

4. Conclusion

The energy of single discharge q_i given by the formula (4) is a function of both the apparent charge value and its phase. In degradation processes, there are changes in both the apparent charge distribution and the phase distribution, and also their phase location range is changed.

Thus, the energy distributions are complicated and they do not correspond to changes of charge distributions. During the long time ageing process of thermosetting insulation at complex thermal and electrical stresses, changes of partial discharge energy were observed. Decrease in maximum energy values and the appearance of extremum of Dn(w) distribution at a low energy value were effects of ageing processes. The discharges with a high energy at the beginning of the test, identified with a streamer form, are responsible for surface changes in discharge source, and they lead to increase in surface conductivity γ_s. It, in turn, causes the increase of extinction voltage and diminishing of PD charges and, in consequence, it leads after the ageing process to pulseless discharges. They can be proceeded by multiple tiny pulses which show some common features with swarming micropartial discharges. The measurements of discharge energy give additional information on this matter. If the time constant of surface charge leakage $\tau_s = \varepsilon_d\varepsilon_0 r_c / 2\gamma_s$ (where r_c - inclusion radius) is decreased because of increased surface conductivity [7], then, it can be concluded that the disappearance of high energy pulses and the decrease of average energy value in Dn(w) distribution will be the result. The appearance of an extremal value at Dn(w) distribution can be interpreted as a „stable" stage discharge activity with the energy corresponding to the modal value, i.e. approx. 50 nJ. At this stage, the effective surface of discharges was formed. It is always the question whether or not a correlation between the cumulative activity of single discharges and their cumulative effect exist, or at least, an energy range of such a correlation. Degradation processes can cause the decrease in pulse energy as an effect of changes occurring in discharge sources.

Acknowledgement

The work presented in this paper was sponsored by the State Committee for Scientific Research, Warsaw, (Contract No : 18.18.120.151).

Literature

[1] B. Florkowska - "Application of partial discharge patterns for assessment of multistress synergy in thermosetting stator bar insulation", Proc.9th ISH, paper 5648, Graz, 1995.

[2] B. Florkowska - "Assessment of temperature influence on partial discharges in epoxy/mica insulation", Proc. 5th ICSD, pp.356-360, Leicester, 1995.

[3] J. Sletbak - "The influence of cavity shape and sample geometry on partial discharge behaviour", IEEE Trans. on DEI, Vol. DEI-3, No. 1, pp. 126-130, 1996.

[4] J. M. Wetzer, A.Pemen, P.C.T. van der Laan - "Experimental study of the mechanism of partial discharges in voids in polyethylene", Proc.7th ISH, paper 71.02, Dresden, 1991.

[5] F.H. Kreuger - "Discharge detection in high-voltage equipment", Butterworth & Co., London, 1989.

[6] B. Florkowska, M. Florkowski, P. Zydron, R.Wlodek - "Partial discharge acquisition and data processing", Advances in Modelling & Analysis, Periodicals B, AMSE Press, vol. 30, No.4, pp. 7-15, 1994.

[7] C. Hudon, R. Bartnikas, M.R.Wertheimer - "The physics of partial discharges in solid dielectrics", IEEE Trans. on EI, Vol. EI-28, No. 1, pp.1-8, 1993.

INFLUENCE QUANTITIES IN THE CALIBRATION OF PD CALIBRATORS CONTRIBUTION TO THE UNCERTAINTY ESTIMATE

R. Gobbo, G. Pesavento

A. Sardi, G. Varetto

C. Cherbaucich, G. Rizzi

Dept. of Electrical Engineering -
University of Padova, Italy

IEN "G. Ferraris", Torino, Italy

CESI - Milano, Italy

Abstract - The numerical integration of the current pulse delivered from a PD calibrator is the easiest way to carry out a performance test of the calibrator itself by using readily available experimental apparatus. The paper discusses the role and the relative importance of the various influence quantities which intervene in the measurement and can contribute to degrade the uncertainty figure. It is shown that type and length of connecting leads as well as the bandwidth of the scope can have only a second order influence, whereas the correct estimate of the vertical sensitivity and the procedure used to carry out the integral, namely the definition of the base line and the choice of the integration limits, can play a more significant role.

Introduction

A PD calibrator consists essentially of a step voltage generator coupled through a high quality low value capacitor to the terminals of the object under test [1]. In principle its characteristic quantity is the charge being delivered which, under circumstances usually met in practice, is related to the product of the step voltage by the capacitance. Consideration of these simple elements would in fact establish a connection between traceable quantities such as DC voltage and capacitance and the charge being obtained: however there is a variety of elements intervening in this process which can contribute the degrade the relation between the two sets of quantities. There are of course stray parameters, of both inductive and capacitive type, associated with the circuit components, the finite rise time of the step and the overall behaviour of the device used to integrate the current pulse. When used in circuit of practical interest the question of the lead length in case of equipment having terminals far away is again a potential source of error. Several studies have been already carried out and in some cases the conclusions are to some extent contradictory [2÷6]. The aim of the work was intended to address two key points: on one side the relation existing between the base quantities (DC voltage and capacitance) and the value of the charge being obtained in various conditions of practical interest and, on the other, the effect of the interaction of a commercial calibrator with the load.

1. Procedure

Most of the tests were carried out starting from a Reference Calibrator developed by using a stabilised DC voltage being connected to the coupling capacitor by means of a mercury wetted relay. The nominal charge based upon the measured value of applied voltage and capacitance was estimated to be 49.7 pC. Pulse measurements were taken by using a digital oscilloscope Tektronix Type DSA 601 with a maximum sampling rate of 500 MS/s; however, when operating in Equivalent Time, sampling rates up to 5 GS/s could be obtained.

Tests were carried out independently in the Laboratories participating to the project; integration algorithms were mainly home made to allow a more flexible control of the various parameters being considered. The schematic diagram of the calibrator is shown in Fig. 1 where possible connections between step generator, coupling capacitor and measuring resistor are indicated together with stray elements associated to the various components. R_g indicates the internal resistance of the step generator. A whole series of tests was also carried out by using a commercial calibrator: in this case the effects associated to the connection to the load and the value of the load itself were analysed in detail for different values of the charge being delivered.

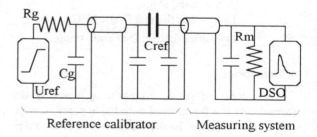

Reference calibrator   Measuring system

Figure 1 Basic structure of a PD calibrator coupled to a measuring system

The following influence quantities were considered in detail:

1. Scope bandwidth
2. Internal resistance of the generator
3. Length and type of connecting leads between generator, capacitor and digital scope
4. Value of the measuring resistor
5. Dynamic behaviour of the scope channel input.

It has to be noticed that the effects which can derive from the elements described in points 1,2 and 4 are to some extent equivalent because, with respect to the current pulse being recorded, all can introduce a low pass filtering, thus resulting in voltage signals

High Voltage Engineering Symposium, 22–27 August 1999
Conference Publication No. 467, © IEE, 1999

with increased time duration. High ohmic loads result in higher peak value whereas BW limitation and increase of the internal resistance of the generator act in the opposite way. Due consideration was paid to the integration procedure and namely to the choice of the base line and of the time limits of the window being considered.

l.1 Influence of the scope bandwidth

The 50 Ohm resistance of the channel input was used as the measuring resistor; all connections between the various parts of the calibrator were reduced to a minimum and in practice only the connectors were included. Equivalent Time sampling and averaging were used; the overall length of the record was around 500 ns. Scope bandwidth was varied between 100 and 20 MHz, thus resulting in different shapes of the pulse as shown in Fig. 2 which reports also the time pattern of the computed integrals. The data are summarised in the following table and each value refers to groups of 10 measurements.

Scope setting [mV/div]	BW [MHz]	Charge [pC]	Std. Dev. [%]	Pulse duration [ns]
50	100	49.757	0.15	15.05
50	20	49.718	0.23	33.65
20	20	49.376	0.09	33.50

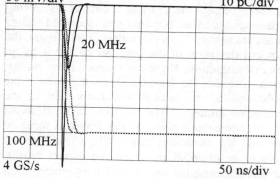

Fig. 2 Pulse modification and integral time pattern with various scope bandwidths

1.2 Influence of the internal resistance Rg

The variation of the internal resistance results in the production of steps having different rise times and this, in case of comparison, could better adapt the pulse being generated to that of the calibrator under test. Relevant parameters are reported in the following. The internal resistance was varied over a broad range without experiencing significant variations as shown in Fig. 3 and related table.

R_g [Ω]	BW [MHz]	Charge [pC]	Std. Dev. [%]	Impulse duration [ns]
82	100	49.908	0.19	14.80
150	100	50.048	0.40	20.20
1500	100	49.915	0.37	38.08

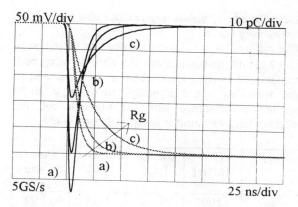

Fig. 3 Modification of the pulse shape for various R_g

1.3 Influence of connecting leads between generator, capacitor and digital scope

In the same measuring conditions listed above, the components of the calibrator were split and tests repeated with full and reduced scope bandwidth. Relatively long coaxial cables were inserted along the chain without appreciable variation in the charge value. The only effect is a modification of the shape of the current pulse due to a mismatch of the line.

Sensitivity 50 mV/div - Bandwidth 100 MHz			
Config.	Charge [pC]	Std. Dev. [%]	Impulse duration [ns]
a)	49.757	0.15	15.05
b)	49.773	0.13	15.75
c)	49.864	0.13	20.50
Sensitivity 20 mV/div - Bandwidth 20 MHz			
Config.	Charge [pC]	Std. Dev. [%]	Impulse duration [ns]
a)	49.376	0.09	33.50
b)	49.402	0.08	34.05
c)	49.563	0.04	36.25

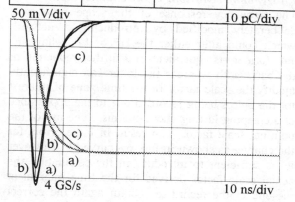

Fig. 4 Pulse shapes and related integral for various connections of the circuit components
a) Connection between different calibrator components and scope as short as possible
b) Coaxial cable ($Z_o = 50\ \Omega$, $l = 1.5$ m) between scope and coupling capacitor
c) Coaxial cables ($Z_o = 50\ \Omega$, $l = 1.5$ m) between scope and coupling capacitor and between this and step generator

1.4 Influence of measuring resistor and type of connecting leads

The value of the measuring resistor was increased to 185.7 Ω and the connections extended by using both coaxial and twisted leads. Results are presented in the following and although there is a slight systematic difference with the previous set, the effect of lead variation is completely negligible.

Config.	BW [MHz]	Charge [pC]	Std. Dev. [%]	Pulse duration [ns]
a)	100	50.028	0.14	37.90
b)	100	50.044	0.15	41.40
b)	20	50.128	0.13	62.65
c)	100	50.040	0.06	55.00
d)	100	50.005	0.16	42.05
e)	100	50.076	0.14	57.15

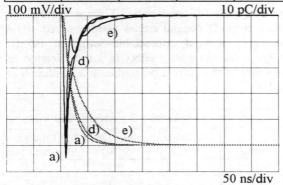

100 mV/div   10 pC/div

50 ns/div

Fig. 5 Pulse modification due to different types of leads
a) Connection between U_{ref}, C_{ref}, R_m and scope as short as possible
b) Couple of twisted leads ($l = 0.3$ m) between R_m and scope input
c) Coaxial cable ($Z_o = 50\,\Omega$, $l = 1.5$ m) between R_m and scope input
d) Twisted leads ($l = 0.3$ m) between C_{ref} and R_m
e) Coaxial cable ($Z_o = 50\,\Omega$, $l = 1.5$ m) between C_{ref} and R_m

1.5 Dynamic behaviour of the channel input

The frequency response of the scope input was deliberately modified by introducing a network which could alter either the HF or LF behaviour. The former is equivalent to a further reduction of the bandwidth whereas the latter can significantly modify the scale factor in the time zone of interest for the pulses being recorded due to the appearance of a creeping lasting about one ms. The use of the nominal scale factor can result in large errors for the charge, as shown in Fig. 6 - curve b). However it is sufficient to introduce in the calculation the actual scale factor derived for the time interval at the end of the record to obtain again the correct value, as shown in the following table.

Config.	BW [MHz]	Charge [pC]	Std. Dev. [%]	Impulse duration [ns]
a)	100	49.994	0.06	37.95
b)	100	49.959	0.05	37.60
c)	100	49.800	0.17	51.25

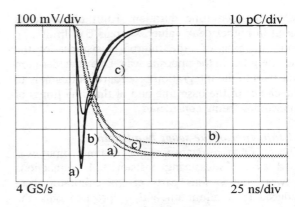

100 mV/div   10 pC/div

4 GS/s   25 ns/div

Fig. 6 Pulse shapes with different scope compensations
a) Compensated input
b) Heavy long lasting creeping
c) Mismatch in the HF range

2. Discussion

The examination of the available set of data indicates that under controlled conditions it is possible to obtain a very good agreement between measured and computed charge value. Despite the variety of cases and the different combinations it is readily seen that measurements are compatible. In most cases it can be noticed that the same modification in the circuit is more effective when the setting of the scope is changed. The presence of stray components and even the connections which were kept deliberately long do not introduce significant errors. Examination of all the cases indicates that the largest differences being experienced between nominal and measured charge values are below 1%. With respect to this figure, the standard deviation of the sets can be lower by an order of magnitude. However, to reach these results, much care has to be placed to the scope, the calibration of its vertical sensitivity and to the acquisition strategy. For the time being it seems that the indications given by the Standard focus on the need of a minimum bandwidth, which does not seem to be essential to obtain the correct results, without stressing the matter of the extremely short time duration of the pulses which requires a very high sampling rate which is not common for that category of scopes: for single shot applications at least 500 MS/s are necessary to reproduce correctly the current shape and also in this way the number of points in some cases is very small. Two aspects which are of paramount importance are related to the procedure used to integrate the signal and namely the definition of the base line and the time limit at which the process can be stopped. It has been found that there is a clear correlation between the confidence limits which can be estimated for the zero level of the base line and the standard deviation which is found for the samples. The integration carried out over short time intervals is less scattered, due to the fact that the possibility to build up an error due to the action of no white noise, often coupled from the screen deflection system into the measuring channel, is less effective.

All these factors can be effectively counteracted by averaging the numerical records with a reduction of both the scatter and of the contribution of noise not strictly correlated to the signal being measured. If these elements are considered, also in the case of industrial calibrators, other factors do not seem to be of particular importance also in extreme cases: Fig. 7 and 8 report the relative deviations which were obtained when using leads of various type and length and also different values of the measuring resistors. The only quantity which has been found to change is the standard deviation associated to the measurements sets; its value becomes overwhelming for very long connections, particularly when not shielded, and this stresses again the role of the noise in the integration process.

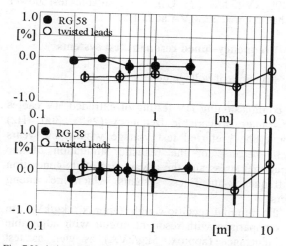

Fig. 7 Variation of the charge for two types of connections as a function of length
Nominal charge: Upper diagram 100 pC, lower diagram 20 pC

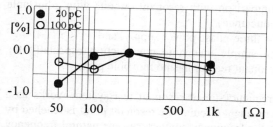

Fig. 8 Variation of the pulse charge for different values of the measuring resistor

3. Conclusions

From the whole series of data it can be concluded that, in principle, the calibration procedure based on the numerical integration can be very robust due to the limited effects which has been demonstrated for a variety of possible influence quantities. However its implementation is not as straightforward as it could appear: it is true, in fact, that scope bandwidth, lead length and value of the measuring resistor do not affect appreciably the result but, at the same time, the necessary sampling rate, the estimate of the base line and the definition of the integration limits are not easily determined.

The scope must have, either in real or equivalent time, a sampling rate of the order of 1 GS/s to reconstruct the current pulses and this is not always associated to the minimum bandwidth being specified by the Standard.

The noise associated to the A/D converter would suggest in any case the use of averaging techniques which result in a better estimate of the base line of the record: it has to be noticed that any error in this value would contribute to the charge in a way which is proportional to the upper time integration limit.

Calibration of the vertical scope setting is essential and has to be done under dynamic conditions to account for possible creeping in the response.

The initial part of the step response contributes to the modification of the pulse shape but does not modify the value of its integral, provided it has settled when the integration process is stopped: the scale factor associated to this time epoch has to be used to evaluate the charge value.

This procedure can become more difficult for low amplitude calibration pulses which would often require scope settings near the lower sensitivity limit where signal to noise ratio is usually unfavourable.

REFERENCES

1. IEC Publication 60270: "Partial discharge measurements", 1981.
2. Lemke, E. et al.: "Experience in the calibration technique for PD calibrators", 3rd European Conference on "High Voltage Measurements and Calibration, Paper 6.1, Milan 1996.
3. Cherbaucich, C. et al.: "Partial discharge calibrators: Practical experience in their calibration", 4th European Conference on "High Voltage Measurements and Calibration, Paper 3.2, London 1998.
4. Benda-Berlijn, S. M. et al.: "Performance test of PD calibrators, calculating its uncertainty and establishing its record of performance according to IEC 60270 CD 1997-06-13", 4th European Conference on "High Voltage Measurements and Calibration, Paper 3.1, London 1998.
5. Drazba, K. et al.: "Performance tests of PD measuring instruments and calibrators", 4th European Conference on "High Voltage Measurements and Calibration, Paper 3.3, London 1998.
6. Lukas, W. et al. : "Comparison of two techniques for calibrating PD calibrators", Proc. 10th ISH, Montreal, 1997, Paper 3106.

Address of authors:

R. Gobbo, G. Pesavento
Dept. of Electrical Engineering
University of Padua
35100 Padua - Italy

Fax +39 049 8277599
E-Mail gobbo@light.dei.unipd.it

AFTER LAYING TESTS OF 400 kV XLPE CABLE SYSTEMS FOR BEWAG BERLIN

R. Plath; U. Herrmann

K. Polster

J. Spiegelberg; P. Coors

Institut „Prüffeld für elektrische
Hochleistungstechnik" GmbH, Berlin

Bewag-AG,
Berlin

HIGHVOLT Prüftechnik
Dresden GmbH

Abstract. The after laying tests carried out on a 400 kV XLPE cable double-system (length 6.3 km) used AC test voltage combined with partial discharges (PD) measurement on joints and sealing ends for the first time. The approx. 30 MVA test power was delivered by the two-stage mobile resonant test system of variable frequency which was specifically designed for this duty. The requirements to be met by the test system as well as the test arrangement and its connection technology are explained. For the PD measurements, directional couplers were integrated into each joint. The data measured on all joints were simultaneously recorded and centrally evaluated. The voltage tests carried out up to 400 kV and the high-sensitive PD measurements done on the joints with a PD detection level of 2 pC confirm that such test and measuring systems allow to achieve informative commissioning and diagnostic tests on extra high-voltage cables also under difficult on-site conditions

1.   400-kV cable installation

To enhance the long-term secure and economical power supply of Berlin, Bewag is building a 400 kV diagonal interconnection through Berlin's load centres.

For the section between the substations of Mitte and Friedrichshain, Bewag decided to use 400 kV XLPE insulated cables for the first time /1/. They were assembled in the continuous cable tunnel running at a depth of approx. 25 m beneath the surface. The cable suppliers were to successfully pass a one-year long-time test and a type test.

To assure the reliable and safe service life of the cable installation as a whole, quality assurance measures were an additional part of the project. They included tests during the manufacturing process and routine tests conducted on every essential component of the cable installation. Quality assurance was completed by commissioning tests carried out after assembling. It was checked whether there were any defects produced during on-site assembling.

Therefore, the XLPE-insulated cable installation was subjected to AC testing after assembling and at the same time high-sensitive partial discharge (PD) measurements were done on all accessories simultaneously.

Due to the very high capacitive load, on-site AC testing of the 6.3 km long cable link was only possible by using the world-wide most powerful mobile resonant test system. At the maximum test voltage level of 400 kV RMS ($\sqrt{3}$ U_0), the available test power reached approx. 32 MVA at 25 Hz.

2.   Frequency-tuned resonant test systems

2.1   Principle of voltage generation

For commissioning HV testing of extruded HV cables AC voltage of variable frequency (25 ... 300 Hz) became the preferred test voltage within few years /2/,/3/, because of the realistic simulation of the operational stress and the advantages of its generation by frequency-tuned test systems, which are among others:

- the minimal specific weight (approx. 1kg/kVA) in comparison with resonant circuit with adjustable inductance (approx. 5kg/kVA) as well as test transformers with compensation (more than 10kg/kVA),
- the lower feeding power (at least factor two in comparison with resonant circuits with adjustable inductance),
- the three-phase power supply, and
- the simple mechanic construction especially the resonant reactor has no movable parts.

The frequency-tuned resonant test system works as a series resonant circuit. The resonant point is reached by tuning the frequency converter to the natural frequency of the series oscillation circuit (resonant reactor and capacitive load). The frequency range of 30 ... 300 Hz (1 :10) is generally accepted which means a permissible load range between minimum and maximum test object capacitance of 1 : 100. The load range can be increased up to 1 : 144 if the minimal frequency of 25 Hz is permitted. In comparison with them the resonant systems with adjustable inductance have only an operating range of 1 : 20. Furthermore an important characteristic is the frequency-dependent quality factor which can be described as the comparison of the capacitive test power P_P to the active power P_L. The power loss is mainly caused by the losses of the reactor and the exciter transformer. By using a suitable construction of the resonant reactor a Q factor greater than 100 can be achieved at least. In this case the feeding power is lower than one percent of the test power.

High Voltage Engineering Symposium, 22–27 August 1999
Conference Publication No. 467, © IEE, 1999

2.2 Rating of the test systems

The technical data and the mechanic design of the resonant reactor affect the parameters and characteristics of the test systems. The test conditions of the 400 kV cable installation of BEWAG have been determined the nominal parameter of the test systems which are as follows:

V_{test} = 1.73 U_0 = 400 kV
C = 0.190 μF/km 6.3 km ≈ 2.2 μF
f_{min} = 25 Hz
I = 80 A, 8 h ON/16 h OFF
P = 32 MVA (!)

Additionally the test system should also suitable for testing of cables in a rated voltage range of 110 kV to 220 kV as well as for 400 kV gas insulated transmission lines (GIL) at 80% of the routine test level V_{test} = 0.8*630 kV = 504 kV. Furthermore the weight of the transportation unit including trailer and truck shall be taken into consideration because a weight of not more than 40 tons avoids a special permission for the road traffic. These conditions require to separate the resonant reactor into two units. Consequently, the frequency converter and the exciter transformer are separated too. As a result, there are two independent test systems which can operate separately or be connected in parallel (higher test power) or in series (higher test voltage) as well. Hence, the second resonant reactor is installed on a insulating construction which is rated for 254 kV (see also Fig. 5). Thereby, the covered test cases are shown in Table 1 and Fig.1.

	rated volt. kV	test voltage kV	capa-citance μF	cable length km	test current A	test power MVA
a)	110	160 (2.5U_0)	2.5	12.5	64	10.3
	132	190	2.5	12.5	75	14.3
	150	220	2.1	10.5	80	17.6
	220	254 (2.0U_0)	1.6	8.0	80	20.3
b)	110	160 (2.5U_0)	5.0	25	128	20.5
	132	190	5.0	25	150	28.5
	150	220	4.2	21	160	32.2
	220	254 (2.0U_0)	3.2	16	160	40.6
c)	400	400 (1.73U_0)	1.25	6.5	78.5	31.4
		504 (0.8U_p)	0.8	11.5*	80	40.3

* 70 nF/km

Table 1: Test system 2 x 80 A / 254 kV
Maximum parameters of cable testing
a) single test system
b) two test systems in parallel
c) two test systems in series

The oil insulated resonant reactor is especially designed for outdoor conditions with a steel tank and an oil-to-air bushing of composite insulation (fibreglass tube with silicon rubber sheds). Its nearly load independent inductance of 16 H is realised by an iron core with multiple gaps which enable a linear magnetization. The high quality factor is caused by a special design of this core and of the winding to guarantee minimum losses. During the 1 hour test operation the dissipation heat can usually drawed off by self-cooling from the surfaces of vessel and radia-

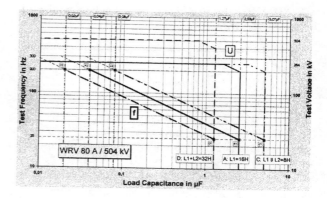

Fig. 1: Load-frequency and voltage characteristics

tors (ONAN). According to the BEWAG test conditions which request a test duration of 8 hours it is necessary to cool the radiators with a ventilator (ONAF).

A minimum quality factor (design quality factor) of more than 100 at a frequency f = 25 Hz needs a feeding power of only 200 kVA for a test power of 20 MVA. This is equal to the rated power of the frequency converter. Therefore the two frequency converters work in a master-slave operation if the two resonant reactors are connected in a series or parallel circuit. Additionally the two exciter transformers are circuited on the primary side in parallel (Fig. 2) and on the secondary side in series.

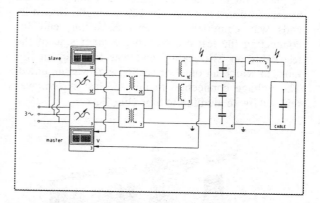

Fig. 2: Test circuit with 2 reactors in series

2.3 Acceptance test of the resonant test system

Before the resonant test system was used for the on-site tests, it had been carefully tested at its rated voltage 254 kV (1 hour including PD measurement) at voltages up to 120% of this value, at test object punctures and so on. These tests were repeated for the series and parallel connection of two test systems.

By means of switchable load capacitors and therefore different resonant frequencies the semiautomatic operation up to the test voltage was proven. The voltage measuring system was calibrated for the operating frequency range from 25 Hz to 300 Hz. The harmonic content of the test voltage meets the requirements of the international standard IEC 60060-1 and is always less than 5 percent. The measured quality factor confirms the calculated frequency response (Fig. 3) and exceeds its design value (Q = 100) remarkably. Therefore the requested test power of approximately

32 MVA can be generated by a feeding power of approx. 200 kVA.

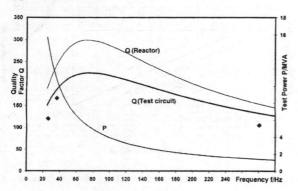

Fig. 3: Quality factor Q and test power P characteristics (design value Q = 100)
♦ measured values

3. Tests

3.1 Test arrangement

The AC voltage test and the PD measurement on the 400-kV cable installation was awarded to IPH, the high-voltage and high-power test laboratory situated at Berlin-Marzahn. The tests took place at the substation (GIS) of Friedrichsfelde.

Due to its large dimensions the AC test system had to be placed outside the cable installation and connected by an additional test cable of 120 m length. One of its ends was equipped with an SF_6-sealing end for connecting the test object. On the other side of the test cable an SF_6 sealing end was fitted in an SF_6 tank, to which an outdoor sealing end was mounted. The voltage connection to the AC test system was realised by an aluminium tube (Fig. 4).

Fig. 4: Outdoor sealing end with test cable

One resonant reactor was used for the test up to 254 kV. It was mounted onto a trailer. Between resonant reactor and test cable terminal a blocking impedance was arranged to reduce interferences for the PD measurement and to protect the resonant reactor

against fast transients in case of breakdown. The blocking impedance was mounted to a capacitive voltage divider on one side and to a support on the other side. The voltage signal of the voltage divider is used by the control and supply unit for voltage regulation and simultaneously for voltage measurement. The test voltage was furthermore measured by the integrated capacitive pick-up at the outgoing bushing of the resonant reactor. The safety for the tested cables was improved by providing a second independent switch-off device. To synchronise the PD measurement system to the test voltage phase, the AC signal was transmitted via an optical link to the control room.

The resonant reactor was supplied by an excitation transformer placed next to the resonant reactor on the trailer.

The 400 kV tests need two AC test systems. A 254 kV test system identical to that of IPH was therefore provided by KEMA. One of the two resonant reactors was mounted to an insulating construction and both AC test systems were series-connected (Fig. 2 and 5).

Fig. 5: Arrangement for 400-kV test

The voltage connection was arranged from the resonant reactor on high-voltage potential to the test cable via the blocking impedance which, in this case, was mounted on a capacitive 500 kV voltage divider and a 500 kV support.

3.2 AC voltage tests

Since the tests anyway included the observation of the PD level, the test voltage was increased by steps of 50 kV/5 min until the final test voltage was reached.

In the first part of the tests, the individual phases of each system were tested one after the other at the test voltage of 254 kV requested by Bewag over a period of 15 min and after that at 230 kV over a 45-min period. With the cable capacitance being approx. 0.190 µF/km, the resulting resonant frequency of the test circuit was approx. 37 Hz.

The second part of the commissioning test was a load test for conditioning the cable installation with elevated operating current (1685 - 1900 A). Over a 20-day period a cyclic load was applied to simulate potential service conditions that may occur in future service life. Furthermore, the load test served to verify the effectiveness of the cable installation's forced air cooling.

In the last part of the tests, the test voltages of $1,73 \cdot U_0 = 400$ kV during 15 minutes and of 350 kV during 45 minutes were applied one after the other to each of the cable conductors. The result was a resonant frequency of approx. 26 Hz for the test voltage. Fig. 6 shows the time characteristic of the 400-kV voltage test, delivered by the computer control of the test system.

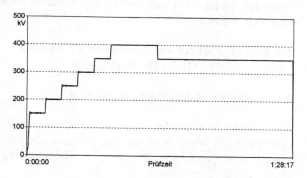

Fig. 6: Time characteristic

3.3 PD measurement

PD measurement was carried out only on the accessories, because all cables were already tested during the routine test.

Each phase of both three-phase systems of the 6.3-km cable installation included 8 joints and 2 GIS sealing ends. On all of these accessories directional couplers /4/ for the decoupling of PD impulses were additionally fitted.

Bewag required simultaneous PD measurements on all these accessories to detect any PD as early as possible. This was to minimise the risk of breakdown. During voltage testing no staff was needed inside the tunnel because the distributed measurement system was under remote control by optical links.

Preliminary investigation showed that the noise level in the tunnel was low as compared to typical on-site conditions. The tunnel depth of 25 m and the large length of cable between two joints attenuated high frequency noise components very well. IPH designed and applied a PD measurement system fulfilling Bewag requirements including a PD detection level of ≤ 2 pC on site. Furthermore, two directional couplers left- and right-hand side of each joint enabled on-site calibration and even high-precision PD location.

Capacitive sensors instead of directional couplers were used for the GIS sealing ends of one of the two three-phase cable systems. These sensors did not reach the same selectivity like the directional couplers. Even at the far end, outer noise (e.g. PD from rain) superimposed the PD measurement at the GIS sealing end.

3.4 Test results

No breakdown occurred during all voltage tests. PD measurements on 60 accessories (12 GIS sealing ends and 48 joints) showed PD in only one single

case. Due to the low PD level (approx. 5 pC, single pulses up to 7 pC) additional measurements were carried out in order to be very sure before opening the joint. After dismantling the joint, the PD causing failure was found. The PD location executed before the opening showed a displacement to the real failure location of approx. 2 cm, confirming the high accuracy of UHF measurement techniques. After repair, the joint reassembling was done exactly in the same place, because practically no damage on the cable insulation occurred due to early fault recognition.

4. Conclusions

The experience from after laying tests of 400 kV XLEP cables with frequency-tuned resonant test systems confirms their simple erection on site, their easy handling as well as their high reliability. Such test systems for test voltages up to 500 kV and test power up to 40 MVA are available.

The test results confirm also the high overall quality level of the 400-kV-XLPE cable system. Nevertheless, besides all routine tests on cables before laying and the use of prefabricated and pretested accessories, a small risk remains due to assembling work on accessories on site. On-site AC testing combined with sensitive PD measurements reduces this risk noticeably.

5. References

/1/ Henningsen, C. H., Polster, K., Müller, K. B., Schroth, R. G.:
New 400 kV XLEP long distance cable systems, their first application for the power supply of Berlin,
CIGRE Session (1998), paper 21-109

/2/ Schufft, W. et al.:
Powerful frequency-tuned resonant test systems for after-laying tests of 110 kV XLPE cables.
9th ISH Graz 1995, paper 49.86

/3/ Hauschild, W., Schufft, W., Spiegelberg J.:
Alternating voltage on-site testing of XLPE cables: The parameter selection of frequency-tuned resonant test systems.
10th ISH Montreal (1997),Vol. 4, pp. 75-78

/4/ Pommerenke, D., Strehl, T., Kalkner, W.:
Directional coupler sensor for partial discharge recognition in high voltage cable systems.
10. ISH Montreal 1997

Address of author:

Dr. Ing. R. Plath
Institut „Prüffeld für elektrische Hochleistungstechnik" GmbH (IPH)
Landsberger Allee 378
12681 Berlin

Applying a Voice recognition system for Substation inspection / maintenance services

S.Nakano, T.Tsubaki and S.Hironaka

Hitachi Engineering & Services, Ltd.

Abstracts: We have developed a new set of portable voice recognition system which is compact, lightweight, and has two-way communication(Sending & Receiving) for the purpose of inspection/maintenance services for substation equipment. We are using these newly developed systems for daily visual inspection, periodic inspection and for commissioning tests of substation equipment. By applying this system with the added guidance and judgment functions, unskilled personnel is possible to carry out inspection/maintenance services of substation equipment. And the data collected at site is possible to be fed to the computer directly, we can easily do the trend control management. Also we are applying this system for education and self-training of unskilled personnel who will be engaged in the maintenance of equipment. This paper describes the outline of voice recognition system, actual examples and achieved results.

Keywords: Voice recognition ,Substation Maintenance, Trend control ,Voice guidance ,Training

1. INTRODUCTION

With the increase in dependence on electric power, effects of electrical faults will have an impact on the social life. Preventive maintenance and diagnosis of incipient faults is becoming more important. The modern substation system is based on SF6 Gas insulated switchgear (GIS), Vacuum circuit breakers (VCB) and other molded electrical products. The detection of incipient stage of faults/failures at an early stage is becoming difficult by the old type of visual inspection and regular examination methods, which largely rely on the inspectors' five senses. The reliability of substation-equipment are becoming up-grade, so the number of operators and maintenance personnel who lack in experience with serious accidents, troubles or failures is increasing. Although requirement of preventive maintenance is increasing, the practice of that is more difficult.

On the other hand personal computers have started to be widely used in the field of inspection/maintenance services as a result of their dramatic advancement. It is very useful to fully utilize the data processing capacity of the personal computer for preventive-maintenance. Because of the difficulty in inputting data directly to the personal computer at site, the data collected by operators and maintenance personnel at site are often fed to the personal computer by office personnel or by operators themselves when they get back to their office.

In order to accomplish good preventive-maintenance, we have studied the possibility of using voice interface between operator/maintenance personnel and personal computers so that they can directly input data to personal computer at site.

2. DEVELOPMENT OF A TWO-WAY AUDIO RECOGNITION SYSTEM.

2.1 Concepts of developing the system

In order to implement audio communication between operator/maintenance personnel and personal computer directly at substation site , a system with the following characteristics is required:

① It is a hands-free system.

② It has a high degree of recognition and is not inputting by inference.

③ It is an interactive (two-way) system due to the necessity of data input to computer & inspection guidance from computer.

The two-way audio-recognition system has been developed on the basis of above requirements.

2.2 Hardware specifications

The newly developed two-way audio-recognition system (HESCOM/V) is connected to the serial port (Com-port) of the personal computer (DOS/V compatible machine) with a dedicated cable for processing audio data. The specification of voice recognition unit is shown on Table 1, the block diagram in Fig. 1 and the exterior view in Fig 2.

Table1 Specification of Voice Recognition Unit

Number of recognizable characters	100,000 words (100 dictionaries)
Dictionary unit	1,000 words (Average 5 syllables per word)
Talker	Unspecified number of talkers.
Sampling	11.025 kHz
Data format	16 bits PCM
Interface	RS232C
Input	Exclusively designed headset
Weight	500g

High Voltage Engineering Symposium, 22–27 August 1999
Conference Publication No. 467, © IEE, 1999

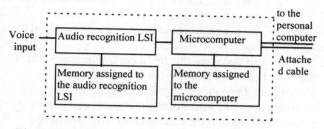

Fig. 1 Block diagram of voice recognition unit

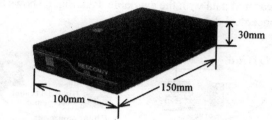

Fig. 2   Exterior view of voice recognition unit

2.3   Outline of the two-way audio communication system

The newly developed audio recognition system is constructed as a compact two-way portable system. It enables carrying out inspection/maintenance service in the following manner with help of the audio guidance (interactive) function:

(Personal computer) gives audio guidance on an inspection item.
(Inspector) inputs inspected details by voice.
(Personal computer) repeats the input data by voice.
(Inspector) verifies the repeated data.
(Personal computer) gives audio guidance on the next inspection item.
The inspector is able to carry out the inspection work in a hands-free manner using a PHS-system.(PHS ; Personal Handyphone System) In addition, the inspector's input data preparation of inspection reports and other services have been greatly facilitated with this system. An outline of the system is shown in Fig. 3.

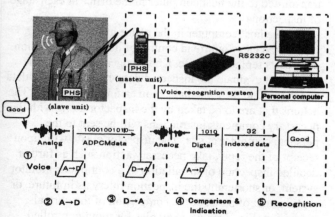

Fig. 3 Two-way communication system

3. EXAMPLES OF THE APPLICATION OF THIS SYSTEM AND IT'S RESULTS

3.1   Improvement of the inspection capability by the guidance function

It will be necessary to carry either the check sheet or the personal computer to the site and these procedures often require the inspector to read out each item for inspection and use both hands to record data either by pen or keyboard. So it has some problems as it hampers mobility and safety. The use of the new two-way communication system helps solve both problems and improve the efficiency of inspection/maintenance services with the added guidance and judgment functions. In the case of new two-way communication system with PHS , the inspector carrying out inspection needs to be equipped with only a microphone/earphone set and a PHS slave unit. PHS is constructed with master unit and slave unit. It keeps the inspector is hands free and improves mobility and safety. The daily visual inspection practice of outside equipment at a rainy day are improved especially. Inspection will be carried out by following the personal computer guidance as demonstrated in the following example:
(Personal computer) What is the gas pressure of No. 1 breaker?
(Inspector) 5.3
(Personal computer) Is it OK at 5.3?
(Inspector) OK (5.3 will be recorded in the PC)
(Personal computer) What is the gas pressure of No. 2 breaker?
(Inspector) 4.6
(Personal computer) 4.6 is below the lower limit of 4.8. Is that OK?
(Inspector) Previous value.
(Personal computer) Previous value was 5.1. What is the gas pressure of No. 2 breaker?
(Inspector) 4.6
(Personal computer) 4.6 is below the lower limit of 4.8. Is that OK?
(Inspector) Yes. (4.6 is recorded in PC together with a mark indicating that it is below the specified value).

The inspection is carried out by the guidance from personal computer, so the inspector concentrates on the actual inspection and the possibility of erroneous data entry is none. By applying this system with the added guidance and judgment functions, unskilled personnel is possible to carry out inspection/maintenance services of substation equipment. And after carrying out inspection and maintenance services, records of those services are automatically fed to the computer. Fig. 4 shows how the inspection is actually carried out.

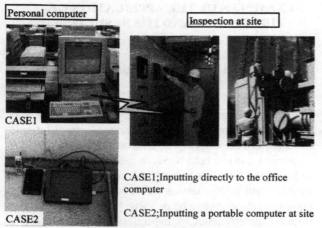

Fig. 4   Inspection using voice recognition system

CASE1;Inputting directly to the office computer

CASE2;Inputting a portable computer at site

3.2   Early detection of failures through trend control of inspected data

The existing procedure for judging by the fixing criteria for healthiness of the equipment was inadequate to determine deterioration or potential irregularities at an early stage. Because such fixing criteria was mostly based on ample margins to allow for variation of equipment characteristics. The examples shown in Fig. 5 are the results of evaluation of the measurement of the closing time of VCB by relying on the judgment of whether the value was below or above the specified limit. The measured results of VCB1, VCB2, and VCB3 based on the above judgment criteria indicated that the closing times of all the VCB were within limits and were all determined to be "good". However, through the trend control of the actual situation, VCB3 showed a sign of irregularity even though it was within the limit and indicated that some action has to be taken. Actually in this case it was discovered that the lubricating oil in the closing mechanisms of VCB3 was hardened and as a result the closing time was delayed. If left as it was, it would have led to further hardening of lubricant, increase the friction in the mechanism, and finally make further closing impossible before the next inspection.

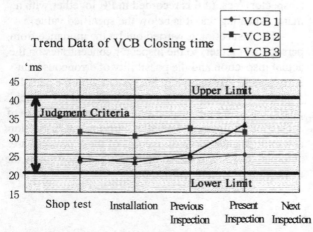

Fig. 5   Example of trend control of VCB Closing time

This was one of the examples to indicate that trend control of the data was the only possibility to help determine the irregularities in the system especially at the incipient stage.

Generally speaking this trend-control management is supported by many persons due to its usefulness.[1][2] By applying this voice recognition system, the data collected at site by operator/maintenance personnel is possible to be fed to the computer directly. We have started monitoring this data to determine the preliminary signs of failures by comparing the changes in the measured data with the available data. Fig. 6 shows a conceptual diagram of this system

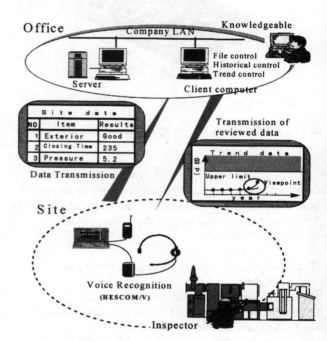

Fig. 6   Conceptual system diagram

The inspector at site will input data directly into the portable personal computer through the voice recognition system. He will check the previous data and evaluate if there are any changes from the last data. The inspection will thus be continued and the inspected data will be transmitted to the host computer in the office at each stage of inspections.

The host computer in the office has a software program to carry out trend control of historical inspection data, parts replacement etc. Knowledgeable staffs in the office will evaluate the inspected data, historical trend data and determine the remedial action as found necessary. The actions that are to be taken by the inspector along with the necessary comments/instructions will be sent to the personal computer at site. The inspector at site will, upon receipt of the reply, take necessary action such as further detailed inspection or installation of a continuous monitor system, or shortening the inspection intervals in future, or replacement of parts, or the completion of the normal inspection. Fig. 7 shows examples for trend-controlled data of GIS .

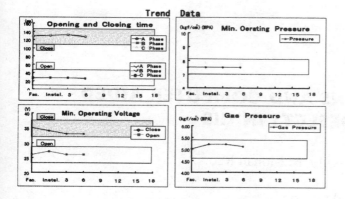

Fig. 7 examples of trend-controlled data for GIS

3.3. Training of maintenance personnel by using voice guidance

In recent times, mainly SF6 gas circuit breakers and vacuum circuit breakers are produced. Although other types of breakers such as Oil, Air-Blast and Magnetic-Blast breakers are scarcely produced, large numbers of these breakers are still used in the field. On the other hand Personnel working on the maintenance and inspection of these breakers are aging and retiring rapidly. So educating and quickly training of younger inexperienced personnel for taking their place in the field is necessary. The urgency for training younger personnel is demanding before the retirement of experienced technicians. Education with the use of manuals and comparison with actual items is not efficient and lecturers by experienced technicians can not educate so many personnel at one time.

To overcome these problems a new audio interactive training system has been developed ; by combining the two-way audio recognition system and personal computers for the short-term training. With this training system, trainees can be trained all by themselves at anytime. This educational system is composed in a step by step manner with audio guidance and trainee's response. This guidance conforms with the standard process.

By following the step ; by step audio guidance, trainees can learn and carry out the inspection/maintenance with ease. Trainees are also supported by other references such as "Drawing", "Photograph" , "Check sheet" , etc. For example, with voice input, the requested items such as "Drawing", "Photograph" and "Check sheet" will be displayed on the computer screen. This training system is found to be very effective as it allows even newly employed personnel to quickly and fully train themselves in any of the inspection/maintenance processes. Fig. 8 shows how the training is carried out.

4. CONCLUSION

The newly developed two-way (interactive) voice recognition system has proved to be very useful in the inspection and maintenance of substations. By applying this system with the added guidance and judgment functions, unskilled personnel is possible to carry out more widely inspection/maintenance services of substation equipment. And the data collected at site is possible to be fed to the computer directly, we can easily do the trend control management. Also we are applying this system for education and self-training of unskilled personnel who will be engaged in the maintenance of equipment. However, to facilitate more advance and efficient maintenance technology, we will continue to develop the system by further integrating the generic functions of personal computers such as calculation, tabulation, graphing.

5. REFERENCES

(1) K.Maruyama, : "Techniques of the Failure Diagnosis for Main Equipment Recieving and Distribution Equipment" , **J.IEIE** Japan, Vol.13 No.12
(2) K.Maruyama, T.Yamagiwa, H.Yamada, S.Sato, : "Preventive Maintenance Technology for Substations " , The Hitachi Hyoron Japan, Vol.75 No.12

Address of author
Seizou Nakano
Hitachi Engineering & Services ,Ltd.,
3-2-2,Saiwai-cho,Hitachi-Shi,Ibaraki-Ken ,317-0073 Japan
E-Mail : snakano@hesco.hitachi.co.jp

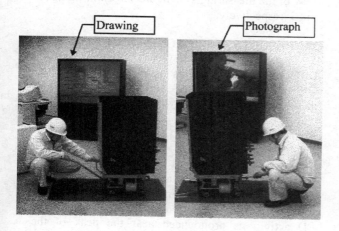

Fig. 8 Training of Inspector by using Voice guidance .

GENERATION OF PD PATTERNS AND COMPUTER AIDED ANALYSIS FOR DEFECTS IN GIS

S.C.Gupta, A.K.Adikesavulu, M.Mohan Rao

BHEL R&D Division
Hyderabad
INDIA

ABSTRACT

Partial discharge (PD) detection is one of the most important tests for assessing the quality and viability of an insulating system. The present paper deals with PD detection in 145kV gas insulated substation (GIS) module by utilizing modern computer aided discharge analyser system (CDA).

Laboratory tests have been carried out on a model busbar, with special reference to the simulation of various types of defects and generation of PD patterns to identify different discharge sources. PD test results reveal a good correlation between the discharge patterns and the type of defects.

1. INTRODUCTION

The majority of defects in GIS can cause deterioration in the insulating strength of SF_6 gas and solid insulation due to accelerated ageing which may result in reduced life time. The most commonly encountered defects [1-3] in GIS are metal protrusions, mobile and fixed particles, internal defects (voids) in the support insulator etc. Since discharges from these defects exhibit typical patterns, the signature produced from them can serve as a good tool to ensure quality during manufacture and reliability in service [4].

The use of the conventional method [5] for PD detection is cumbersome and subjective in nature due to dependence on operator's skill and expertise. The advent of computer aided measuring system [6 -7] has paved the way for processing a large amount of data and to transform this information into useful output in terms of important PD quantities, like the maximum and average magnitude of pulses, number of discharges of different levels, besides several statistical parameters.

The present investigation places emphasis on the generation of PD patterns emerging from the different sources inside GIS model busbar using the CDA system. The results have been analysed with a view to correlate discharge patterns with the type of defects. This may be used as a guiding factor for quality control at different stages of GIS development.

2. TEST SET-UP

Tests were carried out on a 145kV GIS module. This module, with a stainless steel enclosure (ID 264mm), centrally located aluminium conductor (dia 89mm), included a 325 kV gas insulated metal clad transformer and cone insulator at both the ends. The test enclosure was filled with SF6 gas at a pressure of 3 to 4 bar. The test set-up is shown in figure 1.

The measuring system consisted of PD detector with gated output facility (Type 700) to drive the computerised discharge analyser (Type (CDA3). The schematic diagram of circuit connection and a view of the measuring system are shown in figures 2 and 3 respectively.

3. RESULTS AND DISCUSSIONS

Different types of defects have been simulated in GIS module. The results and trends exhibited in their PD patterns are briefly discussed below:

Sharp protrusion

This defect was simulated by projecting sharp pins of 5-10mm length and of approximately 0.2mm dia on the HT conductor. Figure 4 shows the phase resolved plot for 10mm long pin. It is seen that the PD activity is pronounced near the peak of the negative half cycle (270° phase angle). The number

High Voltage Engineering Symposium, 22–27 August 1999
Conference Publication No. 467, © IEE, 1999

Fig.1 Test set-up for pd measurement

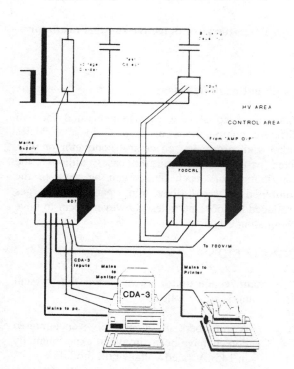

Fig.2 Circuit connections with pd detector and CDA

Fig.3 PD measuring system (CDA and Detector)

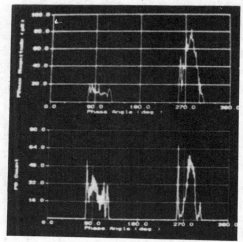

Fig.4 PD pattern for sharp protrusion on HV conductor

of PD pulses and their magnitudes are much lesser near the positive half cycle (90° phase angle) as compared to those near the negative half cycle. Further, the change of gas pressure from 3 to 4 bar and size of sharp pin did not reflect any noticeable change in the PD pattern.

Mobile particles

For simulation of this defect, particles of different sizes and materials were kept in the enclosure. Figure 5 shows the PD pattern for a single loose particle. It is observed that the PD pulses appear almost throughout the AC cycle. However, there is cluster of discharges during the rising part of the positive and negative half cycles.

Particles attached to the support insulator

Two particles were positioned at diagonally opposite locations on the insulator surface near the HT conductor. A typical PD pattern is shown in figure 6. It is noticed that the discharges initiate after the positive and negative peaks of the AC cycle and show symmetrical pattern as regards their number and magnitude.

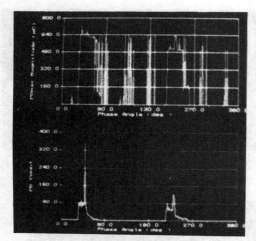

Fig.5 PD pattern for mobile particles

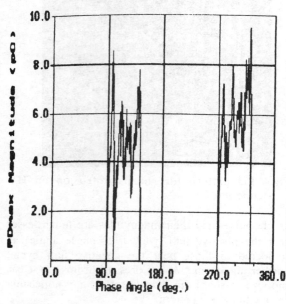

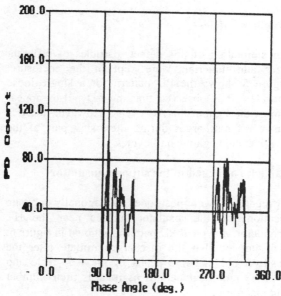

Fig.6 PD pattern for fixed particles on insulator surface.

Defective cone insulator

A defective cone insulator, with minute metallic inclusions and having small voids (reflected by high PC level of around 100pc), was selected for this investigation. The behaviour of partial discharges is shown in figure 7. In this case, discharges seem to begin in the rising portion of positive and negative half cycles approaching the respective peaks. The PD pattern, however, is very asymmetrical since discharges of higher magnitude occur in the negative half cycle.

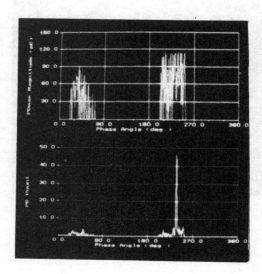

Fig.7 PD pattern for defective support insulator

Combination of defects

Phase resolved plots were also generated for two different combinations of defects, loose particles either with protrusion on central conductor or with fixed particles. In both the cases, although the contribution of each defect can be seen in the combined phase resolved plot, partial discharges produced by loose particles, however, seem to have dominating effect.

CONCLUSIONS

The main conclusions drawn from the present investigations are as follows:

i)   A number of commonly encountered defects have been simulated experimentally in 145 kV rated voltage GIS module.

ii)   PD signatures have been obtained for individual as well as combination of two type of defects at a time.

iii) In most cases, it is possible to find correlation between PD signatures and type of defect. This can be used gainfully as a quality control tool at the time of development, manufacture and shop floor testing of different GIS modules.

ACKNOWLEDGEMENTS

The authors thank Sri D.Suryanarayana, General Manager for his keen interest and encouragement and BHEL Management for granting permission to publish this work. We would also like to express our deep appreciation to Smt.K.Sarojini Subhashini for careful preparation of this manuscript.

REFERENCES

[1] R.Baumgartner, B.Fruth, W.Lanz and K.Petterson, "PD in gas insulated substations measurements and particle considerations," IEEE Electrical Insulation magazine, Jan/Feb. 1992, Vol 8, No.1, pp 16-27.

[2] O.Celi, W.Koltunowicz, A.Pigini, B.S.Manjunath, M.C.Ratra, G.Basile and S.Girolomoni, "Study of diagnostic methods for identification of defects inside GIS", proceedings of ISH, New Orleans, 1989, No.32.04.

[3] A.Bargigia, W.Koltunowicz, A.Pigini "Detection of partial discharges in gas insulated substations" , IEEE Trans on power delivery, vol 7, No.3, July 92, pp 1239-49.

[4] K.Petterson, R.Baumgartner, W.Lanz, D.Cramer and M.Albiez, "Development of methods for on-site testing as part of the total quality assurance testing program for GIS", CIGRE 1992 Session, Aug./Sept. 1992 pp 23-202.

[5] R.E.James, B.T.Phung and T.R.Blalckburn, "Computer aided digital techniques for partial discharge measurements and analysis", proc. Of International Symposium on digital techniques in high voltage measurements, Toronto, Canaa, 28-30 Oct., 1991, pp 2-7, 2-11.

[6] E.Gulskie and F.H.Kreuger, "Computer aided recognition of discharge sources", IEEE Trans. on electrical insulation, vol 27, No.1, Feb.92, pp 82-92.

[7] S.D.Neilson and J.P.Reynders," The use of computer-aided partial discharge pulse height analysis for diagnostics in GIS," proceedings of ISH Yokohama, Japan, Aug. 1993, No.66.02 pp, 149-52.

Address of Authors:

High Voltage Engineering Dept.
BHEL, Corporate R&D Division
Vikasnagar, Hyderabad – 500 093
INDIA.

ACOUSTIC PARTIAL DISCHARGE MEASUREMENTS FOR TRANSFORMER INSULATION – AN EXPERIMENTAL VALIDATION

S.Rengarajan, A.Bhoomaiah, K.Krishna Kishore
High Voltage Laboratory
Bharat Heavy Electricals Limited
Corporate R & D
Hyderabad
INDIA

ABSTRACT

Acoustic partial discharge (PD) measurement is a useful method of identifying and locating PD in a power transformer, in service. In this paper, an experimental validation has been done on a transformer insulation model using an acoustic PD detector. While the efficacy of the method is proved, it is observed that the signal gets attenuated significantly when it passes through pressboard barriers. Even though acoustic PD values can be correlated with the maximum discharge magnitude for a single discharge activity in the laboratory simulations, this relationship could not be established during actual PD measurements in a transformer at site, as the acoustic PD signal depends upon many factors.

INTRODUCTION

Partial discharge activity inside a power transformer can generate heat and degrade electrical insulation, leading to the failure of the equipment. Even though, transformers are tested for PD after manufacture, the discharges can develop inside the transformers during service. Hence, on-line monitoring of PD becomes essential to avoid any unplanned outage of equipment. The electrical method of detecting PD cannot be employed at site because: (i) it is an off-line measurement, (ii) the sensitivity of measurement is affected because of the noisy environment at site, and (iii) it cannot locate the PD source, which is essential for taking-up repairs. As an alternative to the electrical method, acoustic method of detecting PD has become popular, as it is very easy to adopt at site. This paper describes experimental experience with regard to the efficacy of the acoustic method of detection.

EXPERIMENTS

For the experimental investigation, an acoustic PD detector has been used. The detector mainly consists

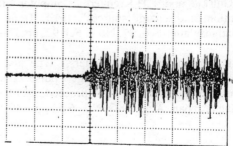

Fig.1: Typical acoustic signal

High Voltage Engineering Symposium, 22–27 August 1999
Conference Publication No. 467, © IEE, 1999

of a sensor, which, when placed on the tank wall of a transformer, will pick up acoustic PD signal and convert at into an electrical signal. In order to avoid the external noise, the sensor is provided with a pre-amplifier and a filter circuit, tuned to the frequency of 150 kHz. The sensor is connected to the main detector, which consists of a narrow band amplifier and a pulse counter circuit. The counter measures the peak of PD pulse train as numbers for a time interval of 1 second and displays. A typical acoustic discharge signal output is shown in figure1.

The experimental set-up to generate the PD pulses consists of a steel tank and two types of electrodes, viz., (i) pin-plane gap and (ii) rod- plane gap. These electrodes are the corona sources and fixed inside the tank at one end, as depicted in figures 2 and 3.

Fig.2: Experimental set-up

Fig.3: Electrode configuration

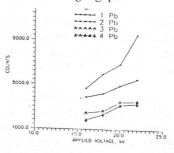

Fig.4: Magnetic holder for acoustic sensor

The gaps between electrodes are adjustable by means of gear mechanisms. The acoustic sensor is fixed on the centre of the tank wall on the opposite side to the region where the electrodes are provided.

This is achieved by means of a suitable magnetic holder specially designed for this purpose as shown in figure 4. A silicone- based coupler is applied on the sensor face, which is just pressed against the tank wall to avoid air film for effective transmission of the signal. Vertical slots of 3mm width are made in the tank to fix metal plates or pressboard barrier between the corona source (a pin of radius <0.5 mm has been used for this purpose) and the sensor. The entire tank is filled with unprocessed transformer oil of BDV ranging from 40 to 50 kV. Though various experiments were done with pin-plane gap and rod – plane gap, the results of pin- plane gap are presented in this paper.

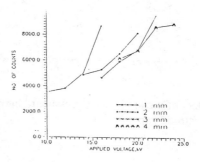

Fig.5: Variation of counts with voltage for different electrode gaps

In all the experiments, the acoustic count has been measured by creating single discharge in the electrode gap by applying high voltage (50 Hz) to the electrodes through an epoxy bushing. This is very important for comparison of results, since repetition of discharges within the set time of 1sec (acoustic detector measuring time) leads to higher acoustic counts.

RESULTS AND DISCUSSIONS

The variation of PD counts with applied voltage for different gaps is depicted in figure 5. In general, as the gap length increases, the count also increases because of increase in discharge current magnitude. The shift of the graph towards right side as the gap length increases can be attributed to the higher PD inception levels for larger gaps.

Fig.6: Variation of counts with no. of pressboards

It is clear from figure 6 that the acoustic signal gets attenuated significantly, when a pressboard barrier is introduced in the signal path. However, attenuation is not witnessed (figure 7) when a metal plate is introduced.

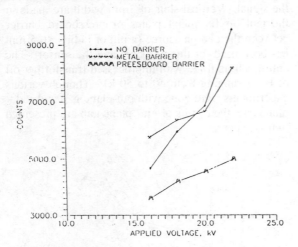

Fig.7: Effect of metal vis-a-vis pressboard barrier - Electrode gap = 3 mm.

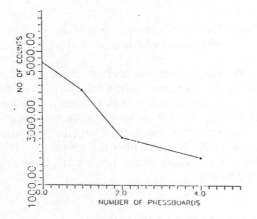

Fig.8: Effect of no. of pressboards on counts

Even though the attenuation increases with the addition of pressboard barriers, the rate of increase of attenuation gets reduced, as indicated by figures 6 and 8. Also, the location of pressboard in the signal path does not alter the degree of attenuation significantly, as depicted in figure 9.

It is also observed that when two discharge sources are operating simultaneously, the effects of these discharges are aiding each other, resulting in an increase in the total acoustic PD count.

Figure 10 shows the correlation between the acoustic PD counts and the maximum discharge magnitude in pC, which was measured simultaneously by a conventional PD detector. However, this type of correlation could be possible only when a single

discharge occurs in one second, which is the time set for acoustic PD counts. Hence, this correlation cannot be established in a transformer at site, where there are a number of discharge sources and repetition of discharges in one second.

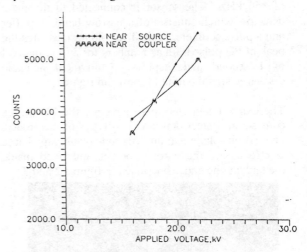

Fig.9: Effect of pressboard location On counts.

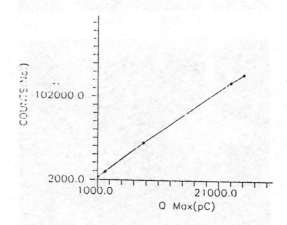

Fig. 10: correlation between maximum discharge And count

SITE MEASUREMENTS

The acoustic detector was also used for on-line PD measurements on many transformers (up to 400kV, 200MVA) at various sites. The measurement locations were decided strategically depending upon the winding configurations, locations of leads and tap changers. Faults could be identified by using the acoustic PD detector and subsequently rectified.

CONCLUSIONS

(a) An acoustic PD detector can detect oil corona discharges inside the transformer accurately, especially when the sensor is directly exposed to the discharge source through the oil medium.

(b) The acoustic PD count depends upon intensity of discharge, number of discharge sources and repetition rate.

(c) Pressboard barriers attenuate the acoustic PD signals significantly. However, the rate of attenuation is not proportional to the number of pressboards barriers added.

(d) A metal barrier does not attenuate signals like a pressboard barrier.

(e) Acoustic signals initiated in a transformer indicate only the total discharge activity but not the maximum discharge magnitude, which is a standard measure of PD.

ACKNOWLEDGEMENT

The authors thank the Management of BHEL, Corporate R&D Division for the permission to publish the paper. The authors are sincerely thankful to Mr. D.Suryanarayana, General Manager (ES) who initiated this project work and for his constant encouragement. The project team wishes to thank Transformer Engineering Division, BHEL, Bhopal for referring the project. Thanks are due to Andhra Pradesh State Electricity Board and National Thermal Power Corporation Limited for providing the transformers for investigation at various sites. The experimental and site support work by Mr.V.K.V.Nair and the paper preparation by Ms. K.Sarojini Subhashini are gratefully acknowledged.

REFERENCES

[1] L.E.Lundgaard, "Partial Discharge-Part XIV Acoustic Partial Discharge Detection-Practical Application", IEEE Electrical Insulation Magazine, Vol.8, No.5. Sept./Oct. 1992.

[2] E.Howells and ET Norton, "Detection of Partial Discharges in Transformers using Acoustic Emission Techniques", IEEE Trans. on PAS, Vol.PAS-97, No.5. Sept./Oct., 1998, p.p. 1538-1549.

ADDRESS

S. RENGARAJAN
Senior Manager
High Voltage Laboratory
Bharat Heavy Electricals Ltd., Corporate R & D
Vikasnagar, Hyderabad-500 093, **INDIA**

APPLICATION OF ELECTRICAL FIELD SENSORS IN GIS FOR MEASURING HIGH VOLTAGE SIGNALS OVER THE FREQUENCY RANGE 10 HZ TO 100 MHZ

M.Seeger, G. Behrmann, B. Coric, R. Pietsch
ABB High Voltage Technologies Ltd., Zürich
Switzerland

Abstract: The measurement accuracy of conventional electrical field sensors (mainly used as Partial Discharge sensors for the UHF-method) in gas insulated switchgear (GIS) was investigated over a frequency range of 10 Hz to 100 MHz. The scale factor of the measuring systems was determined at power frequency voltage only and used for all measurements. This is of special importance for the application to GIS in service, where only power frequency voltage is available. Frequency and step response measurements were performed, yielding an overall accuracy within 5% over the frequency range of interest. Measurements with standard waveforms according [IEC 60-1] were performed. The accuracies for the measurement of power frequency voltage, lightning impulse peak and switching impulse peak were determined to be 1.2%, 3% and 5% respectively, confirming the results of the frequency and step response measurements. Time parameters of the waveforms could be determined with an accuracy within 10%.

1.Introduction

In Gas Insulated Switchgear (GIS), beside the power frequency voltage, stressing voltages and overvoltages over a wide frequency range may occur. They originate from e.g. harmonics, switching transients, and lightning, and can be classified according [IEC71-1] in the following way:

Classes of stressing voltages and overvoltages:
Low frequency: continuous: 50/60 Hz
temporary: 10- 500 Hz
Transients: slow front: $20 \, \mu s < Tp < 5000 \, \mu s$,
$T2 < 20 \, ms$
fast-front: $0.1 \, \mu s < T_1 < 20 \, \mu s$,
$T2 < 300 \, \mu s$
very fast-front: 0.3 MHz – 100 MHz,
30 kHz – 300 kHz

The standard wave shapes for slow front and fast front overvoltages are the standard lightning and switching impulse, respectively, defined by [IEC 60-1]. Very fast front overvoltages (VFFO) originate from fast discharge phenomena inside the GIS, e.g. occurring at disconnector

switching or flashover (see e.g. [1]). The wave shapes and frequencies of VFFO are a complex superposition of travelling waves inside the GIS (0.3 MHz - 100 MHz) and of oscillations of the GIS with the connected apparatus, [IEC1321-1].

Due to the variety of wave shapes, no standard waveshape has yet been defined for such phenomena. The different voltage signals in GIS can be measured accurately using high voltage dividers with appropriate frequency response.

The accuracy and traceability of the calibration for measuring systems of low frequency to fast-front transients is ensured by the calibration procedures of [IEC 60-2], using standard wave shapes. Since no standard wave shape is defined for VFFO, this is not applicable to the measurement systems of VFFO. For these measurement systems which consist of electrical field sensors (e.g. [2],[3]), information about suitability, calibration and accuracy is given in [IEC1321-1] in form of a technical note.

All these calibration and measurement methods are easily applicable in the high voltage laboratory where the equipment for the generation and measurement of the different wave shapes is available. In on-site GIS, measurements of these signals with the same accuracy as in the high voltage laboratory is not a trivial task.

In this contribution, the application of electrical field sensors [2],[3] in GIS for measurement of high voltage signals ranging from 10 Hz to above 100 MHz was investigated, thus covering the range from power frequency voltage up to VFFO. The electrical field sensors were conventional UHF couplers used for the UHF partial discharge measurement. These couplers are increasingly being mounted in GIS for partial discharge measurement, on-site during commissioning tests, and for service/monitoring purposes [4], [5].

The aim of the investigation was to quantify the accuracy of the measurement system over the whole frequency range of interest. The scale factor was determined at power frequency and kept constant for all measurements, which is

High Voltage Engineering Symposium, 22–27 August 1999
Conference Publication No. 467, © IEE, 1999

the usual procedure in the high voltage laboratory and on-site. The accuracy for measurement of peak and time parameters in the different frequency ranges was determined by comparison with conventional high voltage measuring systems applying the procedures of [IEC 60-2].

2. Principle of measurement system

The measurement system consists of an electrical field sensor in the GIS (PD-sensors), two impedance-matching modules (Module 1, Module 2) and a 50Ω coaxial 'thick-wire Ethernet' cable, see figure 1.

Details of the measurement system are described in [6]. In electrical terms, the system is essentially a damped capacitive divider. The signal is measured by an oscilloscope with 1MΩ input impedance. C_1 is the capacitance between the inner conductor of the GIS and the sensor (typically 0.1pF). C_2 is given by the stray capacitance between the sensor and GIS enclosure (typically 20pF). The divider ratio of the system is set to lower the voltage at the oscilloscope input to the order of some volts. This is achieved by choosing an appropriate value for C_4, the capacitance in Module 2.

The total divider ratio is given by:

$$\frac{U_1}{U_2} = \frac{C_1}{C_2 + C_3 + C_4 + C_k}$$

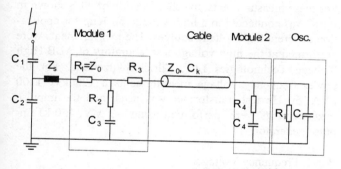

Figure 1: Equivalent circuit diagram of measurement system
C_1 Capacitance between inner conductor of GIS and electrical field sensor
C_2 Stray capacitance between electrical field sensor and GIS enclosure
Z_s Characteristic output impedance of the electrical field sensor
Module 1   RC network determined by the elements R_1, R_2, R_3, C_3 for the impedance matching of sensor connection Z_s and Module 2
Cable   50Ω 'Ethernet' cable with impedance Z_0 and capacitance C_k
Module 2   RC network determined by the elements R_4, C_4 for the impedance matching of cable (50Ω)
Osc.   Input impedance of the oscilloscope (C_i=15pF / R_i=1MΩ)

Typical values of the divider ratio are on the order 10^5. For this application, the divider ratio of the system must be constant over the frequency range 10 Hz to 100 MHz;

therefore the low frequency cut-off f_1 of the system should be less than 10Hz and the high frequency cut-off f_2 should be above 100MHz.

3. Measurements and results

To demonstrate the suitability of the measuring system the frequency and step response of the system were measured and the results given in section 3.1. Secondly, standard wave shapes (AC, LI, SI) were measured. The measured values were compared with the results of calibrated high voltage measuring systems; these are given in chapter 3.2. All measurements were performed simultaneously on three measurement systems of identical design in order to verify material and constructional tolerances. The scale factors were determined at power frequency and kept constant for all measurements.

3.1 Frequency and step response

Direct measurement of the frequency response of the complete measurement system, as proposed in [IEC 1321-1], was not possible, owing to the high divider ratio (ca. 10^5). In order to measure the output signal on the oscilloscope with a reasonable signal-to-noise ratio, a sine-wave source for frequencies below 1Hz to over 100 MHz with an output voltage on the order of some kV into 50Ω would be required which was not available.

Therefore, as a first step, the frequency response of the impedance-matching modules and the 'Ethernet' cable was measured at low voltage (1 to 10 V). The divider ratio for this part of the measurement system is typically 10^3. Second, the step response of the whole system (including the sensor) was measured.

Frequency response of impedance-matching modules

In the first measurement, the frequency response of the impedance-matching modules without electrical field sensor was determined. The capacitive voltage divider, formed by the sensor and the GIS, was simulated by a series capacitance of C_2 (20pF). This is justified by applying the Thévenin's equivalent [7]. The measurement was made with the test circuit shown in figure 2, using conventional sweep/function generators.

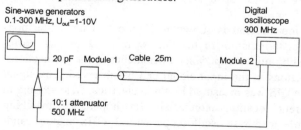

Figure 2: Test set up for the frequency response measurement

The goal was to ascertain the low frequency cut-off and also to show that the divider ratio is constant in the frequency range of interest. The signals were measured with a digital oscilloscope with an analog bandwidth of 300 MHz. Figure 3 shows the results.

The frequency was varied from 0.1Hz to 300MHz in appropriate steps and the output voltage measured. Comparison of input and output voltage versus frequency yields the frequency response of the measurement system. From the figure it can be observed a constant value between 8Hz and 100 MHz. This measurement was performed on all impedance-matching modules. The results all fell within a measurement uncertainty of 4% to that shown in figure 3.

Figure 3: Frequency response of impedance matching modules

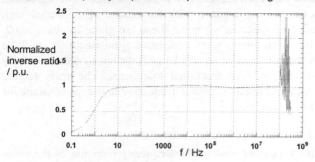

Step response measurement

The test circuit for the step response measurement is shown in figure 4. The electrical field sensors were installed in a short section of GIS with rated voltage of 550 kV. Cone-shaped adapters were installed on both ends of this GIS in order to provide a smooth, reflection-free transition between the GIS-enclosure and the 50 Ω signal connections.

The step signal was generated by a fast pulse generator (rise-time < 320 ps). The duration of the step signal was 2 µs, which was the longest duration achievable.

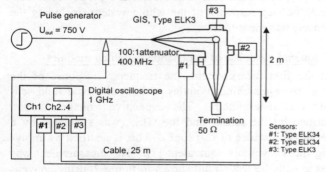

Figure 4: Test set up for the step response measurement

The input signal is shown in Figure 5. Due to the non-perfect termination of the GIS with 50 Ω and the corresponding minor impedance mismatches at the connections, reflections in the input signal can be observed. The signal in the GIS was measured by three electrical field sensors in different locations connected via impedance matching module to a fast oscilloscope with 1 GHz analog bandwidth. The measured step response is shown figure 5.

Figure 5: Step response of measurement systems

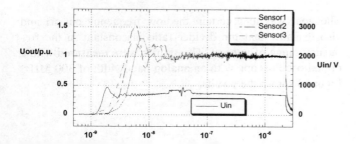

The level of the output signal was only in the range of mV. Noise suppression of the waveforms was achieved by using the "average" function of the oscilloscope over 1000 sweeps. Due to the waveform of the input pulse (e.g. reflections), the analysis methods and criteria given in [IEC1321-1] could not be applied. From figure 5 it can be observed however, that between 20 ns and 2 µs, the unit step is flat within the noise level of about 5%. This is valid for all three measurement systems. Rise times of 2.2 ns +/- 0.7 ns, 1.5 +/-0.7 ns, and 1.5 ns +/-0.7 ns resulted for measuring systems 1, 2, and 3, respectively, with settling times of approx. 20 ns. The high frequency cut-off of the measuring systems can be roughly estimated from the rise times to be higher than 200 MHz. The measuring systems, therefore, fulfil the requirements for the measurement of VFFO. From the frequency response measurement, the low cut-off frequency (-3dB) was determined to 1 Hz.

Thus, the measuring system can be used for the measurement of signals in GIS between 10 Hz and 100 MHz. The accuracy can roughly be estimated from the frequency and step response measurement to be within 5%.

3.2 Measurement of power frequency voltage, lightning and switching impulse

For these measurements, the short section of GIS shown in fig. 4 was connected to a high voltage bushing for application different high voltage sources. The measurements were performed at the high voltage test laboratory of ABB High Voltage Technologies Ltd., Zürich/Switzerland. For the power frequency voltage measurements a 1200kV, 1200 kVA, 50 Hz test transformer was used, and the impulse measurements were performed using a 3 MV, 150 kJ impulse generator.

Power Frequency Voltage:

Power frequency voltage of 50 Hz was applied to the test set up. The rms value of the applied voltage was determined by a conventional, calibrated, 'approved measuring system' (according [IEC 60-2]) acting as a reference measuring system. The total uncertainty of this measuring system amounted to 0.9%. The wave shape of the applied power frequency voltage was simultaneously recorded using the field sensor apparatus (sensors, modules, cables, oscilloscope) as shown in fig. 4. The rms value was determined from the wave shapes using a standard parameter function of the oscilloscope. The scale factor of the measurement systems investigated was determined at an applied voltage of 400 kV to:

Table 1: Scale factors of measurement systems

Meas. system	1	2	3
Scale factor	$4.804 \cdot 10^5$ +/- 0.2%	$4.971 \cdot 10^5$ +/- 0.3%	$3.811 \cdot 10^5$ +/- 0.2 %

Figure 6 shows the results of the measurements at power frequency voltages ranging from 50 kV to 600 kV. Deviations of less than 1% with respect to the reference system were observed, demonstrating the linearity and also the reproducibility of the measuring systems. The systematic variations common to all of measuring systems may be explained by minor non-linearites present in both the investigated measuring systems and the reference. For a confidence level of 95%, assuming a non-linearity of 0.5%, the total measuring uncertainty can be determined to 1.2 % (applying [IEC60-2]).

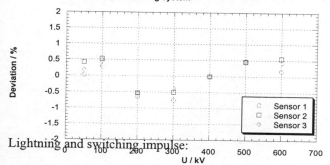

Figure 6: Deviation of measured rms power frequency voltage to reference measuring system

Lightning and switching impulse:

Standard lightning and switching impulses according [IEC 60-1], were applied to the test set up. The wave shapes were simultaneously measured with the field sensors and a conventional 'approved' reference measuring system (calibrated according to see [IEC 60-2]). All wave shapes were analysed by the software of the reference measuring system. The scale factors determined in the power frequency measurement (see table 1) were used. , The procedures given by [IEC 60-2, A1] were used to determine the uncertainty of the measured voltage and time parameters. For each polarity 10 impulses were applied. The parameters of the applied waveform and the results are given in table 2.

Lightning impulse: 750 kVp, T1=1.2 µs T2= 51 µs					
Meas. system	Correction for $\hat{U}$ a)	$\delta\hat{U}$ b)	Correction for T1 a)	Correction for T2 a)	δT1 and δT2 b)
1	2.3 +/- 0.5	1.4 %	-3.2+/- 1.9	3.8+/- 0.9	5.3 %
2	2.6 +/- 0.7	1.4 %	3.8+/- 2.2	8.5+/- 2.2	5.3 %
3	2.8 +/- 0.4	1.4 %	0.4+/- 2.1	5.0+/- 0.9	5.3 %
Switching impulse: 650 kVp, 220 µs / 2440 µs					
Meas. system	Correction for $\hat{U}$ a)	$\delta\hat{U}$ b)	Correction for Tp a)	Correction for T2 a)	δT1 and δT2 b)
1	1.3 +/- 0.3	1.4 %	-1.4+/- 2.8	-3.7+/- 0.6	5.4 %
2	1.4 +/- 0.2	1.4 %	-1.8+/- 3.2	-3.5+/- 0.5	5.4 %
3	1.1 +/- 0.3	1.4 %	-3.4+/- 0.6	-4.6+/- 0.6	%

a) Mean deviation to value of reference measuring system in %. Averaged over 10 positive and 10 negative impulses.

b) Total uncertainty . Calculated according [IEC 60-2, A1] for a confidence level of 95%. The uncertainties of the reference measuring system were 1.1% and 4.5% for peak and time parameter, respectively. A non linearity of 0.5 % was assumed.

Table 2: Result of lightning and switching impulse measurements

The values in table 2 show that peak values were systematically higher than those of the reference measuring system. Deviations up to 2.8 % and 1.4 % for the lightning and switching impulse, respectively, were measured. For the lightning impulse, this is larger than the measuring accuracy. The deviations lie within the measurement accuracy for the switching impulse. The differences of the values

between the measurement systems 1 to 3 are very low, indicating that constructional tolerances can be neglected. The observed deviations are therefore mainly due to the measurement accuracy and the design of the measuring systems. The deviations of the time parameters of the waveforms are mostly within the measurement accuracy of 5-6%. The correction factors could be determined more precisely using a more precise reference measurement system. However, the measurement accuracy was sufficient to show the variation of the scale factor from power frequency voltage to lightning impulse voltage. It can be concluded that using the scale factor determined in the power frequency measurement, lightning and switching impulse peak parameters can be determined within an accuracy within 5 % and 3 %, respectively. For the time parameters, an accuracy within 10% is obtained.

4. Conclusions

The present investigation established the accuracy of measuring high voltage signals in GIS using electrical field sensors, over the range 0.1 Hz to above 200 MHz. Cut-off frequencies of 1 Hz and 200 MHz, respectively, could be determined from the frequency and step response measurements. The frequency response of the impedance matching modules plus the associated cable was flat within 4% over the range 8 Hz to 100 MHz. In the step response, no oscillations or decay could be observed for step durations up to 2 µs. Thus the measurement system is suited for the measurement of high voltage signals ranging from 10 Hz to 100 MHz. The accuracy of measuring peak values and time parameters was determined by the application of standard wave shapes. Calibrated high voltage measuring systems were used as a standard reference. An accuracy within 1.2 % was obtained for measurement of the power frequency voltages (rms). For the peak of switching and lightning impulses, accuracies of 3% and 5%, were determined, confirming the results of the frequency and step response measurement. Time parameters of lightning and switching impulses could be determined within 10%. The accuracy of measured VFFO peak values could not be determined directly, but was estimated from the frequency and step response measurement to be within 5%. The results showed that constructional tolerances are much lower than the measurement uncertainties. Thus, the accuracies given should also be valid for other measurement systems of identical design and construction.

References

[IEC60-1]   IEC60-1, 1989,"High Voltage Test Techniques Part 1:General definitions and test requirements"

[IEC60-2]   IEC60-2, 1994," High Voltage Test Techniques, Part 2: Measuring systems"

[IEC60-2, A1] IEC60-2, Amendment 1, 1996, "High Voltage Test Techniques, Part 2: Measuring systems"

[IEC71-1]   IEC71-1, 1993, "Insulation co-ordination, Part 1: Definitions, principles and rules"

[IEC1321-1] IEC1321-1, 1994, "High Voltage Test Techniques with very fast impulses, Part 1: Measuring systems for very fast front overvoltages generated in gas insulated substations".

[1] IEEE Guide for gas-insulated substations, Std C37.122.1-1993

[2] J. Meppelink; P. Hofer, "Design and calibration of a high voltage divider for measurement of very fast transients in gas insulated switchgear". 5th ISH (1987), 71.08.

[3] M. Albiez, W.Zaengl, K.J.Diederich, J. Meppelink: "Design and calibration of an universal sensor for the measurement of partial discharges and very fast transients in GIS", Proc. of 6th ISH, New Orleans, 1989, 42.28.

[4] G.Behrmann, S.Neuhold, R.Pietsch: "Results of measurements in a 220kV GIS substation during on-site comissioning tests". Proc. of 10th ISH, Montreal, 1997, Vol.4, p451.

[5] J.S. Pearson, B.F. Hampton, A.G. Sellars," A continuous UHF monitor for gas-insulated substations", IEEE Trans. on electrical Insulation, Vol.26 No.3, June 1991.

[6] M.Albiez, PhD thesis, DISS. ETH Nr. 9694, ETH Zürich, 1992

[7] P.Horowitz, W.Hill, "The art of electronics", Cambridge university press, 2nd edition, 1989.

Effect Of Mechanical Pressure And Silicone Grease on Partial Discharge Characteristics for Model XLPE Transmission Cable Joint

Z. Nadolny*, J.M. Braun, and R.J. Densley
Ontario Hydro Technologies, Toronto, Ontario, Canada
Visiting Researcher, Poznan University, Poland.

Abstract

This paper reports on the effects of mechanical pressure and silicone grease on the partial discharge (PD) characteristics at EPR-epoxy interfaces in model cable joints. An ultra wide band PD detection system was used. The PD parameters investigated are: rise time, width, fall time, amplitude, inception voltage and frequency of pulses. PD can cause degradation and eventually lead to breakdown. The presence of air bubbles or contaminants at interfaces enhances the local electric field and initiate PD. The effectiveness of mechanical pressure and/or silicone grease to reduce PD activity was demonstrated.

Introduction

The cable joint is considered to be the weakest part of an extruded cable system and plays a very important role in the reliability of the transmission system [1]. EPR-epoxy interfaces are a sensitive part of the cable joint and if PDs occur along this path, breakdown may result. Figure 1 shows the relationship between breakdown strength and mechanical pressure for the two types of interface [1].

This paper presents data of the PD charactaristics that occur in a modle cable joint. The true slope of the PD current pulses were measured i.e. an ultra wide band PD detection system was used. Mechanical pressure and silicone grease improve the interfacial breakdown strength by about 50 %. Mechanical pressure probably decreases the sizes of

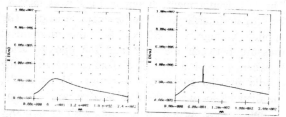

Figure 2. E-field distribution along EPR-epoxy interfaces of a joint without and with an air bubble

air bubbles located at this interface and thus reduce the PD activity.

Computer Simulation of the Electric Field

In order to analyze the distribution of the electric field in a model cable joint, a computer simulation of the electric field based on Maxwell software was calculated for the cable joint geometry discussed in [1]. The maximum interfacial electric field was 2.1 kV/mm along clean EPR-epoxy interfaces, and 4 kV/mm along this interface with an air bubble (see Fig.2). The air bubble has an elliptical shape with diameters D1=1 mm, D2=0.1 mm.

If we define the air bubble shape, **m**, as the bubble width divided by its height, an increase in **m** from 2

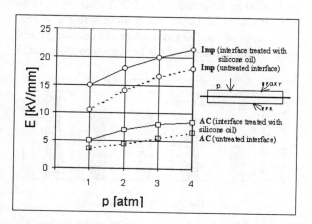

Figure 1. Breakdown characteristics of the interface between the EPR and molded epoxy insulation [1]

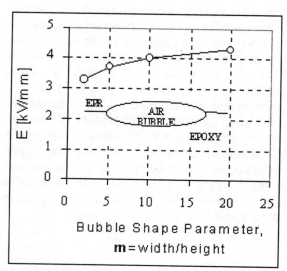

Fig.3. Relationship between bubble shape parameter **m** and maximum electric field along EPR-epoxy surface of XLPE Cable Joint

High Voltage Engineering Symposium, 22–27 August 1999
Conference Publication No. 467, © IEE, 1999

to 20 causes an increase in the maximum interfacial local electric field from 3.4 to 4.3 kV/mm (see Fig.3). The maximum local electric field can thus increase to 4.3 kV/mm along EPR-epoxy interfaces with a very narrow air bubble, which may be sufficient to initiate PD.

Sample And Experimental Setup

The test arrangement shown in Fig.4 was used to investigate interfacial PD pulse characteristics. The applied voltage was 30 kV and the mechanical pressure between the EPR and the epoxy varied between 1 to 4 atm. The distance between the needle and the interface was 1 mm. Horizontally, the distance between the two needles was 20 mm. Samples were tested with and without silicone grease on the EPR-epoxy interfaces. An Ogura needle, 100 μm in tip radius, was used as high voltage electrode in the upper epoxy cylinder and a second Ogura needle, 500 μm in tip radius was used as the ground electrode in the lower cylinder.

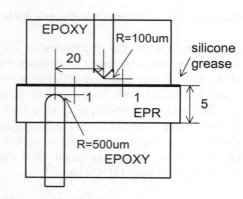

Figure 4. Schematic of the test sample

The PD signals were detected as voltage pulses across a 50 Ω resistance connected between the measuring electrode and ground, and fed to an analog oscilloscope (Tektronix 7104, 1 GHz bandwidth) and a digital oscilloscope (Tektronix DSA 602, 2 Gigasamples per second, 1.1 GHz bandwidth, see Fig.5). Digital measurements of the amplitude, width, rise and fall times to describe a PD pulse were carried out using the digital oscilloscope. Information about the pulses was stored in a PC computer memory via a GPIB port. The tree shape was monitored by a video camera and a long-range microscope.

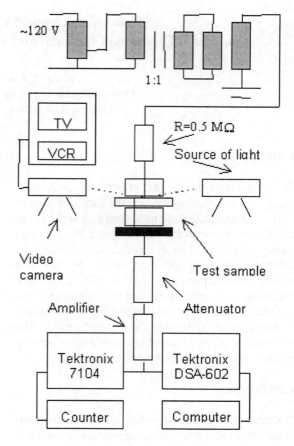

Figure 5. Experimental setup

Result And Discussion

Raising the mechanical pressure from 0 to 4 atm causes an increase in the inception voltage from 7 to 14 kV without silicone grease, and from 8 to 20 kV with silicone grease. Raising the mechanical pressure causes a more rapid increase in PD inception voltage for the coated interface (see Fig.6).

Figure 7 shows the variation in the parameters rise time, width, fall time and amplitude of 100 individual PD pulses recorded at 0 and 4 atm, without silicone grease. The PD is characterized by a large scatter at 0 atm, but is more stable at 4 atm. The values of the PD parameters are also larger at 0 atm. Different sizes of air bubbles probably cause

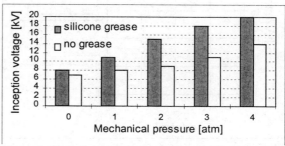

Figure 6. Variation of inception voltage with mechanical pressure for the EPR-epoxy interfaces with and without silicone grease

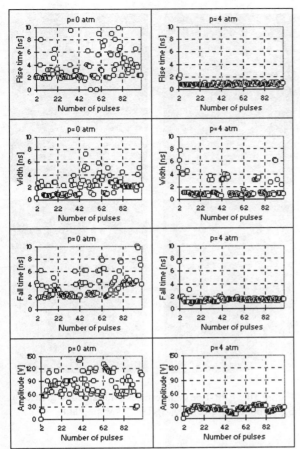

Figure 7. Individual PD at 0 and 4 atm, U=20 kV, no silicone grease

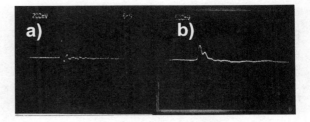

Figure 8. Typical PD pulses (a) within the epoxy, and (b) along the EPR-epoxy interfaces

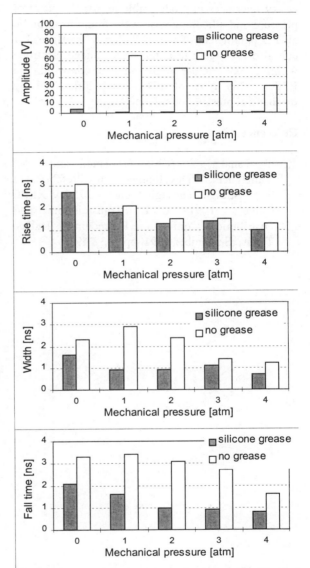

Figure 9. PD parameters as a function of mechanical pressure with and without silicone grease, U=30 kV

the large scatter of the PD parameters recorded by the digital scope. An increase in mechanical pressure decreases the values of the PD parameters, particularly the amplitude. At the higher mechanical pressure, the air bubbles are reduced in size and number with the result that PD along the interface is reduced. However the mechanical pressure does not influence the electrical conditions nor the PD parameters from around the high voltage needle inside the epoxy, characterized by very narrow pulses, as shown in Fig.8a. The interfacial pulses consist of a short rise time pulse with additional smaller amplitude steps (see Fig.8.b). These steps may be due to the ionic component of a single PD or due to additional PD, triggered by the initial PD and which occur almost simultaneously.

Increasing the mechanical pressure causes a decrease in all PD parameters, such as rise time, width, and fall time, and particularly amplitude from 5 to 0.5 V with silicone grease, and from 90 to 30 V without silicone grease, as shown in Fig.9.

The silicone grease decreases all PD parameters such as rise time, width, fall time and amplitude. The silicone grease has the biggest influence on amplitude which decreased about ten times. Figure 10 shows the EPR-epoxy interfaces with and without silicone grease. For the arrangement with silicone grease (a, b) many bubbles are located at the interfaces around high voltage electrode. The bubbles can be gas-filled or residues of PD activity.

Figure 10 (c,d) shows the interface without silicone grease. A white powder has formed at the EPR-epoxy interfaces; this powder may be the result of chemical reactions between the EPR and epoxy materials.

Conclusions

The PD characteristics were measured for EPR-epoxy interfaces in model cable joints. Mechanical pressure and silicone grease were found to improve all PD parameters of the EPR-Epoxy interface. PD can also propagate near the high voltage electrode within epoxy or close to ground electrode within the EPR. This type of PD is not influenced by mechanical pressure or silicone grease.

References

1. T. Tanaka, Y. Sekil, H. Satoh and M. Yamaguchi, "Development of Prefabricated Type Straight Through Joint for 275 kV XLPE Cable", 1991 IEEE Transmission and Distribution Conference Proceedings, Dallas, Texas, September 1991. pp 243-249

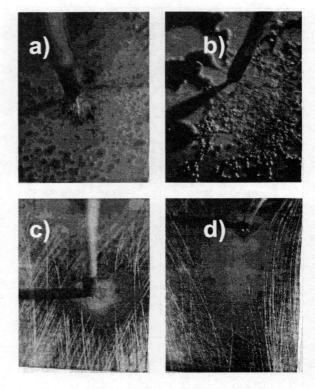

Figure 10. EPR-Epoxy interfaces with silicone grease (a, b), without any grease (c, d), U=20 kV, p=2 atm

APPLICATION OF FWT (FAST WAVELET TRANSFORM) FOR AUTO-DETECTION SYSTEM OF PARTIAL DISCHARGE IN POWER CABLES

Yoh YASUDA[†], Jun MATSUURA[†], Takehisa Hara[†],

CHEN Min[‡], Ken'ichiro HIROTSU[‡] and Siegeki ISOJIMA[‡]

† *Department of Electrical Engineering, Kansai University, Japan*

‡ *Electric Power System Technology Research Laboratories, Sumitomo Electric Industries, Japan*

1. Introduction

Demand of constructing underground power CV cables in cities is being increased as those cities are enormously expanding all over the world. Consequently, the cables' reliability as products is becoming an important role for their maintenance. One of the serious problems that may happen is destruction of insulator in the cables. The best and conventional way to prevent such a crucial accident is generally supposed to ascertain partial corona discharges occurring at small void in organic insulator. However, there are some difficulties to detect those partial discharges because of existence of external noises in detected data, whose patterns are hardly identified at a glance. By the reason of the problem, there have been a number of researches on the way of development recent years to accomplish detecting partial discharges by employing neural network (NN) system, which is widely known as the system for pattern recognition[1].

The authors have been developing the neural network system of the auto-detection for partial discharges[2]-[4], which we actually input numerical data of waveforms to and obtained appropriate performance from. In our former studies we used firstly direct sampling data of waveforms and FFT spectra afterward. The results from the study employing FFT was much improved from the former one, which used the direct data. In spite of the performance, the FFT method couldn't reach our complete satisfaction. The presumable reason is that the FFT spectra excluding time information is unable to show us how partial discharges or noise signals decay as time goes by. Therefore, we applyed Fast Wavelet Transform (FWT) to acquire more detailed transformed data in order to put them into the NN system. Applying FWT, we were able to express the waveform data in time-frequency space, where we easily detect partial discharges when we input them to the NN system. For the first step of the research, it is of importance to confirm the patterns of partial discharges and noise signals. In this paper we present the statistic results by FWT analysis for partial

discharges and noise signals which we obtained experimentally. Moreover, we present results out of the NN system which were dealt with those transformed data.

2. Continuous Wavelet Transform of Partial Discharge

Before showing results after FWT, we would like to discuss a continuous wavelet transform (CWT) because it gives us much better visual comprehension of how wavelet transform works to actual numerical data. Besides, both of the methods are basically not different from each other. Generally, the CWT's equation is defined as following;

$$(\mathbf{W}_\psi f)(b,a) = \int_{-\infty}^{\infty} \frac{1}{\sqrt{|a|}} \overline{\psi\left(\frac{x-b}{a}\right)} f(x)dx \qquad (1)$$

where $\psi(x)$ is a mother wavelet and $f(x)$ is a function that is transformed[5],[6]. The function can be an one-dimensional map like in the case of our study. We selected Gabor wavelet as the mother wavelet. Then, the equation is expressed as following;

$$\Psi(x) = \frac{1}{2\sqrt{\pi}\sigma} e^{\frac{-x^2}{\sigma^2}} e^{-ix} \qquad (2)$$

where σ is 8. The reason why we chose Gabor wavelet as the mother wavelet is that the function is very similar to Window Fourier transform which combines Gaussian function as Window function. Therefore, it is somewhat convenient to realize data by wavelet transform when it comes to comparing with those by Fourier transform. Further, the spectra of them can be easily visualized because the output data are absolute values.

To obtain numerical data of partial discharges and external noise signals, in our labolatory we built up CV cables (26kV, 10m-length) which has an artificial deficit to make partial discharges (see Fig. 1)[4]. By measuring them with a digital storage oscilloscope (DSO), a sampling cycle of the measuring data was 1nsec. and 1000 sampling points were obtained in each pulse waveform. Starting time of the measurement is ordered as 300nsec. before the pulse wave is arisen by pre-trigger function of the DSO. On

High Voltage Engineering Symposium, 22–27 August 1999
Conference Publication No. 467, © IEE, 1999

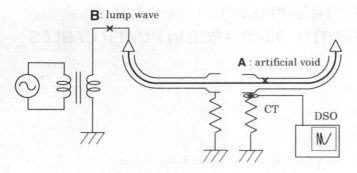

Fig.1 Measurement diagram

the other hand, noise signals were measured at a joint part of the cables, while we were generating air discharges at the end of the cable. We considered these noises as external noises to use. The number of 10 of partial discharges and 16 of external noise signals are selected as typical types to treat them for this study. Then, by employing wavelet transform to them, we analyzed how distinctive they appear and if we can distinguish useful properties for the auto-detection of partial discharges.

Figure 2 shows an actual partial discharge

waveform measured in the way mentioned above. Figure 3 shows spectra of the partial discharge that is transformed by CWT expressed in the Eqs. 1 and 2 (the vertical axis in the figure is chosen as $1/f$ dimension, so the axis scale is not linear.). The black-and-white contour map contrasts the spectrum strength in time-frequency space of the partial discharge. The figure indicates that the strong spectrum values lay over a wide extend of frequency at around 300nsec., when the pulse is severely arisen.

Contrarily, one of the typical external noise waveforms we obtained and its contour picture are shown in Figs. 4 and 5, respectively. Compared with the picture of the partial discharge in Fig.3, that of the noise signal has its strong spectrum values at the lower side of the picture, which means the strength of noise spectrum exists on lower frequency and doesn't decrease for a long period.

Finally, by employing CWT, we successfully obtained more detailed properties of each of the waveforms than those by FFT. Moreover, the visualized properties of them make us easier to recognize their shapes, which means NN system can

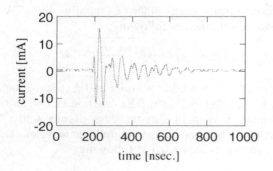

Fig.2 Waveform of a Partial Discharge

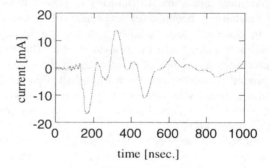

Fig.4 Waveform of a Noise Signal

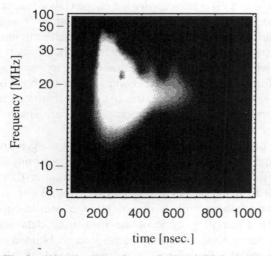

Fig.3 Wavelet Transform of a Partial Discharge in Fig.2

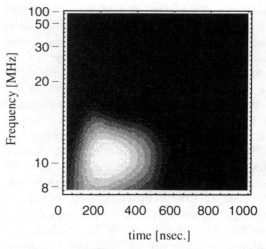

Fig.5 Wavelet Transform of a Noise Signal in Fig.4

be also expected to have better performance for the detection of partial discharges. It is significant that we obtained the spectra of each of the waveforms including time information with such ease.

3. FWT analysis for partial discharges

So far we have described the performance of CWT. It is extremely useful to visualize the waveforms in order to confirm the properties of them. However, the transformed numerical data are not efficient for signal processing because the integral equation (see Eq. 2) has some difficulties to be computed with precise resolution. Compared with it, discrete wavelet transform is treat with dispersed data, so it is highly expected for an effective analysis of time-frequency property to be carried out. FWT, for that reason,

should be the one that is applied for signal processing so as to adjust data to NN system.

We employed an one-dimensional discrete orthogonal wavelet transform with Harr basis[7]. Here, we consider a 2^n of measured data group as one-dimensional vector $x=\{x_1, x_2, \ldots x_{2^n}\}$, and the discrete wavelet transform is expressed as following;

$$x' = Wx \qquad (3)$$

where W is a wavelet transform matrix. Here is an example of the wavelet transform matrix with Harr basis as following;

$$W = \left(\frac{1}{\sqrt{2}}\right)^3 \begin{bmatrix} 1 & 1 & 1 & 1 & 1 & 1 & 1 & 1 \\ 1 & 1 & 1 & 1 & -1 & -1 & -1 & -1 \\ \sqrt{2} & -\sqrt{2} & \sqrt{2} & -\sqrt{2} & & & & \\ & & & & \sqrt{2} & -\sqrt{2} & \sqrt{2} & -\sqrt{2} \\ \sqrt{2}^3 & -\sqrt{2}^3 & & & & & & \\ & & \sqrt{2}^3 & -\sqrt{2}^3 & & & & \\ & & & & \sqrt{2}^3 & -\sqrt{2}^3 & & \\ & & & & & & \sqrt{2}^3 & -\sqrt{2}^3 \end{bmatrix} \qquad (4)$$

where the data amount x is 8 ($2^3=8$). The matrix is created submitting a certain rule, and always keeps the condition of $2^n \times 2^n$ matrix. The data x' transformed with the Eq.3 is also able to be expressed in time-frequency space as the same way as shown in Figs. 3 and 5.

There were 1000 sampling points of measured data as mentioned earlier of this paper, but the amount of sampling points for FWT to transform them is restricted to be exactly the number of 2^n. What was done for the limitation in the study is that we manipulated the data by adding the value of 0 to the rest of 24 points from the end of it. Then, we could treat the data having the amount of $2^{10}=1024$ with discrete wavelet transform. To obtain the characteristics of the waveforms of the partial discharges and the noise signals, we divided a

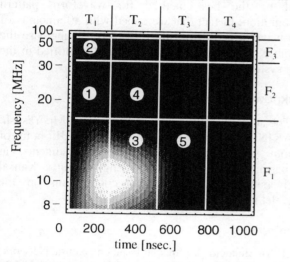

Fig.6 Conceptual Illustration for Division of Time-Frequency Space of Wavelet Transform

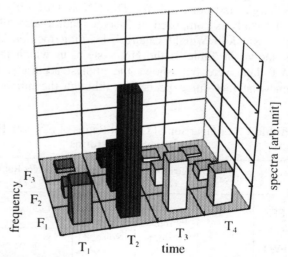

Fig.7 Averaged Spectra of Fast Wavelet Transform of Partial Discharges and Noise Signals
(left: partial discharge, right: noise signal)

spectrum picture into 12 sections as shown in Fig.6 (for visual comprehension, Fig.6, whose data are transformed by CWT, is shown as the conceptual illustration), and calculated an average value from each of the sections.

We also computed average values at each of the 12 sections from the results of FWT analysis for the overall data of 14 partial discharges and 16 external noises. The results of them are shown in Figs.7. Apparently, the difference between the average of the partial discharges and the noise signals is observed in the figures. The average spectra of the partial discharges spread out vertically with intensive values from lower to higher frequency at the early period of the measurement, meanwhile the noise signals' ones have strong values in lower frequency and the strength stays higher than certain amounts all time long. The illustrations also reveal the property of decrement of both of the waveforms, which we couldn't obtain by FFT before.

Thus, the FWT result by both of the partial discharges and noise signals are worth being put in the NN system for auto-detection of partial discharges. The reason why the averages of the overall waveforms were calculated (see Figs.7) is that we can determine which spectrum parts should be put in the NN system. As doing so, fast and efficient performance of the NN system will be expected.

4. NN System For Auto-Detection of Partial Discharges

We actually put both of the CWT and the FWT result of the waveforms into the NN system for auto-detection of partial discharges. Several sections in spectrum pictures were selected for the input data. The selected areas were numbered in Fig.6, whose properties seem to be distinctive between the partial discharges and noise signals. The NN system utilized in the study is composed of 3-layer feedforward NN with "Random Weight Change" learning method[3]. As shown in Table 1, the NN system, in which the FWT data were put, shows the highest accuracy. It somewhat met our expectation, although the number

of the data dealt with were not so large.

5. Conclusion

Conclusively, the statistic results of the data of partial discharges and noise signals transformed by employing FWT (Fast Wavelet Transform) revealed the characteristics of the waveforms in time-frequency space. We effectively succeeded to obtain those data that show us the distinguished properties. The significance of this analysis is that we can observe not only the frequency distribution of the spectrum, but also the time distribution of it at the same time. We expect that it will lead the NN system to executing the identification of partial discharges in underground power CV cables. The results in employing the NN system with FWT show some aspects that the FWT will be the best method for waveform pattern recognition. In fact that we coped with 30 amounts of the waveforms in this study, we have to increase the number of the data to be learned and identified in the NN system.

Acknowledgement

The authors would like to acknowledge Mr. Yasushi MAKINO and Mr. Eiyoshi NAKATSUJI, both of whom are currently undergraduate students at Department of Electrical Engineering, Kansai University, for their helpful cooperation on the wavelet and the NN calculation.

References

[1] E.S-. Sinencio *ed.*: Special Issue on "Neural Network Circuit Implementations", *IEEE Trans. on NN*, **Vol.2**, No.2, 1991.

[2] T. Hara, A. Ito, K. Yatsuka, K. Kishi and K. Hirotsu : "Application of the Neural Network to Detecting Corona Discharge Occurring in Power Cables", *Proc. of 2nd Int. Forum on Appl. of NN to Power Systems*, pp.259-264, 1993.

[3] K. Hirotsu and M.A. Brooke: "An Analog Neural Network Chip with Random Weight Change Learning Algorithm", *Proc. of Int. Joint Conf. on NN*, pp.3031-3034, 1993.

[4] Y. Yasuda, T. Hara, M. Chen, K. Hirotsu and S. Isojima: "A Neural Network System for Auto-Detection of Partial Discharges in Power Cables using an Analog Parallel Circuit", *Proc. of Int. Conf. on Elec. Eng. '98*, pp.610-613, 1998.

[5] C.K. Chui: "AN INTRODUCTION TO WAVELETS", Academic Press, Inc. (1992)

[6] "An Introduction to Wavelets" homepage, http://www.amara.com/IEEEwave/IEEEwavelet.html

Table 1 Comparison of Correct Ratios of NN with FFT and Wavelet Transforms

	Layer	Learning Data	Detected Data	Correct Ratio [%]
FFT		Partial: 7 Noise: 8	Partial: 7 Noise: 8	99.64
FWT	5 : 5 : 1			99.94

A STUDY OF MEMORY EFFECT OF PARTIAL DISCHARGES

YL Madhavi and MN Narayanachar

Department of High Voltage Engineering
Indian Institute of Science
Bangalore-560 012, INDIA

Abstract : From the computed conditional pulse amplitude and phase distributions it is shown that, as the gap spacing is reduced the charging of dielectric surface affects the corona pulse-time distribution which becomes significant at lower gap spacings. Most probable values are found to be the best statistical indicators for interpreting the results.

Introduction

Partial discharge (PD) phenomena which occur both in the presence and absence of solid dielectrics are inherently stochastic processes that exhibit significant statistical variability in such characteristics as pulse amplitude and phase of occurrence. The statistical behaviour of PD is governed primarily by memory effects such as associated with charge deposited by PD on the dielectric surface. When ac generated PD occur near the dielectric surface the predominant memory effect may be due to surface charge deposition [1-3].Much work has been done in interpreting the stochastic nature of PD phenomena. It is shown that the sum amplitude of the discharges on the positive half cycle Q^+ affect the phase of occurrence ϕ^- of the first pulse in the negative half cycle. This is shown to be true in the case of second negative pulse too. The unconditional PD amplitude and phase distributions are shown to be sensitive to relatively small physical or chemical changes occuring in the gap spacing, such as might result from interaction of the discharge with the surface. The sensitivity to non-stationary behaviour is much less evident in conditional distributions. In the present work the relative effect of the first negative pulse q_1^- on the second negative pulse q_2^- has been studied by considering their magnitude and the time separation Δt distribuions.

Experimental arrangement and procedure

Experiments were performed in a controlled environment using point-plane configuration, with and without a perspex dielectric at various voltage levels and at different gap spacings. Straight detection method of PD detection was used. The sample was placed in a normal oven wherein standard ambient conditions were maintained following the methods given in IS : 2260-1973, Appendix A, (Clause 3.1.1). The temperature was $27 \pm 2^0 C$ and the humidity $70 \pm 5\%$ during the course of experiments.

The high voltage 50Hz ac setup was found to be discharge free upto 10kV. Voltages of 1.05, 1.1 and 1.2 times the inception voltage were applied. The gap spacings varied from 1 mm to 5 mm with perspex dielectric. Phase resolved pulse height analyser was used for measuring the amplitude and phase of the discharge pulses. Each record was taken for a time duration of 1 min and 10 such records were taken at 2 min interval. The procedure was repeated for point-plane at 5 mm spacing without the perspex dielectric.

Experimental results and analysis

In order to understand the basic phenomena, it is essential to study the stochastic nature of the PD behaviour at different gap spacings and at different voltage levels. As very few pulses occurred at inception voltage level, the experiments were conducted at overvoltages and data collected for a sufficient time duration so that a proper statistical analysis could be made. In the case of perspex dielectric surface there were a small number of pulses in the positive half cycle followed by a large number of pulses in the negative half cycle, whereas in the case of point-plane, the pulses in the positive half cycle were negligible. From the recorded distributions it was found that the magnitude and phase of occurrence of all the pulses occurring in the negative half cycle are distributed quantities. For the purpose of analysis only the distributions of the magnitude and phase of occurrence of the first and second pulses in the negative half cycle were considered. From the computed conditional amplitude and phase of occurrence distributions the correlations between successive pulses can be determined. The parameters chosen in the present work are q_1^-, q_2^- and Δt. From the data obtained the following conditional distributions were computed,

$P1(q_2^-/q_1^-)$: gives the probability of occurrence of q_2^- given q_1^-.

$P1(\Delta t/q_1^-)$: gives the probability of occurrence of q_2^- in a time Δt given q_1^-.

High Voltage Engineering Symposium, 22–27 August 1999
Conference Publication No. 467, © IEE, 1999

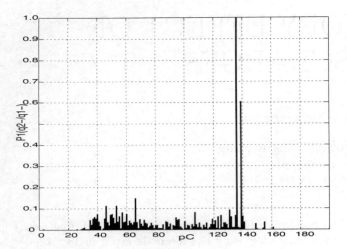

Fig. 1. $P1(q_2^-/q_1^-)$ for Point-Plane 5mm GAP, 1.1Vi

$P1(q_2^-/\Delta t)$: probability of q_2^- occurring, following a given Δt.

A number of distributions were computed for all the gaps and at all levels of voltages. Typical distributions are presented in Figures 1-9. The probabilities are normalised with respect to the maximum occurring in the particular distribution. The distributions are best characterised and explained by considering only the most probable values. Figures 1-3 and Figures 4-6 show the conditional distributions $P1(q_2^-/q_1^-)$ and $P1(\Delta t/q_1^-)$ respectively, for the most probable value of q_1^-. Figures 7-9 show the conditional distributions $P1(q_2^-/\Delta t)$ for the most probable Δt. We get a number of distributions $P1(q_2^-/q_1^-)$ and $P1(\Delta t/q_1^-)$ for different windows of q_1^-. Similarly we obtain a number of conditional distributions $P1(q_2^-/\Delta t)$ for different windows of Δt. From Figures 1-6, we can pick out the most probable q_2^- and the most probable Δt given the most probable q_1^-. These values are given in Table-I for all the cases. From the distributions shown in Figures 7-9 one can find the most proabable q_2^- for the most probable q_2^-. The product of $P1_{max}(q_2^-/\Delta t)$ and $P(\Delta t)$ is a joint probability which indicates the probability of the most probable q_2^- occurring in a particular window Δt where, $P(\Delta t)$ is the unconditional distribution. Such joint probabilities were worked out for all the gaps and at all voltage levels. Table-II gives a typical case where joint probabilities are given for gap spacings of 1 mm and 5 mm with perspex dielectric and for 5 mm without dielectric at the same voltage level(1.1Vi). Table-III gives only the maximum of the joint probabilities and the respective window Δt. Note that Δt of μ s corresponds to 3.14×10^{-4} rad.

Discussion

Interpretation of the results obtained from the analysis forms an important part of this discussion. From Figures 1-6 it can be seen that; following a most probable discharge q_1^-, there is a most probable q_2^- and

TABLE I

$V_{5pl}=3.92$кV , $V_5=3.69$кV , $V_4=4.12$кV , $V_3=3.32$кV , $V_2=3.84$кV , $V_1=2.55$кV

gap mm	q_1^- pC	q_2^- pC	Δt μs
5, pt-plane			
1.05 V_{5pl}	65	70	80
1.1 V_{5pl}	135	140	80
5, pt-pspx			
1.05 V_5	115	110	160
1.1 V_5	165	175	80
1.2 V_5	165	170	80
4, pt-pspx			
1.05 V_4	95	90	160
1.1 V_4	115	120	120
1.2 V_4	115	120	80
3, pt-pspx			
1.05 V_3	115	120	160
1.1 V_3	115	120	120
1.2 V_3	105	110	120
2, pt-pspx			
1.05 V_2	125	130	120
1.1 V_2	125	130	120
1.2 V_2	125	130	120
1, pt-pspx			
1.05 V_1	135	130	440
1.1 V_1	115	130	320
1.2 V_1	125	130	440

TABLE II

gap ,mm $\rightarrow$	1 perspex	5	5 pt-plane
joint probability $\times 10^{-2}$ $\downarrow$	0.33	2.1	2.5
	1.4	1.8	1.3
	3.1	0.43	0.72
	3.6	0.25	0.4
	2.6	0.17	0.55
	1.3	0.24	0.4
	0.37	0.18	0.29
	0.13	0.33	0.22
	0.01	0.23	0.19
	0.073	0.28	0.15
	0.056	0.16	0.076
	0.04	0.14	0.11

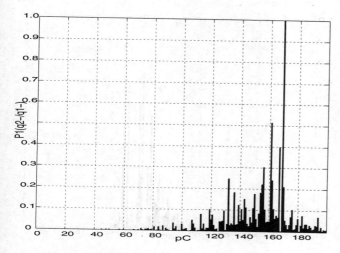

Fig. 2.   P1(q_2^-/q_1^-) for Point-Perspex 5mm GAP, 1.1Vi

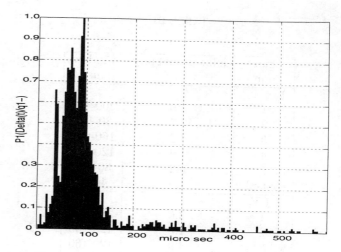

Fig. 5.   P1($\Delta t/q_1^-$) for Point-Perspex 5mm GAP, 1.1Vi

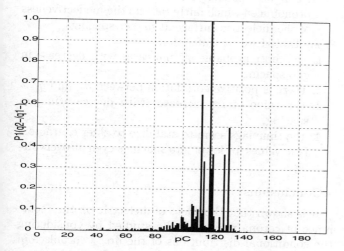

Fig. 3.   P1(q_2^-/q_1^-) for Point-Perspex 1mm GAP, 1.1Vi

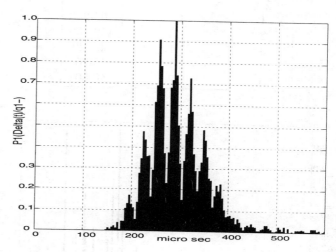

Fig. 6.   P1($\Delta t/q_1^-$) for Point-Perspex 1mm GAP, 1.1Vi

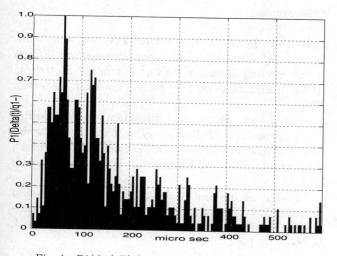

Fig. 4.   P1($\Delta t/q_1^-$) for Point-Plane 5mm GAP, 1.1Vi

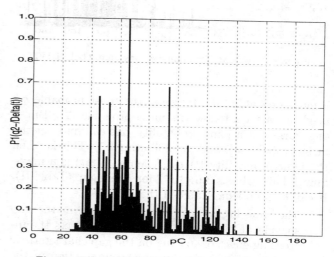

Fig. 7.   P1($q_2^-/\Delta t$) for Point-Plane 5mm GAP, 1.1Vi

TABLE III

gap mm	joint probability $\times 10^{-2}$	window μs
1.05 V_i,		
1	1.3	400-440
2	2.7	80-120
3	4.6	160-200
4	1.9	120-160
5	6.2	80-120
5pt-pl	1.5	40-80
1.1 V_i,		
1	3.6	280-320
2	1.5	120-160
3	4.9	120-160
4	2.3	80-120
5	2.1	40-80
5pt-pl	2.5	40-80
1.2 V_i,		
1	4.6	160-200
2	1.2	120-160
3	3.1	80-120
4	2.1	40-80
5	2.2	40-80

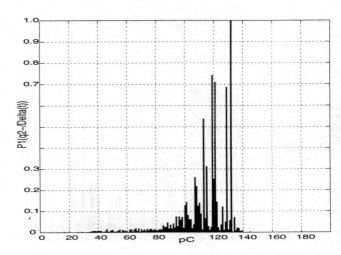

Fig. 9. $P1(q_2^-/\Delta t)$ for Point-Perspex 1mm GAP, 1.1Vi

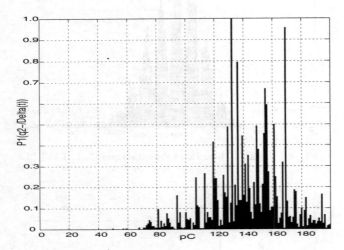

Fig. 8. $P1(q_2^-/\Delta t)$ for Point-Perspex 5mm GAP, 1.1Vi

a most probable Δt. However the most probable q_2^- need not always follow the most probable Δt. But if one works out a joint probability, one can find out the most probable q_2^- and Δt, following q_1^-. This is when the joint probability $P1(q_2^- / \Delta t) \times P(\Delta t)$ is a maximum. This value is 3.6×10^{-2} for 1 mm gap (Table-II). It can be seen from Table-II that,

1. the time lag increases with decrease in the gap spacing at all voltage levels. This reflects the increasing effect of spacing on the gap characteristics that might occur due to the interaction of the discharges on the surface of the dielectric.

2. the overstressing of the gap tends to decrease the time lag in the range 1 to 3 mm of the gap spacing.

At higher spacing (4 and 5 mm) the effect is not pronounced which fairly reflects the ineffectiveness of the dielectric surface at higher spacings.

From Table-I we can infer that,

1. the most probable q_1^- increases with decrease in gap spacing.

2. there is positive correlation between q_2^- and q_1^-.

3. the value of Δt increases with decrease in gap spacing.

4. the time lags seem to stabilize to thier characteristic values for lower gap spacings even at overvoltages.

Conclusions

It is shown that the effect of dielectric on the recurrence of corona pulses can be studied by considering the conditional distributions of the first two pulses in the negative half cycle and the time difference between them. The analyses of the parameters show the effect of the previous discharges on the following events.

References

1. R.J.Van Brunt, E.W.Cernyar, P.Von Glahn, "Importance of Unravelling Memory Propagation Effects in Interpreting Data on Partial Discharge Statistics", IEEE transactions on Elec.Insul., Vol.28, pp.905-916, 1993.

2. R.J.Van Brunt, M.Misakian, S.V.Kulkarni, V.K.Lakadawala, "Influence of a Dielectric Barrier on the Stochastic Behavior of Trichel-pulse Corona", IEEE transactions on Elec.Insul., Vol.26, pp.405-415, 1991.

3. P.Von Glahn and R.J.Van Brunt, "Continuous Recording and Stochastic Analysis of PD", IEEE transactions on Dielectrics and Elec.Insul., Vol.2, No.4, pp.590-601, 1995.

IN-SERVICE MEASUREMENTS OF UHF ATTENUATION IN A GAS INSULATED SUBSTATION

M D Judd[1], J S Pearson[1], O Farish[1] and B M Pryor[2]

[1]University of Strathclyde, UK   [2]ScottishPower, UK

Abstract

An attenuation survey was carried out for UHF signals in a 400 kV gas insulated substation (GIS). The GIS is fitted with UHF couplers that form part of a partial discharge monitoring system. An impulse train generator was used to inject 15 V impulses into the couplers at a repetition rate of 100 MHz. The output signals at couplers further along the GIS were recorded using a spectrum analyser. Signal attenuation was calculated by summing the input and output powers for each harmonic. Differences in attenuation between the three phases are evaluated. Results confirm that the losses cannot be accounted for simply by assigning a fixed attenuation to each GIS component. However, the spread of the measurements was less than might be expected for such a wide variation in signal paths.

1. Introduction

The GIS in question was fitted with a continuous UHF monitor 14 years ago and the system has a long and proven track record. Therefore, we planned to gather data on the signal attenuation in this GIS for comparison with results from other substations and with laboratory measurements. Constraints on GIS construction mean that couplers cannot always be located at equal intervals along the busbar. For this reason it is interesting to measure the differences in signal attenuation between adjacent couplers. To perform the measurements, signals were injected into one coupler and recorded at the next coupler using a spectrum analyser, as in [1]. In all, eight different signal paths were tested on each of the three power phases.

The input signal was generated using an impulse train generator (ITG), which is a resonant device incorporating a step recovery diode [2]. This small module is driven using a 100 MHz sinusoid and it generates a 15 V pulse on every cycle of the driving signal. The pulse width is about 130 ps and the output spectrum contains harmonics of the 100 MHz input signal that extend to frequencies beyond 18 GHz.

This method of pulse injection was investigated because of the relatively low voltage that must be injected into the coupler to obtain a measurable output on the spectrum analyser when compared with pulse units based on avalanche transistors or mercury switches, which have a lower repetition rate. As pointed out in [1], the use of a high repetition rate does alter the amplitude of the measured signal

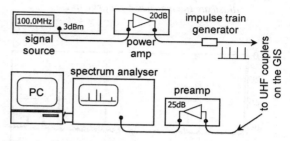

Fig. 1   Equipment for signal injection and detection.

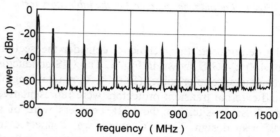

Fig. 2   Output spectrum of the impulse train generator (attenuated by 30 dB).

spectrum. For this reason we will focus primarily on relative attenuation measurements by comparing the signals on different phases.

2. Experimental procedure

The test equipment used to carry out the measurements is shown in Fig. 1. The impulse train generator requires a 100 MHz sine wave input with a power of 200 - 500 mW into 50 Ω. In these experiments, the driving signal was generated using a programmable signal source and a medium power UHF amplifier. However, the same signal could in principle be generated with a much simpler dedicated unit.

The ITG output is shown in Fig. 2, as recorded using a spectrum analyser with a 30 dB calibrated microwave attenuator in series to protect the analyser. The attenuator calibration data was then used to determine the signal power at the ITG output. The input power P_i was defined as the sum of the individual powers of the harmonics from 300 MHz to 1400 MHz inclusive. Based on the measurement shown in Fig. 2, we find that

$$P_i = 11.4\,\text{mW} \qquad (1)$$

High Voltage Engineering Symposium, 22–27 August 1999
Conference Publication No. 467, © IEE, 1999

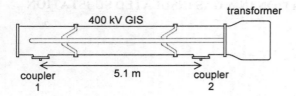

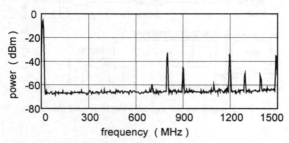

Fig. 3   Section of GIS used in the laboratory to illustrate the measurement principle.

Fig. 4   Output from coupler 2 in the laboratory test.

Fig. 5   Test equipment in use during the on-site attenuation survey.

To demonstrate the technique, consider the section of 400 kV GIS shown in Fig. 3. The GIS conductor diameters, coupler cables (RG214, 36 m total length) and external transient protectors were representative of those on-site, but the UHF couplers were of a different design. The ITG output was connected to the cable from coupler 1 and the spectrum analyser was connected to the cable from coupler 2. The resulting spectrum is shown in Fig. 4, from which the output power was determined by summing the contributions from all the harmonics in the range 300 - 1400 MHz that are visible above the noise floor, giving

$$P_o = 0.91 \ \mu W \qquad (2)$$

From (1) and (2), the attenuation A is calculated as

$$A = -10 \log\left(\frac{P_o}{P_i}\right) = 41 \ dB \qquad (3)$$

Note that this result does *not* imply that the UHF attenuation of 5.1 m of GIS is 41 dB. Included in this figure is a large offset that represents fixed losses, the major one being the inefficiency of the coupler at radiating the signals. In fact, previous experiments in this particular test rig have shown an attenuation of only about 5 dB when the UHF signal is excited by a PD in the chamber containing coupler 1. However, if we now changed the layout of the GIS to which the couplers were attached and repeated the test, we could measure *changes* in signal attenuation because the fixed losses would remain the same.

In effect, the measurement involves sampling the frequency-domain response of the GIS at 100 MHz intervals. A broadband UHF signal detection system

Table 1 - List of experiment numbers and details of the intervening discontinuities

Expt. No.	Circuit breakers (CB)	90° bends (excluding 2 per CB)	Branches
1	0	0	0
2	0	5	2
3	1	0	1
4	1	0	2
5	1	0	1
6	1	0	1
7	0	2	2
8*	0	4	1

* there was one coupler in between the two couplers used for this test on a longer section.

is assumed, so that if a signal appears above the noise, it will be detected regardless of its frequency as long as it lies within the specified range.

3. Results of on-site tests

The test equipment and a node of the UHF monitoring system inside the substation building are shown in Fig. 5. Most of the UHF couplers are mounted at a high level, so it is necessary to inject signals into the cables leading from the node boxes to the couplers. All the coupler cables have a standard length of 15 m to simplify time-of-flight measurements for locating defects. Therefore, any attenuation introduced by the cables will be common to all measurements. The configuration of the GIS section between each pair of couplers was recorded in terms of the numbers of corners, circuit breakers and branches that were present. This information is summarised in Table 1. Due to the method of GIS construction, positions of the internal insulating spacers were not apparent. However, when

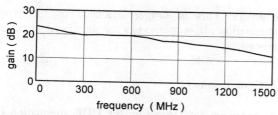

Fig. 6   Frequency response of the UHF preamplifier and the 24 m length of RG214 cable.

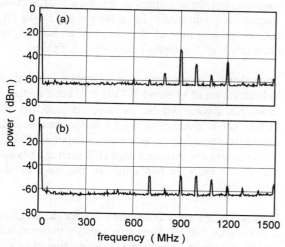

Fig. 7   Output from couplers recorded on site. (a) Expt. No. 2, yellow phase. (b) Expt. No. 5, blue phase.

Table 2 - List of experiment numbers with intervening distances ℓ and the measured signal attenuation A.

Expt. No.	PHASE: RED		BLUE		YELLOW	
	ℓ (m)	A (dB)	ℓ (m)	A (dB)	ℓ (m)	A (dB)
1	6.9	51.0	6.9	51.6	6.9	52.4
2	22.5	68.4	15.5	58.0	19.7	60.6
3	20.3	63.9	20.3	69.6	20.3	65.5
4	16.9	69.7	19.9	79.3	18.3	75.3
5	12.5	65.3	9.5	70.5	11.0	69.1
6	19.0	68.2	19.0	70.0	19.0	65.2
7	9.0	62.1	15.1	71.3	12.0	63.8
8*	33.7	75.6	34.6	75.4	33.4	78.9

* there was one coupler in between the two couplers used for this test on a longer section.

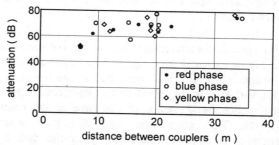

Fig. 8   Measured signal attenuation as a function of the length of GIS busbar.

comparing measurements between different phases, it is likely that the intervening components will be the same even though their spacing may be different.

The measurements involve high levels of attenuation, so a preamplifier was used to boost the signals emerging from the second coupler before they were displayed on the spectrum analyser. To make a connection to the distant coupler, it was necessary to use a 24 m length of RG214 cable in addition to the preamplifier. The frequency response of the cable and preamplifier combination was measured using a network analyser. The overall gain, shown in Fig. 6, was subtracted from the measured results to recover the true signal levels emerging from the output coupler's cable at the node box.

With the preamplifier connected, the noise floor of the signal on the spectrum analyser was about -63 dBm. No interference signals were visible above this level at frequencies up to 1500 MHz. The spectrum of the coupler output signal was recorded for each of the 24 coupler pairs and the data files were stored on a PC for processing. Two examples of the measured signals are shown in Fig. 7.

The output power P_o for each measurement was calculated as follows: If the amplified coupler output

at a particular harmonic was below -60 dBm, this was regarded as being too close to the background noise for reliable signal detection. Otherwise, the peak amplitude of the harmonic was noted and the gain of the preamplifier/cable combination at that harmonic was subtracted. The power contributed by all the harmonics from 300 to 1400 MHz was summed and divided by P_i to give the measure of attenuation that is used here. By this means, attenuation levels of up to 80 dB could be measured.

The lengths ℓ of the busbars between each pair of couplers were determined with the aid of a scale drawing. The measured attenuation values and lengths are listed for each phase in Table 2. Fig. 8 shows the attenuation figures plotted against the length of busbar separating the couplers. This takes no account of the complexity of the signal path in terms of discontinuities but does indicate a general trend of increasing attenuation with ℓ.

4. Analysis and discussion

The results are presented in a more useful way in Fig. 9, where the all values for A and ℓ are plotted relative to the other two phases. Thus all the results for GIS sections where the lengths of each phase were identical appear on the x-axis at $d = 0$. The best linear fit to this data indicates an overall attenuation

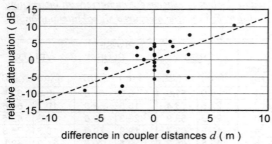

Fig. 9 Inter-phase comparison of the attenuation data, showing the best linear regression fit.

rate of 1.3 dB m^{-1}, but clearly the situation is more complicated than can be adequately represented by such a figure.

The spread of attenuation values on the vertical scale in Fig. 9 indicates how signal levels can vary even for apparently identical signal paths. The reason for this variation is the effect that components beyond the section being measured can have on the result [1,3,4]. This conclusion was confirmed when attempts to improve the linearity of the data by assigning fixed attenuation values to each of the discontinuities listed in Table 1 proved unsuccessful. Excluding Expt. No. 1, which will be discussed separately, the maximum difference in attenuation between adjacent couplers on any phase is 21 dB. This includes geometry variations from 0 to 5 corners, 0 to 2 busbar branches and 0 or 1 circuit breakers. Thus, while a simple interpretation of the attenuation figures is not possible, they do indicate that none of these GIS components is introducing a blind spot into the PD detection system.

The measurement that most closely resembles the laboratory one is Expt. No. 1, where the couplers are 6.9 m apart in a straight section of GIS. The average attenuation here is about 10 dB more than was measured in the laboratory. This difference may in part be due to the different coupler designs; however, it is also likely that the much longer busbars on either side of the couplers on-site results in a greater leakage of signal energy away from the compartments containing the couplers.

A brief comment with regard to the result of Expt. No. 8 (see Table 2) is called for. In this instance, the second couplers were not at the next node along the busbar, but at the one beyond it. Although the attenuation is greater than for any of the other results, it is certainly not twice the typical attenuation of 65 dB measured between adjacent couplers. This is further evidence of the fixed attenuation offset that exists for the signal injection procedure.

5. Conclusions

A technique for measuring the UHF attenuation of GIS using a 15 V impulse train generator has been presented. The results show a typical total attenuation in the range 60 - 70 dB between adjacent couplers, but these figures do include a large fixed attenuation (estimated to be in the range 35 - 40 dB) due to the inability of the couplers to radiate the necessary high frequency components efficiently.

Inter-phase comparison of signal attenuation indicates a trend of about 1.3 dB m^{-1}, but deviations from this figure can be as large as ± 6 dB for differences in distance of only a few metres, despite the similarity of the intervening components. This demonstrates the influence that GIS sections beyond the section between two couplers can have on the UHF signal.

In practice, the least favourable site for detection of a PD source will be when it is located between two couplers. Our measurements indicate that typically the signal emitted by a defect would be subject to an attenuation of no more than 15 dB before it reached the nearest UHF coupler.

6. Acknowledgements

M. Judd acknowledges the support of the EPSRC under research grant No. GR/L34785. The authors would like to thank Scottish Power plc for permission to publish this material and Mr J Mesa for assistance with the measurements.

7. References

[1] S Meijer, E Gulski, J J Smit and A J L M Kanters, "Determination of PD sensitivity of GIS using signal reduction measurements", Conf. Record IEEE Int. Symp. on Electrical Insulation (Washington), Vol. 1, pp. 65-68, 1998

[2] Hewlett-Packard Co., *Communications Components*, Part No. 33002A, 1993

[3] R Kurrer and K Feser, "Attenuation measurements of UHF partial discharge signals in GIS", Proc. 10th ISH (Montreal), Vol. 2, pp. 161-164, 1997

[4] M D Judd, B F Hampton and W L Brown, "UHF partial discharge monitoring for 132 kV GIS", Proc. 10th ISH (Montreal), Vol. 4, pp. 227-230, 1997

PD-Pattern of defects in XLPE Cable Insulation at different Test Voltage shapes

D. Pepper, W. Kalkner

Department of High Voltage Engineering, Technical University of Berlin, Germany

Abstract

There is a growing interest in onsite PD mesurement of XLPE-insulated medium voltage cable systems. Due to the high capacitive load, these cable systems are voltage tested using either resonant circuits or very low frequency (VLF) generators with sine or cosine-rectangular voltage shape.

The aspect of PD behaviour at such voltages is investigated. PD measurements on XLPE test samples are presented at test voltages with variable voltage shape and frequency. In addition to recent work, the main focus is on voids in XLPE as a typical defect. PD fingerprint as well as PD pulse rates are considered.

1. Introduction

Onsite testing of medium voltage cable systems is still not an easy task. Today there are mainly two testing techniques used for XLPE insulated medium voltage cable systems, the measurement of dielectric properties of the insulation (e.g. measurement of loss factor at VLF, return voltage or depolarization current) and the measurement of partial discharges.

PD measurement allows the detection and localization of small defects (e.g. voids) inside the insulation of the cable or its accessories. The test voltage is, in respect to the high capacitive load, either generated by a resonant circuit or very low frequency (VLF 0,1Hz) with sine or cosine-rectangular voltage shape is used. Therfore in this paper the PD measurement at different test voltage shapes is considered. Parameters of PD activity (level of apparent charge, pulse rate, fingerprints) are observed while varying parameters of the test voltage (shape and frequency). Furthermore, by varying the test voltage it is attempted to gain more insight in physical parameters of PD activity.

2. Test samples

Continuing the work of [1], where the PD behaviour of cable defects like a not assembled cable termination or a defect of the outer cable shield was observed, focus in this paper is on voids in XLPE. Specimen are disks of 1mm thickness and 60mm diameter which contain a spherical void of 0.5mm to 0.75mm diameter. The XLPE disks are placed between two electrodes of 40mm diameter forming a homogeneous electrical field. The whole arrangement is put into insulating oil to prevent surface discharges. The samples are very suitable to simulate voids in real XLPE cables since the voids are spherical.

In addition, the samples are relatively robust against electrical treeing which gives the possibility to perform PD measurements on samples which change only little during the measuring time.

3. System for PD measurements

The measurement setup is described in [1] and mainly consists of three parts. First part is a very low noise high voltage source, capable of generating test voltages up to $35kV_{peak}$ with a maximum current of $30mA_{peak}$ at variable frequencies in the range from DC to 1kHz. Any voltage shape can be generated depending on the software for the controlling digital signal processor (DSP). This high voltage source is described in more detail in [2].

The second part consists of a high voltage coupling capacitor, a PD coupling unit (bandwidth 100kHz to 10MHz) an (optional) amplifier which broadens and rectifies the PD pulses, and an oscilloscope (LeCroy 9310) which digitizes the PD signal and provides a GPIB interface to a personal computer.

Finally the PD activity is stored and visualized on a personal computer using the software tool 'TEmess' which was developed at the Technical University of Berlin as well as the High Voltage Generator. Among other options (e.g. controlling of the oscilloscope, localization of PD, partly automatic calibration, ...) the software is able to show PD activity in three different ways: 1) apparent charge versus test voltage phase (used in this paper); 2) apparent charge versus time (indicates change of specimen); 3) apparent charge by time versus phase of test voltage (indicates change of specimen, too).

4. Measurements

All measurements were performed using test voltage shapes as already known respectively generated also of commercially available high voltage sources (despite the ability of the system to generate other, even unusual test voltage shapes, too). Among others, the sinusoidal and cosine-rectangular voltage shapes were chosen for measurements. At cos.-rect. voltage shape two parameters were varied, the fundamental frequency and the time function of the change of voltage polarity (usually part of a sinusoidal from $\frac{1}{4}\pi$ to $\frac{3}{4}\pi$ for the falling edge), here

High Voltage Engineering Symposium, 22–27 August 1999
Conference Publication No. 467, © IEE, 1999

called slope frequency. At sinusoidal test voltage the frequency is varied, too.

4.1. Fingerprints

Examples for the measured PD patterns are presented in fig. 1, 2, 3 and 4. All figures show the apparent charge versus the phase of the test voltage. The fingerprints are displayed in grayscale where a darker area indicates a higher number of pulses with an equal amplitude and phase position. In figure 1 and 2 the parameter is the fundamental frequency, in figure 3 and 4 the slope frequency is varied. It can be seen, that the higher slope frequency leads to a higher level of apparent charge while the fundamental frequency has almost no influence on the PD level.

The fingerprints with variable slope frequency were measured with the amplifier/rectifier, the fingerprints with variable fundamental frequency could be measured without the amplifier. The reason is the limited memory of the oscilloscope which makes the amplifier with its feature of pulse broadening necessary at low fundamental frequencies. Therefore, the shape of the fingerprints look a little different depending on the use of the amplifier but are in general very similar to each other.

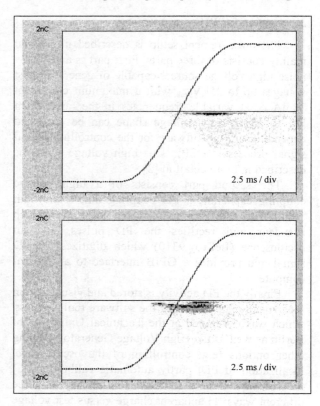

Fig. 1: *Fingerprint of a sample with a void in XLPE (apparent charge versus phase)*
*test voltage: cos.-rect. (**rising edge**) 15kV$_{peak}$*
upper: 0.1Hz / 50Hz
lower: 10Hz / 50Hz
slope frequency constant 50Hz
fundamental frequency varied

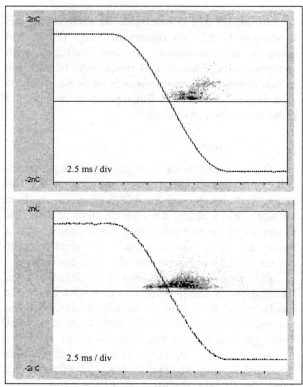

Fig. 2: *Fingerprint of a sample with a void in XLPE (apparent charge versus phase)*
*test voltage: cos.-rect. (**falling edge**) 15kV$_{peak}$*
upper: 0.1Hz / 50Hz
lower: 10Hz / 50Hz
slope frequency constant 50Hz
fundamental frequency varied

4.2. PD pulse rates

The number of PD pulses are recorded for a void in XLPE at variable frequencies of the sinusoidal test voltage as well as for cos.-rect. test voltage at variable fundamental and slope frequency. The results are shown in fig. 5, 6 and 7.

At sinusoidal test voltage the number of PD pulses per period is decreasing with rising frequency as well as for rising slope frequency at cos.-rect. test voltage. Frequencies above 100Hz are not displayed here since there was hardly any PD activity at higher frequencies.

When the fundamental frequency of the cos.-rect. test voltage is varied, there is a maximum of PD pulses around 1Hz. Rising the fundamental frequency close to the slope frequency (slope frequency is two times higher than fundamental frequency) leads to a character of PD activity similar to sinusoidal test voltage since the voltage shape becomes almost sinusoidal. Therefore the PD activity almost stopped

at a test voltage with a fundamental frequency of 40Hz and a slope frequency of 50Hz. To obtain a pulse rate similar to cos.-rect. test voltage the voltage of the sinusoidal test voltage was set to $30kV_{peak}$.

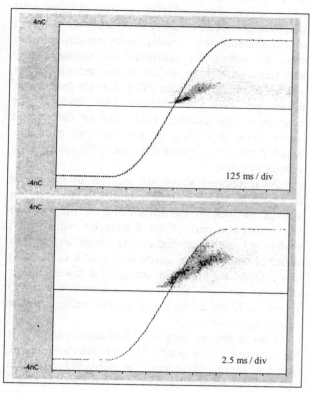

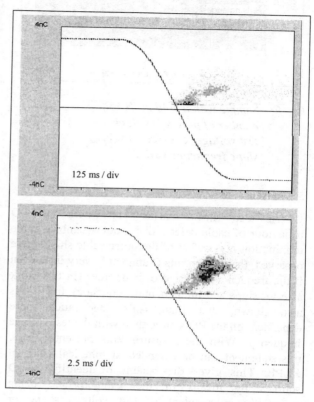

Fig. 3: Fingerprint of a sample with a void in XLPE (apparent charge versus phase)
test voltage: cos.-rect. (**rising edge**) $25kV_{peak}$
upper: 0.1Hz / 1Hz
lower: 0.1Hz / 50Hz
fundamental frequency constant 0.1Hz
slope frequency varied

Fig. 4: Fingerprint of a sample with a void in XLPE (apparent charge versus phase)
test voltage: cos.-rect. (**falling edge**) $25kV_{peak}$
upper: 0.1Hz / 1Hz
lower: 0.1Hz / 50Hz
fundamental frequency constant 0.1Hz
slope frequency varied

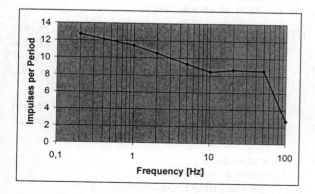

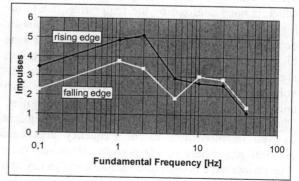

Fig. 5: Sample with a void in XLPE
number of pulses per period
test voltage: Sin $30kV_{peak}$

Fig. 6: Sample with a void in XLPE
number of pulses per slope
test voltage: cos.-rect. $15kV_{peak}$
fundamental frequency varied

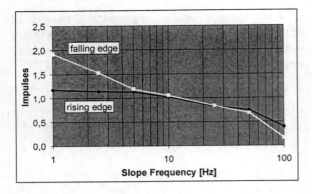

*Fig. 7: Sample with a void in XLPE
number of pulses per slope
test voltage: cos.-rect. 15kV$_{peak}$*
slope frequency varied

5. Discussion

As also shown earlier [1] where the PD behaviour of cable defects like a not assembled cable termination or a defect of the outer cable shield were observed, the fingerprints change only very little even if the frequency is varied over decades (in this work also for slope and fundamental frequency). In [1] it is also shown, that fingerprints alter much more dependent on the PD source than with the test voltage frequency. With the measurements presented here, this statement can be extended to spherical voids in XLPE. This gives the benefit of identifying PD sources by means of their fingerprints, being essentially independent of test voltage shape or frequency (for cosine-rectangular and sinusoidal test voltages).

Also it could be deducted that the physical parameters of PD activity which are important for the shape of fingerprints are frequency independent (even if this seems not to be probable). During measurements it could be observed, that PD pulses occur with a certain time-distance between them which is not visualized by the fingerprints due to the high amount of PD pulses and their stochastic occurence. Also other researchers come to a new kind of visualization of PD activity [3] where PD pulses are visualized in dependency of the change in test voltage (Δu). This is supposed in [3] to be more exact and clear when PD sources have to be identified. Thus, fingerprints of the kind used in this work might as well be insensitive to the frequency dependent physical parameters of PD activity. However, these fingerprints have proven to be suitable to identify and distinguish PD sources from each other. It seems that fingerprints can be used for this purpose furthermore even with varying the frequency and shape of test voltages.

On the other hand there is a variation of the rate of PD pulses and the level of apparent charge with the variation of frequency. In general, there is a lower number of PD per period or slope in XLPE voids with higher (slope-) frequencies which was expected due to the rising gas pressure level inside the void with rising test voltage frequency. The higher gas pressure level strengthen the insulation capabilities of the void. This means that resonant test voltage generators which supply frequencies up to 300Hz (sinusoidal shape) might not be effective to detect voids by means of PD measurement. During measurements for this work the voltage for sinusoidal test voltage shapes was twice as high as for the cosine-rectangular test voltage to obtain a similar PD pulse rate (number of pulses per period or slopes). This leads to the conclusion that existing voids can be detected at lower stress depending on the test voltage shape, which could be an important aspect for testing high voltage systems.

As mentioned above, the number of PD pulses decreases for higher slope frequencies at cos.-rect. test voltage. A look at the fingerprints in fig. 3 and 4 show, that the part of the fingerprint with a high probability occurs at higher PD levels and is thus easier to detect during onsite testing of a cable. This can compensate the disadvantage of a lower number of pulses per slope and make the detection of a PD source as likely as for cos.-rect. test voltage with a lower slope frequency.

One important aspect of frequency variation is the measuring time onsite. If fingerpints are needed, one has to record several thousand PD pulses to obtain a meaningful fingerprint. Therefore in general one would have to face longer measuring times for lower test voltage frequencies. Partly this is compensated by the higher number of PD pulses at lower frequencies.

References

[1] D. Pepper, W. Kalkner: PD-measurements on typical defects on XLPE-insulated cables at variable frequencies, Int. Symp. on High Voltage Engineering, ISH 1997, Montreal, Canada

[2] H. Emanuel, M. Kuschel, C. Steinecke, D. Pepper, R. Plath, W. Kalkner: A New High Voltage Dielectric Test System for Insulation Diagnosis and Partial Discharge Measurements, Nordic Insulation Symp., Bergen, Norway, 1996

[3] H. Martin: Impulsfolgen-Analyse: Ein neues Verfahren der Teilentladungsdiagnostik, Dissertation Universität-Gesamthochschule Siegen, Germany, 1997

Address of authors:
Technische Universität Berlin
Institut für Elektrische Energietechnik,
Fachgebiet Hochspannungstechnik (Sekretariat HT3)
Einsteinufer 11, D-10587 Berlin

Tel.: ++ 49 30 - 314 23470
Fax: ++ 49 30 - 314 21142
e-mail: Pepper@ihs.ee.tu-berlin.de
WWW: http://ihs.ee.tu-berlin.de

SENSITIVE ON-SITE PD MEASUREMENT AND LOCATION USING DIRECTIONAL COUPLER SENSORS IN 110 kV PREFABRICATED JOINTS

P.Craatz

Bewag
Net Center, Section
CableTechnique
Berlin

R.Plath

IPH
High Voltage Lab
Berlin

R.Heinrich, W.Kalkner

Technical University Berlin,
Institute for Electrical Power
Engineering

Abstract

To verify the reliability of 110 kV XLPE cable joints a long term prequalification test was carried out at IPH. 12 joints of various manufacturers were integrated in a circular 110 kV XLPE cable test loop. The short distances between the joints required a selective PD measurement and localisation location to measure PD from each joint separately. Therefore directional coupler sensors (DCS) were integrated at each joint to provide a sensitive and unequivocal PD measurement and localisation.

1. Introduction

Based on the experience with prequalification tests on 400 kV XLPE cable systems [1], which showed partly bad performance of lapped joints compared to prefabricated and pretested joints, Bewag decided to use only prefabricated joints also in its 110 kV cable network in future. In order to verify the reliability of available 110 kV joints, Bewag Berlin charged the IPH test lab in Berlin to carry out a prequalification test on prefabricated joints for 110 kV XLPE cables under on-site conditions.

One aim of this prequalification test was to check the overall performance of the different types of joints to choose the preferred joint suppliers. Besides, Bewag intends to assemble the 110 kV joints using its own staff. So the second aim of the test was to prequalify Bewag staff for assembling these types of joints. Before starting the prequalification test, the joint manufacturer qualified Bewag jointers. Bewag teams supervised by manufacturers experts assembled all joints in the prequalification test at IPH. So Bewag gained experience about the different assembling procedures and the handling.

2. On-site test set-up and test parameters

12 joints of 6 different manufacturers were installed on a cable loop of 200 m length. The cable loop consisted of two 110 kV XLPE cables from different manufacturers but with the same dimensions and construction. The conductor cross section was 630 mm^2 (copper), the insulation thickness was 18 mm. Besides, the cables offered longitudinal water tightness in the screen area and a laminated aluminium sheath with outer PE sheath. The tested joints were installed each on both cables (2 x 6) to check, whether there is any influence of the cable type. The cable length between two joints was 15 m. The complete cable loop was laid in the ground.

During the commissioning test, an AC voltage up to 3 U$_0$ (U$_0$ = 64 kV) was applied. Sensitive PD measurements (sensitivity < 2 pC under on-site conditions, depending on the location of the joint) were carried out on all joints using directional coupler sensors, which were installed left and right hand side of each joint. So it was easy to distinguish between PD in two neighboured joints even close to the sealing ends where outer interference is not damped due to the short cable length in between.

After successful commissioning tests, the long-time test (180 load cycles, 2.5 U$_0$) was started to simulate the stress of 50 years of operational service. An 8/16h load cycle was chosen, according to thermal time constants of the cable. The maximum cable conductor temperature during load cycles was 90...95 °C. A few load cycles were carried out to reach even higher temperatures (100...105 °C). Two different temperature measurement systems were used to measure cable sheath and screen temperature. The screen temperature was measured by optical fibres included in the cable screen, enabling hot spot detection for the complete loop. PE sheath temperature was measured with Pt1000-based active sensors with ultra low power consumption and connected via optical links to avoid any interference in case of breakdown.

Calculations for the layout of the cable heating circuit gave relatively high impedance for the big test loop, which would have required a large number of heating transformers and compensation capacitors for AC heating. Therefore a special DC heating circuit was installed to minimise the number of heating transformers and reduce the costs. A very short 110 kV cable loop feeds into a bridge rectifier consisting of 8 high current diodes. The rectifier has to operate on test voltage potential (160 kV). The output of the rectifier is connected to the 200 m test loop. Thus, no compensation capacitors were needed and the number of required heating transformers was reduced by factor of 8 compared to AC heating. The reduction factor depends mainly on the impedance relationship of the two cable loops. In case of DC heating, the impedance of the big test loop leads to a smoother current. On the other hand, two additional outdoor sealing ends were required for the short (AC) loop. Bewag used four different outdoor sealing ends in the test set-up in order to gain more experience with different termination constructions, too.

To determine the thermal drop between conductor and screen (fibre-optic temperature measurement system) resp. between conductor and outer sheath (Pt1000 measurements) a dummy test was carried out, too.

The test set-up, especially the high voltage connection between transformer and cable loop provides unfortunately a good antenna for various noise sources.

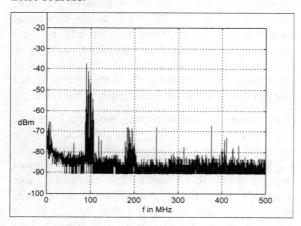

Fig. 1: Spectral characteristics of distortions

The most severe distortions were identified as radio broadcasting at around 100 MHz, short wave at 10 MHz and TV broadcasting in the range from 170 MHz to 230 MHz and above 400 MHz (fig. 1).

3. Directional Coupler Principle and Properties

A directional coupler is a passive device, which is commonly used in a large variety of RF applications [2]. Typical operating frequencies are in the higher MHz to GHz range.

The directional coupler principle can be applied for signal detection (e.g. partial discharges [3]) on high voltage cables using a special sensor with 2 outputs [4], [5]. Depending on the pulse travelling direction, energy is coupled to different output ports. Thus the origin of the PD pulses can be determined.

During installation of the cable line the correct mounting and the properties of each directional coupler have been checked. It was found that the characteristics of the directional coupler sensor are almost independent from the construction of the joint. However, the properties of the high voltage cable have significant influence on the characteristics of the sensor, especially on the directivity. In order to achieve maximum sensitivity and directivity, the directional coupler sensor is usually adjusted to the properties of the cable. However, this optimisation was only done for one of the two different cables used in the test. Thus, a part of the sensors had limited performance.

To test the characteristics of a directional coupler sensor, a calibration signal was injected at port B (see fig. 2) of the directional coupler left from the joint.

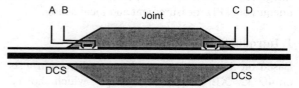

Fig. 2: Directional coupler sensor (DCS) in a joint

The signal coupled through the joint to the second directional coupler was measured using a broadband oscilloscope (LeCroy LC574AL). Fig. 3 shows, as an example, typical output signals of a directional coupler.

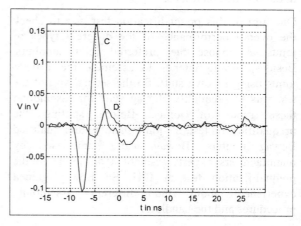

Fig. 3: Typical output signals of a directional coupler PD sensor (signal injected at port B, fig. 2)

The calibration signal, injected at port B propagates through the joint and is decoupled at port C (close to the joint, see fig. 2), whereas almost no signal is coupled out at port D. This indicates that the pulse came from the left of this coupler (joint). A signal from the right (e.g. an external distortion) would be coupled at the ports D and B. The possible coupling opportunities are summarised in the table below.

	A	B	C	D
Signal from left	x		x	
Signal from right		x		x
Pd from the joint		x	x	

Thus an unequivocal and reliable discrimination of PD from external noise and from PD of adjacent joints is possible. This is entirely important for the given test set-up to differentiate between the joints.

4. PD measurement and results

4.1 PD measurement

Since no broadband measurement system with multiple channels and remote control was available at that time, when the PD measurement was carried out, each joint was measured separately using broadband amplifiers and an oscilloscope.

The measurement was calibrated using a calibrator with less than 1 ns rise time. Taking the coupling attenuation of the directional coupler sensor and the amplification of the broadband amplifier into account the measured output voltage at the oscilloscope was 30 mV per pC in the cable. This relation is only valid, if the pulse width of the calibration signal corresponds to the PD pulses. However, the calibrator used for this measurement fulfilled this condition.

The PD measurements at each joint separately may be regarded as not suited because not all joints can be monitored at the same time. However, because of the short distance between the joints, severe PD faults from adjacent joints would be discovered in spite of the attenuation of the high voltage cable and joints. The origin of PD can be determined using the directional coupler technique, which allows to separate PD of the measured joint from other signals from right or left.

Initially, all joints were measured at U_0 ($U_0 = 64kV$) No PD of more than 2 pC, depending on the location of the joint, was measured. However, PD was measured at higher voltage levels.

The realised sensitivity was depending slightly on the location of the joints. The high noise coupling due to the high voltage test set-up lead to a noise floor, which reduced the sensitivity in spite of the selective coupling properties of the directional coupler sensors. At one end of the cable loop, close to the high voltage connection, the maximum sensitivity was about 6 pC at the first joint, increasing to < 1 pC in the middle of the cable loop due to the attenuation properties of the high voltage cable and the high operating frequency of the directional coupler sensors. At the other end of the cable loop, close to the heating circuit, the noise coupling reduced the sensitivity only up to 2 pC, because of the attenuation through the cable of the heating circuit.

Fig. 4 shows measured PD at one joint at 2.5 U_0. The peak value was about 54 pC at 2.5 U_0. In spite of the limited directivity of the not optimised directional coupler sensors at this joint it can be seen, that the output signal at port B is larger than at port A resp. output at port C is larger than at D. This indicates PD from the joint.

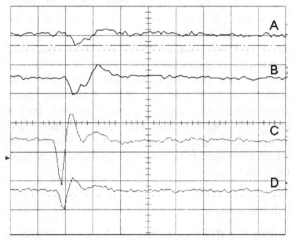

Fig. 4: PD at 2.5 U_0 (Amplitude: 500mV/div, time 10ns/div)

An additional opportunity to discriminate between PD from the joint and external signals is the propagation time consideration. The maximum at port C occurs shortly before D resp. B before A. The time shift between ports B and C can also be used for determination of the exact location of PD in the joint resp. between 2 sensors.

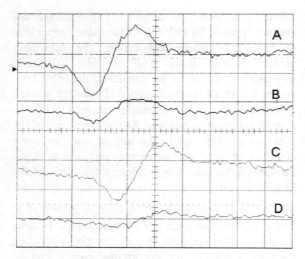

Fig. 5: PD signal from left at 2.5 U_0, (Amplitude: 50mV/div, time 10ns/div)

Fig. 5 shows PD from the joint discussed above measured at a joint located at the other end of the cable loop. In this case, the PD signals are attenuated and widened due to attenuation and dispersion of travelling through 4x15 m cable and 3 joints. The joint itself, however, is PD free: The output signals at port A and C of the directional couplers indicate clearly a signal from left. Port B and D show almost no output signal. The time shift of the output signals at port A and C is about 10 ns, which corresponds to the propagation time through the joint.

4.2 Test results

During the first two month of the prequalification test, three breakdowns on joints occurred. The first two breakdowns were probably caused by problems with the compression connectors of the conductor, leading to very high temperatures inside the joint. In the first case, joint inspection show irregular mechanical stress on the connector and cable, leading to increased contact resistance. In the second case, the cause of the fault was too low pressure on the conductor connector during joint assembling, also with the consequence of increased contact resistance. Unfortunately, the optical fibre temperature measurement system was not able to detect these hot spots clearly before breakdown, because the path of the fibre was positioned close to the joint enclosure and therefore not sensitive enough to detect hot spots inside the joints. Besides, no drop curves (difference temperature between conductor and fibre position) are available by now.

The third breakdown was evaluated as a pure electric one. In this case, PD was detected several days before the breakdown occurred, but not during first commissioning of this joint. In case of the first

two faults, a standard type test (20 cycles, 2.0 U_0) would not have been able to detect them, because they occurred after a longer period with higher voltage (2.5 U_0). No further breakdowns occurred up to the end of the load cycle test.

After the load cycle test, PD measurements at 2.5 U_0 were repeated on all joints. No PD was detected.

During final impulse testing, one joint failed at 90% of 550 kV impulse test level. After repair, the impulse test was continued (550 kV lightning impulse, 550 kV switching impulse test, first positive, then negative polarity). One sealing end failed at 550 kV lightning impulse, positive polarity. Fortunately, this sealing end was part of the AC loop of the heating circuit, so it was disconnected from the main loop to go on with impulse testing without any repair work. Then, a second joint failed after the third 100 % shot, switching impulse, negative polarity.

Conclusions

The prequalification test on 110 kV joints confirms the experience from the former 400 kV tests: running a test under real-life conditions (civil work, assembling and testing on site, cable laying in ground etc.) results in much more experience - and sometimes not expected effects - for the customer than conventional type tests in the laboratory do. Thus a higher level of confidence for the final user can be achieved.

Additional measurements (e.g. sensitive PD detection, hot spot monitoring) lead to experience, which may be helpful for the future to avoid or to predict problems in accessories before a breakdown occurs. Nevertheless, the final impulse testing laid bare weak points in the accessories, which remained uncovered even by sensitive PD measurements.

References

[1] Henningsen C.G., Polster K., Obst D.: Berlin Creates 380-kV Connection with Europe, Transmission & Distribution World, July 1998

[2] Oliver B. M.: Directional Electromagnetic Couplers, Proc. IRE 42 (1954) 1686-1692

[3] Sedding H.G., Campbell S. R., Stone G. C., Klempner G.S.: A new sensor for detecting partial discharge in operating turbine generators, IEEE Trans. On Energy Conversion, Vol. 6, No.4, Dec. 1991

[4] Pommerenke D., Krage I., Kalkner W., Lemke E., Schmiegel P.: On-site PD measurement on high voltage cable accessories using integrated

sensors, International Symp. on High Voltage Engineering, ISH 1995, Graz, Austria

[5]   Pommerenke D., Strehl T., Kalkner W.: Directional Coupler Sensor for partial discharge recognition on high voltage cable systems, International Symp. on High Voltage Engineering, ISH 1997, Montreal, Canada

Address of authors

Bewag AG
Puschkinallee 52, 12435 Berlin, Germany
Fax: ++49 30 267-17843

IPH Institut "Prüffeld für elektrische Hochleistungstechnik" GmbH
Landsberger Allee 378, 12681 Berlin, Germany
Fax: ++49 30 54960-267
e-Mail: plath@iph.de

Technical University Berlin, Institute for Electrical Power Engineering
Einsteinufer 11, 10587 Berlin, Germany
Fax: ++49 30 314 21142
e-mail: rheinrich@ihs.ee.tu-berlin.de

Advanced Measuring System for the Analysis of Dielectric Parameters including PD Events

Lemke E. / Strehl T.

LEMKE DIAGNOSTICS GmbH, GERMANY

ABSTRACT: For preventive diagnosis of HV equipment different measuring procedures are in use, such as the detection and analysis of PD phenomena as well as the measurement of the capacitance and loss factor. The submitted paper reports on the integration of such different measuring systems to a common, compact and computer based device. This offers for the first time the possibility of a simultaneous measurement of both, impedance-parameters of HV insulation and partial discharges. Simultaneously PD faults can be located. For analyzing characteristic PD types a database expert system is integrated. So the global insulation condition can be assessed in a complex manner.

1. INTRODUCTION

Both, the loss factor / impedance measurement and the partial discharge measurement are accepted techniques for investigations on the dielectric properties of high voltage insulation materials. For this purpose, the presented new developed system combines the technology of both proved methods.

Caused by the difference of these methods regarding to the appropriate usage most often are both [1] applied in order to get comprehensive test results. On the one hand the integral overall state of a system and on the other hand the differential local insulation system fault analysis is of interest to estimate the condition of HV systems [2]. Only, when applying both diagnostic methods, a meaningful criterion of the tested insulation system can be found [1]. For obvious reasons, the combination of both measurement techniques into one integrated system gains a high benefit.

2. SYSTEM ARCHITECTURE

The system is characterized by a general modular conception. It is designed to implement most of the functionality in digital components, exclusively. Analog parts are used only on a small indispensable scale.

The platform of the user interface for both systems is a conventional iX86-based computer system. The signal preprocessing is managed by separate independent DSP units. These units feed the main system bus of the computer. Parallel running real-time NT kernel drivers transport the preprocessed compressed data to a Windows-based user front-end. Different input units with selectable digitizing characteristics are realized. The implementation of different acquisition units is cascadable.

2.1. LOSS FACTOR / IMPEDANCE MEASURING TECHNOLOGY

Obliged by many inconveniences of the classical and traditional bridge technology based on the Schering-idea (1919) and it's improvement in the 1960-ies by Kuster-Peterson using the current comparator principle, a new system to measure loss factor and other impedance quantities was developed. The measuring principle [3] is schematically shown in figure 1. The magnitude and phase relation of the two currents flowing in the measurement- and reference branch are continuously measured by two independent potential free, fibre-optical connected and battery supplied active current sensors.

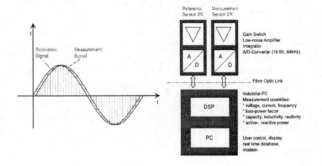

Figure 1: **Concept of the loss / impedance measuring system**

Signal Acquisition, Sampling, Digitizing

The voltage which drops across the equivalent input-capacitance of the two sensors is acquired and compensated by a fine graduate integrating amplifier (see figure 1). The modulation range is automatically controlled by a fast under- overload-recognition system. Following the high-impedance, low-noise amplifier a 2-channel 16 bit A/D-converter and an electro-optical interface is running in each sensor. With a sampling rate of 64 kHz of both channels of the two sensors the digitized signals are transmitted via a fibre optic link to the receiving unit in the main-workstation. For potential-free signal acquisition the sensors are battery powered. Moreover, the battery charge state as well as the internal sensor temperature is monitored.

High Voltage Engineering Symposium, 22–27 August 1999
Conference Publication No. 467, © IEE, 1999

Signal Processing, Signal Compression

The continuously transmitted data stream of the two sensors is buffered and feeds one of the digital signal processing units. The processor performs continuously the spectral dispersion running the DFT (Discrete Fourier Transformation) of the signals of each sensor. The vectorial resolved and spectral dispersed quantities of the two high accurate measured current to voltage converted signals are transmitted to evaluate the requested impedance values.

A real time database on the computer is continuously refreshed with the frequency selective quantities of the complex current values.

For each period of the applied test voltage a data set is added to the measuring data base. From the windows based program platform the user controls on demand which interesting impedance values are displayed, stored or embedded to other application software. The quantities under measurement comprise all sorts of interesting impedance parameters and derived mathematical and statistical parameters.

As shown in figure 2 loss / power factor, capacitance, inductivity, current, voltage, resistance, active / reactive power and charge can be displayed as a function of the frequency. But also drift and standard deviation are put out. All quantities can be displayed graphically as a function of each other (see figure 2) or can be displayed and stored in an EXCEL spreadsheet embedment.

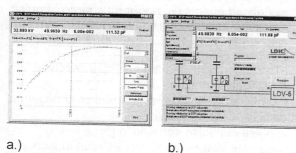

a.)   b.)

Figure 2a: Main window, selection of the displayed quantities
2b: Graph, tan δ vs. voltage

To run a preconfigured measurement procedure [5], set points like timing and interruption commands can be adjusted. Furthermore, these set point variables can also be used to run, e.g., a impedance vs. frequency scan for certain nodes in order to perform a distortion factor measurement.

2.2. PARTIAL DISCHARGE MEASURING TECHNOLOGY

The electrical PD detection is an indispensable tool for quality inspection tests in laboratory after manufacturing and for diagnostic field tests after installation and maintenance. Due to the enormous wide range of applications of PD measuring systems the instrumentation arrangement can be flexible composed according to the particular measuring situation. Hence, the PD-measuring system consists of different package modules.

Hardware Concept

The schematic diagram is shown in figure 3. The following functional units can be distinguished.

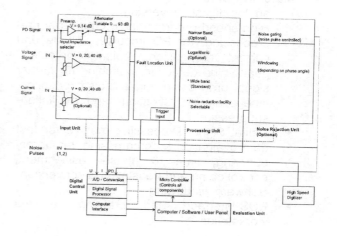

Figure 3: Schematic diagram of the PD-measuring unit

- **Input Unit:**
  - Receiving the signal from the coupling unit.
    (A wide range of different types of coupling units can be connected)
  - adjustable wide band pre-amplification
  - software controlled filter bank
  - tunable attenuator to get a very high dynamic range (93 dB)
  - Voltage input to acquire the instantaneous value of the test voltage, for phase resolved PD-assignment
  - High impedance input to acquire the instantaneous value of the current, for PD-phenomena which appear in dependence of the current values

- **Processing Unit (PU)**
  different types of plug-in units can be assembled and combined
  - wide band PU for very sensitive PD-measuring, including special features to use the wide bandwidth for noise suppression
  - logarithmic wide band PU of a very high dynamic magnitude resolution for processing the „real" individual PD pulses
  - frequency selective PU for narrow band PD-detection, tunable in a wide frequency

range or for several selected fixed frequencies

- **Noise Rejection Unit**
  - phase angle resolved hardware windowing
  - noise impulse gating, triggered by external events or by corresponding gating input units, for masking of the noise signals, sensor / antenna inputs for noise impulse suppression

- **Matching Unit**
  - analog digital conversion
  - High-speed programmable gate array pre-processing, for PD impulse, voltage and current recognition, polarity detection, peak value detection, matching for further signal processing, digital noise reduction, timegating for reflection- and oscillation-suppression
  - Digital Signal Processing module for fast data evaluation & compressing, temporary buffering, time- & phase- assignment, control command translation & conversion
  - Software windowing
  - BUS-Interface and bridging to the workstation

- **Control Unit**
  - µ-Controller based control bus for all devices & components
  - auxiliary ports for additional operation functions

- **external Multiplexing Unit**
  - Software controlled switch over module for multichannel systems
  - remote controlling & routing of the switch-over-device

- **Evaluation Unit**
  - Workstation to perform the complete diagnostic evaluation
  - Real-time displaying of all measured PD-parameters with a pulse repetition rate higher than 100 kHz
  - storing, reporting and user adjustable protocol generator
  - measurement process automation
  - database link, Object link and embedment interface,
  - enhanced analysis & statistical toolbox (Phase resolved 3D Pattern evaluation, etc.)
  - implemented PD-Expert System

Software Concept

The software concept is characterized by strict partitioning of the numeric data processing into different programmable hardware units.
The first numerical data processing is realized in a field programmable gate array (FPGA). After that, the digital signal processing unit (DSP) is responsible to do the PD-pulse to voltage-phase assignment, the control command translation and the precompression for the main processor subsequent treatment. Based on all this progressive software grading the user interface program in the workstation runs exclusively for the front-end application.

Moreover, all control commands for the hardware management are operated by a separate internal µ-controller.

Analysis & Statistical Features

The front-end software is running under a Windows NT operation system. A comprehensive PD-evaluation toolbox [4] is implemented. It serves for all types of analyzing. Typical routines and evaluation-processes for quality tests in a production environment, on site tests or even for periodical or continuous monitoring are already realized. Also features for a wide range of scientific PD-investigations are available [4].

The analysis and statistical toolbox covers the following functionality:

- replay of all PD quantities in compliance with the specifications of the standards (IEC, VDE, AEIC, IPCEA, ASTM, ANSI) and derived quantities (q, q^2, $qxU(t)$) using an operation panel similar to a audio-, video-player [4]
- display types:
  traditional presentation in time and elliptic mode
  time- and voltage dependent presentation of all defined quantities: apparent charge q; pulse repetition rate and frequency n, N; average discharge current I, discharge power P, quadratic charge rate D, etc.
  PD-frequency distribution, phase resolved 2D & 3D representation
- PD-pulse distribution, segmental or continuous display
- impulse / impulse correlation [5]
- water fall diagrams of the PD-distribution vs. charge
- average PD-current vs. phase and time
- all phase- and polarity resolved statistical PD-parameters (q, $H(q)$):
  maximum, minimum, mean value, standard deviation, skewness, kurtosis, and cross correlation

Software Automation & Control & Protocol Interface

The basic and enhanced extension capability of the analysis features is suitable to generate all types of requested protocols. The PD-data can be transferred on-line to spread sheet or data base software. Based on preinstalled custom designed document-templates all types of protocols and scoring sheets can be compiled and modified at any time. Additional evaluation procedures can be executed using other application software.

An OLE (object link and embedment) interface presumes the implementation and hosting of other software. Using this common exchange platform complete software based remote control is possible. Exemplary, the embedment of control units (PLC-Components) form High voltage test field automatization is a typical application to use object links in order to combine stored program controllers with measurement devices.

A bi-directional control signal and data transfer between different software packages is performed in the same way as using local or wide area network data transmission. When connecting the system with a digital or analog telecommunication network it is immediately possible to operate telecontrolled monitoring of Partial Discharge and Loss Measurement.

Partial Discharge Expert system

A common mathematical modeling of all occurring PD failures is not available up to now. Only for exceptional cases exists a mathematical model which is suitable to describe a subclass of PD problems. Therefore, a automatized diagnosis system for PD failures is limited to the recognition of specific symptoms in PD measurement data records.

In this connection it must be noted that the characteristic feature extraction of the data record is cut out for the key position [6] in the quality of the diagnosis result. In the scientific field of the PD-fault recognition exits a wide range of formulations about the suitability of different features to be extracted.

Realized for the presented PD expert system is a combination of two independent feature detectors. The Fourier correlation coefficient of the phase resolved charge signal is normalized to the number of the test voltage periods. To describe the phase resolved PD distribution is only a limited number of coefficients of the Fourier series necessary [6] and used for the feature extraction array. Additionally, the variation of the coefficients versus the test periods is inserted to the feature pool.

Furthermore, the classical statistical operators [7] of the derived histogram functions of the PD-frequency distribution are included in the feature extraction matrix.

After the extraction, the two resulting feature arrays are subjected by a classification schedule. The classification is effected by means of comparison of feature extraction arrays of the actual measured PD-data with feature objects of all existing PD failure records stored in a reference database. As the classification result, the qualified probability of the classmembership of the classified object array related to identified PD fault is

evaluated and after a mutual coincidence check displayed on the computer screen of the PD-measuring system. A typical classification result is shown in figure 4.

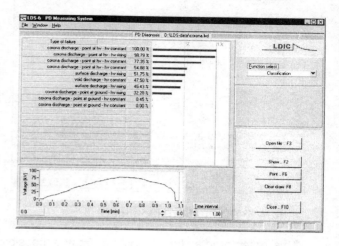

Figure 4: PD fault identification

2.3. PARTIAL DISCHARGE LOCATION TECHNOLOGY

It is obviously of high importance not only to know the values of the PD quantities like apparent charge level, repetition rate and phase correlation of a measured PD-signal, but also the source position of the detected pulses. Mainly, in spatial extended objects like power cables the PD-fault localization is of highest interest.

To locate PD-faults in cables the traveling wave principle is commonly used. A wide band digitizing unit for time-domain reflectometry is therefore installed in the system (see figure 3).

Reflectometer Method

For the PD-pulses the cable appears to be a dielectric waveguide. Therefore the PD-pulse at the point of origin is divided into two equal parts in accordance to the differential characteristic impedance of the cable. A separate fault location amplifier supplies the highspeed digitizer with the received RF-PD-signal. The signals are converted with a digital dynamic of 10 or optional 12 Bit and a sampling rate of 100 Ms/s.

The time delay between the direct measured signal and the first subsequent reflected impulse echo is evaluated in the fault localization unit

The computer based system is able to run the impulse echo evaluation for the complete corresponding PD-impulse-reflectograms. Hence, not only the PD-fault with the highest magnitude is detectable but also localization of multiple faults [8] is possible. All localized PD-faults are extracted, evaluated and mapped automatically. Thus, the PD-mapping diagram represents the PD-pulse magnitude as a function of its particular position vs. the number of the localized PD-pulses. A typical PD-position map is shown in figure 5.

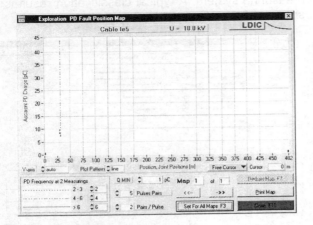

Figure 5: PD fault position map

This reflectometer functionality is easy accessible from the same common LDS-6 software platform.

Enhancement of the Location Sensitivity

To improve the location sensitivity and accuracy a number of high sophisticated features are realized additionally:

- A continuous pulse averaging and an adjustable threshold level can be applied to reduce continuous interference's, impulse oscillations and background noise.
- A FFT feature assists the user finding harmonic radio interference's. Supported by this harmonic analysis, a selection of digital filters can be adjusted optimally.
- The rise time of the pulses is used to discriminate between near and far-end PD-sources.
- For an exact position-independent determination of the real PD-impulse magnitude a transmission-loss-adjustment is automatically executed for all located PD-pulses.

3. SUMMARY

The presented system provides the possibility to combine together a great variety of measurement techniques to one system. A high accurate and frequency independent impedance- and loss factor analyzer can run parallel with a complete digital partial discharge analyzing system. With respect to the particular measurement task and the corresponding disturbance situation, several different processing units can be used. In addition, a reflectometer module for PD-fault location can be implemented.

The developed software for all available measurement devices complements the capability with regard to the noise suppression facility and therefore the improvement of the sensitivity. Comprehensive analysis and diagnosis software tools and a PD-expert system permit a user-friendly PD-fault recognition and identification.

The application range covers the complete spectrum: Scientific and industrial research purposes, routine tests in a production environment, on-site diagnostic tests and installations for condition monitoring.

References

[1] Holboll, J.T., Edin, H., "PD-Detection vs. Loss Measurement at High Voltages with Variable Frequencies", 10th ISH, Montréal, Canada, 1997

[2] Lemke, E., Schmiegel, P., „Analysis of the Properties of Dielectrics on the Basis of a Non-conventional Measuring Procedure", 6th DMMA Conference, Manchester / UK, 1992

[3] Kaul, G., Plath, R., Kalkner, W., "Development of a Computerized Loss Factor Measurement System for Different Frequencies, Including 0.1 Hz and 50/60 Hz", 8th ISH, Yokohama, Japan, 1993, paper 56.04

[4] Lemke, E., Rußwurm, D., Schellenberger, L., Zieschang, R., "Computergestütztes Teilentladungs-Diagnosesystem" 7 Tagung Techn. Diagnostik, Merseburg, 1996

[5] Hoof, M., Patsch, R., „Pulse-sequence-Analysis: A New Method to Investigate the Physics of PD-Induced Aging", IEE Proc. Sci. Meas. Techn., Vol. 142,1 1995

[6] Hücker, T., "UHF Partial Discharge Expert System Diagnosis", 10th ISH, Montréal, Canada, 1997

[7] Gulski, E., Digital analysis of partial discharges", IEEE Trans. On Dielectrics and Elec. Insulation, Vol. 2, No. 5

[8] Strehl, T., Kalkner, W., "Measurement and Location of Partial Discharges During On-Site Testing of XLPE Cables with Oscillating Voltages", International Symposium on High Voltage, 9th ISH, Graz, Austria, 1995

Author(s)

LEMKE DIAGNOSTICS GmbH

Prof. Dr.-Ing. E. Lemke
Dipl.-Ing. T. Strehl

Radeburger Str. 47
01468 Volkersdorf
Germany

Voice ++49 35207 863 0
Fax ++49 35207 863 11
mailto: info@ldic.de

INFLUENCE OF MODIFICATIONS IN PARTIAL DISCHARGE PATTERNS ON A NEURAL NETWORK'S RECOGNITION ABILITY

O. Rudolph, W. Zierhut

R. Badent

ABB Calor Emag Schaltanlagen AG, Germany

IEH, University of Karlsruhe, Germany

ABSTRACT

For several years the interest in automated analysis of Partial Discharge (PD) Measurement has increased, because manual PD pattern interpretation by human experts is one of the most powerful diagnostic methods for design and improvement of electrical hv equipment.

The main aspect is the interpretation of the phase resolved PD patterns. A database of 4000 pd patterns of on-site measurements has now been used for basic investigations with neural networks. In a first step 500 of them have been pre-selected and classified by a human expert. These files have been classified into 44 classes. For a better understanding of the general recognition abilities of N.N.s, patterns of on-site measurements have been artificially modified by splitting of pd matrices, using picture smoothing algorithms and zero value adjustment.

INTRODUCTION

On-site measurements of partial discharges on all type of high voltage equipment is becoming state-of-the-art. The interpretation of „living" pd impulses on oscilloscopic displays is restricted to few human experts so far. Using digital recording and data storing allows generation of so-called phase resolved partial discharge patterns. On the one hand basic investigations on analytic interpretation have been done (Niemeyer (1), Fruth and Gross (2), and Irmisch (3)). But PD patterns may also be used as input files for neural networks. Their classification shows generally promising results (Gramlich (4), and Kehl (5)).

Automated reliable type classification of any pd pattern would allow best interpretation of the condition of hv insulations. Due to different reasons there still exist erroneous results of N.N. which have not yet been described.

PD PATTERNS

The patterns have been stored from pd measurements system type ICM as a vector of 256x256 long integers. In order to reduce the data storage size, the files have been reduced to a 128x128 vector and stored in a compressed file format. The investigated patterns are from measurement circuits, which have been set up according to IEC270. There are also measurement data available from various other pd measurements with different techniques (acoustic, electric/magnetic sensors or antennas). The data have been acquired from more than 50 on-site measurements on several types of hv equipment since 1995, mainly gas-insulated / metal-enclosed switchgears (10-170kV), capacitive voltage transformers (400kV), and hv cable accessories (10-400kV). The pd patterns used with this investigation are all from bipolar acquisition mode and of single source type.

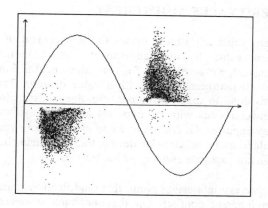

Fig. 1: Typical pd pattern

PARAMETERS OF USED NEURAL NETWORKS

A basic backpropagation neural network has been used (Badent et al (7)). The network has been set up with an input vector of 128x128 input neurons, 30 to 210 hidden neurons, and, according to the number of tested classes, an amount of output neurons. The amount of testing cycles was varied from 200 to 4500.

BASIC NETWORK

In prior intensive investigations on neural networks it has been tried to optimize them for the use with pd patterns. Parameters to be optimized were the number of hidden neurons, the learning rate, the activation function, the number of training cycles, and the initialisation of the network before training (7).

High Voltage Engineering Symposium, 22–27 August 1999
Conference Publication No. 467, © IEE, 1999

These investigations gave the basic network for the present paper. It proved 120 hidden neurons to be sufficient for successful training of at least 50 pd patterns. Also the learning rate was fixed to 0.1 according to prior good experiences with this value. Each pattern was trained 200 times to the network and initialisation was done by an empirically found fixed distribution with excellent properties for this task.

For all tests within this investigation the same set of pd patterns was used. This set consisted of seven classes with five training patterns each.

To get a deeper understanding of the network and to learn about the training process, so far mainly network parameters were studied. Also preprocessing of pd matrices with some promising results have been realized (5). Now in a new approach, three new „parameters" of pd matrices were defined: Zero value adjustment, picture smoothing, and splitting of pd matrices. All three parameters are methods to modify pd patterns artificially.

ZERO VALUE ADJUSTMENT

Zero value adjustment means the replacement of a zero value by the adjustment value (another constant, e.g. a value of –5). This adjustment is done for all matrix elements with a value of zero. This was done to make a sharper, non-linear distinction between areas with pd impulses and areas without any impulses. Of course, this also means to give zero values more importance during training since they can now activate and trigger hidden neurons.

As a very interesting point, this modification causes an inherent conflict. On the one hand it can be expected that zero value adjustment helps the network during training, due to the enhanced information in the matrices. On the other hand it probably decreases the recognition ability of distorted or even multiple source patterns, because now the former zero values have almost the same influence on the result as the pd impulses themselves. The examination of this conflict can give principal information about the influence of zero value matrix elements to the general recognition abilities of a neural network.

PATTERN SMOOTHING ALGORITHM

$$b_{m,n} = \sum_{i=-1}^{2} \sum_{j=-1}^{2} \left(k_{ij} \cdot a_{m+i,\,n+j} \right) \qquad (1)$$

$k_{i,j}$: smoothing factor
$a_{m,n}$: matrix element

The smoothing algorithm used in this paper sums up the active and fifteen neighbouring elements. The new value serves now as the input value for the network training. The target of pattern smoothing is to bring a little fuzziness into the patterns in order to „generalize" them to some extend. This generalizing effect while training a neural network is usually generated by the use of a high number of training patterns per class. Therefore, the use of generalized patterns should lead to the same result.

The principal effect of generalization by smoothing is quite simple. Assuming that an area of chequered patterns within the matrix is shifted by a single element in any direction leads to a matching difference value (mdv) of 1 according to Rudolph (9), which represents complete mismatch (Fig. 2a).

Fig. 2a: 100% different patterns before smoothing

Such a phase or amplitude shift can be seen in real pd patterns due to the statistical behaviour of pd pulses. If there are not all possible statistical effects are represented in the training patterns, the recognition ability of unknown patterns by the neural network will be reduced. By using smoothing algorithms like the one mentioned above (1), the two different patterns of Fig. 2a result in the same smoothed pattern (mdv of 0, see Fig. 2b).

Fig. 2b: Example of patterns after smoothing

An example of the smoothing algorithm used on a real pd pattern can be seen in Fig. 3a and Fig. 3b.

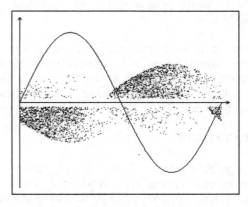

Fig. 3a: Pd pattern before smoothing

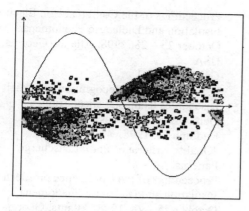

Fig. 3b: Pd pattern after smoothing

SPLITTING OF PD PATTERNS

As it is commonly known, symmetric pd sources, e.g. point-point arrangements, round bubbles, etc., lead to symmetric pd distributions while asymmetric pd sources, e.g. point-plane arrangements, delamination, etc., cause asymmetric distributions. This asymmetry has already been studied by splitting the pd matrix, evaluating the pd quantities for each of the positive and negative half-cycles separately and classifying the matrices by using the results from both parts (as done by Gulski (8)). Following this argumentation one has to consider whether the use of the described methods (zero value adjustment and smoothing) will behave differently if used on split matrices.

TRAINING SERIES

Four neural networks of the described basic type have been trained, one for each combination of the two first methods described above. Tab. 1 shows these combinations together with the IDs of the resulting networks.

Network No.	ID	Smoothing	Zero Value
1	ZMT	No	0
2	SiMT	Yes	0
3	EMT	No	-5
4	DMT	Yes	-5

Tab. 1: First training series

In the neural network DMT, in which the two methods of zero value adjustment and pattern smoothing are used at the same time, the pattern smoothing is done before the zero value adjustment. Therefore small holes (squares of four elements) within clouds of pd activity are closed before they will be given too high an importance by the zero adjustment method.

In addition to this series, four more networks have been trained using only reduced pd patterns.

Network No.	ID	Smoothing	Zero Value
5	SMT	No	0
6	FMT	Yes	0
7	VMT	No	-5
8	ZMF	Yes	-5

Tab. 2: Second training series

The performance of the resulting networks was tested by three test series:

- First, the networks were tested with their own training patterns to check if the learning was sufficient and successful.
- Second, the networks were tested using either the positive and negative half of the training patterns.
- In the third and last test, the networks were tested exemplary with a set of 30 unknown patterns of one trained class.

RESULTS

According to the table below (Tab 3.), the expected benefit of each method for complete patterns can be seen. Surprisingly, the use of both methods at the same time is even better than just the linear addition of the single benefits, so there might be some interesting synergy effects. In fact there were two cases where only the combined use of smoothing and zero value adjustment led to correct classification.

On the other hand there were cases of misclassifications caused by using just one of the two methods, which did not happen without preprocessing. This negative effect might be eliminated by optimizing the smoothing and/or adjustment value.

ID	Smoothing	Zero Value	Recognition
ZMT	No	0	63,30%
SiMT	Yes	0	70,00%
EMT	No	-5	66,70%
DMT	Yes	-5	93,30%

Tab. 3: Recognition of complete pd patterns

In contrast to this result, there is an unexpected disadvantage if the methods are used on reduced patterns (Tab. 4), which needs further examinations.

ID	Smoothing	Zero Value	Recognition
SMT	No	0	70,00%
FMT	Yes	0	66,70%
VMT	No	-5	53,30%
ZMF	Yes	-5	60,00%

Tab. 4: Recognition of reduced pd patterns

CONCLUSION

Concluding one can say that:

1. the use of smoothing increases the recognition rate of complete patterns by ~ 7%,
2. the use of zero value adjustment increases the recognition rate of complete patterns by ~ 3%,
3. the use of both methods increases the recognition rate by ~ 30% and
4. with split patterns there is a reduction of recognition by 5 - 10%, which is subject for further studies.

In a next step there will be investigations with the training of artificially produced pd patterns according to Niemeyer (11) with test patterns from real measurements. Due to this, it will be possible to pre-train N.N.´s to just proposed pd defects without any real reference data.

REFERENCES

1. Niemeyer L., 1993,
    "Interpretation of PD Measurements in GIS",
    ABB Corporate Research, Baden, ISH, Switzerland
2. Fruth B. A., Gross D W, 1995,
    "Modeling of Streamer Discharges between Insulating and Conducting Surfaces"
    5th Int. Conf. on Conduction and Breakdown in Solid Dielectrics, July 10-13
3. Irmisch M., 1995,
    "Hochspannungs-Isolationsdiagnostik an supraleitenden Großmagneten"
    Forschungszentrum Karlsruhe Technik und Umwelt, Wissenschaftliche Berichte FZKA 5615, Dezember
4. Gramlich J., 1995,
    "Analyse von TE-Fehler-Matrizen und Entwicklung eines Klassifizierungssystems",
    TH Karlsruhe, IEH, Germany
5. Kehl A., 1995,
    "Aufbereitung und Normierung von TE-Matrizen zur besseren Klassifizierung mittels Bapst",
    TH Karlsruhe, IEH, Germany
6. Rudolph O., 1992,
    "Optimierung eines rechnergestützten Teilentladungsmeßsystems",
    TH Karlsruhe, IEH, Germany
7. Badent R., Rudolph O., Zierhut W., Breuer A., Schwab A. J., 1998,
    "PD-Pattern Recognition of On-Site-Measurement Data",
    Proceedings of the Conference on Electrical Insulation and Dielectric Phenomena, October 25 – 28, 1998, Atlanta, Georgia, USA
8. Gulski E., 1991,
    "Computer-aided Recognition of partial Discharges using statistical Tools",
    Delft University Press, USA
9. Rudolph O., 1998,
    "Quality Criteria of Partial Discharge Patterns",
    Proceedings of the Conference on Electrical Insulation and Dielectric Phenomena, October 25 – 28, 1998, Atlanta, Georgia, USA.
10. Saur S., 1995,
    "Isolationsfehlererkennung aus der TE-Matrix mittels Neuronaler Netze",
    TH Karlsruhe, IEH, Germany
11. Niemeyer L., 1995,
    "A Generalized Approach to Partial Discharge Modeling",
    ABB Corporate Research, Baden
    IEEE Transactions on Dielectrics and Electrical Insulation, Vol. 2 No. 4, Aug.

DETECTING DELAMINATED STATOR WINDING INSULATION USING THE RAMPED HIGH-VOLTAGE DC TEST METHOD

L.M. Rux
United States Department of Interior
Bureau of Reclamation
Denver, Colorado USA

S. Grzybowski
Mississippi State University
Department of Computer and Electrical Eng.
Mississippi State, MS USA

Introduction

High-voltage dc tests have traditionally been used to detect cracks and fissures, surface contamination, and moisture in rotating machine insulation systems. Not generally recognized, however, is the ability of dc testing to diagnose delaminated groundwall insulation. This impaired insulation condition is generally evaluated using ac tests, such as dissipation factor tip up and partial discharge measurements [1-3]. In this paper, the ramped high-voltage dc test method [4,5] is described, and its ability to detect and monitor delaminated stator winding insulation is demonstrated. Supporting field data are presented and analyzed.

Delamination

Delaminated insulation can significantly shorten the service life of generator and motor stator windings. Delamination, i.e., separation of adjacent layers, may result from thermal and/or electrical deterioration of the resin binder during machine operation. Delamination can also occur if the winding insulation was not sufficiently impregnated during manufacture.

Delamination can appear at the interface between the copper conductor and the groundwall insulation or between the resin and mica layers which comprise the groundwall insulation. The former is more serious because it may also affect strand and turn insulation, although in both situations, the thermal conductivity of the insulation is reduced. Poor heat transfer between the copper conductor and the stator iron causes the winding insulation to operate at higher than normal temperatures, leading to accelerated aging and further degradation of the resin binder. In addition, gas-filled voids in the insulation may result in internal electrical discharges which can further erode the insulation, particularly the strand insulation and the turn insulation in multi-turn coil designs. Eventually, normal operating bar forces can initiate mechanical failure of the winding. This is a potentially catastrophic situation as turn-to-turn insulation failures can cause extensive stator winding and iron damage, resulting in a costly unscheduled outage.

Under both normal and extreme operating conditions, delaminated insulation is more susceptible to the combination of thermal, electrical, and mechanical deterioration effects. Although the degradation process may be slow, it can result in a significant reduction in useful life. A high-quality generator stator winding can be expected to provide 30 to 40 years of reliable service. In situations where the insulation is delaminated, however, the useful life may be reduced by 10 years or more—a substantial decrease in terms of return on investment.

High-Voltage DC Testing

Diagnostic tests are vital for assessing insulation condition and preventing in-service failures. Although no single test is ideal for detecting all types of insulation problems, variations in measured insulation current versus applied dc potential have been shown to be useful in detecting and analyzing certain modes of stator winding deterioration.

During high-voltage dc tests, the three primary components of measured stator insulation current are capacitance charging current, dielectric polarization current, and conduction current. The charging current is attributable to the winding-to-ground capacitance and is equal to the rate of change of applied voltage times the capacitance. Stator winding capacitance depends on the conductor-to-core geometry (i.e., area, shape, and spacing) and dielectric material used. Following an applied voltage step, the capacitance charging current decays exponentially to zero in seconds. Dielectric polarization results from polarized atoms and molecules becoming displaced and aligned when placed in an electric field. The polarization current response depends on the dielectric material used in the stator insulation system. Polarization current is similar to capacitance charging current except that it typically takes minutes to hours to decay to a negligible value in response to an applied voltage step. The conduction component of stator insulation current passes through an insulation's volume, over its surfaces, and through defects. It is a continuous, irreversible current which results from a voltage being applied across the insulation. The conduction current of high-quality insulation will, in general, be small and proportional to the applied voltage.

The stepped dc voltage test involves application of the direct high voltage in a series of steps. Current readings are taken at each interval, and the current versus voltage curve is plotted. The curve is examined for increases or other variations in the measured current response, a possible indication of

High Voltage Engineering Symposium, 22–27 August 1999
Conference Publication No. 467, © IEE, 1999

stator insulation weakness. Dielectric absorption currents may dominate the measurements and mask subtle but significant variations in insulation current. To minimize dielectric polarization effects, the test voltage may be held at each level long enough to allow the polarization current to decay to a negligible value. This stabilization time may exceed an hour for some machines. Alternatively, the voltage steps may be graded in such a way that the polarization component of measured current is proportional to the applied voltage [4].

When using either stepped voltage method, controlling test time intervals and voltage increments, as well as visually reading and averaging a fluctuating insulation current meter, often result in poor test data accuracy. The faulty data may appear as misleading dips, peaks, and false knees in the plotted current versus voltage insulation curves. Uncertainty decreases confidence in the data, or worse, may even lead to improper assessment of insulation condition.

Ramped High-Voltage DC Tests

The ramp test can be considered a step test in which the voltage steps and time intervals are made very small. As the size of the voltage and time increments approaches zero, a voltage ramp is formed. The application of a ramped voltage (instead of discrete voltage steps) automatically linearizes the polarization component of insulation current so that minor deviations are more easily seen. A programmable high-voltage dc test set is used to automatically ramp the high voltage at a preselected rate, typically 1 kV per minute. Insulation current versus applied voltage is plotted with an *X-Y* plotter, providing continuous observation and analysis of the insulation current response as the test progresses. Therefore, it is no longer necessary to hand plot insulation current versus applied voltage to evaluate an insulation. Insulation integrity can be assessed directly from the automatically recorded insulation current curves because the human factor has been eliminated from the time, voltage, and current parameters. The observed insulation current nonlinearities are directly related to insulation quality and condition.

A typical stator winding response to a ramped dc voltage is shown in Fig. 1. The measured current is a composite of the capacitance charging, polarization, and conduction currents. The charging current is constant because the rate of change of the applied voltage is constant. The polarization component has been linearized by the application of a continuous ramped voltage. The conduction current is negligible up to the maximum value of the applied test voltage.

When the insulation under test has been affected by moisture, surface contamination, cracks, uncured resins, or other insulation problems, the current versus voltage curve will exhibit nonlinear characteristics related to the specific defect and its corresponding influence on insulation resistance, capacitance, and/or polarization [5].

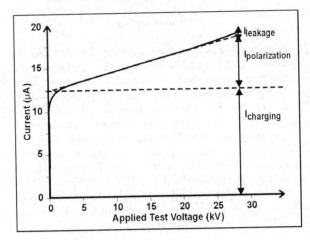

Fig. 1. Characteristic response of normal stator winding insulation to an applied ramped dc voltage.

Capacitance Increase with Voltage

Capacitance is normally described as a function of geometry and dielectric constant [6]. For example, a parallel-plate condenser of area A and plate separation d has the capacitance

$$C = \frac{A}{d} \varepsilon_0 \varepsilon$$

Where ε_0 is the dielectric constant or permittivity of vacuum and ε is the relative permittivity of the dielectric material between the electrodes.

The dielectric permittivity of most homogeneous materials is independent of the magnitude of the applied electric field. In the case of composite insulations, however, this may not always hold true. For example, when delaminated epoxy-mica insulation is placed between the electrodes, internal gas spaces or voids are introduced. Given a sufficient voltage gradient across the void, it becomes short-circuited and causes a temporary increase in stator winding capacitance. The electrical discharge phenomenon also produces a rise in dissipation factor with increasing voltage. Dissipation factor and capacitance increases due to discharge of internal gas spaces have been reported by Dakin and others [7-9]. Dissipation factor tip up has become a standard quality control check of insulation integrity. In comparison, relatively little emphasis has been placed on analyzing the concurrent increase in measured capacitance.

A severely delaminated insulation system will contain numerous voids of varying size and depth having different breakdown gradients. As the applied voltage is increased, the whole gas space within the insulation does not discharge at once. The voltage across each individual gas space must reach its breakdown voltage, and an initiatory electron must be present for ionization to occur [10].

The generation of initiatory electrons is a statistical process. Thus, there will be a time lag between the moment the gas space breakdown voltage is reached and the discharge actually takes place. The existence of different depth voids having different breakdown gradients coupled with the statistical nature of the ionization process results in discharges across internal gas spaces occurring in a number of small steps over a range of applied voltage. Each discharge short-circuits a small area of gas space, thereby increasing the capacitance of the insulation by a small increment. This phenomena is also known as high-voltage polarization.

Ramped Voltage Test Results

High-voltage dc ramp tests have been shown to detect delaminated stator winding insulation. Fig. 2(a) shows the benchmark ramp test results of a new 13.8 kV hydroelectric generator stator winding with polyester-mica groundwall insulation. The current versus voltage (*I-V*) response is linear up to the maximum applied test voltage of 30 kV dc conductor-to-ground, an indication of normal insulation. The next test was performed after 4 years of operation. At this time, the *I-V* response exhibits a distinct increase in measured insulation current at approximately 8.5 kV, as displayed in Fig. 2(b).

Subsequent ramp tests consistently exhibit the nonlinear increase in measured current. Test results after 12 years of machine operation are shown in Fig. 2(c). The curve appears normal when the applied test voltage is below a few kilovolts. But as the voltage approaches approximately 7 kV, the slope of the curve sharply increases. As the applied test voltage continues to ramp up, the slope of the *I-V* response levels off again at approximately 11 kV. The final slope of the current trace is slightly greater than the initial slope. After 16 years of operation, the generator failed at 22.4 kV dc conductor-to-ground during a routine ramped voltage test. The test curve of the failed phase is shown in Fig. 2(d) and is essentially identical to that of the 12-year test.

Removal and examination of the failed coil revealed a crack through the groundwall which appears to have been caused by bending stresses from coil installation. (The generator operated without failure for 16 years despite the cracked insulation.) Dissection of the failed coil also exposed delaminated groundwall insulation over the entire length of the coil; the layers of insulation were

separated from each other and from the copper conductor. A thorough evaluation of the stator winding was performed. Additional insulation failures occurred and several more coils were removed for dissection. Evidence of widespread delamination was found in neutral-end as well as line-end coils. The decision was made to rewind rather than repair the generator.

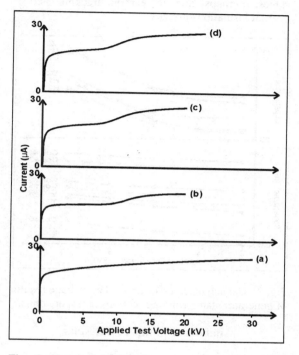

Fig. 2. Ramped high-voltage dc test results of a delaminated polyester-mica stator winding. (a) Initial results. Results after (b) 4 years, (c) 12 years, and (d) 16 years of operation

Capacitance Measurements

Before the failed stator winding was removed from the core, capacitance versus applied ac voltage measurements were made on several sections of the winding. Two parallel circuits from a phase which was not involved in the failure were sectionalized into three parts, each having five series-connected coils. The sections which operated at the line-end were designated H1 and H2, the middle sections were designated M1 and M2, and the neutral-end sections were designated L1 and L2. Capacitance versus voltage measurements were also made on seven unused spare coils which were originally supplied with the stator winding. The test voltage was applied to the winding specimens in a series of steps equal to 2, 4, 6, 8, and 10 kV rms conductor-to-ground. These voltage steps represent 25, 50, 75, 100, and 125 percent of the rated phase-to-neutral operating voltage. The capacitance versus voltage measurements are presented in Fig. 3.

The spare coil test data showed that prior to operation, the average capacitance increase was approximately 1.2 percent over the 25 to 125 percent voltage range. After 16 years of service, the stator winding sections exhibited an average capacitance increase of approximately 7.1 percent over the same voltage range. The capacitance increase was roughly the same for all the stator winding sections, regardless of their position in the parallel circuit. However, those parts which operated at the higher voltages did exhibit lower overall capacitance values, perhaps due to carbon tracking across internal insulation surfaces.

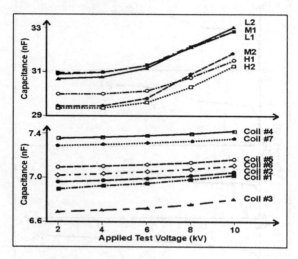

Fig. 3. Capacitance versus applied ac voltage results of generator stator winding sections and spare coils.

Analysis of Ramp Test Results

The nonlinear increase observed on the ramp test current traces appears to be related to the breakdown of gas-filled voids in the delaminated stator winding insulation. As described above, the applied test voltage is raised causing the internal gas spaces to ionize and the winding capacitance to increase. The winding capacitance C, connected to a voltage source V, stores a charge

$$Q = CV$$

As the capacitance increases during discharge of internal gas spaces, charge must flow from the supply to the electrodes to maintain the applied potential; hence, the observed increase in measured insulation current.

As the applied voltage continues to increase, the more active voids will have ionized and the less active ones will begin to discharge. Eventually, all available gas spaces will become ionized and the maximum winding capacitance is achieved. The slope of the insulation current versus voltage curve is once again dominated by dielectric polarization effects rather than changes in winding capacitance.

Conclusions

The test results presented herein demonstrate the ability of the ramped high-voltage dc method to detect and monitor insulation delamination. The results indicate the example generator stator winding insulation began to delaminate soon after the unit was put into service. Over time, the delamination became more severe, as evidenced by the extension of the voltage range over which the capacitance increase was observed as well as by the overall capacitance increase.

Several other generators owned by Reclamation exhibit similar evidence of delamination. Research will be undertaken to relate the change in capacitance observed in the ramp test results to insulation void volume. Also, partial discharge analysis of one or more of the affected units will be performed in the near future. Discharge measurements will be compared to the ramped voltage test results. Trends will be studied over time to evaluate the progression of delamination and its effect on insulation life.

References

[1] G.C. Stone, "Use of Partial Discharge Measurements to Assess the Condition of Rotating Machine Insulation," IEEE Electrical Insulation Magazine, Vol. 12, No. 4, July/August 1996, pp. 23-27.

[2] Y.J. Kim and J.K. Nelson, "Assessment of Deterioration in Epoxy/Mica Machine Insulation," IEEE Transactions on Electrical Insulation, Vol. 27, No. 5, October 1992.

[3] J. C. Botts, "Corona in High Voltage Rotating Machine Windings," Vol. 4, No. 4, July/August 1988, pp. 29-34.

[4] L.M. Rux, "High-Voltage DC Tests for Evaluating Stator Winding Insulation: Uniform Step, Graded Step, and Ramped Test Methods," IEEE Conference on Electrical Insulation and Dielectric Phenomena, 1997 Annual Report, Vol. 1, pp. 258-262.

[5] L.M. Rux, "The Ramped High-Voltage DC Method of Evaluating Stator Winding Insulation," Iris Rotating Machine Conference, Dallas, TX, March 1998.

[6] A. von Hippel, Dielectrics and Waves, Wiley, New York, 1954.

[7] T.W. Dakin, "The Relation of Capacitance Increase with High Voltages to Internal Electric Discharges and Discharging Void Volume," AIEE Transactions, Paper 59-151, October 1959, pp. 790-795.

[8] R. Seeberger, "Capacitance and Dissipation Factor Measurements," IEEE Electrical Insulation, Vol. 2, No. 1, January 1986, pp. 27-36.

[9] S. Boggs, "Partial Discharge: Overview and Signal Generation," IEEE Electrical Insulation, Vol. 6, No. 4, July/August 1990, p. 37.

[10] P. Morshuis, M. Jeroense, and J. Beyer, "Partial Discharge Part XXIV: The Analysis of PD in HVDC Equipment," IEEE Electrical Insulation, Vol. 13, No. 2, March/April 1997, pp. 6-16.

FREQUENCY-TUNED RESONANT TEST SYSTEMS FOR ON-SITE TESTING AND DIAGNOSTICS OF EXTRUDED CABLES

W. Schufft, P. Coors, W. Hauschild, J. Spiegelberg,

HIGHVOLT Prüftechnik Dresden GmbH, Germany

Abstract: Frequency-tuned resonant test systems are, meanwhile, state of the art for on-site testing and diagnostics of high-voltage extruded cables. After experience with some realised systems the technical data, especially the specific weight, and the performance have been further optimised. A specially adapted diagnostic technique has been developed for this application. Basic research on cable samples with different failures has qualified AC voltage near to the power frequency to be the optimum test voltage wave shape. Resulting from these it is logical to apply this test voltage shape also on medium-voltage cable systems. An example is introduced in this paper.

1. Introduction and history

Frequency-tuned resonant test systems were introduced at the end of the seventies for GIS on-site testing /1/ and later on also applied for after-laying tests of high-voltage (HV) cables /2/. These test systems consist mainly of a frequency converter, an exciter transformer and a resonant reactor with fixed inductance. Resonant reactors with fixed inductance were designed in the beginning as cylindrical modules in an insulating case.

Because of the limited heat dissipation through the insulating tube this early cylinder-type design enables test currents of some Amps in a short-time duty cycle only, which may be sufficient for testing short cables of a few hundred meters. The testing of HV cables of typical lengths between 4 and 15km requires test currents between 100 and 200A.

First steps to meet this test current requirement were done with two powerful frequency-tuned resonant test systems in the beginning of the nineties, using larger cylinder-type resonant reactors, see systems no. 1 /3/ and 2 /4/ in table 1. A remarkable short-time power has been be reached.

Another important step was the introduction of a test system with a first tank-type reactor (no. 3 /5/).

The heat dissipation through the metallic vessel which can be optimally enforced by external radiators with fans is much better and allows long-time tests, for instance 8 hrs. Other essential advantages are the better resistance against mechanical shocks and the option of a cable plug-in connection to lead the test voltage to the cable to be tested, see also chapter 2.2. . Meanwhile this design has proven to be optimum, the systems 5 – 9 are realised in this way. An exception is the system no. 4 using a cylinder-type resonant reactor /6/. The remarkable test power of limited duty cycle can be reached by an external heat exchanger.

2. Progress in the design of frequency-tuned resonant test systems

After frequency-tuned resonant test systems have been recognised to be the most effective and only practicable solution for the cable on-site testing of HV cables, there was an uncertainty to define a test frequency range larger than that of power frequency for laboratory testing, which is defined to be within 45 and 65Hz /7/. The permissible frequency range (f_{min} - f_{max}) determines the obtainable load capacitance range (C_{min} - C_{max}) /8/:

$$(f_{max}/f_{min})^2 = C_{max}/C_{min} \qquad (1)$$

With f_{min}=45Hz and f_{max}=65Hz the load range, i.e. the ratio C_{max}/C_{min} is approx. 2, which is not sufficient to design a practicable frequency-tuned resonant test system. An extended frequency range is absolutely essential. So there were more or less cautious approaches for the frequency range in the beginning, see systems no. 1, 2, 3 in table 1. An essential step was done by the CIGRE WG 21.09 /9/ defining for cable on-site tests a frequency range from 30 to 300 Hz to be "near to power frequency". All following realised systems were based on this

No.	Year	Origin	Rated voltage	Rated current	Test power (related to 50 Hz)	Frequency range	Specific weight of the system
1	1992	Switzerland	250 kV	37 A	23 MVA	20 – 160 Hz	unknown
2	1993	South Africa	132 kV	122 A	16 MVA	50 – 100 Hz	unknown
3	1996	Germany	160 kV	40 A	8 MVA	35 – 71 Hz	1.9 kg/kVA
4	1996	Switzerland	220/440[1] kV	66/133[2] A	24/48 MVA	30 – 300 Hz	0.9 kg/kVA[3]
5 + 6	1998	Germany	150/300[4] kV	90/180[5] A	22.3/44.3 MVA	30 – 300 Hz	0.97 kg/kVA
7 + 8	1998	Germany	254/400[4][6] kV	80/160[5] A	31.4/62.8 MVA	25 – 300 Hz	0.8 kg/kVA
9	1998	Germany	160/320[1] kV	50/100[2] A	26.5 MVA	30 – 200 Hz	1.1 kg/kVA

[1] 2 reactors in series;   [2] 2 reactors in parallel,   [3] specific reactor weight only;
[4] 2 systems in series;   [5] 2 systems in parallel,   [6] 504kV for GIL testing

Table 1: Survey of realised systems for HV cable on-site testing.

High Voltage Engineering Symposium, 22–27 August 1999
Conference Publication No. 467, © IEE, 1999

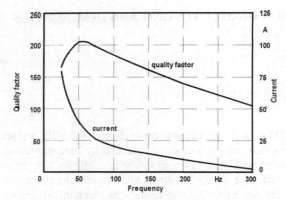

Fig. 1:Quality factor and current vs. frequency for a resonant test system 80A, 254kV, no. 7, 8 in table 1

recommendation, see systems no. 4 - 9 in table 1. Basic investigations on the breakdown behaviour of polyethylene samples with typical failure pattern have confirmed that there is no significant difference in the breakdown behaviour over a wide frequency range /10/.

2.1. Frequency range and quality factor

With respect to a maximum test power and a lowest specific weight (kg/kVA) the frequency range is the essential for the design of a frequency-tuned resonant test system. The lower the minimum permitted frequency f_{min}, the lower is the necessary test power P to test a given capacitance C with a given test voltage V:

$$P = 2\,\pi\,f_{min} \cdot C \cdot V^2 \qquad (2)$$

But for the specific weight this tendency is practically limited by a larger cross section of the iron circuits of the resonant reactor and of the exciter transformer, which is required to avoid the saturation of the iron core at lower frequencies. The enlargement of the magnetic circuit leads to an essential higher weight. So the specific weight of the system, i.e. the weight of the system related to the 50Hz-equivalent test power given in kg/kVA, would go up. Table 1 contains as far as known the specific weights of the realised systems. It is obvious that the specific weight came down after choosing 30Hz as a minimum frequency. The optimum for the minimum frequency regarding to a minimum specific weight is even below 30Hz, see the systems no. 7 and 8 having a minimum

frequency of 25Hz and a lowest specific weight of 0.8 kg/kVA. Consequently a minimum frequency between 20 and 25Hz is recommended.

The maximum test frequency f_{max} determines the load range as given in equation (1). A frequency range 30 - 300Hz leads to a load range of the factor 100, which seems to be sufficient. With the increase of the frequency the frequency-depending supplementary losses (hysteresis losses, skin effect etc.) in the resonant reactor and exciter transformer decrease also. Besides these an increased polarisation heat in the measuring capacitor has to be considered. From this point of view a higher frequency than 300Hz is disadvantageous.

The quality factor q of a resonant test system is the ratio between test power and required feeding power to cover all ohmic losses in the test circuit. Because the polyethylene insulation of an extruded cable has a very low tan δ (ca. $3 \cdot 10^{-4}$) the quality factor is determined by the resonant reactor and exciter transformer losses. A maximum quality factor would be desirable regarding a minimum feeding power, but it would require larger cross sections of the iron core and of the copper winding wire, which results in a higher specific weight. For low frequencies (and therefore high currents) the quality factor is determined mainly by the pure ohmic losses in the copper wire, for higher frequencies by all frequency-depending supplementary losses. Both influences result in a maximum of the quality factor over frequency. As an example fig. 1 shows the quality factor and current versus frequency for the resonant test systems no. 7, 8 in table 1. The quality factor of ca. 160 is connected with the specific weight of only 0.8kg/kVA.

Both effects, i.e. the lowering of the test power by a lower frequency and a sufficiently high quality factor, which is much higher than for resonant reactors with variable inductance, enable the generation of some 10MVA test power with an on-site available feeding power of some 100 kVA.

2.2. Connection technique and accessories

In practice it is sometimes impossible to bring the heavy test system to the immediate vicinity of the cable to be tested. So it is necessary to lead the test voltage over some 10m. This can be done by a bar wire, which is supported by insulating posts. Thereby the entire transmission line must be

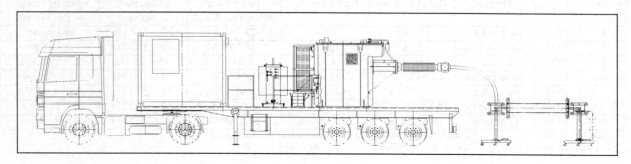

Fig. 2: Frequency-tuned resonant test system 90 A, 150 kV on a trailer with HV filter for PD measurement

Fig. 3: Parallel connection of two identical test system 90A, 150kV to extend the available test current to 180A (By courtesy of BICC Cables Ltd., U.K. and Pirelli Construction Company Ltd., U.K.)

protected by a safety loop with warning lamps and emergency-off switches. A more convenient solution is to use a connection cable with two flexible terminations at the ends - one is connected to the HV terminal of the resonant reactor and the other to the cable to be tested. Such cables are available as so-called "emergency cables" originally meant for provisional connections in substations. By means of this cable the test voltage can be lead easily into indoor substations or down to underground cable facilities.

Tank-type resonant reactors enable the direct plug-in connection of such a connection cable. Two solutions for this direct connection cable plug-in have been realised meanwhile:

- The plug-in connection can be made directly by an SF_6-immersed sealing end for transformers. The other end of the cable has a standard air cable termination or a sealing end for GIS acc. to IEC 60859, to be plugged in a GIS, which terminates the cable to be tested. In the latter case there is a closed encapsulation between the resonant reactor and the cable under test, which provides best preconditions to eliminate outer electromagnetic disturbances (noise) and to enable sensitive PD measurements.

- The resonant reactor is fitted with an oil-SF_6-bushing. The SF_6-part of this bushing projects into a cylindrical SF_6-vessel. A SF_6-air-bushing projects into the other end of the vessel (fig. 2). Both bushings are connected by a plug contact inside the vessel. The SF_6-air bushing can be replaced by a connection cable with a standardised sealing end as described above.

2.3. Transportation system

In spite modern cable on-site test systems have a lowest specific weight the resulting total weight of such a system is about 30 tons.

Such systems can be handled by customary truck trailers being modified for this purpose. fig. 2 shows a trailer for the transportation of test system no. 5 in tab. 1. The control and feeding unit, including the inverter is located in an air-conditioned and illuminated 10ft container at the front side of the trailer. This container serves for the operation of

the system. It has a door, windows and a board mains. The resonant reactor is located above the trailer axles, its bushing projects to the rear side. The exciter transformer is standing between the container and the resonant reactor. A foldable stairs serves for the access to the trailer platform, which is surrounded by a safety railing. The trailer has a foldable roof and side canvas to protect the system against bad weather during transport and parking.

Alternatively, with reference to future ship transportation all control equipment and the exciter transformer can be arranged in a 20ft container. The resonant reactor is designed to fit on a so-called flat rack container, for its fixing it has counterparts for container twist locks at the bottom plate and for lifting container fittings at the top plate. By storing the accessories in the container the test system is split up into two units for sea transportation – the 20 ft container and the resonant reactor in a container flat rack.

Fig. 3 shows the parallel operation of a containerised and a test system on a trailer, to extend the load range, systems no. 5 and 6 in table 1.

2.4. Software for control and protocol

The control and feeding unit contains all power electronics and control modules required for operating a frequency-tuned resonant test system.

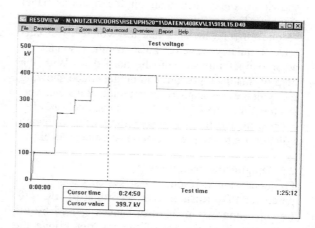

Fig. 4: Test record voltage vs. test time

The entire system is controlled by a PLC, type SIMATIC S5-95U. An operator panel COROS OP 15 is used for the input of the test data and to display the measured values (voltage, frequency, current) and other necessary information concerning the state of the system.

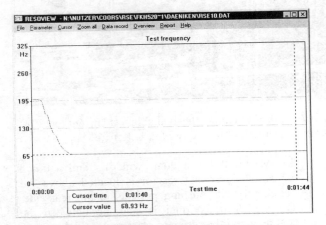

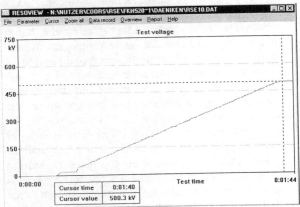

Fig. 5: Records for searching resonance frequency (upper screen) and following voltage rise (lower screen)

For a remote control the SIMATIC unit can be linked to an external PC, preferably a laptop, via an RS 232 interface. The laptop enables by means of a special software a more comfortable operation of the system, because much more information can be shown on the screen than on the operator panel display. This software stores during a test every second all relevant data like test voltage, test current, frequency, inverter pulse width, resonant reactor temperature etc. on the hard disk. With the help of this data a record can be generated showing test voltage, resonant reactor temperature etc. versus test time, see fig. 4 with a real test cycle and fig. 5 showing the frequency searching with following voltage rise to the pre-selected test voltage level.

2.5. Diagnostics technique
Frequency-tuned resonant test systems can be prepared for PD measurements according to IEC 270. For the suppression of disturbances caused by the steep switching flanges of the inverter bridge the control and feeding unit generates a signal to trigger the gating unit of an especially modified PD detector /11/. Additionally HV filters consisting of measuring capacitor, blocking impedance and coupling capacitor can be applied for PD measurement purposes, see fig. 2 right side. A PD sensitivity below 10pC can be reached. The obtainable PD sensitivity under on-site conditions depends on the environmental conditions (external noise, earthing conditions, etc.) and from the damping conditions and length of the cable. It is estimated that the IEC 270 method can be applied up to a cable length of max. 2.5km, when the PD measurement is executed only from one end, and max. 5km from both ends.

For longer cables the sensitivity of PD measurements according to IEC 270 becomes too low and non-conventional methods using sensors in the cable joints must be applied /12/. An encapsulated cable plug-in connection between resonant reactor and cable under test (as described in chapter 2.2. before) provides a totally screened circuit with optimal preconditions for sensitive PD measurements.

3. Frequency-tuned resonant test systems for medium-voltage cable testing
After frequency-tuned resonant test systems have been introduced successfully for the on-site testing of HV cables there is a certain logic to apply this principle also on the testing of medium-voltage cables, especially under the point of view to enable PD and tan δ diagnostics. Before VLF (very-low-frequency) test systems for medium-voltage cables became applicable /13/, there was indeed the attempt to introduce a frequency-tuned resonant test system for this purpose /14/. But it was to weak in power to master water-tree-damaged and older oil-paper cables with their bad tan δ value.

3.1. Selection of technical data
There are some special aspects to select the technical data for future frequency-tuned systems for medium-voltage cables.

In opposite to the HV cables there is till now no recommendation related to the test level, i.e. the multitude of the phase-to-earth voltage U_0 for the different system voltages. At the other side the maximum length of the cable to be tested is not so clear. For the test system a rated voltage of 36kV meeting 3 U_0 of the 20kV class and 2 U_0 of the 30kV class and a rated current of 17A representing a cable capacitance of 2.5µF (ca. 10km) at 30Hz have been chosen. The triple 36kV, 17A, 30Hz results in a 50Hz-equivalent power of 1MVA. Different to HV extruded cables a worse tan δ has to be considered for medium-voltage cables, which will determine the quality factor of the resonant circuit now. So it makes no sense to design exciter transformer and resonant reactor for an extremely high quality factor. This is possible also, because a much lower feeding power related to HV cable test systems must be supplied.

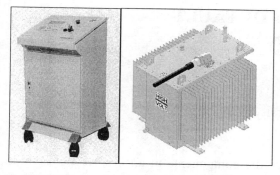

Fig. 6: Control and feeding unit and resonant reactor 17A, 36kV for testing cables up to 10 km

3.2. Design example

The frequency-tuned resonant test system for medium-voltage cable on-site testing consists of a control and feeding unit, an exciter transformer and a resonant reactor, see fig. 5. The resonant reactor is realised in a conventional power transformer design, i.e. oil-immersed in a metal tank. It contains besides the active part (coil and core) a capacitor for the voltage measurement. The test voltage is lead out via a plug-in connection and a 20m-connection cable with a air cable termination at the other end. To adapt the output voltage of the control and feeding unit to the resonant circuit there is a dry-type exciter transformer. The control and feeding unit is desk type and contains all power electronic and control components including a peakvoltmeter, see also chapter 2.4 before. The medium-voltage test system has the following technical data:

Nominal voltage: 36kV
Nominal Current: 17A
Frequency range: 30 – 300Hz
Capacitance range: 25nF – 2.5µF (ca.100m – 10km)
Duty cycle (at 20°C):
for nominal current 17A: 30 min ON - 45 min OFF
for reduced current 11A: continuously

Weights:	Resonant reactor	1100 kg
	Exciter transformer	295 kg
	Control and feeding unit	95 kg
	Total	1490 kg

4. Conclusions

Marked progress has been made in the design and performance of frequency-tuned resonant test systems for on-site testing and diagnostics of HV extruded cables:

- A frequency range beginning at 20 ... 25 Hz and ending at 300 Hz enables to reduce the specific weight of such test systems to 0.8 kg/kVA or lower.
- A tank-type reactor with radiators is the optimum solution related to the permissible duty cycle and a plug-in connection technique to the cable under test.
- There are optimised solutions for the transportation of such a system based on containers and trailers.

- An adapted PD technique and comfortable PC software for control and protocol are available.

The application of frequency-tuned resonant test systems for the on-site testing of medium-voltage cables is a logic conclusion. An example has been introduced.

5. References

/1/ Bernasconi, F., Zaengl, W.S., Vonwiller, K.: A new HV series resonant circuit for dielectric tests. 3rd ISH Milan 1979, paper 43.02

/2/ Aschwanden, T.: Vor-Ort-Prüfung von Hochspannungs-Kabelanlagen. Bulletin SEV/VSE, Vol. 83, 1992, S. 31 - 40

/3/ Onodi, T.: Neue Methoden und Erkenntnisse in der Kabeldiagnostik. 37. Internationales Wissenschaftliches Kolloquium TU Ilmenau, Sept. 1992

/4/ Lang, M. A. I. et.al.: A variable frequency series resonant test set for after-laying tests on XLPE cables. CIGRE Session (1994), paper 21-105

/5/ Schufft, W. et al.: Powerful frequency-tuned resonant test systems for after-laying tests of 110 kV XLPE cables. 9th ISH Graz 1995, paper 49.86

/6/ Mohaupt, P. et al.: High Voltage Testing using Series resonance with variable frequency. 10th ISH Montreal (1997),Vol. 4, pp. 351 - 354

/7/ IEC-Publ. 60060-1 (1989): „High-voltage test techniques Part 1: General definitions and test requirements

/8/ Hauschild, W., Schufft, W., Spiegelberg J.: Alternating voltage on-site testing of XLPE cables: The parameter selection of frequency-tuned resonant test systems. 10th ISH Montreal (1997),Vol. 4, pp. 75-78

/9/ Working Group 21.09 (J. Becker et. al): After laying tests on high-voltage extruded insulation cable systems. ELECTRA No. 173 (1997) pp.33-41

/10/ Schiller, G.: Das Durchschlagverhalten von vernetztem Polyäthylen (VPE) bei unterschiedlichen Spannungsformen und Vorbeanspruchungen. Thesis, TU Hanover 1996

/11/ Hauschild, W., Spiegelberg J., Lemke, E.: Frequency-tuned resonant test systems for HV on-site testing of SF₆-insulated apparatus. 10th ISH Montreal (1997),Vol. 4, pp. 75 - 78

/12/ Pommerenke, D., Strehl, T., Kalkner, W.: Directional coupler sensor for partial discharge recognition in high voltage cable systems. 10. ISH Montreal 1997

/13/ Boone, W. et al.: VLF HV generators for testing cables after laying. 5th ISH Braunschweig (1987), paper 62-04

/14/ Jäckle, E.: Prüfung von Kabelanlagen mit Resonanz-Prüfgeräten. Elektrizitätswirtschaft 7(1987)86, S. 245 – 300

Address of author:

Dr. Wolfgang Schufft
HIGHVOLT Prüftechnik Dresden GmbH
Marie-Curie-Str. 10
D-O1139 Dresden, GERMANY

ON-LINE MONITORING INSTRUMENT OF FAULT DISCHARGE IN LARGE GENERATORS

Guangning Wu Dae-Hee Park

School of Electrical Engineering, Wonkwang University, 570-749, Korea

Abstract

This paper describes a newly developed partial discharge (PD) on-line monitoring system used for large turbine generators. The system consists of a broadband current transducer, a computer-aided PD measurement system. By using a programmable fabricate band pass filter and an adaptive digital filter, the system can suppress the noise and extract PD signal from the intense noise surroundings successfully. At the end of this paper, some field test results, obtained from a 200MW generating set, were presented and discussed.

1. Introduction

Recently the development tendency of power equipment is high voltage, large capacity. During operation, the electrical, thermal, and mechanical stresses, along with environmental factors, can combine to degrade the electrical insulation. Therefore the insulation system is deteriorated during equipment operation. Thus the measured trend in PD activity from periodic testing over the life of the equipment can indicate if insulation deterioration is occurred. This facilitates scheduling of preventive maintenance prior to failure. The study of PD test especially for on-line monitoring becomes important and popular [1][2][3].

Generally, electrical noise can be categorized into two types: thermal and external. Thermal noise is the more fundamental and results from thermally induced current fluctuations in amplifiers and detection impedance. However, external noise tends to be much more severe during the PD testing of HV equipment. In many cases, external noise can cause false indications, reducing the credibility of on-line and off-line PD tests [4]. The external noise sources include follows [5][6]: (1) PD and corona from the power system which can be coupled directly to the apparatus under test(in on-line test) or radioactively coupled (in on-line or off-line test); (2) radio transmissions and power line carrier communication systems; (3) arcing from slip ring and shaft grounding brushes in rotating machinery, arcing between adjacent metallic components in an electric field where some of the components are poorly bonded to ground or high voltage; (4) thyristor switching; (5) other pulse interference, for example, arc welding, relay switching, thunder.

A lot of PD on-line monitoring systems were developed since 1980s. Because of the complexity and difficulties of interference suppression in the generator, most of their research concentrated on how to suppress noise and extract PD signal correctly. The most famous among them are PDA [7] and TGA [8]. PDA was developed by M. Kurtz in the early years of 1980s. The sensor used in this system was coupling capacitor. The coupling capacitor was fixed at the output terminal of a differential amplifier. It could detect PD in hydraulic generator with suppressing the noise successfully. However, when it was used for turbine generator, it could not be applied satisfactorily. Because the noise inside turbine generator is much higher compare to hydraulic generator, Sometimes noises are hundreds of times than discharge signal. The capability for PDA to suppress noise is only 10 times. Meanwhile, PDA require circle bus at least 2 meters or longer. As for the turbine generator, there is no such structure. Therefore, PDA can not be applied to PD on-line monitoring of large turbine generator. Another system which was called TGA was developed successfully in the beginning of 1990s. The sensor used for TGA is called Stator Slot coupler (SSC). The principle for TGA to recognize discharge and noise is the pulse shape. The system was on the basis of high-speed data acquisition and waveform analysis. The disadvantage of TGA is that SSC must be installed in the generator stator slot. In this case, the safety problem of generator itself must be considered seriously. This is the reason why the system can not be applied widespread.

In this paper, a new PD on-line monitoring system was developed and presented. It is a computer-aided PD on-line monitoring system. The system is composed of a broadband current transducer system, and a digital partial discharge analyzer. It can be used for both on-line and offline test of large turbine generators. Some typical defects, which often exist in the large generator stator bar, were simulated and detected by using new system. Some field test results are also given and discussed at the end of this paper.

2. System Description

The block diagram of the system is shown in Fig.1, which was mainly composed of two parts: PD coupling circuit and PD detection circuit. PD coupling circuit is used for coupling the PD signal. PD detection system is the core of total system. It is used for acquisition of the signal which coupled by the sensor and displaying the detection results after signal processing.

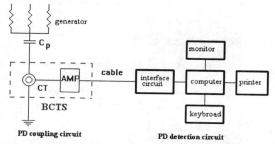

Fig.1: Test circuit of PD on-line monitoring system for large generators

2.1 PD coupling circuit

The sensor used in this system was a broadband current transducer system (BCTS). This transducer system was composed of a high-frequency current transducer, a preamplifier. The ground lead was chosen as a good measurement location because it was at a low potential about ground and PD current at any location in the test sample caused current pulse flowing in the lead. This kind of transducer was worked by way of magnetic coupling. Therefore the measurement circuit and the HV circuit had no direct electrical contact. It was safe to use especially for HV equipment on-line monitoring.

As we have known, the stator winding of generator is a distribution parameter component. Stator winding are embedded deeply in the stator slot. As for the electromagnetic coupling among each coil and circle, it is so weak that the capacitor among different windings can be ignored. Therefore, the transmission line theory can be applied to analyze the characteristics of PD pulse along the stator winding [9][10]. Former research results have shown that high frequency component of

the PD signal has obviously attenuated. The attenuation also depends on the transmission distance. Therefore, the main frequency component of the signal at the neutral point of generator is beyond the frequency range of 10MHz. Therefore, the upper limit of BCTS, which is applied, is 20MHz. As for the lower limit, the frequency should be much higher than the frequency of power supply in order that the resonant component of power supply could be avoided. But it should not too high. Otherwise, it will affect the reappear of impulse waveform. Therefore, bandwidth of BCTS that is applied in this system is from 10kHz to 20 MHz.

This transducer system also met the following requirements:

- Fast transient response (establishing time is in the range of a few ten nanoseconds);
- High sensibility (the minimum PD pulse magnitude which can be coupled is ±2mV);
- Linearity within the range of detection PD pulse currents (in the dynamic range of ±2mV to ±2000mV, no distortion);
- Stability in use (installed at the generator neutral point which is in zero potential, the transducer has good fixing and shielding).

Therefore, the new developed BCTS can be applied for coupling of PD signals of large generators.

2.2 PD detection circuit

Block diagram of PD detection circuit is shown in Fig.2.

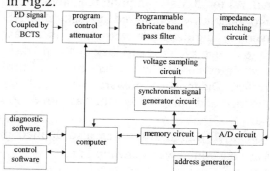

Fig.2 Block diagram of PD detection circuit

To extract the discharge parameters for computer analysis, some form of signal processing and digital conversion was necessary. In this system, the hardware consisted of a program controlled attenuator, a programmable fabricate band pass filter, a impedance matching circuit, and an A/D converter to digitize the PD pulse.

Program controlled attenuator was designed for judging the signal scope. According to the obtained original signal, computer gave the corresponding response to adjust signal magnitude, thus the PD signal can be enlarged (or attenuated) to a suitable scope. The dynamic range of this circuit is 40dB. Then the signal entered a programmable fabricate band pass filter. In this system, three filter band was designed, which was from 10kHz to 100kHz, from 10kHz to 2MHz, and from 10kHz to 10MHz, respectively. As for this system, different filter band was designed in order to satisfy different kinds of requirements of field test. After attenuation, filtering and impedance matching circuit, PD signal was flowing into the AD conversion circuit. Meanwhile, the voltage signal was sampled and converted to square wave, which could provide the information of phase position. In this system, the high speed AD converter worked by way of parallel comparison method. The maximum sample rate is 20Ms/s. Demultiplier circuit was designed so that the low sample rate could be obtained up to 25ks/s. AD conversion bits is 8. Hardware logic generation circuit was also adopted to support hardware address code and control logic. Due to the high-speed conversion, high-speed memory circuit was also required. Therefore, high-speed static flash memory was applied. Bi-address buffer was adopted to detract the requirement for memory speed. In order to observe the PD signal in one cycle of power supply, AD conversion length was required at least one cycle of power supply signal. As for 20Ms/s sampling rate, 400k Bytes memory caches was required at least. After AD conversion, digitized data was stored in a mass storage device such as hard disk or floppy disk. System inter control was operated by control software. Diagnostic software was used for analyzing and displaying the acquired PD signal.

An adaptive digital filter was also applied in this system which was shown in Fig.3.

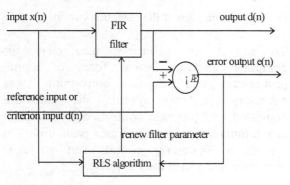

Fig.3 RLS adaptive digital filter

It was a FIR structure, RLS algorithm adaptive digital filter. This filter has the following features: (1) Structure is simple, phase shift is steady and linear, waveform would not distort. (2) RLS has fast convergence speed, it can follow the tracks of PD signal well when the signal appeared, then adjust its FIR structure. The noise-suppressing ratio of this filter was 40dB.

3. System Application

The PD on-line monitoring system was applied to a 200MW generating set in a power plant in order to obtain filed test result. PD monitoring results of time domain and frequency domain was shown in Fig.4a, Fig.4b, respectively.

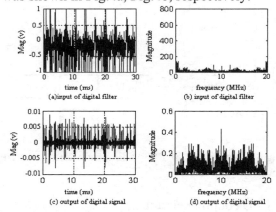

Fig.4 Field test results

We can conclude from Fig.4b that the central frequency of the discharge signal is less than 1MHz. The above filter system, which contains both hardware and software filters, was applied to suppress the noises. PD signal, which was after filtering, is shown in Fig.4c and Fig.4d. From Fig.4d, we can conclude that each frequency component was similar with that of the whole frequency domain. Therefore, there was no hazard discharge sources existed.

From the above on-line monitoring test results, we can also conclude that rather large noise was existed in the generator. Magnitude of the noise sometimes a few hundreds of times than discharge signal.

4. Conclusions

(1) A broadband PD on-line monitoring system was developed which can be applied for PD on-line monitoring test effectively and tracing insulation degrading procedure.
(2) Two typical discharge sources were simulated: Internal discharge in the stator bar;

Slot discharge between insulation and stator core. Characteristic spectrum was obtained, which can be used as the basis of generator PD on-line monitoring and diagnostic technique.

(3) The filter system which was applied contained a programmable fabricate filter and an adaptive digital filter. The programmable filter consists of three bands, which is from 10kHz to 100kHz, from 10kHz to 2MHz, from 10kHz to 10MHz. The adaptive digital filter is a FIR structure, RLS algorithm digital filter. With the above filter system, the noise can be suppressed up to 40dB. Discharge signal can be extracted from rather large noise surroundings. Simulated test and field test results show that this filter system can be a great help to suppress noise in PD measurement especially for on-line monitoring system.

References

1. G. C. Stone, "partial discharge part VII: Practical techniques for measuring PD in operating equipment," IEEE Electrical Insulation Magazine, Vol.7, No.4, pp. 9-19, 1991.

2. Wilfried Hutter, "partial discharge - Part XII: Partial Discharge Detection in Rotating Electrical Machines," IEEE Electrical Insulation Magazine, Vol.3. No.8, pp. 21-32, 1992.

3. M.Kurtz, J.F.Lyles, G. C. Stone, "Application of partial discharge testing to hydro generator maintenance," IEEE Trans. on Power Apparatus and Systems, Vol. PAS-103, No.8, pp. 2148-2156, 1984.

4. U.Kopf, K.Feser, "Rejection of narrow-band noise and repetitive pulses in on-site PD measurements," IEEE Trans. on DEI, Vol.2, No.3, pp. 433-447, 1995.

5. K.Itoh, Y.Kaneda, S.Kitamura, K.Kimura, "New noise rejection techniques on pulse-by-pulse basis for on-line partial discharge measurement of turbine generators," IEEE Trans. on Energy Conversion, Vol.11, No.3, pp.585-594, 1996.

6. Sher Zaman Khan, Zhu Deheng, "A new adaptive technique for on-line partial discharge monitoring," IEEE Trans. on DEI, Vol.2, No.4, pp.700-707, 1995

7. G.C.Stone et al, "The ability of diagnostic tests to estimate the remaining life of stator insulation," IEEE Trans. on Energy Conversion, Vol.EC-3, No.4, pp.833-841, 1988.

8. H.G.Sedding, S.R.Campbell, G.C.Stone, G.S.Klempner, "A new sensor for detecting partial discharges in operating turbine generators," IEEE Trans. on Energy Convertion, Vol.6, No.4, pp.700-706, 1991.

9. Wilson, R.J.Jackson, N.Wang, "Discharge detection techniques for stator windings," IEE Proceedings-B, Vol.32, No.5, pp.234-244, 1985.

10. J.W.Wood, H.G.Sedding, W.K.Hogg, I.J.Kemp, H.Zhu, "Partial discharges in HV machines; initial considerations for a PD specification," IEE Proceedings-A, Vol.140, No.5, pp.409-416, 1993.

11. F. H. Kreuger, "Partial Discharge Detection in High-Voltage Equipment," Butterworth, 1989.

EVALUATION OF PARTIAL DISCHARGE ACTIVITY BY EXPERT SYSTEMS

K. Záliš

Czech Technical University, Faculty of Electrical Engineering, Dept. of Electrical Power Engineering, High Voltage Laboratory, Czech Republic

Abstract: Several rule-based expert systems were developed for diagnostics of high voltage insulation systems, especially for the evaluation of partial discharge activity. The complex project for the evaluation of a partial discharge measurement on high voltage insulation systems has also been made. This complex evaluating system includes two parallel expert systems for the evaluation of a partial discharge activity on high voltage electrical machines.

1. INTRODUCTION

Due to the rise of cost and power of large electrical machines and equipment in electrical power network, it is necessary to pay attention to the operating reliability. The evaluation of the state of large electrical machines and equipment and the estimation of their behavior in future operation is executed by special diagnostic methods. The partial discharge (PD) measurement is one of the most effective methods for the evaluation of modern resin insulating systems and PD measurement is also more and more used for the indication of vibrating bars in stators of rotating electrical machines.

Usually there are not many problems with the PD measurement itself. There are many high quality commercial devices for this purpose on the market. The problems may rise in the calibration process of the measuring circuit. Most standards, regulations and directions specify a galvanic connection of an external calibrator with a complete measuring circuit, including the object under the tests. In many cases, the object has a high value of capacitance, e.g. in large hydroelectric alternators. Test calibration signal is then very damped down and misrepresenting. In this case it is necessary to leave the tested object out of the calibration process. Another big problem is the one of signal interference. The internal interference (a corona in a measuring circuit, ungrounded metallic parts, etc.) can be eliminated by special filters, arrangements and other techniques or special devices. Random disturbances can be removed by the statistic processing of measured PD data. But the elimination of the external periodic interferences spreading usually from the network is more complicated. The typical example is the interference impulses from the electrical equipment controlled by thyristors. Their impulse phase shift depends on the load of the electrical machine. In this case, PD data must be evaluated by the top specialist in this branch.

However, main problems are with the evaluation of measured the PD data, determination of critical levels of PD activity for different kinds of machines and the decision making of their future operation behavior.

2. EVALUATION OF PD DATA BY EXPERT SYSTEMS

For the purposes of the evaluation of measured PD data, the best way is to apply special expert systems with the elements of artificial intelligence [1]. The evaluation process usually begins with the selection of the input data. Measured PD data are statistically processed to remove wrong PD data. Then these „adjusted" data are processed by the special expert system. The results of the expert system processing are not only in the form of processed values of PD diagnostic parameters, but also in addition, in the form of the recommendations for the future operation

High Voltage Engineering Symposium, 22–27 August 1999
Conference Publication No. 467, © IEE, 1999

of the machine or equipment. The biggest advantage of these expert systems lies in the fact that the expert systems enable, even to a non-expert user, to determine the risks of further operation of the device without the necessity of consulting the matter with top experts. The expert systems are also very good and effective tools for the education of maintenance men and specialists in the electrical power engineering branch. It helps them to get experience and skill in evaluating high voltage insulation systems.

At the High Voltage Laboratory of the Czech Technical University (CTU) in Prague, Faculty of Electrical Engineering (FEE), Dept. of Electrical Power Engineering, several rule-based expert systems have been developed in the cooperation of top diagnostic workplaces of the Czech Republic for the diagnostics of high voltage insulation systems and for the evaluation of PD data. The IZOLEX expert system [2] evaluates diagnostic measurement data from commonly used off-line diagnostic methods for the diagnostics of high voltage insulation of rotating machines, non-rotating machines and insulating oils. The CVEX expert system evaluates the discharge activity on high voltage electrical machines and equipment by means of an off-line PD measurement. The CVEXON expert system is for the evaluation of the discharge activity by on-line measurement and the ALTONEX expert system is the expert system for on-line monitoring of rotating machines. Expert systems for the evaluation of an off-line measurement (CVEX, IZOLEX) are also in practice, while expert systems for an on-line measurement (CVEXON, ALTONEX) are under testing, and on the basis of requirements of testing workplaces, corrections of their knowledge bases are being performed.

All these developed expert systems are regularly updated with regards to the latest results of scientific research and practice. The evaluation of individual diagnostic methods is mainly stressed. The possibilities of their further development lie in the automation of measuring processes together with the mutual linking of individual sources of knowledge (databases, computer programs, etc.) in such a way that the expert systems can provide the user with a broader overview of the past and present states of the device and its grading according to other experts' criteria. Another possible area for the development can be found in incorporating the diagnostics of other electric components like capacitors, insulators, cables together with other kinds of diagnostics, e.g. with mechanical vibrations and heat stress.

3. COMPLEX SYSTEM FOR THE PD DATA EVALUATION

In these days, the complex system for the evaluation of PD measurement has been developed in the High Voltage Laboratory of the CTU. Two parallel expert systems for the evaluation of PD activity on HV electrical machines and equipment work in this complex evaluating system: a rule-based expert system performs an amplitude analysis of PD impulses for determining the damage of the insulation system, and a neuron network is used for the recognition of PD patterns (a phase analysis of PD impulses) to determine the kind of PD activity and location of the resource of PD activity. Both expert systems, including the unit for a standard evaluation of the discharge activity, operate simultaneously, and special software ensures coordination between them. The connection between a computer and a measuring unit enables to load the digitized measurement data directly into the computer. This procedure enables to evaluate diagnostic parameters immediately and enables to use this diagnostic method as an on-line measuring method.

PD data from the measuring probe is processed in the measuring unit. The values of the area, and the phase shift of each PD impulse are digitized. The set of this digitized data per 20 ms (one period of the supply network) is then transferred into computer, into the special software central unit where the digitized PD data are further processed. The disturbances and random data are discarded and the rest of the PD data is processed for the visualization and for inputs into expert systems.

The construction of expert systems requires the construction of a "bank of experts", the determination of limit values of diagnostic parameters and the development of a suitable training set for the training process of the neuron network.

The rule-based expert system for the amplitude analysis of PD data is based on the proved CVEXON expert system. The amplitude analysis of PD data determines the extent of the damage of the high voltage insulation system. After preliminary processing, the data from the central unit are delivered, into a small input database. The expert system evaluates this data and saves the results of the consultation (probabilities of hypotheses) into a small output database. After subsequent processing, the results are displayed on the screen of the monitor of the computer.

A neural network is being developed for the PD data evaluation by the phase analysis of PD impulses, i.e. by the recognition of PD patterns. Both input and output of the data are practically similar, as in case of the rule-based expert system. They are stored in the form of text files.

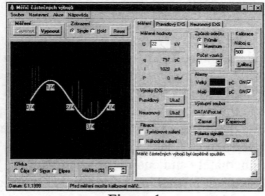

Figure 1

The visualization of all results is done in an accessible form for the user, i.e. all results are shown on the virtual front panel of the standard measuring instrument (see Figure 1). Besides a standard visualization of PD data impulses during the period of the supply network, the results of expert systems evaluation, modes of filtering and

the results of statistical processing are also continuously displayed. The level alarms are visualized as well.

The complex system has its own archive database. It is possible to determine the different period of data saving for the different state of an operating system – a normal operation, enhanced PD activity, overfullfilment of alarm levels, important decisions of expert systems, etc. These records of PD data are very important for the consequent analysis of the defect state of the observed machine.

In the first stage of the work on this research project, in 1997, the measuring unit and rule-based expert system were developed. Based on our experience with commercial measuring units, the special measuring unit, including a calibration equipment, has been made in the cooperation with the Development Laboratory of the CTU in Poděbrady. This unit enables quick and non-distortionless scanning of PD impulses.

Our work in 1998 was focused on the increase of the efficiency of the data flow in the system, the development of the central database for the measured data and the design and production of better multipurpose measuring unit including a new type of a calibration equipment. The rule-based expert system for the amplitude analysis of PD data was developed and successfully tested.

The final stage of our research activity, in 1999, lies in the development of a neural network for the recognition of PD patterns (a phase analysis of PD impulses). The construction of this expert system requires the development of a suitable training set. We decided the neural network will be taught in three steps: the first, the elements of the training set will be composed of the artificial data from the scientific literature [3, 4], the second, PD data from the well-known simple real PD arrangements, and the third, PD data from the real devices with known defects.

The newly developed evaluating system for the PD stand has several advantages in comparison with commercially produced

PD devices:

- The digitization of PD data directly in the measuring unit, the transfer of the digitized data into the computer via a standard serial (RS232) line, and the minimization of interference impulses by means of PD data processing.
- The possibility of SW modification following the specific conditions of singular tested machines or equipment in operation.
- Low price of this stand (complex system for PD measurement in operation) in comparison with commercially produced PD devices.
- The improvement of the mechanical resistance and the operational reliability of the PD device considering the fact that the new PD stand has minimum of mechanical and analog parts.
- This complex system can be modified for purposes of education, and the testing of knowledge and experience of technicians or students in the electrical power engineering branch.

4. CONCLUSIONS

Between 1994 and 1998, the new principle of PD device has been developed at the High Voltage Laboratory of the CTU, Faculty of Electrical Engineering in Prague in the collaboration with the Development Laboratory of the CTU in Poděbrady. In these days, the stable measuring stand (measuring workplace) for PD measurement under operational conditions in on-line (non-interruptive) mode is created. This new conception of PD data evaluation has several advantages in comparison with the conceptions of standard commercial instruments. The digitization of PD data in the measuring unit directly, the transmission of digitized data into the computer and their further processing by software reduce the influence of an external interference into minimum. In addition, the processing of PD data by software enables the data processing without any limitation, in accordance to the latest research. The modification of all parts of this complex system is very easy. Low price is in comparison with commercially produced instruments an indispensable advantage of this complex system.

On the base of diagnostic PD monitoring (on-line measurement and evaluation during the operation) by the proposed PD stand (a measuring workplace), the operational reliability and safety will be rapidly increase.

Results of this project will increase the safety and reliability of the operation of large electrical machines (alternators and transformers) and HV equipment working in the Czech National Network System. This complex PD stand will also find the special applications at the nuclear electroenergetics workplaces.

The author gratefully acknowledges the contributions of colleagues and students of the Electrical Power Engineering Department of the CTU for their cooperation, help and stimulating discussions. This research is financially supported by the Grant Agency of the Czech Republic (grant No. 102/97/0481) and by the CTU.

REFERENCES

[1] Záliš K.: *Database and Expert Systems for Dielectric Diagnosis*. In: Proceedings of the 9th International Symposium on High Voltage Engineering, August 28 - September 1, 1995, Graz (Austria), ref. 5593.

[2] Záliš K.: *IZOLEX - Expert System for Diagnostics of Insulating Systems*. In: Acta Polytechnica, Vol. 36, No. 3, 1996, CTU Prague, p. 55-68.

[3] Florkowska B: *Vyladovania niezupełne w ukladach izolacyjnych wysokiego napięcia – analiza mechanizmów, form i obrazów (Partial Discharges in High Voltage Insulation Equipment – Analysis of Mechanisms, Forms and Patterns)*. Wydawnictwo IPPT PAN, Warszawa, 1997.

[4] König D., Rao Y.N.: *Partial Discharges in Electrical Power Apparatus*. Vde-Verlag GmbH, Berlin and Offenbach, 1993.

A NEW MULTI-PURPOSE PARTIAL DISCHARGE ANALYSER FOR ON-SITE AND ON-LINE DIAGNOSIS OF HIGH VOLTAGE COMPONENTS

J.P. Zondervan[*], E. Gulski[*], J.J. Smit[*], T. Grun[**], M. Turner[**]

[*] High Voltage Laboratory, Delft University of Technology, The Netherlands
[**] Haefely Trench AG, Tettex Instruments Division, Switzerland

ABSTRACT

This contribution reports about a solution for on-line partial discharge (PD) testing. The possibilities of a standard digital detector for off-line PD testing have been extended by tools for on-line PD recording according to the VHF/UHF PD detection technique. The development of the measuring system aimed at realization of modern technologies in a flexible solution, while keeping the familiarity of the functions of the original detector intact.

1. INTRODUCTION

As the title of the paper implies, this contribution introduces a new partial discharge (PD) detector. Given the number of contributions on new, novel or innovative PD detectors that have been published in the last few years, one might ask:
- Is there need for yet another new type of PD detector?
Or, in more applicable terms:
- What can this new PD detector add to the range of existing PD detectors?
In order to respond to this question satisfactorily, the contribution will emphasize on two key features of the system:
a)   The detector is suitable for both on-line and off-line discharge measurements.
b)   The system is able to operate with a variety of common types of coupler.
The keynotes that characterize the innovation in the system are
i)   Following modern trends
ii)   Familiarity
iii) Flexibility
All the effort resulted in a system that does not only satisfy the user's demand of utilizing modern technologies in practical devices, but also responds to the general prevailing preference for upgrading a common device to modern standards instead of developing a completely new system.

2.   ON-LINE PD DETECTION TECHNOLOGY

PD detection is a very popular tool for assessing the insulation state of any HV type equipment (eg. power transformers, rotating machines, GIS). PD testing was originally utilized for, and is still most applied as, type and routine tests to ensure the quality of HV equipment during manufacture and immediately after delivery.

However, since issues such as diagnostic maintenance and lifetime management became of more and more importance in the efficiency-focussed perceptions of the power utilities, PD testing and analysis have also been found to serve as input for lifetime management systems [ref.1]. Although standards for PD detection have not yet caught up with those for the more 'classic' inspection techniques, such as loss-factor for rotating machines and gas-in-oil-analysis for power transformers, it is generally believed that PD detection is the most sensitive and therefore most powerful technique for revealing defects that jeopardize the lifetime or lead to unavailability of HV-equipment [ref. 2,3].

For quality checks during service of HV equipment, periodic PD measurements are usually performed off-line, during outages and maintenance overhauls. As the time period between two scheduled outages can be up to a couple of years, there exists a reasonable possibility for an in-service failure. To reduce this possibility, efforts are made to develop tools that provide early warnings for possible failures. Conversely, utilities tend to minimize their overall maintenance costs by extending the time interval between two revision dates (e.g. from 4 to 6 years). This is only realized when efficacious tools are applied to monitor the system's condition in order to ensure a safe and reliable in-service duration.

The technology of on-line PD detection has been recognized as a powerful tool to assess the quality of the insulation condition during regular operation of HV equipment [ref. 13]. Several measuring techniques are in use or in development to detect discharges in HVAC equipment during their regular operation. The **VHF/UHF PD detection** technique was reported earlier to be a fruitful means to measure discharge patterns in GIS, turbogenerators and power transformers [ref. 4-7,14]. This technique uses a spectrum analyser as a tunable filter in combination with a digital PD detector. By selective tuning of the spectrum analyser, PD that emanate from defects in the insulation are recorded, after which they are digitally processed

High Voltage Engineering Symposium, 22–27 August 1999
Conference Publication No. 467, © IEE, 1999

for analysis and interpretation. More background on this technique can be found in [ref. 3,4].

3. VHF/UHF PD DETECTOR

For a practical implementation of the VHF/UHF detection technique, the industrial TE-571 detector by Haefely Trench Tettex has been adapted. A software interface has been developed to establish a communication link between the TE-571 and a spectrum analyser. The communication runs through a serial RS-232 port and is controlled from the TE-571. As for the spectrum analyser, two types of programmable spectrum analysers were found to be suitable for the application: the 2711 by Tektronix and the HP-8590 by Hewlett Packard. In figure 1 the detection circuit of the VHF/UHF PD detector in an on-line PD test is schematically depicted.

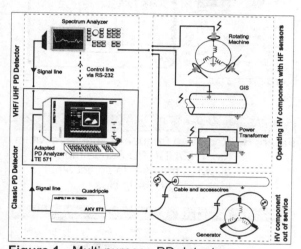

Figure 1 Multi purpose PD detector:
- classic detector for off-line PD tests (lower)
- VHF/UHF PD detector for on-line PD tests (upper)

The TE-571, a standard detector for off-line PD tests [ref. 8], has been extended with tools to enable on-line PD detection. The modified system automatically detects if it should operate in the classic or VHF/UHF detection mode, by checking for the presence of a spectrum analyser.
Figure 2 depicts the flow chart of the employment of the VHF/UHF PD detection technique for on-line PD tests. First, the spectrum of the signal response at the coupler's site is recorded, see figure 3, and analysed in order to find possible frequencies that are suitable for a PD measurement. In this scope, the term suitable refers to highest sensitivity and least interference as has been reported in [ref. 4,6,14].

After one or more suitable frequencies are found, all manually and automatically generated parameters, see figure 4, are stored in the proposed CIGRÉ data format [ref. 9]. Then the spectrum analyser is set in *zero-span mode* [ref. 3,4] to one of the selected frequencies and the system proceeds to the VHF/UHF PD Scope mode where the PD are recorded and displayed.

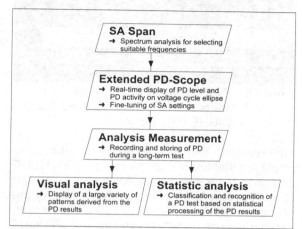

Figure 2 Flow chart for on-line PD test and analysis with the presented VHF/UHF PD detector

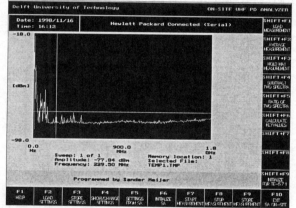

Figure 3 Example of a recorded spectrum from which suitable measuring frequencies are selected

4. VHF/UHF PD SCOPE

In the VHF/UHF PD Scope mode, the discharges are recorded and displayed on the familiar ellipse time base, see figure 5. As there is not yet a calibration standard available for on-line PD detection, the discharge height or PD level is shown in Volt-unit (corresponding to the reading of the spectrum analyser) instead of the Coulomb-unit.

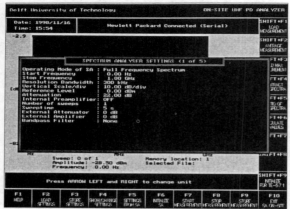

Figure 4 Example of part of the parameter sheet that is stored in the proposed CIGRÉ data format

Furthermore, two additional parameters are displayed that are of importance to the VHF/UHF detection technique: The *centre frequency* and *resolution bandwidth*. If some fine tuning of the spectrum analyser is necessary, then both parameters can be adjusted within the VHF/UHF PD Scope mode. As there may have been more than one suitable frequency that resulted from the spectral analysis (as discussed in the former section), the user can re-display the former recorded spectrum as a guide to select a different frequency, see figure 6.

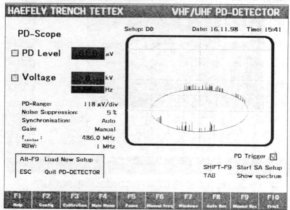

Figure 5 The VHF/UHF PD Scope measuring screen. The discharges on the ellipse were observed at a 420 kV GIS test-installation

5. MEASUREMENT AND ANALYSIS

In addition to the VHF/UHF PD Scope mode for real-time recording and display of PD activity, an analysis measurement can be started in order to process a large variety of basic and deduced quantities characterizing the properties of a PD test [ref. 10]. The test time is variable from 2 minutes to 100 hours. After a measurement has finished, the data is stored and can be displayed or printed whenever required for visual analysis. As an example, in figure 7 some basic quantities

(maximum charge, mean charge and discharge intensity) related to the phase position of the voltage cycle are depicted, the so-called phase-resolved patterns. Figure 8 shows an example of the $H_n(\varphi, q)$-distribution, a deduced pattern that visualizes the three-dimensional relationship between discharge magnitude, discharge intensity and phase angle. Furthermore, discharge intensity distributions, time behaviour graphs of several parameters (e.g. discharge magnitude, quadratic rate), phase-resolved distributions along the test time and a coloured phase-magnitude-intensity graph are available.

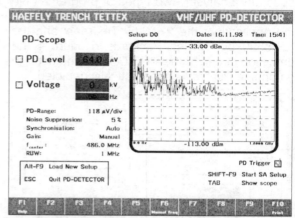

Figure 6 Recorded frequency spectrum within the VHF/UHF PD Scope mode, allowing the user to select other measuring frequencies

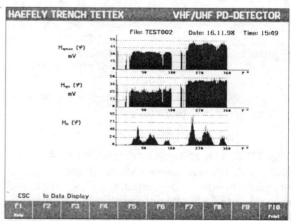

Figure 7 After a measurement is performed, several basic and deduced quantities are stored for display; e.g. phase-resolved patterns

As visual analysis usually requires highly skilled test engineers, the VHF/UHF PD detector offers a computerized method of analysis, which is less subjective. The properties of a PD test are characterized by a set of statistically processed parameters, earlier reported as a *fingerprint* [ref. 11]. With this fingerprint, a test of an unknown PD source can be compared with a databank consisting of fingerprints of known PD sources, in order to classify or recognize the unknown PD source.

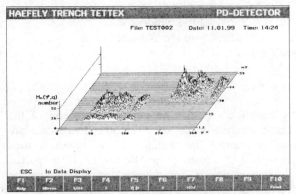

Figure 8 Example of the three-dimensional $H_n(\varphi,q)$-distribution as measured at a 420 kV GIS test setup

6. CONCLUSIONS

Besides the comprehensive features of the detector as discussed in the former sections, its main feature is its versatility. As mentioned in section 3, the detector operates for both off-line (according to IEC 270 recommendations) and on-line PD tests. As for the latter, the VHF/UHF PD detection technique does not require a specific type of coupler. Any coupler that operates in the VHF- or UHF-region is suitable to operate with the VHF/UHF PD detector. The detector has been proven to operate successful with VHF capacitive couplers and Rogowski coils (turbogenerator, [ref. 4,5]), stator slot couplers (turbogenerator, [ref. 12]), VHF current transformers (power transformer [ref. 7]) and UHF disc couplers (GIS, [ref. 6]).

Furthermore, the detector can be deployed for routine and type testing during the manufacturing stage as well as for in-service tests.

REFERENCES

1. J.J. Smit, *High Voltage Insulation Diagnosis for Lifetime Management*, Proc. ISEIM, Toyohashi, 1998, pp. 23-28

2. E. Binder et al., *Predictive Maintenance of Generators,* Cigré session, Paris, 1992, pap. 11-305

3. H.D. Schlemper, R. Kurrer, K. Feser, *Sensitivity of On-site Partial Discharge Detection in GIS*, ISH, Yokohama, 1993, Vol. 3, pp. 157-160

4. J.P. Zondervan, E. Gulski, J.J. Smit, R. Brooks, *PD Pattern Analysis of On-line Measurements on Rotating Machines*, Proc. CEIDP, Minneapolis, 1997, pp. 242-245

5. J.P. Zondervan, E. Gulski, J.J. Smit, R. Brooks, *Use of the VHF Detection Technique for PD Pattern Analysis on Turbogenerators*, Proc. ISEI, Washington, 1998, pp. 274-277

6. S. Meijer, E. Gulski, W.R. Rutgers, *Acquisition of Partial Discharges in SF_6 Insulation*, CEIDP, San Fransisco, 1996, pp. 581-584

7. J.P. van Bolhuis, E. Gulski, J.J. Smit, T. Grun, M. Turner, *Comparison of Conventional and VHF Partial Discharge Detection Methods for Power Transformers*, ISH, London, 1999

8. E. Gulski, R. Oehler, *New Generation of Computer-Aided Partial Discharge Measuring Systems*, ISH, Graz, 1995, Subj. 4, paper 4570

9. Task Force 15.03.08, *CIGRÉ Data Format for GIS Partial Discharge Software Applications*, Electra no.177, April 1998, pp. 87-93

10. E. Gulski, P. Seitz, *Computer-Aided Registration and Analysis of PD in HV Equipment*, ISH, Yokohama, 1993, pap. 60.04

11. F.H. Kreuger, E. Gulski, A. Krivda, *Classification of PD*, IEEE Tr. on EI., 1993, Vol. 28, No. 6, 917-931

12. J.P. Zondervan, E. Gulski, J.J. Smit, M.G. Krieg-Wezelenburg, *Comparison of On-line and Off-line PD Analysis on Stator Insulation of Turbogenerators*, Proc. ISEI, Washington, 1998, pp. 270-273

13. J. Fuhr et al, *Generic Procedure for Classification of Aged Insulating System*, Proc. 3rd Int. Conf. on Prop.and Appl. of Diel. Mat., Tokyo, 1991, pp. 35-38

14. S. Meijer, E. Gulski, J.J. Smit, F.J. Wester, T. Grun, M. Turner, *Interpretation of PD in GIS Using Spectral Analysis*, ISH, London, 1999

EXPERIENCE WITH INSULATION CONDITION MONITORING ON A 1050 MW GENERATOR

S.A. Higgins
Eskom

J.P. Reynders
University of the Witwatersrand

Johannesburg, South Africa

ABSTRACT

There are a number of off-line condition monitoring tools that have historically been used to establish the state of machine insulation. On-line insulation condition monitoring tools are a relatively new development and have not yet established themselves as an industry standard.

Eskom was presented with an insulation problem, which on first analysis indicated that an immediate repair should be undertaken. By integration of the information provided by the classical off-line and newly introduced on-line techniques it was decided to continue running the machine.

The latest detailed off-line investigation vindicated this decision.

1. INTRODUCTION

Eskom has actively started to apply a combination of off- and on-line insulation condition monitoring tools to the management of its generators. An insulation problem presented on a 1050 MW Unit provided an ideal opportunity to test how the results from these techniques could be integrated to facilitate the decision to keep the machine on the grid. It also allowed some meaningful comparisons to be drawn between the test methods.

2. DESCRIPTION OF THE INSULATION TEST TECHNIQUES USED

Eskom has evaluated a number of insulation test techniques in an attempt to formulate the best long-term insulation care strategy for its machines. Over a period of years the quadratic rate, TVA probe, IEC 270 and on-line techniques have been evaluated.

2.1 Quadratic rate partial discharge detection

Quadratic rate is the sum of the squares of the apparent charges, during a certain time interval divided by this interval [1]. The instrumentation used by Eskom displayed the quadratic rate value in a logarithmic format.

The instrument measures partial discharge as follows:

A coupling capacitor is tied to either end of the phase winding under test. The other two phases are tied to earth. The winding to be tested is then energised by means of an HV test set. The partial discharge signals are detected at the output the voltage divider formed by the coupling capacitor and the coupling tied to earth. A capacitive/inductive voltage divider is used. A block diagram of the coupling of the quadratic rate instrumentation is shown in Figure. 1.

Once the signal has been detected by the voltage divider the input to the instrumentation is switched via a switching matrix. The signal is presented to the switching matrix unprocessed.

The signal from the switching matrix is processed by one of three selectable narrow band filters centred at 10 kHz, 30 kHz, and 130 kHz. The quadratic rate of the filtered signal is calculated by analogue circuitry. The result of this calculation is presented as an analogue value on a meter in dB format. The signal is at all times transient within the analogue domain, and no long-term storage of the quadratic rate measurement is possible.

An absolute reading of 36 dB or less at 0.6 of Line voltage is the Eskom acceptance criteria.

2.2 IEC 270 partial discharge detection

The method of coupling to the phase under test is very similar to the quadratic rate method. The major difference is that the IEC 270 method uses a capacitive/resistive divider. The method of coupling is shown in figure 1.

Once the signal has been detected by the voltage divider, the input to the instrumentation is switched via a switching matrix.

Once the switching matrix has selected the signal the IEC 270 system passes the signal through a selectable wide-band band-pass filter. The low-end cut-off of this filter is selectable between 40kHz, 80kHz and 100kHz. The high-end cut-off is selectable between 250kHz, 400kHz and 800kHz. The peak of this filtered signal is detected, and digitised. The values of the individual peaks of the discharges are recorded and displayed. These values are stored digitally for post processing. From the

High Voltage Engineering Symposium, 22–27 August 1999
Conference Publication No. 467, © IEE, 1999

individual discharges the total discharge current is calculated. A block diagram of the instrument is shown in figure 2.

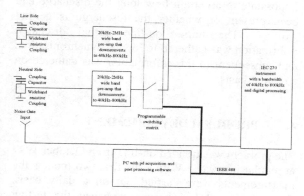

Figure 1. Coupling of the partial discharge test equipment to the machine

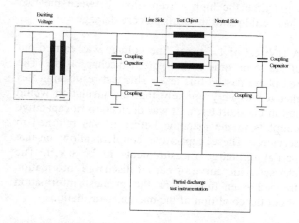

Figure 2. Block diagram of the IEC 270 instrument

2.3   TVA probe partial discharge detection

The apparatus for this test consists of a probe and hand held peak pulse meter (PPM). The probe is moved along the slot sections of an energised stator, and the measured peaks are read from the meter. The TVA probe used is of the type designed for use by British Columbia (B C) Hydro [4]. The probe and meter are commercially available.

The TVA probe head consists of a series resonant Resistive Inductive Capacitive circuit, designed to resonate at a frequency between 6 and 7 MHz, enclosed within epoxy resin. The probe head captures a pulse made up of a narrow band of frequencies generated by the discharge sites [4]. The peak of this pulse is directly proportional to the discharge magnitude [2]. At this resonant frequency the partial discharge pulse spectra are attenuated very rapidly as they propagate away from the discharge site. This allows localisation of the discharge site to within a few centimetres [3]. The resonant frequency is low enough to allow a very simple circuit to be used to display the pulse height value. The TVA probe is shown in Figure 4.

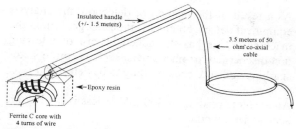

Figure 4. TVA probe

The acceptance criteria vary between insulation types, but a reading above 10 mA is considered cause for concern.

2.4   On-line partial discharge detection

Digital instrumentation is used to gather, process and display partial discharge information. The instrumentation is attached to permanently mounted coupling capacitors via connections housed in a terminal box. Two sets of coupling capacitors are used on each phase. The first set is mounted at the phase output terminals of the machine. The second set is mounted at least two meters away from the first set toward the generator transformer.

A block diagram of the on-line detection method is shown in Figure 5.

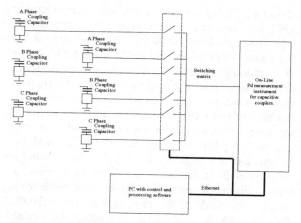

Figure 5. Block diagram of the on-line partial discharge detection instrument

The instrumentation performs the data processing at the same time as it takes the measurement. The partial discharges on the positive and negative sections of the mains cycle are separated, and displayed on a graph.

A predominance of discharges in the positive cycle indicates that there is a higher concentration of sites closer to the copper conductor and the groundwall. A predominance of discharges in the negative cycle indicates that there is a higher concentration of sites closer to the core slot wall. An equal spread of discharges across the positive and negative cycles indicates void discharges.

As a result of the method of detection, the analysis equipment detects all electrical discharges within the machine. This has the added benefit of allowing non-discharge type phenomena such as coil to stator sparking and core back burning to be detected. [5].

An absolute reading of 400 mV or less is the Eskom acceptance criteria.

3. PROBLEM IDENTIFICATION

Up to the start of 1997 Eskom only had the quadratic rate instrumentation available for insulation testing.

A routine quadratic rate test carried out in July of 1996 showed that the red phase line-side partial discharge measurement was 3 – 4 dB's higher than the other two phases. This type of discrepancy between phases occurs frequently, and was not considered cause for concern. The absolute reading was 22 dB, which is 14 dB below the acceptance criteria. The machine was returned to service with a clean bill of health.

During 1997 the quadratic rate equipment was retired, and both the TVA and on-line techniques became available to Eskom.

A routine TVA probe test carried out during October 1997 indicated two discharge sites with abnormally high readings. Both readings were recorded on the red phase. A disadvantage of the TVA probe test is that readings cannot be localised between top and bottom bars. The first reading was 20 mA, which is double what would be considered acceptable for a machine of this type. It was detected in the centre of the core 16 wedges from the exciter end. The top bar at this point is energised at 8.4 kV, and the bottom bar is energised at 2.4 kV. This discharge site, taken in isolation, is not seen as a problem. Partial discharge measurement is very subjective, and the tolerance of the reading can be +/- 100%. This reading is only twice the acceptable value. Partial discharge inception in a hydrogen environment pressurised to 3-bar pressure is in the region of 8kV. Only the top bar is stressed above this value, and only for a very small part of the power frequency cycle.

The second high reading was detected at the end of the core one wedge from the exciter end. This reading was 80 mA. The top bar at this point is energised at 12.3 kV, and the bottom bar is energised at 13.8 kV. The discharge is eight times the acceptable value. This is high even within the tolerance of partial discharge testing. Both top and bottom bars are both stressed above 8 kV for a considerable portion of the power frequency cycle. This discharge site is seen as a problem.

The combination of two weaknesses in the insulation system of a single phase is cause for concern. If both weak insulation points failed to earth at the same time the high current flow between the two sites could result in localised core damage.

A HV test of 1.1 times line voltage was carried out at the same time as the partial discharge test. The red phase passed this test. This test result indicates that it was very unlikely that the insulation would fail in the short term. It was decided to fit capacitive couplers to the machine before it was returned to service. These capacitive couplers allow on-line partial discharge measurements to be taken. The characteristic curve of partial discharge information trended over time gives the greatest information about the condition of the machines insulation.

The TVA test was the first of this kind to be carried out on this machine. From a single test it is impossible to ascertain how long the discharge had been present or whether the discharge is getting worse. The initial on-line partial discharge information was gathered following synchronisation, and subsequent to this information was gathered on a regular basis.

4. PROBLEM DEVELOPMENT

The problem was trended on-line from October 1997 at regular intervals. Within the first two months the measurements were carried out on a daily basis. Once the trend had been established the testing interval was first increased to monthly and, when the trend was confirmed, to three-monthly.

The initial peak discharge measurement at synchronisation was 250mV. There was a greater concentration of discharges in the positive cycle indicating that the discharge site was closer to the groundwall.

Over the next few days the peak discharge increased rapidly to 400mV. All the tests indicated that the discharge site was close to the groundwall.

Over the next two months the discharge slowly increased, until it peaked at 850mV. This was very worrying, but the insulation did not fail at this value. Following this peak, the value levelled out, and then slowly decreased to 700mV. The value has fluctuated over time, but has remained in a range around 700mV.

A full off-line insulation investigation was carried out in January 1999. Both a TVA probe and IEC 270 tests were undertaken. The IEC 270 test indicated delamination near the line end. The TVA probe indicated that the peak discharge at both sites had decreased. The TVA probe only detects sites that discharge at greater than 20 times a second. A small delamination may discharge at this rate but as the delamination increases in length it discharges less and less frequently. A large delamination may discharge as infrequently as once every few seconds, and this would be a large discharge. It is likely that the large discharges measure during the first TVA probe test were the start of a delamination. As the machine ran the delimitation increased in length. This would explain the steady increase in the on-line readings.

5.   PROBLEM MANAGEMENT

Both the TVA reading and the on-line results indicated a very serious insulation problem. If the results were taken on face value the machine should have been repaired immediately. It is, however, in Eskom's interest to keep the machine running, since it plays an important role in the stability of the grid. An outage to carry out an immediate bar repair would have been unplanned and lead to significant additional costs in comparison with a planned outage.

When the TVA probe test first detected the problem, a HV test was performed on the machine. The insulation did not break down at 1.1 times line voltage, which corresponds, to 26.4 kV. The discharge site would only be subject to a maximum of 13.8 kV. On the strength of this test result the machine was returned to service. It was, however, necessary to monitor the development, if any, of the void in the insulation. For this reason the capacitive couplers were fitted, and the on-line testing was performed.

The peak partial discharge was seen to increase, over the period of a few months, to alarming levels. However the readings indicated that the site was closer to the conductor than the slot. At no time did a reading indicate that the discharge site was treeing toward the grounded slot wall. If the discharge site remained close to the copper there is only a very slight chance of a stator earth fault developing.

The latest off-line tests indicate that the fault may be a delamination. This information does not change the way the problem is being managed. As this on-line measurement indicates that the peak discharge has levelled out, it is likely that the insulation will remain stable and there is very little chance of a failure occurring.

The machine will run with on-line monitoring of the machine insulation being performed at regular intervals.

6.   CONCLUSIONS

A combination of information from more than one condition monitoring technique was used to make a decision regarding an insulation fault. The information from the quadratic rate method did not identify the severity of the fault. This is because this technique integrates the partial discharges to give a global indication of the state of the insulation. However when the information from the quadratic rate test is combined with the other three tests, which all indicate peak discharge values, it becomes easier to identify possible problems and localise the possible source of the high discharge sites.

Using the information from various insulation condition monitoring techniques a machine that ordinarily would have had to been repaired has been run with a known fault.

Following the latest inspection the machine will continue to run in its closely monitored state for at least one more outage before the insulation problem is repaired.

7.   ACKNOWLEDGEMENTS

We would like to thank the Eskom Technology Group Research Division for providing the funding that made this research work possible.

8.   REFERENCES

1. IEC, (1981), "Partial Discharge Measurements", International Electrotechnical Commission, IEC Standard, Publication 270, Second Edition, 1981.

2. Kreuger, F.H. (1964), "Discharge Detection in High Voltage Equipment", Heywood, 1964.

3. Smith, J.W. (1996), "Some Comments on Application of TVA Probe Testing", CEA Fourth Motor and Generator Partial Discharge Conference, Houston, Texas, 22-24 May 1996.

4. Smith, L.E. "A Peak Pulse Ammeter-Voltmeter Suitable for Ionisation (Corona) Measurement in Electrical Equipment, Doble Paper 37AIC70.

5. Stone, G.C. Lloyd, B.A. Campbell, S.R. (1996), "Development of an Automatic Continuous On-Line Partial Discharge Monitor for Generators", Fourth CEA/EPRI International Conference on Generator and Motor Partial Discharge Testing, Houston, May 1996.

PHYSICAL MODELLING AND MATHEMATICAL DESCRIPTION OF PD PROCESSES WITH APPLICATION TO PRPD PATTERNS

C Heitz

ABB Corporate Research, Switzerland

ABSTRACT

A general framework for the physical description of partial discharge (PD) processes is presented that holds for different types of PD defects. A PD process is a special stochastic process consisting of point-like discharges and drift/recombination phases between the discharges where only the fast point-like discharges can be measured as PD signals. It is determined by few basic physical parameters. Mathematically, a PD process can be formulated either in a point process framework or in a stochastic process framework. Both approaches are presented. The link between the process and observable quantities like Phase Resolved PD (PRPD) patterns is derived.

With the new approach a much easier interpretation and analysis of PD patterns is possible. For all of the three discharge types (internal, surface, corona) examples are presented that demonstrate the applicability of the model.

INTRODUCTION

In the present contribution a generalized framework for the physical description of partial discharges is presented. It can be used for a large variety of PD phenomena, including internal discharges, surface and corona discharges in gases or liquids.

A PD process is considered as a non-observable stochastic process consisting of fast transient discharges (PD events) and drift/recombination phases between the discharges (i.e. slow discharges). Only the fast discharges can be observed leading to a restricted possibility of observing the PD process. Thus, the modelling consists of two steps: In a first stage the dynamics of the PD process itself is formulated which is governed by two processes, namely a continuous slow discharge process and a quasi-discontinuous fast discharge process. The interaction of these processes can be described in a closed mathematical framework. In a second step the link between the process dynamics and observable quantities like, e.g., PRPD patterns is made.

In the following a fixed discharge geometry is assumed for simplicity. Discharges caused by, e.g., moving particles are not considered. However, they can be described in the same framework with only slight modifications (see Section DISCUSSION).

High Voltage Engineering Symposium, 22–27 August 1999
Conference Publication No. 467, © IEE, 1999

PHYSICAL MODELLING I: DYNAMICS

A PD is a localized discharge phenomenon that is restricted to a limited space and usually starts at a point with high electrical stress, called "defect" in the following. Conceptually one can divide the total electric system into a small localized "PD system" where discharges take place, and an embedding system. The *PD system* is defined by the defect geometry (void in insulator, particle in gas, protrusion in oil, ...). Its time-dependent state is described by the surface or space charges deployed in the vicinity of the defect by previous discharges. These charges change during the PD activity, leading to a specific dynamics. In the following, instead of these charges we consider the "internal" field $E_i(t)$ being produced by them at the defect site. The *embedding insulation system* does not take part directly at the PD activity. It acts on the PD system by supplying an "external" field $E_0(t)$ (the applied field without any space or surface charges) at the defect which is proportional to the applied voltage.

The dynamics of a PD system consists in the temporal evolution of the field $E_i(t)$ in the presence of the driving external field $E_0(t)$. It is determined by two sorts of processes. First, there is a sequence of point-like quasi-discontinuous discharges (the so-called *PD events*). Between these events, the charge distribution near the defect changes by a slow continuous process. Both processes are regarded in the following.

PD events (fast discharges)

During a discharge at time t the total electric field $E_{tot}(t)$

$$E_{tot}(t) = E_0(t) + E_i(t) \qquad (1)$$

drops to a residual field E_{res}:

$$E_{tot} \mapsto E'_{tot} = \pm E_{res} \qquad (2)$$

The positive sign is chosen if $E_{tot}(t)>0$ and vice versa. The value E_{res} of the residual field is defect specific, e.g. for air-filled voids $E_{res} \approx (0.2 \ldots 0.5)\, E_{cr}$ where E_{cr} is the critical field of air (Gutfleisch and Niemeyer (1)), for a streamer corona discharges in a strongly attaching gas E_{res} is about the critical field of the gas. It is assumed that the residual field has a constant value for each discharge.

With Eq. (2) a discharge leads to a sudden jump in E_i:

$$E_i(t) \mapsto E'_i(t) = \pm E_{res} - E_0(t) \qquad (3)$$

Drift and recombination of charge carriers (slow discharge)

The bipolar charge distribution deployed in the vicinity of the PD defect may change under the effect of the electric field. In particular it may vanish by drift and recombination. This, in turn, changes the field E_i in a way that is specific for the drift/recombination process. As examples, the change of E_i may be proportional to E_{tot} (e.g. for drift) or to E_i (for recombination, since E_i is proportional to the number of charge carriers). In general, it can be described by a differential equation

$$\dot{E}_i(t) = f(E_i, E_0(t)) \qquad (4)$$

where $\dot{E}$ means the temporal derivation of E. In contrast to the jump process during a PD event, this field change is continuous and deterministic until the next PD event takes place.

The function $f(E_i, E_0(t))$ may depend both on E_i and on $E_0(t)$. Often, however, the exact way of how E_i decays is not very important. On the other hand, the decay time always is an important parameter. A simplification consists in setting $f = -E_i/\tau$ leading to an exponential decay with time constant τ.

Interaction of both processes

Both processes, discharge (jump of E_i) and drift/recombination (continuous change of E_i), interact in a real PD process. They are coupled by the discharge probability in the following way: Let $c \cdot dt$ be the probability that a fast discharge occurs in the time interval $[t,t+dt]$. This probability may depend on the internal field E_i as well as on the external field $E_0(t)$:

$$c = c(E_i, E_0(t)) \qquad (5)$$

It is zero for $E_{tot} = E_i + E_0(t) < E_{inc}$ with a typical inception field E_{inc} that depends on the defect geometry and the surrounding insulation material. The form of the function $c(E_i, E_0(t))$ for $E_{tot} > E_{inc}$ depends on the mechanism of first electron supply. Some possible forms are given in Heitz (2). For example, for voids one often can assume a constant c, whereas in a corona c strongly increases with E_{tot}.

For small dt the time evolution of $E_i(t)$ is shown schematically in Fig. 1. The whole PD process is completely determined if the parameters c, E_{res} and f are known for

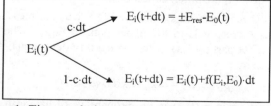

Fig. 1: Time evolution of $E_i(t)$ to $E_i(t+dt)$ for small dt. The upper path with probability $c \cdot dt$ represents the change of $E_i(t)$ due to a discharge, the lower path with probability $1-c \cdot dt$ is the time development for the case of no fast discharge.

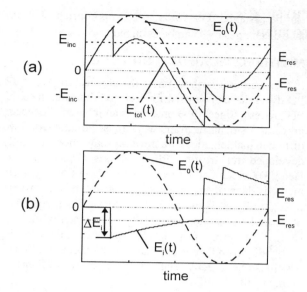

Fig. 2: Typical trajectories under AC (a) for $E_{tot}(t) = E_a(t) + E_i(t)$, (b) for $E_i(t)$.

all t and E_i. In Fig. 2 typical (random) trajectories of $E_{tot}(t)$ and $E_i(t)$ for the case of AC voltage are shown. Note that the same process realization is shown in Figs. 2(a) and (b).

PHYSICAL MODELLING II: OBSERVATION

The dynamics of the system has been formulated in terms of the internal field $E_i(t)$. The PD events manifest themselves as sudden reductions of the total field E_{tot} or, equivalently, in sudden changes of the internal field E_i. These changes, however, cannot be observed by the usual PD measurements. Instead, the bipolar charge quantity $\pm q$ being deployed during the discharge can be measured. In this section the link between field reduction and measurable charge q is made.

The jump ΔE in the internal field during a fast discharge according to Eq. (3) is given by

$$\Delta E = E_i' - E_i = \pm E_{res} - E_i - E_0(t) \qquad (6)$$

This field change is accomplished by a bipolar charge quantity $\pm q$ which is a function of ΔE. Additionally, due to geometric fluctuations of the discharge structure (e.g. different streamer paths in a streamer corona) the charge may fluctuate around a mean value q_m. In general, the dependance of q on the field reduction ΔE can be described by a probability density function $p_{ch}(q|\Delta E)$ where $p_{ch}(q|\Delta E) \cdot dq$ is the probability that the charge lies in the interval $[q, d+dq]$ for a discharge with given field reduction ΔE. The smaller the fluctuations, the more this function is peaked at q_m.

The functional dependence of $q_m(\Delta E)$ on the field reduction ΔE is defect specific. Some possible forms for different defects have been given in Heitz (2). Often a functional form $q \propto \Delta E^\alpha$ with a defect specific exponent α

can be assumed. For voids $\alpha=1$ is appropriate, for surface discharges $\alpha=3$ can be assumed.

MATHEMATICAL DESCRIPTION

In this section it is shown how the physical model of the PD dynamics can be used for a closed description of the PD process. Since a PD process is an inherent stochastic process, a stochastic framework must be adopted. Two different mathematical frameworks have been recently developed for treating PD processes (Heitz (2) and Heitz(3)) which will be presented in the following. Both methods lead to a probability density $p_d(E_i,t)$ where $p_d(E_i,t)\cdot dE_i\cdot dt$ is the probability that a discharge takes place in the time interval $[t,t+dt]$ and an internal field in $[E_i,E_i+dE_i]$. From this density one can calculate the probability $p_d(q,t)$ of a discharge with charge q at time t which can be directly related to time resolved PD measurements like PRPDs.

Description as a stochastic point process

The sequence of PD events can be treated as a stochastic point process. Each event (i.e. each fast discharge) is specified by a time point t and an internal field E_i. It has been shown in Heitz (2) that this point process is completely described by the first conditional probability $p_d(t,E_i|t',E_i')$ specifying the probability of a discharge at time t with internal field E_i if the last discharge was at time t' with internal field E_i'.

Suppose that the last discharge has been at time t' with internal field E_i'. For times $t>t'$ the internal field evolves deterministically according to Eq. (4). We denote by $\hat{E}_i(t)$ the solution of Eq. (4) with initial condition given by Eq. (2)

$$\hat{E}_i(t=t') = \begin{cases} E_{res} - E_0(t'), & \text{if } E_0(t') + E_i' > 0 \\ -E_{res} - E_0(t'), & \text{if } E_0(t') + E_i' < 0 \end{cases}$$

Thus, $\hat{E}_i(t)$ depends on t' and E_i'. For each $t>t'$, the discharge probability $p_d(t|E_i',t')$ is specified by the usual waiting time distribution

$$p_d(t\,|\,E_i',t') = c\left(\hat{E}_i(t), E_0(t)\right)$$
$$\cdot \exp\left[-\int_{t'}^{t} c\left(\hat{E}_i(t''), E_0(t'')\right)dt''\right]$$

Since the internal field at time t is given by $\hat{E}_i(t)$, we find

$$p_d(E_i,t\,|\,E_i',t') = c\left(\hat{E}_i(t), E_0(t)\right)$$
$$\cdot \exp\left[-\int_{t'}^{t} c\left(\hat{E}_i(t''), E_0(t'')\right)dt''\right] \cdot \delta\left(E_i - \hat{E}_i(t)\right)$$

where $\delta(\cdot)$ is the Kronecker delta function.

Thus, the conditional probability function $p_d(E_i,t|E_i',t')$ can be calculated from the basic physical parameters $c(E_i, E_0)$, $f(E_i,E_0)$ and E_{res}, yielding a complete descrip-

tion of the point process. All interesting properties of the process can be drawn from the knowledge of $p(E_i,t|E_i',t')$. Especially the unconditional probability densitiy $p(E_i,t)$ of having a discharge at time t and field E_i without any knowledge about prior discharges can be obtained by solving the equation

$$p_d(E_i,t) = \int_{-\infty}^{t} dt' \int dE_i' p_d(E_i',t') p_d(E_i,t|E_i',t') \quad (7)$$

which is an integral equation for $p_d(E_i,t)$. It can be solved numerically.

Description as a piecewise deterministic Markov process

The temporal evolution of the state of the PD system (given by $E_i(t)$) is deterministic if no discharge occurs and shows a sudden change ("jump") in the case of a discharge. The probability of a discharge at time t only depends on the system's state and the external field $E_0(t)$. Thus the dynamics of the internal field forms a piecewise deterministic Markov process (Gardiner (4)).

Such processes are generally described by a dynamical equation for the probability density $p(E_i;t)$, the so-called master equation. Here $p(E_i;t)\cdot dE_i$ is the probability of finding the system in the interval $[E_i,E_i+dE_i]$ at time t. A large $p(E_i,t)$ means that, when repeating the experiment many times and measuring E_i at time t, $E_i(t)$ lies in the interval $[E_i, E_i+dE_i]$ with a high probability, and vice versa. Note that this is a different probability as the discharge probability $p_d(E_i,t)$ used in the last section.

The master equation of a PD system can be constructed from knowledge of the short-time dynamics as in Fig.1 and results in (see Heitz (3))

$$\frac{\partial}{\partial t} p(E_i;t) = -c(E_i, E_0(t))\cdot p(E_i;t)$$
$$-\frac{\partial}{\partial E_i}\left(f(E_i, E_0(t))\cdot p(E_i;t)\right)$$
$$+\left(\int_{E_i'+E_o(t)>0} c(E_i', E_0(t))\cdot p(E_i';t)\,dE_i'\right)\delta\left(E_i - (E_{res} - E_o(t))\right)$$
$$+\left(\int_{E_i'+E_o(t)<0} c(E_i', E_0(t))\cdot p(E_i';t)\,dE_i'\right)\delta\left(E_i - (-E_{res} - E_o(t))\right) \quad (8)$$

For periodic $E_0(t)$, $p(E_i;t)$ is periodic as well. Integrating Eq. (8) over some periods leads to this solution independent on the initial conditions.

In PD measurements, only the PD events are measured. This corresponds to the upper path in Fig. 1. The observable part $p_d(E_i,t)$ of the probability density $p(E_i,t)$ is given by

$$p_d(E_i,t) = c(E_i, E_0(t))\cdot p(E_i;t)$$

where again $p_d(E_i,t)$ is the probability of having a discharge at time t and internal field E_i and is the same function as have been used in the point process description.

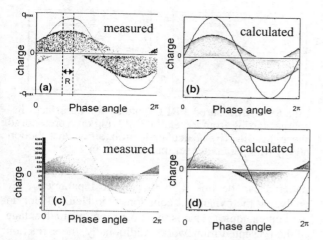

Fig. 3: (a) and (c): Measured PRPD pattern for a void (b) and (d): Calculated probability distribution p(φ,q) with parameter settings as given in Tab. 1.

Again, as within the point process model, the breakdown probability density p_d can be derived from the basic physical parameters c, f and E_{res}.

Transformation from $p_d(E_i,t)$ to $p_d(q,t)$, relation to PRPD patterns

Within both presented mathematical frameworks the dynamics of the PD process can be treated. The probability $p_d(E_i,t)$ can be derived. From this density the observable density $p_d(q,t)$ can be easily calculated using Eq. (6) and the charge function $p_{ch}(q|\Delta E)$:

$$p_{bd}(q,t) = \int p_{bd}(E_i,t) \; p_{ch}(q|\Delta E) \; d\Delta E$$

$$= \int p_{bd}(E_i,t) \; p_{ch}(q|\pm E_{res} - E_0(t) - E_i) \; dE_i$$

The positive sign of $\pm E_{res}$ must be used for those E_i that lead to $E_{tot} > 0$ and vice versa.

This density is an observable quantity. It can be estimated from a set of measured PD events (q_i, t_i), i=1,2,... .

For AC voltages, the so-called Phase Resolved Partial Discharge (PRPD) patterns correspond to $p_d(q,t)$ for t $\in [0,T]$ or $p_d(q, \varphi)$ for $\varphi \in [0,2\pi]$, respectively. A large probability $p_d(q,t)$ means that many PD events are likely to be recorded around the point (q,t). Thus, the point density of a PRPD pattern directly corresponds to the probability density $p_d(q,t)$.

EXAMPLES

In the following, measured PRPD patterns are compared with calculated probabilities $p_d(t,q)$. Eq.(8) has been solved numerically. Note however, that equivalently Eq. (7) could be used for determining $p_d(E_i;t)$. The resulting $p_d(E_i,t)$ was transformed to $p_d(q,t)$ as explained in the last section. For all examples, the function $f(E_i,E_0)$ in Eq. (4) has been set to $-E_i/\tau$. The external field $E_0(t)$ was chosen as a 50 Hz AC field with peak amplitude $\hat{E}_0$. The charge function $p_{ch}(q|\Delta E)$ has been taken to be a

Gaussian distribution with mean $q_m(\Delta E)$ and standard deviation σ All fields are given in units E_{inc}. The measured PRPD patterns are compared with the calculated distributions $p_d(q,t)$ where t corresponds to the phase angle in the interval $[0,2\pi]$.

Internal discharge: Void

Voids where the first electrons are mainly supplied by external radiation are specified by c=0 for $E<E_{inc}$ and c=const. for $E>E_{inc}$ [1]. The relation between q and ΔE can be assumed to be linear: $q_m(\Delta E) \propto \Delta E$. In Fig. 3(a) a measured PRPD pattern for an air filled void in epoxy is shown. In Fig. 1(b) the calculated probabilty distribution p(t,q) is displayed with model parameters as indicated in Tab. 1. The higher the probability p(t,q) for a given location (t,q), the darker displayed is the corresponding location. Thus, a dark region in p(t,q) means a large density of PD events at this location of a measured PD pattern. It can be seen that the calculated probability distribution coincides well with the measured pattern.

In Fig. 3(c) another measured void pattern is shown. The corresponding calculated probability distribution in Fig. 3(d) was made with the same parameter settings as in Fig. 3(b), but with a larger breakdown probability. This larger probability leads to a more structured pattern.

It is important to note that the agreement between model and measurement is quantitative, not just qualitative. As another demonstation of the quantitative agrrement, the PD events in the time interval R around the voltage maximum (see Fig. 3(a)) was examined. In Fig. 4 the measured counts per AC cycle in each charge bin is indicated as a solid line. The corresponding theoretical quantity, calculated from $p_d(t,q)$, is

$$cpc_{R,calc}(q) = \int\limits_{R} dt \int\limits_{q}^{q+\Delta q} dq \; \widetilde{p}_d(q,t)$$

where Δq is the width of the charge bin. This curve is shown as a dashed line in Fig. 4. It can be seen that in both cases the theoretical curve fits the measured data quite well. Since the PD process is a stochastic process,

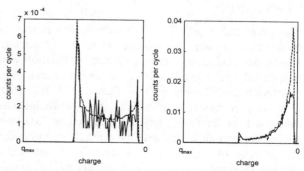

Fig. 4: Void discharges: charge dependence of PD probability at phase angle $\varphi = \pi/2$. Solid line: measured counts per cycle in time interval R (see Fig. 3(a)). Dashed line: Calculated counts per cycle. (a) Pattern from Fig. 3(a). (b) Pattern from Fig. 3(b).

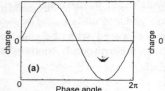
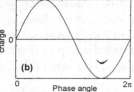
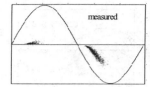
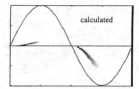

Fig. 5: (a) Measured PRPD pattern for a point-plane configuration in air. (b) Calculated probability distribution $p(\varphi,q)$ with parameter settings as given Tab. 1.

Fig. 6: (a) Measured PRPD pattern for a surface discharge starting from a metallic foil lying on a pressboard sheet in oil. (b) Calculated probability distribution $p(\varphi,q)$ with parameter settings as given in Tab. 1.

the measured curve must be regarded as a stochastic realization of the underlying probability distribution which is given by the calculated curve. The stochastic bahaviour of the measured data can be seen clearly seen in the figure.

Corona discharge: Needle in air

For a point-plane configuration in air the PRPD pattern (see Fig. 5(a)) was measured for a voltage slightly above the inception voltage. For the calculations, $c(E_i,E_0) \propto (E_{tot}-E_{inc})^3$ was used, and a small decay time constant $\tau=120$ µs. The charge was chosen as $q_m(\Delta E) \propto \Delta E^2$.

Surface discharge in oil

A PRPD pattern for a surface discharge in oil along an insulating surface is shown in Fig. 6(a). The discharge was produced by a metallic foil lying on a 2 mm pressboard sheet. Under the pressboard sheet a flat counterelectrode was situated. The metallic foil was on ground potential, thus a positive applied voltage corresponds to a negative surface discharge and vice versa.

The used parameters for Fig. 6(b) are indicated in Tab. 1. According to Atten and Saker (5) we took $q_m(\Delta E) \propto E^3$. Since positive discharges are longer than negative ones (Atten and Saker (5)) $q_m(\Delta E)$ for positive discharges have been set 5 times larger than for negative discharge. The discharge probability for positive discharges has been set twice the probability for negative discharges. As in the other examples, a good agreement between the measured pattern and the calculated probability distribution $p(\varphi,q)$ is achieved.

DISCUSSION

The presented approach for describing PD processes uses basic physical knowledge about the discharge mechanism and charge carrier drift/recombination that has already been used partially in earlier works (e.g. Gutfleisch and Niemeyer (1), Patsch and Hoof(6)). However, in former publications this knowledge has been used for simulating trajectories $E_{tot}(t)$ or $E_i(t)$ over successive periods of the external voltage, imitating the PD process. In contrast, in this paper in the stochastic process formulation a probability density $p_d(E_i,t)$ is calculated either in a point process framework or in a master equation approach. With its help the PD process can be analyzed analytically, which is a much more powerful method than pure simulation. Typical features of PD

patterns can be linked to the physical parameters by analyzing the dynamical equations (see Heitz (3) for an explicit example). This is the most prominent advantage of the presented framework. Additionally, the extraction of the model parameters from the PRPD or other measured PD patterns seems to be possible which is very difficult if only a sequential simulation algorithm is available. Thus the interpretation and classification of such patterns on the basis of the underlying physics could be possible with the presented approach. Future work will address this topic.

Conceptually, the simplification of the process description is a result of separating the inherent process dynamics from the observation process which only transforms the solution $p_d(E_i,t)$. This concept is new and has not been used in former approaches.

The dynamics of the system in a given external field $E_0(t)$ is controlled by a small number of basic physical quantities:

- $c(E_i,E_0(t))$: the probability of a discharge given the internal field E_i at the defect at time t which often has the special form $c(E_{tot})$ with $E_{tot}=E_i+E_0(t)$.

- the field decay function $f(E_i,E_0(t))$ which specifies how E_i decays due to drift or recombination of charge carriers. Often this function can be reduced to a single decay time constant τ, setting $f=-E_i/\tau$.

- the residual field E_{res}.

- the function $p_{ch}(q|\Delta E)$ that specifies how much charge must be deployed for getting a field reduction of ΔE at the defect location. This function is peaked at a mean charge $q_m(\Delta E)$ with a scatter σ

For a given defect these quantities can be derived from the physics of the discharge. They are well defined and can be determined quantitatively if the physics of the discharge is known. Often, however, rough estimates are sufficient that can be given without exact knowledge of all details of the defect geometry and the discharge physics. It has been shown that inspite of its simplicity, the model is able to describe PD processes of different types such as internal discharges, corona and surface discharges in different insulation media.

In the present contribution, the state of the PD system is assumed to be given by a scalar field E_i. This is appropriate for voids and is still a reasonable approximation for other PD systems with fixed geometry, such as co-

Fig.	c(sec^{-1})	τ (sec)	$\hat{E}_0$	E_{res}	q_m	σ
3(a)	9.5	10	5	0.3	$\propto\Delta E$	0
3(b)	600	10	5	0.3	$\propto\Delta E$	0
5	$\propto(E_{tot}-E_{inc})^3$ for pos. point, = 0 for neg. point	$120 \cdot 10^{-6}$	1.08	0.5	$\propto\Delta E^2$	0.015 $\cdot q_m$
6	$\propto dE_{tot}/dt$ for increasing field, =0 for decreasing field	0.2	3	0	$\propto\Delta E^3$	0.3 $\cdot q_m$

Tab. 1: Parameter settings for the calculation of the theoretical probability distributions $p_d(q,t)$. The fields $\hat{E}_0$ and E_{res} are given in units E_{inc}.

rona and surface discharges. For other PD systems the state must be described with more than one variable. An example are hopping particles between conducting surfaces. Instead of $E_i(t)$, the system state is then given by the position x of the particle and its charge q. Both quantities are time-dependent: the position changes due to the electrostatic and gravitational forces (continuous process), the charge changes when the particle starts a discharge (jump process). However, again a probability distribution $p(x,q_{part},t)$ can be introduced and its dynamics can be expressed in a master equation similar to Eq. (8). Alternatively, a point process forumlation can be chosen. Thus the presented approach can easily be generalized to more complicated systems.

Patsch and Hoof (6) have recently developed a model which also has been shown to be valid for different defect types. There are some similarities between their model and our model. However, the model of Patsch and Hoof does not account for the stochastic nature of PD processes. Furthermore it only describes the temporal behaviour of PD processes, the charges of the PD events are not considered. Therefore it is difficult to relate the model to charge resolved measurements like, e.g., PRPD patterns. The presented model goes further than the model of Patsch and Hoof since first it incorporates statistical effects (first electron statistics), and second it describes both the time-of-occurence and the charge of the PD events. Therefore the link to two-dimensional representations like PRPD patterns can be made easily, and the model can be used for interpreting and understanding such patterns.

SUMMARY

In the present investigation a general theoretical framework for describing PD processes has been presented. A PD system is treated as a small discharging system in a large embedding insulation system. The embedding system supplies a driving field $E_0(t)$ that leads to a dynamics in the internal field E_i of the PD system. The dynamics has been shown to be a piecewise deterministic Markov process and can be formulated either in a master equation framework or in a point process approach. It is governed by three physical quantities, the discharge probability $c(E_i,E_0(t))$, the decay function $f(E_i,E_0(t))$, and the residual field E_{res}. Given those parameters, the process is determined unambiguously, resulting in a time-dependent probability density $p(E_i,t)$. Two different mathematical approaches for calculating $p(E_i,t)$ have been proposed.

The PD process by itself being not directly observable, only the time and charge of PD events can be measured. The link between the PD process and observable quantities like a PRPD pattern has been made. It is heavily influenced by a defect specific function $p_{ch}(q|\Delta E)$ which describes how much charge has to be deployed for a given field reduction during discharge.

The presented model is general and can be used for different PD types. The physical process parameters are well defined and can be specified quantitatively if the physics of the discharge is known. Often, however, rough estimates can already be sufficient.

The advantage of the presented approach lies in the closed mathematical representaion of the process which is not possible with the usual simulation algorithms. Pattern features can be linked to the process parameters analytically. Thus, with the presented approach, a new understanding of PD patterns seems possible.

REFERENCES

1. F. Gutfleisch, L. Niemeyer, 1995, "Measurement and Simulation of PD in Epoxy Voids", IEEE Trans. On Dielectrics and El. Insul., 2, No. 5, pp.729-743.

2. C. Heitz, 1998, "A General Stochastic Approach to Partial Discharge Processes", IEEE 6th Intern. Conf. On Conduction and Breakdown in Solids ICSD'98, Västeras, Sweden, pp. 139-144.

3. C. Heitz, 1998, "A New Stochastic Model for Partial Discharge Processes", CEIDP 98, Atlanta, Georgia, pp. 400-406.

4. C.W. Gardiner, 1985, "Handbook of Stochastic Methods", Springer Series in Syergetics. 13. Springer Verlag, Berlin, Heidelberg.

5. P. Atten, A. Saker, 1993, "Streamer Propagation over a Liquid/Solid Interface", IEEE Trans. On Electrical Insulation, 28, No. 2, pp.230-242.

6. R. Patsch, M. Hoof, 1998, "Physical Modelling of Partial Discharge Patterns", IEEE 6th Intern. Conf. On Conduction and Breakdown in Solids ICSD'98, Västeras, Sweden, pp. 114-118.

TRANSFORMER MONITORING USING THE UHF TECHNIQUE

M D Judd[1], B M Pryor[2], S C Kelly[2] and B F Hampton[3]

[1]University of Strathclyde, UK [2]ScottishPower, UK [3]Diagnostic Monitoring Systems Ltd, UK

Abstract

The viability of applying UHF partial discharge monitoring techniques to an in-service oil-filled power transformer is established. Discharges in oil are shown to radiate the high frequency signals necessary to allow detection. On-site pulse injection is used to test the operation of a UHF coupler fitted to the transformer. The effects of on-load tap changing operations are also investigated.

1. Introduction

In fitting a UHF coupler to a power transformer, our intention was to use it for detecting partial discharges (PD), relying on techniques developed for monitoring gas insulated substations (GIS). The transformer is a 40 MVA oil-filled unit owned by ScottishPower, which in recent years has been prone to tripping out during warm summer weather. The cause is abnormal gas production, which may be due to internal discharges. By using a UHF monitor to continuously log data, we hope to establish whether any discharges are evident prior to a transformer shutdown.

Fundamental to the success of the UHF technique in GIS is the extremely short risetime (<100 ps) of PD current pulses in SF_6 [1,2]. If the technique is to be applied to monitoring oil insulated equipment such as transformers, we need to know whether discharge pulses in oil are capable of exciting UHF resonances inside the transformer tank. Previous studies have yielded different conclusions in this respect. Measurements carried out by Okubo [3] indicate that the risetimes of PD pulses in oil are too long to generate the necessary high frequency signals. However, Rutgers [4] has measured UHF signals from oil-filled PD test cells with amplitudes comparable to those from defects in SF_6. A preliminary experiment was therefore carried out to investigate the frequency range of signals radiated by discharges in transformer oil.

2. Preliminary experiment

The experimental arrangement is shown in Fig. 1. The screened Aluminium chamber contains a spark plug as a discharge source. The centre pin of the spark plug was tapered to a flat tip with a diameter of 0.5 mm and a wire of diameter 1 mm was used to form an earth loop above it. The spark plug was suspended in a glass dish that would later be filled with transformer oil. A conical electric field sensor detects electromagnetic signals inside the chamber.

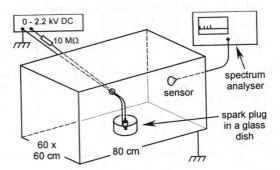

Fig. 1 Arrangement used to compare the signals radiated by discharges in air and oil.

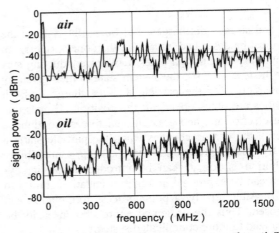

Fig. 2 Spectrum of the discharges in air and in oil. The background noise floor was -64 dBm.

The output from this sensor was displayed on a spectrum analyser.

The DC power supply was connected so that the spark plug electrode had a negative polarity. The 10 MΩ series resistor limits the power supply current. The discharge gap was set to 0.1 mm. Inception occurred at about -0.9 kV in air. In oil, inception occurred at -1.5 kV but the discharge quickly extinguished and the inception voltage began to rise. At the maximum voltage of -2.2 kV, intermittent discharges occurred. Results were recorded at this voltage level. Fig. 2 compares the spectra of the signals radiated by discharges in air and oil. These results confirm that discharges in oil are capable of generating quite large signal amplitudes at frequencies up to at least 1.5 GHz.

3. UHF coupler

Having established that discharges in oil can be detected, a UHF coupler was designed to fit a spare

High Voltage Engineering Symposium, 22–27 August 1999
Conference Publication No. 467, © IEE, 1999

port at the top of the transformer tank. Fig. 3 shows the coupler, which consists of a disc sensor mounted on pillars so that its face lies just below the inner wall of the transformer tank. Sensitivity of the coupler was measured (Fig. 4) using the calibration system [5] developed for GIS couplers. The coupler was designed primarily for mechanical integrity. Consequently, its average sensitivity over the 500 - 1500 MHz band is about 4 dB below that of a normal GIS coupler.

4. Installation

Fitting the Coupler

Fig. 5 is a side view of the transformer tank. To allow the coupler to be installed at port C, the oil was drained to a level slightly below this port. The existing flange was then replaced by the coupler assembly. The installed coupler is shown in Fig. 6. Covers on two smaller ports, A and B, were also removed to permit insertion of a probe to simulate low-level PD.

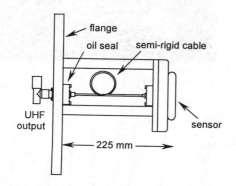

Fig. 3   Side view of the UHF coupler.

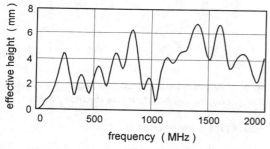

Fig. 4   Measured sensitivity of the coupler.

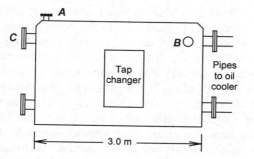

Fig. 5   View of the transformer showing the positions of the test ports (labelled A, B and C).

Pulse Injection Technique

To confirm that the transformer coupler was working, the test equipment shown in Fig. 7 was used. The output of the signal generator is a 10 V step with a risetime of ≈ 50 ps. This signal is attenuated to about 3 V and drives a 40 mm monopole probe. The probe is initially uncharged, but when the voltage step reaches it, a charge proportional to its capacitance is transferred onto the probe, simulating a PD. The amount of charge transferred to the probe for a step voltage of 3 V is estimated to be in the region of 1 pC.

Pulse Injection Tests in Air

Signals from the transformer coupler were amplified using a 25 dB gain, 0.3 - 1700 MHz amplifier and recorded using a spectrum analyser. The probe was first inserted into port A and positioned 0.8 m from the coupler. The main peaks in the amplified coupler output signal were noted and are summarised in Table 1. The probe was then inserted into port B, 3.5 m from the coupler. Signals at the coupler output under these conditions were approximately 4 dB below those measured when the probe was inserted into port A. The received signal frequencies were not significantly altered. These results show that signal attenuation inside the chamber is relatively low.

Pulse Injection Tests in Oil

Port B was then closed and the oil level raised to the top of the tank, immersing the coupler. Under these conditions, pulses could still be injected by lowering the probe into the oil through port A. The signal levels excited (Table 2) were greater in the oil than those recorded for port A in air. The larger signal peaks occurred at lower frequencies, probably due to the increased permittivity of the medium surrounding the coupler.

Fig. 6   UHF coupler on the transformer.

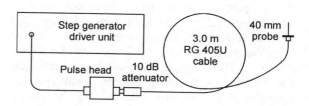

Fig. 7   The equipment used to inject test pulses.

Table 1 - Principal signals with pulse injection at port A. Probe and coupler both in air above the oil.

frequency range (MHz)	peak signal power (dBm)
250	-53
820 - 890	-54
1180 - 1320	-56
1360 - 1440	-56

Table 2 - Principal signals with pulse injection at port A. Probe and coupler both immersed in oil.

frequency range (MHz)	peak signal power (dBm)
390 - 470	-50
510 - 580	-47
800 - 860	-46
900 - 1500	-59

Comparative Pulse Injection Tests in GIS

To compare the received signal levels with those obtained in a GIS, the pulse injection equipment of Fig. 7 was used in a 420 kV GIS test vessel. The GIS was equipped with a standard internal UHF coupler. Pulses were injected by suspending the probe centrally in an open hatch cover. Using the same pre-amplifier, the amplitude of the peaks in the signal spectrum was only marginally greater than those recorded in the transformer. The difference can be accounted for by the 4 dB difference in coupler sensitivities.

A PD test cell containing a 2 mm diameter Aluminium sphere in SF_6 was placed in the GIS to generate UHF signals for comparison with those excited by the injected pulses. This test cell generates PD with an apparent charge of about 50 pC. At PD inception, the UHF signals were 10 - 20 dB larger than those excited using the probe. The fact that even small signals injected using the probe could be detected in the transformer implies that the transformer monitor should be capable of detecting low-level discharges.

5. Monitoring equipment

Fig. 8 shows the transformer with the coupler fitted. The armoured output cable from the coupler was fed to the control room - a distance of about 15 m. To mitigate the effect of cable attenuation, and in view of the need to use an amplifier to detect the smallest injected signals, a 10 dB gain preamplifier was installed at the coupler (housed in the sealed box visible in Fig. 6).

In the control room, signals from the coupler can either be displayed on a spectrum analyser or captured in the time-domain using a portable UHF partial discharge monitor (PDM). To test the complete system, the spark plug used in the preliminary experiment was attached to the end of a steel bar and lowered into the transformer oil through port A. The maximum depth that could be reached was 1.5 m. The spark plug was energised and the resulting discharge signals were recorded using the PDM. A typical discharge pattern is shown in Fig. 9. These signals represent large discharges. Calculations and measurements both yielded estimates in the region of 50 nC for the charge transferred by each discharge of the spark plug in oil.

Fig. 8   The transformer.

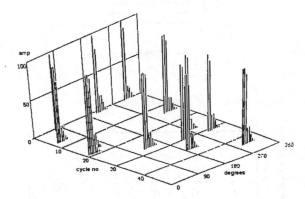

Fig. 9   Control room PDM display showing the spark plug oil discharges inside the transformer.

6. In-service measurements

Energising the Transformer

To achieve a high sensitivity during the initial tests, the 25 dB gain UHF amplifier was used in addition to the coupler preamplifier. In this configuration, the background noise floor on the spectrum analyser was at -60 dBm. Fig. 10 compares the coupler output signals before and after the transformer was returned to load. All of the signal peaks can be attributed to broadcast or communication channels. Although these signals appear quite large, this is due to the very high gain of the measurement system. At these levels the signals would not interfere with normal operation of the UHF PDM system. Despite the high gain, no discharge type signals were observed.

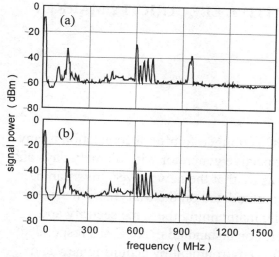

Fig. 10   Signal spectrum from the transformer coupler (a) before energising, and (b) energised and returned to a full load current of 200 A.

The transformer tap changer (visible in Fig. 8) operates on the primary winding. The potential difference between adjacent tap positions is about 500 V. We had anticipated that operation of the transformer tap changer might generate detectable signals at the coupler output, due to sparking at the current carrying contacts. To investigate this, the tap position was manually adjusted up or down by one or two positions with the transformer on load. The majority of tap change operations did not generate UHF signals. On about 30% of tap changes a small spike appeared on the spectrum analyser sweep, but without a consistent amplitude or frequency.

The coupler output was then connected to the portable PDM system, which can reliably detect very fast transient signals. While the PDM captured a more substantial signal, it still did not record a discharge on every tap change operation. Fig. 11 shows a typical PDM display when a tap changing event was detected. A few large spikes appear, which are confined to two or three power cycles.

7. Future work

Installation of the coupler took place early in the year at a time when the transformer is not usually problematic. The tests did not detect discharges inside the transformer. A continuous UHF monitor is now being installed, which will be fitted with a modem to allow remote access to the discharge data via a telephone line. By gathering data over several months we hope to determine whether any discharge activity appears that could provide an early warning of a developing defect.

On the basis of the promising results obtained so far, an autotransformer (400 kV, 1000 MVA) is also being fitted with two UHF couplers to facilitate a further in-service study of this potential application for UHF discharge monitoring.

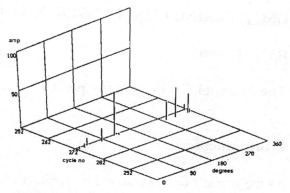

Fig. 11 Control room PDM display showing signals from a tap changing operation.

8. Conclusions

A simple experiment has demonstrated that discharges in oil are capable of radiating signals with significant energy at frequencies up to at least 1.5 GHz. However, electrical discharges in liquids are complex phenomena and this finding may depend on whether the discharge is of sufficient intensity to vaporise the oil.

A UHF coupler has been installed successfully on a 40 MVA oil-filled power transformer. Pulse injection tests revealed minimal attenuation of UHF signals inside the oil tank. When the coupler was immersed in oil the amplitude of the UHF signals increased. Comparative tests in a GIS have shown that the pulses injected to test the transformer coupler radiate considerably less energy than a 50 pC PD in SF_6. We can therefore be confident of detecting relatively small discharges inside the transformer.

9. Acknowledgement

M Judd thanks the EPSRC for their support.

10. References

[1]   M D Judd, O Farish and B F Hampton, "Excitation of UHF signals by partial discharges in GIS", IEEE Trans. DEI, 1996, Vol. 3, No. 2, pp. 213-228

[2]   M D Judd and O Farish, "High bandwidth measurement of partial discharge current pulses", Conf. Record Int. Symp. Elec. Ins. (Washington), 1998, Vol. 2, pp. 436-439

[3]   H Okubo, A Suzuki, T Kato, N Hayakawa and M Hikita, "Frequency component of current pulse waveform in partial discharge measurement", Proc. 9th ISH (Graz), 1995, Vol. 5, pp. 5634.1-5634.4

[4]   W R Rutgers and Y H Fu, "UHF PD-detection in a power transformer", Proc. 10th ISH (Montreal), 1997, Vol. 4, pp. 219-222

[5]   M D Judd, O Farish and P F Coventry, "UHF couplers for GIS - sensitivity and specification", Proc. 10th ISH (Montreal), 1997, Vol. 6

ENGINEERING CHALLENGES IN A COMPETITIVE ELECTRICITY MARKET

Roger Urwin

The National Grid Company plc, UK

1. Introduction

As the world's first privatised independent transmission company, The National Grid Company plc (NGC), has almost a decade's experience at the heart of the radically changing electricity market in England and Wales. NGC owns and operates 7,000 route-km of overhead lines, 650 circuit-km of underground cables and around 300 substations, operating primarily at 275 kV and 400 kV. NGC also has major interconnections with Scotland and France.

After 40 years of Government ownership, the Central Electricity Generating Board (CEGB) was vertically integrated; driven by Government financial regimes and energy policy imperatives; operated within a cost-plus framework; was engineering led and had a strong public service ethos. In 1989 the CEGB was divided into several competing generators and an independent transmission company, namely NGC, while the 12 Area Boards were transformed into regional electrical companies (RECs) responsible for distribution and supply. Additionally, the Office of Electricity Regulation (OFFER) was established to promote competition, oversee prices and protect customers' interests in the areas where a monopoly remained.

Restructuring has led to a more dynamic and demanding generation market. NGC has no control over the planning or siting of future generation, or the closure of existing generation for which only six months notice may be given. However,

NGC is obliged to offer a connection to any prospective generator within ninety days and to ensure that the new power flows can be accommodated on the transmission system, while maintaining the same security and operational standards. NGC has responded to this environment by acquiring new and improving operating skills, applying and developing new technology, and optimising system design.

Now that the electricity market is open and more competitive, the costs associated with network downtime are better understood and can be controlled. For example, if an outage (for maintenance or because of a fault) restricts a generator from running when it otherwise would have done, that generator must be compensated; additionally another generator must be paid to run to cover the shortfall in generation. Controlling these costs, and others imposed by the patterns of power flow on the network, subject to voltage, stability and thermal constraints, has been one of the key areas where NGC's research and development (R&D) programme has helped to reduce costs. These costs form a major part of the Transmission Service agreement between OFFER and NGC.

To meet the challenges of operating in a competitive electricity market NGC has adopted an engineering life cycle analysis (Figure 1), which focuses on the four stages of plant and system life, namely, design, buy and build, operate, and monitor and maintain. NGC's R&D programme uses the same principles in focussing on day-to-day engineering issues so as to improve system reliability and flexibility, through the

High Voltage Engineering Symposium, 22–27 August 1999
Conference Publication No. 467, © IEE, 1999

incorporation of new plant and technology, as well as, increasing the capability and availability of existing plant, and planning for the future.

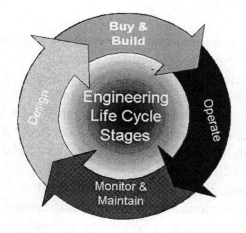

Figure 1: The four stages of the engineering life cycle

Set out below are some examples, from all four stages of the plant/system life cycle curve, in which NGC is actively pursuing research to meet the challenges it faces operating in an ever changing and competitive electricity market.

2. System design and planning for the future

Planning for the future within NGC covers many different areas; designing a flexible transmission system to allow future developments, maximizing the current assets on the system, and using new technology to NGC's best advantage. The present transmission system has evolved during the past 40 plus years to its current state and over this period many new technologies have emerged and are now implemented routinely on the system. New methods of operation and technologies continue to arise and NGC is constantly assessing the applicability of these developments and their implications

for the transmission system. These processes are supported by an active R&D programme which focuses on the needs of the company within the four stages of the engineering life cycle shown in Figure 1.

The move to a competitive electricity market requires NGC to reassess the present and future requirements of the system on a continuous basis. Software tools are important in this process and are used for example to plan and predict the likely capital expenditure programme over the next 25 years (ALERT), and to predict changes in the pattern of generation (PACE). ALERT calculates the predicted annual capital expenditure for the next 25 years as a result of plant currently on the system coming to the end of its life, given the best estimate of the failure probability density function of all the plant on the system. Such data come from routine monitoring of the plant on the system (see section 5.2) and the examination of plant which has been removed from service due to age or poor reliability. In contrast, PACE tries to model the changing pattern of generation, by predicting which generators will close, based on factors such as the generators' location, fuel source and age: the model can also place potential new generators on the system. Tools such as this have helped to predict the need for relocatable compensation equipment, eg. relocatable static var compensators (see section 3.1). Additionally, changing power flow patterns on the system, which could cause thermal constraint are foreseen and appropriate steps taken to avoid the problem (see section 4.1). Both programs model many possible scenarios and therefore help to understand and manage, the changing risk to our transmission system in the future.

3. Creating a climate for innovative design by suppliers

Since NGC does not have sufficient knowledge or any desire to design or manufacture transmission equipment, it cannot directly effect the design of the next generation of HV plant. Nevertheless, through close working relationships with manufacturers and a detailed knowledge of its system requirements, NGC is in a position to create a climate for innovation by providing challenging functional specifications for new equipment. Functional specifications detail the operating requirements for equipment without overly specifying technical details or design, and are a move away from previous practice. This allows manufacturers to meet our requirements through novel solutions without being restricted by detailed, prescriptive, specifications.

3.1 Responding to changing patterns of generation

The changing pattern of generation within England and Wales, and the need to respond to changing network topologies within tight timescales, has led to the need for NGC to install significant amounts of reactive compensation equipment. Since the mid-1980s Static Var Compensators (SVCs) and Mechanically Switched Capacitors (MSCs) have been installed to improve stability margins and voltage control. This ensures that system security and quality standards are maintained. Since 1990 ~6000 Mvar of MSCs and ~3300 Mvar of SVCs have been installed on the transmission system.

Until 1994 our compensation equipment installations were designed to tight technical specifications on the assumption that the equipment would be permanently located at a host site. The changing generation pattern during the early 1990s led to an increased risk that compensation equipment would no longer be required at the site at which it was installed.

Movement of conventional plant was considered and some MSCs were moved, however SVCs were found to be too expensive to move. In view of this, the concept of relocatable SVCs (RSVCs) was introduced.

The need for relocatability provided specific challenges in terms of equipment specification, the definition of the transmission system needs, and the performance requirements of the RSVCs to cover a wide range of possible connection points. Nevertheless, such a move was in line with our change of equipment specification from detailed technical specifications to functional specifications for new equipment. This change allowed manufacturers more flexibility in the design of new equipment and the chance to provide novel solutions.

The RSVCs, which can be moved is less than three months, each generate up to 60 Mvar in 10 Mvar steps. The design allows connection directly to the tertiary winding of any supergrid transformer with a rating between 180 MVA and 1000 MVA. To date NGC has installed 7 RSVCs, with another 5 to be commissioned within the next year.

3.2 Maximising plant availability through controlled switching

The deployment of large numbers of capacitor banks, and the continuing need for regular switching of shunt reactors in heavily cabled parts of the network, such as London, has required NGC to develop a strategy to maximise the availability of these assets, whilst minimising the disturbances switching of such equipment can cause. For example, energisation of an uncharged capacitor bank is analogous to applying a short-circuit to the network and is a routine, typically daily, switching duty.

Uncontrolled energisation has two main implications:

- Severe electrical transients are imposed onto the surrounding network
- The circuit-breaker suffers accelerated electrical wear

The principle of controlled switching is to time the closing signal to the circuit-breaker to ensure that the capacitor bank is energised at, or very close to, voltage zero such that inrush currents are minimised.

NGC makes considerable use of controlled closing of capacitor banks and also uses controlled opening of shunt reactors. In each case it is necessary to control the operating instant of the circuit-breaker to within a small fraction of the power frequency voltage; typically times better than 2 ms are achieved. Only in recent years has circuit-breaker and control technology improved sufficiently to achieve these requirements. For controlled closing to be effective circuit-breaker operating times must be consistent and/or predictable to within 2 ms over a wide range of conditions eg. control voltage, temperature, drive condition etc. Similarly, relay technology must be able to predict these operating times with sufficient accuracy. This may require complex self-adjustment features to be incorporated to account for circuit-breaker variations. NGC has been active in adopting this technology through the use of functional specifications and is one of the major users of this type of technology throughout the world. The benefits have been reduced system disturbances and increased interrupter maintenance intervals.

4. Operating the transmission system to better effect

Under the Transmission Service agreement mentioned above OFFER provides a financial incentive for NGC to reduce the various costs in running the system. Since 1994 the annual cost of uplift, which includes the cost of constraints, has fallen by 40% from £570m to £340m. A major contribution has come from innovative ways of operating and maintaining the transmission system including, the additional use of complex circuit-to-station inter-trips, live-line working techniques, and performing substation maintenance with temporary circuit bypasses. Furthermore, the availability of transmission circuits has risen from 91% as an average annual system value to 96% over five years, while planned unavailability for maintenance now varies from 1% in winter to 8% in summer. In addition improved planning of outages to minimise constraints and the utilisation of higher ratings, where available, have made significant contributions.

4.1 Optimising the use of overhead line circuits

During the 1950s many overhead lines were constructed and most were strung with twin 400 mm^2 aluminium conductor steel-reinforced (ACSR) conductor (known as Zebra). In the 1970s, the maximum continuous winter rating for such lines was 1320 MVA at 400 kV, and thermal constraints were starting to become an issue.

In the 1980s, a refurbishment programme was initiated and these lines were replaced with 500 mm^2 all-aluminium-alloy conductor (AAAC) (known as Rubus). This change also enabled the rated temperature to be increased to 75°C compared to the 50°C for the ACSR, providing a winter rating of 1900 MVA. Consequently, this was the maximum continuous winter rating ascribed to these lines when they were transferred from the CEGB to NGC (see Figure 2).

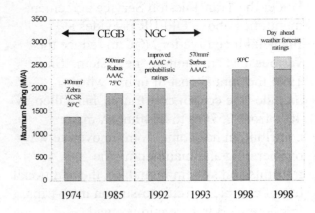

Figure 2: Improvement to overhead line ratings during the last 25 years.

In the1990s, due to the changing pattern of generation some lines running from the north to the south of England became thermally constrained. On NGC's heavily inter-connected system even small increases in the rating have a significant value because, increasing the rating of a constraining circuit also allows more load to be carried by parallel, otherwise unconstrained lines.

Changing the line rating calculation method to include natural convection and the use of lower resistivity AAAC alloy has given small but significant improvements. Also, since the early 1980s, a probabilistic approach to line ratings in which the conductor rated temperature could be exceeded for short periods of time has been used. This approach allows the risk of exceeding the conductor rated temperature to be quantified and managed, whilst increasing the post-fault winter rating to over 2000 MVA. It should be noted that, for the majority of NGC lines, it is the increase in post-fault rating which is required, with only a few circuits being constrained by their pre-fault rating. The principal concern is the provision of adequate post-fault ratings to enable the system to be operated to the accepted security standard with minimal constraints on generation.

Changes to the electricity regulations in 1988 meant that tower design loading could use a probabilistic approach, rather than a simple safety factor. These methods were subsequently used to re-interpret the design strengths of existing towers. As a result, the generic design of towers for twin lines has been found capable of carrying 570 mm^2 AAAC (known as Sorbus) over most of the NGC network without any modification; with a corresponding increase in rating to 2180 MVA. Further increasing the maximum rated temperature from 75°C to 90°C for AAAC on some lines has increased the possible rating to 2400 MVA. However, before such an increase could be considered the implications for loss of conductor strength (by annealing), the effect on conductor grease, the effect on joints, and the implications for conductor creep needed to be studied.

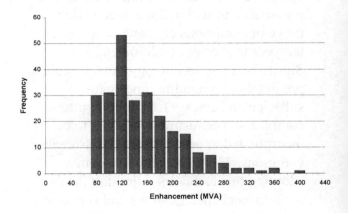

Figure 3: Distribution of rating gains achieved over a one year period.

These ratings are calculated assuming the worst case scenario of poor cooling conditions. There are, however, days on which the weather conditions provide considerably better cooling, and as such, on these days the ratings could potentially be enhanced. NGC now uses day-ahead weather forecasts from the Meteorological Office (MO) to provide 3-hourly values for minimum windspeed 10m above the ground, maximum ambient temperature and solar radiation along the route of

NGC is also investigating the use of conductors that can be operated at even higher temperatures, thus improving the ratings of overhead line circuits still further. On NGC's heavily interconnected system, this could have significant benefits.

Circuit	Conductor Rated Temperature (°C)	Days Upratable (%)	Minimum Uprating (%)	Maximum Uprating (%)	Average Uprating (%)
A	50	72	4	21	11
B	90	63	3	13	6

Table 1: Upratings achievable over a one year period using 24 hour ahead weather forecasting.

particular overhead lines to calculate day-ahead ratings. During the development and trialing of this approach a package of solar power meteorological instruments was installed on a number of towers, and the actual conditions compared to the MO forecasts. The reliability of these data, together with an algorithm that was developed, has meant that weather monitoring has not been required for the circuits on which this approach has been implemented.

So far three circuits have been targeted because they cause constraints, especially during summer outage season. Forecast data are sent electronically to a dedicated computer at National Grid Control Centre. The data are automatically processed, and a rating sheet printed for each circuit, which can then be used in the scheduling of generation for the following day, with potential savings in uplift costs. Table 1 summarizes the enhancements that have been achievable to date on two different circuits, while Figure 3 shows the distribution of rating enhancements over a one year period for Circuit B.

4.2 Improvements to cable circuits

NGC's transmission cables mainly use an oil-filled design and their ratings can be enhanced by:

- Utilising measured ambient temperatures and load variation.
- Measuring cable temperatures and inferring soil thermal resistivity.
- Measuring soil moisture and thermal resistivity.
- Increasing the rated temperature for short periods.

The first three approaches utilise more accurate thermal modelling and all these approaches are used by NGC. Ground temperatures are routinely measured for the Circuit Thermal Monitor (CTM) which fully models the varying load. Rated temperatures have increased generally from 85°C to 90°C, although values of up to 95°C have been used for short-term loading on a few circuits.

Soil resistivity values have been derived from fibre-optic distributed temperature sensing (DTS) measurements or with thermal probes placed close to the cables. The results are encouraging because the soil moisture level is usually high and continuous rating improvements of up to 20% are attainable. The period for which higher load is carried is important in assessing the need to consider soil dry out. Where the rating might only be required for one day the rating increase is comparable with that for an overhead line for the same period and the chance of any soil moisture movement is very low. NGC has used stabilised backfills for most cable circuits installed in the last thirty years which retain a high level of performance even when fully dried out. The use of such stabilized materials and the prevailing load pattern for all circuits suggest that there is minimal chance of soil dry out with current standards of security and plant reliability.

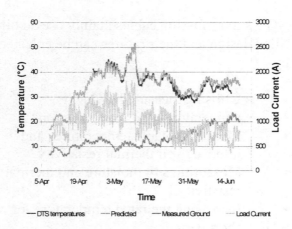

Figure 4: Predicted and actual cable temperature against circuit load.

The circuit thermal monitor (CTM) was developed from an initial electromechanical prototype in the 1970s to a comprehensive system covering 60 circuits by 1996. The main use for the CTM is to provide real-time ratings for strategic cable and transformer circuits within the scheduling and control time scales. The circuits to which it has been applied were selected on the basis of anticipated restrictions due to potential thermal constraints. The CTM uses real-time data at one minute intervals to provide

accurate real-time ratings to engineers at the National Grid Control Centre thus allowing significant scope for cost savings and improved security under both pre-fault and post-fault conditions.

Under normal operating conditions the detailed planning performed by the system operators removes the requirements to utilise the enhanced ratings produced by the CTM. However, when unplanned outages or sudden generator unavailability occur, engineers in our Control Centre can fully utilise the real-time modelling of the CTM. The largest use of the CTM is in planning for post-fault situations, although it has been used pre-fault on many occasions to reduce generation constraints. This has led directly to significant cost savings associated with constraints.

DTS is being installed on all new cable circuits. This not only allows hot spots along the cable route to be readily identified, but can also be used to calculate the effective thermal resistivity of the soil around a buried cable, so enabling more accurate cable ratings to be calculated.

The ability to predict how a cable will respond to an increased load using the CTM and then to monitor this using the DTS can provide significant savings to NGC. For example, where a possible thermal constraint had been identified during a transmission outage at a substation, an improved rating for the affected cable circuit with the new load pattern was modelled by the CTM. The circuit also had a DTS system and Figure 4 shows how the cable surface temperature varied with the daily load on the circuit over the period of the outage. The modelled temperature and the measured temperature are in very good agreement. The temperatures were monitored closely by the DTS over the outage period to provide extra security and the extra rating provided led to significant savings in thermal constraint costs.

4.3 Tools for optimising system performance

In parallel with operating assets more efficiently as discussed above, we are using improved analytical capabilities to develop and maintain an efficient, coordinated and economical system. NGC has developed and adapted a number of software tools to enhance engineering decision making and to improve the management of uplift costs. Programs are used for analysing transmission uplift costs (ESCORT), assessing response and reserve contracts and holding costs (CAVALIER), and evaluating reactive power contracts and investments (SCORPION).

Over operational planning timescales ESCORT is used to evaluate the expected transmission uplift costs of different maintenance plans, while over longer timescales the program is used to assess the expected constraint uplift costs against a range of uncertainties, eg. generator cost or availability. The model includes all circuits and impedances, all generators and typical bid prices, as well as considering constraints, post-fault actions and the effect of quad-boosters. Both ESCORT and CAVALIER can be run post-event using actual system data, so that the output can be compared to the real case, and the models refined further.

Advances in technology allow data to be processed much faster allowing better real time information to be available to Control Centre engineers. GIMS (General Inter-trip Monitoring System) is one such system that provides a simple to understand graphical display, showing the real time security of the transmission network (Figure 5). GIMS developed from two university research projects, one to create a fast system stability analysis program and the other to develop user friendly information displays. Every fifteen minutes GIMS takes the state estimator file (a snapshot of the power flows on the network) and from this calculates whether any parts of the network would not meet operational standards in the event of a fault on the system. GIMS looks for:

- Circuits with power flows above their maximum pre-fault ratings
- Substations with excessive voltage changes
- Generators in danger of pole slipping or sustained oscillations

These are displayed in different colours depending on the level of severity of the potential problem. Additionally, GIMS calculates the requirement for more or less generation to be available on inter-trip.

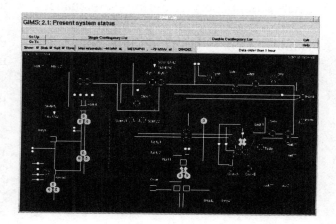

Figure 5: A typical output from GIMS

As part of its R&D programme NGC is continuing to develop links with university departments in a broad range of fields, from power engineering and system control to materials science. Links exist with more than 20 institutions which carry out projects on new technologies, new techniques and assessment of existing systems. These links are further being developed through the appointment of NGC Research Fellows at universities in key technology areas.

5. Increasing network reliability and availability

The business needs of NGC and the constraints imposed on the transmission system necessitate that network reliability and availability are as high as possible. NGC strives to continually improve these measures through better maintenance planning and the use of condition monitoring to assess plant condition.

5.1 Application of software tools to maintenance planning

Regular maintenance of assets aims to improve reliability, however it reduces circuit availability, since circuits must be switched out of service during maintenance for safety reasons. Improved maintenance planning can only be achieved by better understanding of the assets and their maintenance requirements. Much work has, and continues to be performed, within NGC to gain this improved understanding by predicting plant performance and how it is affected by maintenance activities. By optimizing the maintenance intervals and qualifying the risks associated with such changes for individual assets, NGC planning engineers have more flexibility when producing circuit maintenance schedules. However, other factors effecting circuit maintenance schedules to be taken into account include:

- The direct cost associated with planned maintenance activities.
- The downtime costs associated with planned maintenance outages.
- The expected costs associated with unplanned maintenance activities.
- The expected downtime costs associated with unplanned maintenance outages.
- The expected circuit unreliability.
- The expected circuit unavailability.

Furthermore, each circuit may be constructed from up to 40 individual assets eg. circuit breakers, transformers or cable sections, each with its own optimal maintenance interval. Consequently, there will be many instances when the asset maintenance intervals do not coincide and thus over the years many outages will be required. NGC has developed COMPAC, a genetic algorithm based spreadsheet model, which optimizes the maintenance of assets on a particular circuit by grouping together those assets for which maintenance activities can be performed concurrently. This approach has the advantage that the algorithm often finds solutions with better availability and reliability combinations than planning engineers with many years of experience would have found unaided.

By moving to maintenance policies for individual circuits NGC has been able to show substantial reductions in total costs and circuit unavailability. By applying COMPAC to five key circuits it is anticipated that total cost savings between 15-40% could be achieved, with circuit unavailability reduced by up to 30%.

5.2 Condition monitoring to improve asset management

Condition monitoring forms an integral part of NGC's maintenance policy. Through a better understanding of the condition and performance of an asset, NGC can plan a cost effective and efficient operation/ maintenance policy, and also judge the optimum time to replace an asset. Some of the benefits of effective condition monitoring include:

- Improved ability to risk manage assets
- Improved management and/or reduced environmental risks
- Ratings and performance improvements leading to constraint cost avoidance

- Better targeting and prioritisation of assets for reconditioning or replacement

NGC's R&D and engineering programmes, often in collaboration with equipment and plant manufacturers, aims to identify new measurable parameters which can be related to plant health, and may be usefully monitored. This also allows the critical equipment deterioration and failure modes to be determined which indicate the end of asset life or when maintenance must be undertaken.

The condition assessment strategy employed by NGC applies tests in two distinct ways in an effort to acquire information in the most cost-effective way. Routine tests are carried out on a periodic basis for screening to detect incipient failure and indicate general plant condition. Special tests are applied only as required for diagnosis and detailed assessment in response to one of a set of triggering circumstances. Additionally, continuous on-line monitoring of particular targeted assets may be employed as a specific reaction to known or suspected equipment problems.

5.3 Datamining applied to dissolved gas analysis

Dissolved gas analysis (DGA) is a standard technique used routinely worldwide for monitoring the condition of oil-filled plant. Samples are typically taken on an annual basis, with more frequent sampling when a possible problem is identified. NGC (and the CEGB) has undertaken routine DGA from transformer oil samples for more than 30 years, and has amassed an extensive database of records. The gases usually considered in DGA are hydrogen, methane, ethane, ethylene, acetylene, carbon monoxide and carbon dioxide: variations in the levels of individual gases, or ratios of gases, may indicate a problem with the asset.

However, the problem is complicated by the fact that the levels of dissolved gas measured can be affected by the sampling technique and conditions and the laboratory performing the analysis. The analysis of such data is therefore complicated and in general relies on experts using international standards to interpret the data. Methods which simplify the interpretation of DGA data and which may also extract more information from the data would therefore be relatively more useful.

To this end NGC has been investigating datamining as a new method for analysing DGA data. Datamining is a technique which finds patterns in data using data driven techniques such as neural networks, and contrasts with the traditional scientific approach, where data analysis is highly directed by hypothesis and theory. The key point of datamining is to find patterns *not* predicted by established theory and to unearth information hidden within the data. In particular NGC has found success in using Kohonen neural networks to reveal clusters and trends.

The process provides a visual result such as Figure 6, where the position on the map relates to the condition of the transformer. The map shown is generated by a Kohonen neural network from NGC's DGA database without any prior knowledge of the faults. The neural net is driven by correlations in the DGA data to cluster similar sets of data together on the map. These clusters are then examined by NGC experts who assign known faults to particular areas of the map, each of which represents certain DGA patterns. For example, one area of the map corresponds to the presence of hydrogen and acetylene and is indicative of internal arcing, whereas another area is dominated by the presence of methane and ethylene and indicates overheating. Using this approach, past DGA results have been re-examined, and this has revealed some samples which could be reinterpreted as being due to possible transformer problems: these transformers have since been prioritised for further investigation.

Traditionally, trends in certain gas ratios have been followed over time to indicate the condition of a particular transformer. A similar approach can be adopted by plotting successive DGA results on the Kohonen map (Figure 6). The trajectory starts with the earliest sample at the bottom part of the map and moves progressively towards the centre of the cluster associated with overheating. The lateral spike midway along the trajectory corresponds to the transformer being degassed: clearly this operation was only effective in the short term. This transformer was subsequently targeted for more frequent DGA testing to monitor the situation. Although still in the early stages of development, datamining has been successfully applied to transmission, generation and distribution transformers.

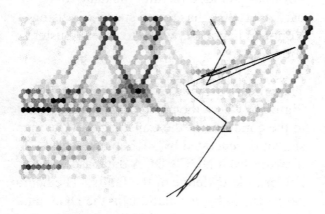

Figure 6: Kohonen map of transformer DGA data

5.4 Preparing for the unexpected

Although routine and directed condition monitoring has produced many benefits, there are occasions when events do not turn out as expected. One such example regards the catastrophic failure of a 30 year old transformer. Until late 1990 the routine yearly DGA results had shown nothing unusual, however, the next routine sampling

in early 1992 revealed a step change in the gas levels with a signature that indicated discharge activity. This change was attributed to a lightning strike on the transformer. Subsequent DGA results showed a steady decline in the fault gas concentrations which did not suggest any continuing activity. Nevertheless, because of the abnormal DGA signature the transformer was recommended for replacement. Problems with similar transformers at other sites led to monthly sampling being instigated in 1997. Again, there was no evidence of fault development, and yet just twelve days after one such DGA test the transformer failed with a massive internal flashover. Although subsequent forensic analysis revealed how the transformer failed, the root cause remains unclear.

Events such as this show that although modern condition monitoring techniques provide many benefits, they are not guaranteed to predict when failures will occur. Only through the continuing research efforts of companies like NGC, the manufacturers of HV equipment, and university groups will progress be made towards even higher levels of reliability through a better understanding of fault initiation and degradation processes in electrical plant.

5.5 Targeted application of on-line monitoring

Where plant items have been identified as having a specific or suspected problem they may be targeted for continuous on-line monitoring. Although this form of monitoring is expensive, the constraint costs which could result from removing an asset from service may be even higher. Consequently, an ability to manage the risk of keeping an asset in service whilst planning remedial work can be extremely beneficial.

In the early 1990s, NGC identified the potential benefits of a robust low maintenance on-line gas analysis (OLGA) system, although no reliable commercial system was available at the time. Through an R&D programme various gas extraction and detection techniques were investigated, and a prototype system was developed and successfully trialed. At the same time, a commercial company was also developing a similar system, based on the different new technology. The knowledge gained by NGC during the development of its own OLGA system has proved invaluable during the subsequent installation of the commercial system. The manufacturer is now interested in commercializing our development as it is more robust than their own. Although not part of NGC's core business, such a collaboration is to the benefit of both companies.

Between 1991 and 1993 NGC installed the majority of 96 bushings of a new design; 69 were installed horizontally and 27 vertically. A few of these horizontally-installed bushings suffered disruptive failures within the first few years of service. Extensive testing by ourselves and the manufacturer concluded that the principal cause of these problems was failure of a filling compound between the exterior porcelain and internal epoxy resin impregnated paper (ERIP), most probably caused by pollution on the porcelain.

A strategy was developed quickly to deal with the situation and enable the rest of bushings to remain in service:

- Hazard zones were imposed around all the bushings immediately and blast protection was erected on some sites
- In the short term, the bushings were greased to mitigate the pollution problem and stress relieving rings were fitted to the HV ends of the bushings to lower the electric field
- In the long term, all the horizontal bushings are to be replaced.

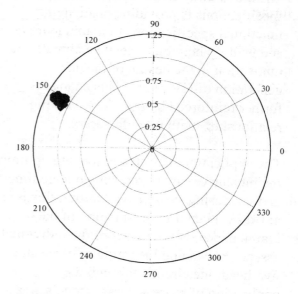

Figure 7: Graphical plot showing stable on-line capacitance measurements

In order to improve our understanding of the degradation process, on-line capacitance monitoring was established on twelve bushings to determine whether partial discharge was involved with the failures. Close interaction with a university research programme has further advanced NGC's understanding of the situation. The initial on-line monitoring led to the fitting of such systems on all such bushings (Figure 7), ensuring any problems could be detected rapidly, without needing an outage for off-line testing. This action has allowed NGC to maintain system integrity whilst managing and understanding the risk.

6. Conclusions

The change to operating in a deregulated market has presented NGC with many new challenges. Through a continually improving understanding of the factors which effect plant performance and reliability NGC is better placed to increase system availability and to control costs.

The move away from detailed technical specifications is providing challenging functional specifications to manufacturers and NGC is helping to create a climate for innovation in the design of the next generation of HV plant. This change is further reinforced by our close working relationships with manufacturers.

The application of software tools to optimise maintenance plans is reducing outage times and increasing system availability; this is aided by a better understanding of the maintenance requirements of the individual assets. Enhancing the rating of cable and overhead line circuits through the application of more accurate models aids this process by creating fewer constraints on the system during maintenance outages. These improvements help NGC cope with the ever changing pattern of generation.

Through a combination of routine testing and targeted tests NGC is learning more about the current condition of it's assets thereby allowing a cost effective and efficient operation, maintenance and replacement policy to be pursued. The examination of historic test data with new computational techniques is providing improved information from the data, and additionally easing the interpretation.

From the specification for new equipment to the decisions on replacement at the end of plant life, NGC is applying technology to meet the demands of operating in today's rapidly changing electricity market in England and Wales.

Novel concept in high voltage generation: Powerformer™

Mats Leijon

ABB Corporate Research, Västerås, Sweden

Abstract

A new generator, Powerformer™, which is able to supply electricity directly to the high voltage grid without the need for a step-up transformer, has been developed by ABB in Sweden. The new generator is suitable for power generation at output voltages of several hundred kilovolts. The new concept is based on circular conductors for the stator winding and it is implemented by using proven high-voltage cable technology. Thus, the upper limit for the output voltage from the generator is set by the state-of-the-art power-cable technology.

Introduction

Until today, electric generators have looked almost the same for nearly a century. Progress has obviously been made in areas such as insulation materials and production technology, but the basic generator concept has remained the same. ABB has developed a radically new electrical machine which principally rests on the fundamentals of nature, as described in the Maxwell equations, rather than resting on traditions emanating from the turn of the century when insulation technology was still in its infancy.

The direct connection of the generator to the grid, leading to the omission of the transformer with its associated losses and maintenance requirements, provides a number of noteworthy advantages for the network operator as well as for the power producer. Such advantages over a conventional system include higher plant efficiency (0.5-2%), increased reactive power production capability, and possibilities for increased overload capability. Altogether, these features sum up to a highly reduced life cycle cost for a power production system implemented with Powerformer as compared to a system using a conventional generator with its step-up transformer.

Throughout the development process of Powerformer, environmental issues have held a central position. This is, for example, reflected in the choice of constructional materials used in the new machine, facilitating recycling at the end of the lifetime of the machine. However, the largest environmental gains come from the increased efficiency of the new energy system compared to that of a conventional system, thus reducing demands made on natural resources.

The first machine to feature the new winding concept is a hydropower generator rated at 45 kV/11 MVA, which is in operation in Porjus in the northern part of Sweden. Moreover, two orders on Powerformer have been placed by Swedish utilities at ABB Generation for one turbo generator rated at 136 kV/42 MVA and one hydropower generator rated at 155 kV/75 MVA.

The immediate consequence of the new solution is the possibility of increasing the output voltage of the generator from today's maximum of about 25 kV to the level of the rating of the available power cables which presently is of the order of 500 kV. The effects of this include the possibility of directly connecting the generator to the power grid without a step-up transformer and thereby eliminating the associated losses and costs. This new machine has been named Powerformer since it incorporates the functionality of both the generator and the transformer.

High Voltage Engineering Symposium, 22–27 August 1999
Conference Publication No. 467, © IEE, 1999

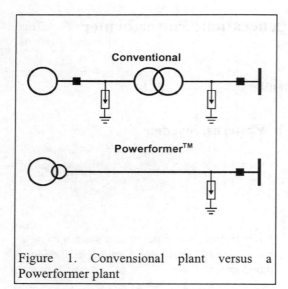

Figure 1. Convensional plant versus a Powerformer plant

Basics

In the mid 19th century, J. C. Maxwell managed to fully describe the phenomena of electricity and magnetism in a single system of equations, commonly known as the Maxwell equations. These equations contain the basis for rotating electric machines, for example Faraday's induction law. Without going into details, some aspects of the different solutions to the Maxwell equations will be briefly discussed here. For a rectangular conductor geometry, solutions to the Maxwell equations for the electric as well as the magnetic field have infinitely high field strengths outside the corners of the conductor. A cylindrical conductor geometry, however, yields a smooth electric and magnetic field distribution which is a prerequisite for a high-voltage electric machine, see Fig. 2. The practical consequence for a rectangular conductor in an electric machine is that the insulation material and the magnetic material of the machine are highly stressed and not uniformly loaded which leads to an uneconomical use of the involved materials. Failures in the machine related to the high stress of the materials are also very likely to occur. Such severe drawbacks are avoided when using circular conductors which allow for the optimal use of the insulation material due to the associated smooth field distribution.

From a physics point of view, round conductors are the natural choice for the stator windings of an electric machine. However, engineering problems have until now led the development in another direction. In order to raise the output power of an electric machine, either the level of the output voltage or the current in the stator windings must be increased. When more output power was required from generators in the beginning of this century engineers, limited by the insulation technology at that time, had to raise the current in the machine instead of increasing the output voltage.

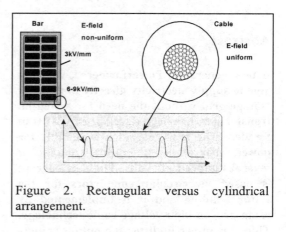

Figure 2. Rectangular versus cylindrical arrangement.

Design of rotating machines

In the design of an electromagnetic system the four main points which can be identified are as follows: electric design, magnetic design, thermal design, and mechanical design. Previously, it was not possible to make an independent treatment of the different parts due to various limitations of the insulation system. With the new winding, however, these points may be treated separately, giving designers by far better possibilities to optimise the performance of the machine. The independent treatment of, for example, the electric design is made possible, on the one hand by the round conductors which result in a smooth field distribution thus minimising the stress on the insulation material and, on the other hand, by the semiconducting layers on each side of the solid insulation which layers confine the electric field to the winding. The thermal design is facilitated since cooling may be accomplished at earth potential. The four traditional design points will be treated briefly below.

Electric design

The insulated conductors used for the stator windings for Powerformer are, as referred to earlier, based on proven high-voltage cable technology. Currently, XLPE insulated cables are available for voltages up to 500 kV. This means that it is possible to design rotating electric machines, using the new cable concept for the stator winding, with output voltages of up to 500 kV. Therefore, Powerformer may be connected directly to the power grid without a step-up transformer. As mentioned above, an increase in the output voltage corresponds to a decrease in the loading current in the machine for a given power rating and a lower current density results in lower resistive losses in the machine.

The winding of Powerformer is a multilayer concentric winding where the potential along the winding increases with each turn starting from the inside of the stator and increasing towards the periphery. Accordingly, the demands made on the insulation thickness increase towards the periphery. It is therefore possible to use thin insulation for the first turns and then increasingly thicker insulation for the subsequent turns. This arrangement makes it possible to optimise the use of the stator core volume. The electric design is further facilitated by the practically zero electric field outside the cable. For example, there is no need to control the electric field in the coil-end region as is the case with conventional machine technology. The personal safety is also increased since the entire outer semiconducting layer of the cable can be held at earth potential.

For jointing and terminating the cables in the winding of Powerformer, readily available standard components are used.

Magnetic design

The stator of Powerformer consists of a laminated core, built up from sheet steel. The stator slots are radially distributed cylindrical bores running axially through the stator and joined by narrow sub-slots. The circular shape of these cable ducts, corresponding to the shape of the cable, is also preferable with respect to the Maxwell equations since the roundness results in an even distribution of the magnetic field. As a consequence, the loss in the generator is reduced and the output voltage contains less harmonics. Since, as already mentioned, the cable diameter decreases towards the rotor due to the decreasing requirements on the insulation thickness, the

cross-section of the cable ducts decreases. This results in a practically constant radial width of the stator teeth (the area between the stator slots) which in turn reduces the losses in the teeth.

In a conventional winding coil the conductor laminations have to be transposed along their length in order to reduce the eddy-current losses. For the new round conductor arrangement, minimisation of the eddy-current losses is achieved by using insulated and twisted strands in the cable forming the winding.

It is possible to install additional windings in the stator of Powerformer which can, for example, be used for supplying the plant with auxiliary power. Moreover, Powerformer can be used as a rotating transformer with possibilities of simultaneous connection to several grid voltages using winding taps.

Thermal design

Inherent to Powerformer is that the loading current for a given output power is considerably lower than in a conventional generator. Thus, the main part of the losses are iron losses in the core and not resistive copper losses in the winding. Consequently, the cooling requirement of Powerformer is concentrated to the core and not to the winding. Since, as already mentioned, the electric field is confined to the winding, the cooling of the stator core is performed at earth potential. Both liquid and gaseous media can be used for cooling.

Mechanical design

Powerformer is fitted with a conventional rotor. Therefore, only the stator related aspects of mechanical design will be covered here.

When designing a generator, current forces exerted onto the end windings during normal operation as well as during short circuits have to be considered in terms of the support of the windings. Due to the lower currents and current densities, the current forces in Powerformer are considerably smaller than in a conventional rotating machine. As a consequence, the support for the end windings can be made simpler in Powerformer.

In a conventional generator the stator winding is typically arranged in two layers, resulting in short stator teeth. The multilayer winding of Powerformer requires deep stator slots and thus the stator teeth are long. Mechanically this leads to lower bending stiffness in the teeth and lower resonance frequency which are unfavourable features. By using a new type of stiff, pre-stressed slot wedge in Powerformer, problems related to the long stator teeth have been eliminated.

Discussion

The clearest evidence that the concept presented above is valid, is that the first machine already is in operation. When it comes to larger machines, it can be seen that the concept is scalable. The cable technology is mature, and cable for several hundred kilovolts is readily available. Possible problems are expected to be easier to solve than for the corresponding conventional system, since the design can be separated into independent parts.

As the generating system is simplified, the distance from generation to the consumer shrinks. This will open up new possibilities, with new features for the power system as a whole. The increased capability of reactive power production is one example of this.

Conclusion

Based upon fundamental laws of nature and proven insulation technology, it has been possible to design a high voltage generator connected directly to the power grid. The cardinal point in this concept is to utilize a power cable for the stator winding. The result is a simpler, cheaper, more efficient and more environment-friendly generating system. The most obvious impact is the elimination of the step-up transformer, which leads to cost cuts and positive effects on the network. Other advantages include improved control over the design process, due to the fact that different parts of the design can be treated separately.

Powerformer, as the machine is called, plays the role of both generator and transformer, and can be used to connect grids with different voltages. The first installation is already in operation, validating the different aspects of this new concept.

Acknowledgment

Dr. Fredrik Owman has provided valuable comments on an early version of this text. Dr. Jan Hemmingsson is commended for reviewing this paper.

References

[1]   Leijon, M.: "Powerformer™ - a radically new rotating machine, ABB Review, no. 2, 1998.

[2]   Leijon, M.; Gertmar, L.; Frank, H.; Martinsson, J; Karlsson, T.; Johansson, B.; Isaksson, K. & Wollström, U.: "Breaking Conventions in Electrical Power Plants", Report 11/37-3, CIGRE Session, Paris, 30th August - 5th September, 1998.

[3]   Leijon, M.; Owman, F. & Karlsson, T.: "Power generation without transformers", Hydropower & Dams, Issue Two, pp. 97-98, 1998.

Effects of Electrode Materials for Arc-firing Probability and Switching Time in UV-laser Triggered Vacuum Gap

Akira Sugawara, Kazunori Satou, Takayuki Itou, Kouichi Itagaki

Department of Electrical and Electric Engineering,
Niigata University, Ikarashi 2-8050, Niigata 950-2181, Japan

Abstract

The switching time and the arc-firing probability of a self-sustaining discharge in UV-laser triggered vacuum gap (ULTVG) were measured. The third harmonic beam of Nd:YAG pulse laser (wavelength 355 nm, energy 5mJ/pulse, power density 4×10^7 W/cm^2) was used as a trigger for ULTVG. The UV-laser pulse injected perpendicularly onto the grounded cathode. The electrode materials were oxygen-free copper and aluminum. The electrodes had 85 mm in diameter. The gap length was 1.6, 1.9, 2.2 mm. The experimental chamber was evacuated to about 1.3×10^{-4} Pa by a turbo molecular pump.

The range of the firing voltage in copper electrodes was higher than in aluminum. The jitter of the switching time in copper electrodes was smaller than in aluminum. The arc-firing probability decreased as the circuit inductance increased.

1. Introduction

The triggered vacuum gap (TVG) has been widely employed as a closing switch in many applications, for instance in pulse power generators at high voltage engineering[1,2]. The TVGs have high dielectric strength, a wide firing range of several kV to tens kV without adjustment of the gap length, and generally rapid dielectric recovery strength. However the TVGs have a polarity effect that switching characteristics are influenced by different voltage polarities of an applied voltage in the TVG. It is necessary that the TVG is triggered at high voltage side of the electrodes, and the triggering circuit is insulated from high voltage. So we study the vacuum gap is triggered by a laser.

The laser triggered vacuum gap was developed by Pendleton and Guenther in 1965[3]. The switching mechanism seems as follows; The laser produces a plasma from the electrode material by an ablation and a photoemission[4,5]. The plasma forms a leader, i.e. a discharge channel. If the gap sufficiently fills plasma then the TVG switches on.

In this study, we use a Nd:YAG pulse laser (wavelength:355nm as a third harmonic beam, the energy of photon 3.5eV, energy 5mJ/pulse, beam diameter 2mmϕ, pulse width 4ns, power density 4×10^7 W/cm^2). The work function of copper or aluminum is 4.48eV or 3.23eV, respectively. It is well known that the electric field decreases the work function, in appearance.

Our study is presented that the ULTVG is influenced by the applied voltage, the gap length, the circuit inductance and the electrode materials.

2. Experimental Setup and Method

The experimental arrangement is shown in Fig.1. The electrodes are 85-mm-diameter and 10-mm-thick, made by oxygen-free copper or aluminum. The anode has a hole 8-mm-diameter at the center for injecting the UV-laser. The cathode connects to the ground. The anode and cathode are mounted in the vacuum tube (Pyrex). The experimental chamber was evacuated to about 1.3×10^{-4} Pa by a turbo molecular pump. The upper flange has a hole and a window is put on it. The transmittance of window and the reflectance of mirror are both about 99% (wavelength:355nm).

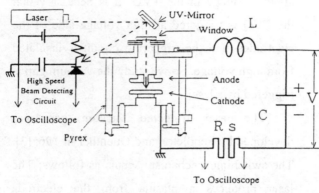

Fig.1　Experimental setup

The time that the UV-laser is injected detects by pin-photodiode. The high speed beam detecting circuit is composed of a battery (9V), a resistor (4.4kΩ), a capacitor (220pF) and a pin-photodiode. The waveform of the rising voltage (9V) is shown in oscilloscope by this circuit.

The experimental circuit is composed of an inductance L and a capacitor C. The circuit parameters are as follows: C=0.25μF, L=3, 107, 217, 791μH, the applied voltage V30kV, the gap length d=1.6mm, and electrode materials Cu or Al.

2.1 The Measurement of
the Arc-firing Probability

In this paper, we define the arc-firing probability as follows: The ULTVG switches on when a sinusoidal current wave flowed by the circuit constant (the angular frequency ω

$=1/(LC)^{1/2})$ is observed. The arc-firing probability shall be introduced as a standard to judge "yes" or "no" of the switching with 30 trials. The trial performed by decreasing voltage (2kV step) from self-breakdown voltage to the range of no-firing.

2.2 The Measurement of the Switching Time

Fig.2 shows waveform of the high speed beam detecting circuit by pin-photodiode and the sinusoidal current waveform measured by shunt resister (Rs=9.7mΩ). In this paper, we define that the switching time is averaged with 30 trials that are time deference between the initiations. We measure the switching time by using the oscilloscope.

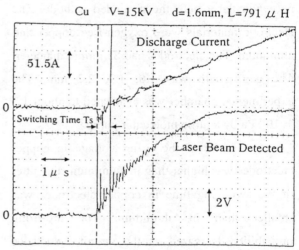

Fig.2　Waveform of discharge current in initiation and laser beam detected

3. Experimental Results and Discussions

An example of the arc-firing probability is shown in Fig.3. The self-breakdown voltages of Cu and Al are about 27.9kV and 20.2kV, respectively. The parameters are L=217μH and d=2.2mm. The arc-firing probability increased as a function of the applied voltage. The pair of copper electrodes for the arc-firing voltage is higher than aluminum. And the curves of arc-firing probability is approximately Gaussian distribution.

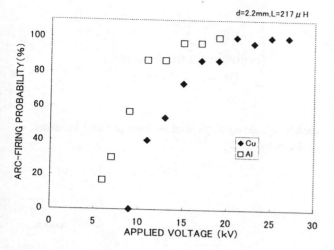

Fig.3. Arc-firing probability

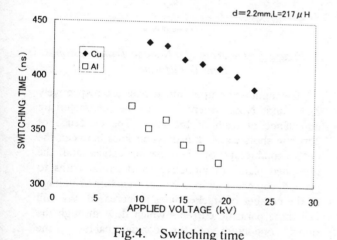

Fig.4. Switching time

Fig.4. shows an example of the switching time. The parameters are L=217 μ H and d=2.2mm. The switching time decreased as a function of the applied voltage. And the switching time for the pair of aluminum electrodes is shorter than copper.

The jitters (variance) of the switching time for a pair of aluminum or copper electrodes are about 50~100 ns, or 20~30 ns, respectively. The jitters for aluminum electrodes are larger than copper.

4. Conclusions

The switching time and the arc-firing probability in UV-laser triggered vacuum gap were investigated. The following results were obtained:

(1) The arc-firing probability increased as a function of the applied voltage.

(2) The self-breakdown voltage and the arc-firing voltage for the pair of copper electrodes were higher than aluminum.

(3) The switching time decreased as a function of the applied voltage.

(4) The switching time for the pair of aluminum electrodes was shorter than copper.

(5) The jitters of the switching time for aluminum electrodes were larger than copper.

References

[1] S.Hazairin, K.Itagaki, et al.:"Switching Characteristics of Two Parallel-Connected Triggered Vacuum Gaps", Trans. I.E.E., Japan, Vol.117B, p.346,1997.

[2] A.Sugawara, H.Samaulah, K.Itagaki and H.Kitamura :"Switching Characteristics of a Triggered Vacuum Gap Employing a Trigger Electrode in the Respective Main Electrodes", Scripta Technica, Inc., Electrical Engineering in Japan, Vol.116, No.6, p.95,1996.

[3] G.Schaefer, et al.:"Gas Discharge Closing Switches", Chapetr 3 Section 3e, P.F.William and A.H.Guenther:"Laser Triggering of a Gas Filled Spark Gaps", Plenum Press, p.145, 1990.

[4] R.L.Boxman, et al.:"Handbook of Vacuum Arc Science and Technology", Noyes Publications, p.57, 1995.

[5] U.Ghera, et al.:"Laser-induced electron source in a vacuum diode", J.Appl.Phys., 66, p.4425, 1989.

OPTIMIZATION OF THE PERFORMANCE OF A TRANSMISSION LINE TRANSFORMER BASED ON THE USE OF FERRITE BEADS

L. Pécastaing, T. Reess,
J. Paillol, A. Gibert, P. Domens

Laboratoire de Génie Electrique
Université de Pau, France

J.P.Brasile

Thomson Shorts Systèmes
Bagneux, France

Abstract

A Transmission Line Transformer (TLT) based on the use of ferrite beads is presented. This work aims at optimizing the performance of the voltage gain and the compactness of the TLT according to the position, the type and the number of ferrites used. The proposed design consists in threading the cables through ferrite beads. These ferrites decrease the effect of parasitic short-circuit transmission line between the outer conductors of the coaxial cables of the TLT. The voltage gain achieved is optimum for a 2, 4 and 10-stage Transmission Line Transformer. The length of each cable is shorter than 1.5 meter.

1- Introduction

During the last few years, the quick improvement in capacitors technology namely their cost, their dimension and the stored energy has permitted the development of pulsed power systems for industry. Pulsed power technologies have a strong potential for pollution control. For example, a non equilibrium plasma (corona discharge) may be used in aqueous based compound. These techniques may also be applied to the treatment of flue gases such as SOx and NOx. The required energy for such applications appears to be very competitive compared with existing methods [1]. High volume of gases treatment is a key issue for industrial acceptance of this technique. This can only be achieved with an energetic corona discharge. Moreover, to produce an energetic corona, the efficiency of these techniques depends closely on the impulse voltage waveform, namely the time to peak, the maximum applied voltage and the repetitive rate of pulses.

Consequently, such methods requires the development of efficient pulse power generators. The specifications for such generators are output voltage ranging from 100 to 500kV, pulse widths from 100 to 500ns, risetime less than 10ns and repetitive rates up to 1kHz. To achieve such performances with a few kilovolts input, an innovative fast pulse transformer is needed.

Figure 1 shows a schematic representation of a N-stage transmission line transformer (TLT). This transformer consists of equal lengths of coaxial cables connected in parallel at the input and in series at the output.

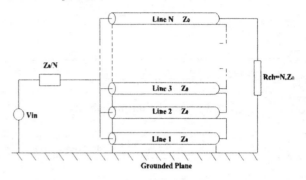

High Voltage Engineering Symposium, 22–27 August 1999
Conference Publication No. 467, © IEE, 1999

Figure 1 : Diagram of a N-stage Transmission Line Transformer

The input and output impedances are respectively Z_0/N and $N.Z_0$ where Z_0 is the characteristic impedance of each of the lines. The existence of parasitic short-circuit transmission lines between the outer conductors of the coaxial cables and the grounded plane, or inductive short-circuit paths to ground, can cause serious loss of transformer voltage as the number of stacked lines is increased. We will call these parasitic currents which flow through the outer conductors of the coaxial cables : the secondary currents.

The choking reactance which isolates the input from the output is usually obtained by coiling the transmission line around a ferrite core or by threading the line through ferrite beads. The aim is to optimize the voltage gain of the TLT without change the time-to-peak of the waveform.

By coiling the coaxial line around ferrite core [2], the voltage yield is 85% and the time-to-peak is about 60ns. The length of the cable is quite long (several ten meters) in order to limit the secondary effect. For example, Wilson and Smith [3] have developed a TLT with 4 cables. The voltage gain achieved is 3.4 and the rise time of the output signal is 50ns. Using 5 cables, the generator of Wilson and al. [4] has a gain of 4.5 and a rise time of 60ns. Some recent works [5]-[6] have allowed to achieve an optimum gain of 10 with a 10-stage transformer. This result with TLT is more efficient than with other types of transformers. Nevertheless, the transformer is constructed using ten lengths of 110m long coaxial cable, and is wound through nine sets of ferrite cores.

As a result the overall height of the transformer is 1.3m.

Our work consists in using the magnetic properties of ferrite beads associated to the TLT characteristics. This method enables to improve the performances and compactness of TLT (decreasing the height and the weight of the generator).

2- Ferrite beads

Ferrites have a great advantage over other types of magnetic material : high electrical resistivity and resultant low eddy current losses over a wide frequency range. Additional characteristics, such as high permeability and time/temperature stability, have expanded ferrite uses into quality filter circuits, high frequency transformers, wide band transformers, adjustable inductors, delay lines, and other high frequency electronic circuitry. As the high frequency performance of other circuit components continues to be improved, ferrites are routinely used into magnetic circuits for both low and high power level applications. Another reason for choosing ferrites is the higher cost of magnetic metals. With low cost, high stability, and lowest volume, ferrites are a good core material choice for frequencies ranging from 10kHz to 50MHz. Ferrites offer an unmatched flexibility in magnetic and mechanical parameters.

Ferrites can be represented by an equivalent series electric scheme as shown on figure 2.

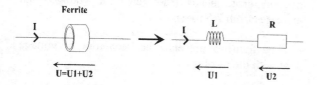

Figure 2 : Equivalent series electric scheme of a ferrite

The resistance and the inductance characteristics depend on the frequency. Consequently, it is not possible to determine easily the electrical characteristic behaviour of ferrites. In order to take into account a large band of frequencies, different ferrites have been used.

Ferrites used during the experiments are :
- Type A : 5MHz Cut-off frequency
- Type B : 2MHz Cut-off frequency
- Type C : 10MHz Cut-off frequency

3- Experimental device

The transformers are constructed using 15Ω or 50Ω coaxial cables. The length of cables depends on the type of test developed (from 1 to 1.5 meter). Voltages are measured with a Tektronix probe (40kV) or a capacitive divider developed in the laboratory (few hundreds of kV). The secondary currents are determined with a Pearson Electronics probe (1V/A).

To drive the TLT, a Blumlein generator (15kV, 12.5Ω, 130ns, 2.3J) is used to begin with. This generator is built with discrete capacitors. The inductances are fully distributed along the transmission lines. In a second step, the input impulse voltage of the TLT is generated by switching 3 parallel coaxial cables, with 15Ω impedance.

4- Experimentations

Two-stage Transmission Line Transformer

In a first time, the aim is to optimize the voltage gain according to the number of ferrites. The system includes the Blumlein generator and a two-stage transformer. It is built using 50Ω coaxial cables, 1 meter long each.

The output of the Blumlein is connected to a carbon composite 25Ω load resistor to match the Blumlein.

The results of these tests are shown in the figure 3 :

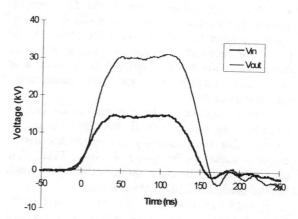

Figure 3 : Input and output voltage of a two stage TLT (8 ferrite beads of each type on each line)

In this configuration, the voltage gain is 2 when the number of ferrite beads of each type is at least 4.

To get a good understanding of effects of ferrites, the secondary current is measured on the outer conductor of the line 1.

When the Blumlein is mismatched (without the 25Ω resistance), the behaviour of voltage and the secondary current measured on the cable 1 is presented on figure 4.

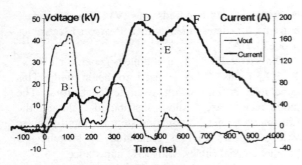

Figure 4 : Output voltage and secondary current measured on the line 1 (8 ferrite beads of each type on each line)

Number of ferrites of each type			Position on the TLT				Voltage Gain
Type A	Type B	Type C	Line 1	Line 2	Line 3	Line 4	
0	0	0					2,5
11	11	10				X	3,4
5	5	5				X	3,3
3	3	3				X	3,1
3	3	3			X	X	3,3
3	3	3		X	X	X	3,3
3	3	3	X	X	X	X	4
3	3	3	X			X	3,6
6	6	6	X			X	4
0	0	6	X			X	3,2
0	6	0	X			X	3,7
6	0	0	X			X	3,9
0	6	6	X			X	3,7
6	0	6	X			X	3,8
6	6	0	X			X	4

Table 1 : Voltage gain in different configurations

The first pulse voltage is the output impulse voltage whereas the following ones are due to the mismatching.

The ferrites lead to the following current equation:

$$I(t) = \frac{U(t)}{R(t)} + \frac{1}{L(t)} \times \int U(t)dt$$

where $R(t)$ and $L(t)$ are respectively the variable resistance and inductance of the ferrites.

From the point A to the point B, the resistive and inductive phases of ferrites occur simultanely : there are only a few amperes during the first impulse voltage (the maximum current is transferred farther). From the point C, only the inductive phase can be taken into account, the resistive phase does not occur. The current is proportional to the integral of the voltage : the points D, E and F correspond to the sign change of the voltage.

Others experiments have been carried out by threading the cables through the ferrite beads. It appears that the secondary current is transferred exclusively after the first pulse of voltage. Finally, the act of threading the cable through the ferrite beads has the same effect than adding ferrites.

It will be interesting to use this technique in the future, but the following experiments (with 4 coaxial cables) aim to determine the best configuration according to the number, the position and the type of ferrite beads used.

Four-stage Transmission Line Transformer

The transformer driven by the Blumlein generator is constructed using 4 lengths of 50Ω coaxial cables, each of which is 1 meter long. The output of the transformer is connected to a carbon composite 220Ω load resistor. The impedance matching is not perfect. This study has allowed to take into account the role of ferrite beads. It appears clearly that the voltage gain is different depending of the quantity, the position and the type of ferrite beads used. Table 1 presents the voltage gains achieved in different configurations.

It is not essential to use long coaxial cables for a good performance of the TLT. The parasite effect of secondary current is minimized thanks to the use of electrical characteristics of ferrite beads. Moreover, the experiments must not be realized above a grounded plane : parasite capacitances are created between the different lines and the grounded plane.

Generally speaking, the voltage gain increases with the number of ferrites used, the voltage gain is also improved when the ferrites are positioned on each cable. Nevertheless, good results can be obtained when the ferrites are located on the lines 1 and 4 (the most important secondary currents flow in the outer conductor of these lines). The contribution of ferrite beads reduces notably theirs effects.

Consequently, this study with a 4 stage TLT has allowed to obtain a perfect efficiency thanks to the use of ferrite beads (the generator voltage gain increases from 2.5 to 4).

The figure 5 presents the behaviour of voltage gain with the ferrites :

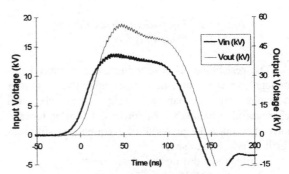

Figure 5 : Input and output voltage for a four-stage TLT (2 ferrite beads of each type on each line)

Ten-stage Transmission Line Transformer

The input impulse voltage of the TLT is generated by switching 3 parallel coaxial cables (20 meters long), each of which has a characteristic impedance of 15Ω. The TLT is constructed using ten lengths of 50Ω coaxial cables, each of which is 1.5

meter long. At the input and the output of the TLT, the impedances-matching are respectively 5Ω and 500Ω.

The aim of this configuration is to study the voltage time-to-peak and the voltage gain.

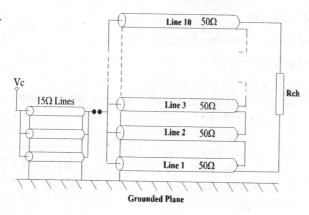

Figure 6 : *System with the 10 stages principle*

In the best system studied, the lines 2 to 9 are threaded 4 times through one ferrite bead of each type (types A, B and C) and the lines 1 and 10 are threaded 3 times through 3 ferrite beads of each type.

The volume of the whole system is only a few dm^3 : the small volume performed by this configuration can be emphasized.

The rise-time of the input voltage pulse is 10ns whereas the output one is about 20ns. This difference is due to the output connections of the TLT. An improvement of these output connections must lead to achieve the same time-to-peak between the input and the output impulse voltage.

The voltage gain of this generator is 10. The gain is optimum.

The comparison between the input and the output impulse voltage is presented on the following figure.

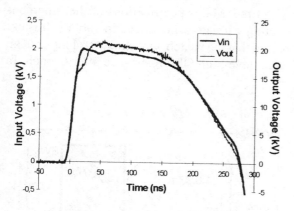

Figure 7 : *Input and output voltage for a 10-stage TLT*

5- Conclusion

Our study aimed essentially at optimizing the performance of a TLT with the position, the type and the number of ferrite beads.

Our technology has allowed to obtain a linear evolution of the voltage gain with a number of cables up to 10 with an undeniable advantage of compactness.

Further works will expand, the range of performance for the output voltage up to 500 kV, the multiple connections at the output will be improved to avoid inopportune breakdowns.

References

[1]-Penetrante B.M.,
"Pollution control applications of pulsed power technology",
Proceedings of the 9th IEEE Pulsed Power Conference, pp. 1-5, Albuquerque, 1993

[2]-Muser A., Zen J., Armbruster R.,
"Transformateurs d'impulsions rapides à lignes de transmission", Onde Electrique,
pp. 520-529, 1964

[3]-Wilson C.R., Smith P.W.,
"Transmission line transformers for high voltage pulsed power generation",
Proceedings 17th Power Modulator Symposium,
pp. 281-285, Seattle, 1986

[4]- Wilson C., Erickson G.A., Smith P.W.,
"Compact, repetitive, pulsed power generators based on transmission line transformers",
Proceedings 7th IEEE International Pulsed Power Conference, pp. 108-112, 1989

[5]- Graneau P.N., Rossi J.O., Brown M.P., Smith P.W.,
"A high voltage transmission line pulse transformer with a very low droop",
Review of Scientific Instruments,
Vol. 67, No 7, pp. 2630-2635, 1996

[6]- Smith P.W., Rossi J.O.,
"The frequency response of transmission line (cable) transformers",
IEEE, pp. 610-615, 1997

Address of authors
Laboratoire de Génie Electrique
Hélioparc Pau Pyrénées
2 Avenue P. Angot
64000 PAU, FRANCE
E-mail : laurent.pecastaing@univ-pau.fr

FAST QUENCH OF SUPERCONDUCTING WIRE BY PULSED CURRENT AND ITS APPLICATION TO PULSED POWER GENERATION

Junya Suehiro, Kouichi Tsutsumi, Daisuke Tsuji, Shinya Ohtsuka, Kiminobu Imasaka and Masanori Hara

Department of Electrical and Electronic Systems Engineering

ISEE, Kyushu University

6-10-1, Hakozaki, Higashi-ku, Fukuoka, 812-8581 Japan

Abstract - An inductive energy storage pulsed power generator needs an opening switch to immediately limit the circuit current and to achieve a high inductive voltage. Quench phenomena of a superconducting wire can be applicable to the opening switch because it provides a fast resistance increase of the wire. In this study, quench characteristics of a superconducting wire were investigated using pulsed current waveforms with various peak values. It has been found that quench of the wire occurred when the current exceeded a critical value. The resistance rise rate of the wire increases with the current peak and wire length. An inductive pulsed power generator, which used a solenoid coil of the superconducting wire as an opening switch, was constructed to demonstrate that the switch could principally work to generate pulsed voltage and power with a fast rising rate.

1. Introduction

Pulsed power technology is finding variety of applications in diverse scientific and engineering fields. Especially, pulsed power generation using inductive energy storage has higher energy storage density than capacitive one and can enhance practical application of this technology. The inductive energy storage pulsed power generator principally needs an opening switch to transfer the inductive energy to an external load. The opening switch should have high OFF resistance which rapidly increases in a short period. Fuses made of thin metal wires are most widely employed as opening switch. Although the fuse has many advantages over other types of switch, it has one disadvantage that it is not suitable for a repetitive operation. Quench phenomena of a superconductor (sudden transition of a superconductor to normal conducting state) accompanying rapid electrical resistance development is potentially applicable to current limiting operation of the opening switch [1]. The phenomena have also been applied to a fault current limiter in electric power

system [2]. Repetitive ON-OFF operation can be realized if the conductor recovers to the original superconducting state by appropriate cooling. In this study, quench characteristics of a superconducting wire were investigated by using pulsed high current. Especially, effects of the current peak value and the wire length were discussed.

2. Experimental Setup and Procedures

An equivalent circuit of experimental setup is depicted in Fig.1. The circuit is basically similar to that of an inductive energy storage pulsed power generator. By closing switch SW_1, energy initially stored in capacitance C is transferred to a series circuit of resistance R_0, inductance L_0 and switch SW_2. Here SW_2 is a superconducting wire whose resistance $R_S(t)$ rapidly increases from zero when quench occurs as a result of higher current over a critical value flowing through the wire. To realize current limiting operation by quench, it is necessary to establish superconducting state during early current rising stage. If the discharge current oscillates at too high frequency, the initial superconductivity may be lost as a result of considerable AC loss. The circuit parameters such as C, R_0 and L_0 were selected so that the circuit current i_S does not oscillate and rises rather slowly with time. Main specifications of the superconducting wire are summarized in Table 1. It is expected that the switch wire should be long enough to achieve high OFF resistance after quench. In

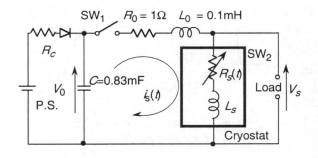

Fig.1 Experimental circuit

High Voltage Engineering Symposium, 22–27 August 1999
Conference Publication No. 467, © IEE, 1999

Table 1 Specifications of superconducting wire

Diameter of wire	0.314 mm
Diameter of filament	7.8 μm
Number of filament	732
CuNi / NbTi	1.21
Twist pitch	9.2 mm
Critical current	224 A (at 1.5 T)
Resistance after quench	4.5 Ω/m (at 15 K)

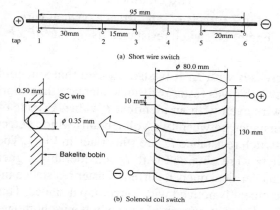

Fig.2 Two types of switch configuration (a) straight short wire (b) solenoid coil

order to examine effects of the wire length on quench characteristics, two types of switch configuration shown in Fig.2 were tested. A short sample is prepared as a straight superconducting wire of 95 mm length on which six voltage measuring taps are connected to investigate spatial distribution of normal zone. The other configuration is a solenoid coil of the superconducting wire of 2 m which also has three voltage taps and is tightened with 10 kg/mm² tension. The superconducting switch was immersed in pool boiling liquid helium and cooled down to 4.2 K. Experiments were conducted changing initial charging voltage V_0 of capacitance C to obtain various peak value I_p of discharge current i_S.

3. Results and Discussion
3.1 Quench characteristics of two types of superconducting switch
Fig.3 illustrates resistance waveform measured between both ends of the short wire switch shown in Fig.2 (a). When the current exceeded about 300 A, quench occurred and resistance R_S abruptly

increased with a rate 2×10^3 Ω/s. When R_S exceeded 200 mΩ, it increased more slowly maybe due to temperature rise of the conductor by Joule heating. Similar experiments were conducted using solenoid switch shown in Fig.2 (b) and a typical result is depicted in Fig. 4. Here quench generated at lower current of 200 A and the resistance increased much faster at 5×10^4 Ω/s and finally exceeded 14 Ω. The higher terminate resistance was obtained simply because of longer wire length and induced current limiting which was not observed with short samples. This rapid current attenuation generated a pulsed inductive voltage V_s between the coil two terminals.

3.2 Switch resistance rising rate
The opening switch needs to have higher resistance increase rate and terminate OFF resistance. The higher OFF resistance of superconducting switch is basically obtained making the switch with longer wire. On the other hand, the resis-

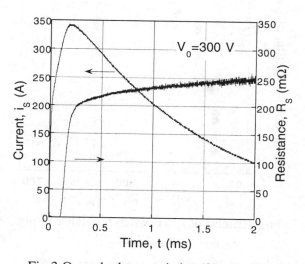

Fig.3 Quench characteristics (Short wire)

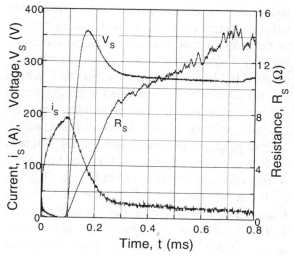

Fig.4 Quench characteristics (Solenoid coil)

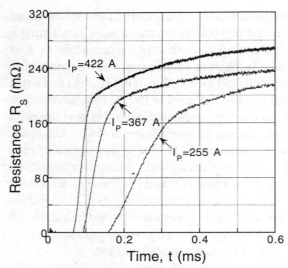

Fig.5 Effects of I_P on switch resistance R_S
(Short wire)

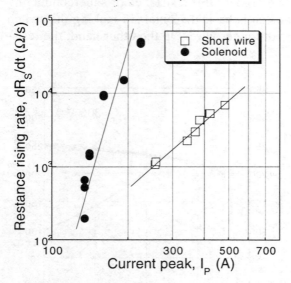

Fig.6 Effects of I_P on switch resistance rising
rate dR_S/dt

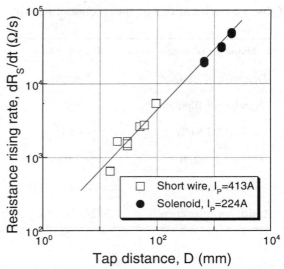

Fig.7 Effects of tap distance (switch wire
length) on switch resistance rising rate dR_S/dt

tance rising rate dR_S/dt is influenced by many factors. Fig.5 shows temporal change of switch resistance R_S measured with a short wire switch varying current peak I_P. The resistance increases just after quench occurrence becomes faster with higher I_P value. Effects of I_P on dR_S/dt are summarized in Fig.6. It should be noted that I_P means current peak which is obtained without quench occurrence and increases linearly with capacitance charging voltage V_0. As shown in Fig.4, actual current peak can be lower than I_P when solenoid coil switch is employed because the high OFF resistance attenuates the current before it reaches I_P. It can be seen that dR_S/dt increases with I_P. Considering that I_P is not actually experienced by solenoid switch as mentioned above, the resistance increase rate essentially depends on current rising rate before quench occurrence which in-

creases with I_P. It is also noticed that solenoid switch gives faster resistance increase than short wire even with an identical I_P value. Effects of distance between two voltage taps which detect switch resistance R_S are illustrates in Fig.7. The data were obtained with I_P considerably higher than critical quench current. Faster resistance increment is noticed with longer tap distance. This implies that quench of the switch wire homogeneously develops over the entire length as long as the switch current rapidly increases to the quench level. A linear relationship between wire length and dR_S/dt found in Fig.7 is employed to theoretically predict how the wire length influences peak values of output power P and wire temperature T due to Joule heating after quench. Theoretical calculations were conducted by using the following equations which describe electrical and thermal behaviors of the experimental circuit and tested switch.

$$\{R_0 + R_S\}\, i_S + L_0\frac{di_S}{dt} + \frac{1}{C}\int i_S\, dt = 0 \qquad (1)$$

$$C_P(T)\, V\, d\theta = \left\{P - h_C\, S\left(T - 4.2\right)\right\} dt \qquad (2)$$

Here $C_p(T)$ is specificheat of the switch wire [3] which is assumed to be same as that of CuNi. h_C is a thermal transfer coefficient [4] from a conductor to the surrounding liquid helium whose temperature is assumed to be constant at 4.2 K even after quench. Volume V and thermal transferring lateral area S of the switch wire are expressed as follows.

$$V = \pi d^2 l\, / \, 4 \qquad (3)$$

$$S = \pi d l \qquad (4)$$

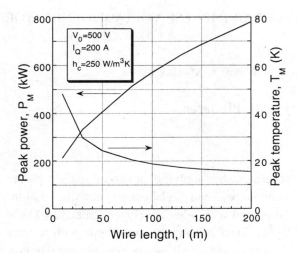

Fig.8 Effects of switch wire length on output pulsed power and switch temperature rise

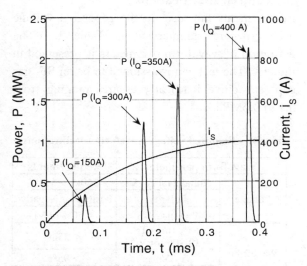

Fig.9 Effects of switch quench level I_Q on output pulsed power

Here l and d are length and diameter of the switch wire respectively.

As shown in Fig.8, it is theoretically predicted that longer switch wire increases output from the pulsed power generator. Lower temperature rise after quench is another benefit of the longer conductor that may shorten quench recovery time and result in higher repetition rate of the switch ON-OFF operation.

From Fig.6, it seems that one can achieve higher output power by increasing I_P and resultant higher dR_s/dt . However, this results in earlier quench occurrence than current peak and lower energy transfer efficiency from inductance to the exter-

nal load. Fig.9 shows theoretical predictions about relationship between quench occurrence timing and output power. As switch quench current I_Q is varied, quench onset time accordingly changes. It is clear that maximum output power is obtained when quench occurs at the current peak where the energy transfer efficiency becomes maximum as well. This result means that higher output power may be obtained by selecting circuit parameters which enables faster current rising and current peak which is close to the switch quench level.

4. Concluding and Remarks

It was proved that quench phenomena of a super-conducting wire is basically applicable to opening switch operation for an inductive energy storage pulsed power generator. Making the switch with longer wire can improve the switch performance with faster current attenuation and lower temperature rise after quench. Faster switch current rising, which is realized by optimization of circuit parameter, is also essential to improve the switch performance.

This work was supported in part by Grant-in-Aid for Scientific Research from the Ministry of Education, Science and Culture, Japan. One of the authors (SO) is a Research Fellow of the Japan Society for the Promotion of Science.

References

[1] W.H. Bostick, V. Nardi and O.S.F. Zucker (Eds.):"Energy Storage, Compression, and Switching", pp.279-305, Plenum Press, New York (1976)

[2] H.J. Boeing and D.A. Paice:"Fault Current Limiter Using a Superconducting Coil", IEEE Trans, on Magnetics, Vol.MAG-19, No.3, pp.1051-1053 (1982)

[3] M. Iwakuma, H. Kanetaka, K. Tasaki, K. Funaki, M. Takeo and K. Yamafuji: "Abnormal Quench Process with Very Fast Elongation of Normal Zone in Multi-strand Superconducting Cables", Cryogenics, Vol 30, pp.686-692 (1990)

[4] N. Amemiya, N. Banno and O. Tsukamoto:"AC Loss and Stability Analysis in Multifilamentary Superconductors Based on Temporal Evolution of Current/Temperature Distribution", IEEE Trans. Magn, Vol.32, No.4, pp.2747-2750 (1996)

DEVELOPMENT OF 1MV STEEP-FRONT RECTANGULAR-IMPULSE VOLTAGE GENERATOR

M. Yashima[*], H. Goshima[*], H. Fujinami[*],
E. Oshita[**], T. Kuwahara[**], K. Tanaka[**], Y. Kawakita[**], Y. Miyai[**], M. Hakoda[**]

  * Central Research Institute of Electric Power Industry (CRIEPI), Japan
** Nissin Electric Co., Ltd., Japan

Abstract

The basic design and performance of a newly developed 1MV steep-front rectangular-impulse voltage generator, named "SPARK", are described. Advanced pulsed-power technology and elaborate circuit design enable an ideal rectangular waveform, in which the rise time of the wavefront is shorter than 20ns and a long flat-topped voltage crest continues up to 10μs, to be generated. By applying this waveform as a test voltage to a model GIS (gas-insulated switchgear) gap, quantitative evaluation of the voltage-time characteristics can be achieved.

1. Introduction

The insulation design of SF₆-insulated power apparatus such as a GIS is principally based on steep-front overvoltages having a rise time of several tens to hundreds of ns caused by lightning surges or disconnector operation. The insulation characteristics of GIS for such overvoltages are basically tested by applying standard test voltages such as a lightning impulse voltage.

Measurement of voltage-time (V-t) characteristics in the short time region of less than 1μs is normally very difficult when using a lightning impulse with a rise time of about 1 to 2μs. To obtain the quantitative evaluation of V-t characteristics in a short time region, an ideal rectangular-impulse voltage is required as the test voltage waveform.

Currently, the ideal rectangular-impulse voltage generator is realized only in a low-voltage range below several kV or several tens of kV. Utilizing recent pulsed-power technology, the generation of a steep-front high-voltage waveform up to the MV order, having a rise time of several tens of ns can be achieved. However, long duration of a flat-topped voltage crest of up to several μs has not yet been realized.

In the present work, we have developed "SPARK", a 1MV steep-front rectangular-impulse voltage

generator, by applying advanced pulsed-power technology and elaborate circuit design. Compared with other developed generators [1], a highly ideal rectangular waveform with a very short rise time, small oscillation and long duration of the flat-topped voltage crest has been realized.

2. Required specifications

Requirements imposed on the generated voltage waveform are summarized in Table 1. The definition of waveform parameters is presented in Fig. 1. The test gap is assumed to be an SF₆ gas gap (capacitive load) at pressures ranging from 0.1 to 0.6MPa.

Table 1. Required voltage waveform specifications.

Waveform parameter		Requirements
Voltage	Voltage range: V_r	200kV - 1MV
	Polarity	Positive or Negative
	Prepulse: V_p	< 30% of V_r
Wavefront	Rise time: T_r (30-90% of V_r)	< 25ns
	V_r / V_{max}	> 95%
	Oscillation: V_{osc}	< 5% of V_{max}
Flat-topped part	Duration	Variable, 10μs maximum
	Droop at 10μs: V_{10} / V_{max}	> 95%

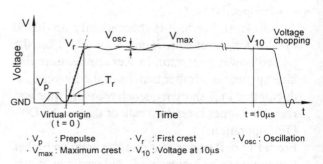

V_p : Prepulse   ·V_r : First crest   ·V_{osc} : Oscillation
·V_{max} : Maximum crest   ·V_{10} : Voltage at 10μs

Fig. 1.   Definition of waveform parameters.

3. Basic system structure and circuit

To obtain a high-voltage rectangular waveform, the PFN (pulse forming network) or PFL (pulse forming line) is usually used in pulsed-power

High Voltage Engineering Symposium, 22–27 August 1999
Conference Publication No. 467, © IEE, 1999

equipment. Although the PFN can generate long-duration pulses ranging from μs to ms, the rise time of the wavefront is not very short (order of μs). The PFL is able to provide a shorter rise time of several tens of ns. The duration is normally limited to several hundred ns due to the physical size limitations of the equipment.

In our plan, an encapsulated, large-capacity Marx generator is used as the main voltage source to enable the generation of a long-duration waveform. The generated voltage is initially stored in the ISC (intermediate storage capacitor) with extremely low inductance, then transferred to a test gap through a series gap in the GS (gap switch) by triggering the gap with a laser beam. The test gap enclosed in the test tank is the final load of the Marx generator. The overall schematic plan of the system and the equivalent circuit are shown in Figs. 2 and 3, respectively.

To restrain overshoot and oscillation from appearing at the wavefront or the flat-topped voltage crest, two damping resistances (RD1, RD2) are distributed on both sides of the GS. The function of the resistance connected to the ground (RD) is to maintain the arc current through the series gap in the GS to provide the waveform duration after charging the test gap.

In the Marx generator, a chopping gap is installed to control the duration of the voltage. This gap is also triggered with a laser beam after the

voltage crest duration of an arbitrarily variable time ranging up to 10μs. The operation of the chopping gap yields a steep voltage drop at the wavetail, which results in a waveform of the ideal rectangular shape.

On both sides of the ISC, oil-filled tanks are installed. These prevent water leakage from the ISC into the oil-insulated Marx generator or into the gas-insulated GS tank in case the ISC spacer is broken due to the surface flashover.

4. Simulation

Based on the equivalent circuit shown in Fig. 3, the voltage waveform applied to the test gap (output voltage waveform) is analyzed using the EMTP (electromagnetic transients program). As the circuit element for each resistance, capacitance and inductance, a lumped constant circuit based on factors such as material, dimension and arrangement, is used. In spite of the short axial length of the ISC, a distributed constant circuit is applied to the ISC for higher accuracy. The analyzed waveforms are shown in Fig. 4.

Time-dependent variable resistance is applied to the arc between the series gap in the GS. Its initial value at the moment the arc bridges the gap is decreased by three orders of magnitude within several tens of ns.

The rise time and the oscillation appearing on the wavefront are analyzed to be 17ns and 4.1%,

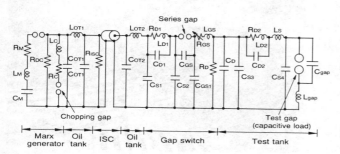

Fig. 2. **Overall schematic plan of "SPARK".**

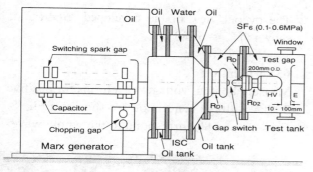

Fig. 3. **Equivalent circuit.**

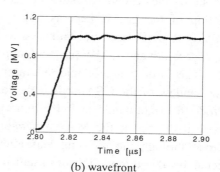

(a) whole waveform

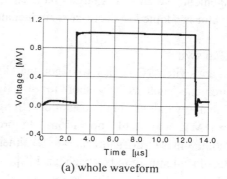

(b) wavefront

Fig. 4. **Analyzed output voltage waveform.**

respectively. The droop of the voltage crest is only 1% at 10μs from the virtual origin of the waveform. The prepulse appearing before the voltage rise, which occurs due to charging of the GS, is only 6% of the voltage crest.

5. Detail of main components

5.1 Marx generator
The oil-insulated encapsulated Marx generator is designed to generate a long wavetail and to reduce internal inductance as much as possible. For these purposes, high ohmic value charging resistors are used, and all components are placed as compactly as possible. Principal specifications of the Marx generator are summarized in Table 2.

Table 2. Principal specifications of Marx generator.

Description	Details
Number of stages	8
Capacitors per generator stage	2 x 1.6μF / 75kV
Impulse capacitance	0.1μF
Maximum total charging voltage	1.1MV
Total charging energy	61kJ

5.2 Intermediate storage capacitor
Conductivity-controlled pure water (resistivity: 1 to 2MΩcm, specific permittivity ε_r: 80) is used as the dielectric medium of the ISC. The coaxial arrangement with a large-diameter inner conductor is adopted to decrease its inductance.

5.3 Gap switch
The gap switch (GS) has to meet the difficult requirements such as very short turn-on time of less than several ns, high insulation strength of above 1MV, and stable operation. In order to overcome these difficulties, an excimer-laser triggered switching gap has been used [2].

Fig. 5 shows the schematic diagram of GS. The insulation of the gap is kept as SF_6 at 0.1 to 0.6MPa. A KrF laser of 248nm wavelength, 15ns pulse width and 15MW peak output power is used. The laser beam is injected through the quartz window of the GS tank into the load-side electrode (the right-hand side in the figure), and is focused at the tip of the Marx generator's side electrode through the focusing lens after being reflected by the turning mirror installed in the load-side electrode. The focused area of the

laser beam at the irradiated point is about 50μm in diameter (energy density: approximately 10^4 J/cm^2). A similar GS is used for the chopping gap in the Marx generator. A laser-triggered GS also enables precise control of voltage generation timing with an extremely small jitter of less than 25ns.

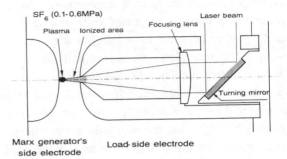

Fig. 5. Schematic diagram of laser-triggered gap switch.

5.4 Damping resistor
To prevent local voltage oscillation, elaborately designed damping resistors should be arranged at appropriate points. Based on the analytical results, two damping resistors are placed on both sides of the GS, and each value is carefully selected. To decrease the stray inductance and capacitance of the resistor, a water resistor is used. An aqueous solution of ammonium chloride (NH_4Cl) is chosen as the resistor material, taking the thermal and insulation properties into account.

The overall view of the developed "*SPARK*" is shown in Figs. 6 and 7.

6. Verification
Measured output voltage waveforms for positive 300kV voltage generation are shown in Fig. 8. Figs. 8a and 8b correspond to the whole wave-

Fig. 6. Overall view of developed "SPARK".

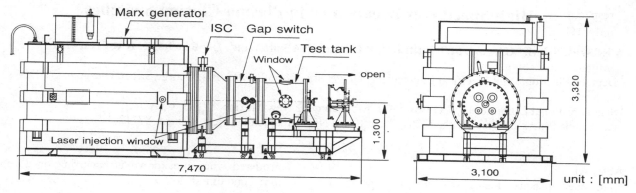

Fig. 7.   Side and front sketches of *"SPARK"*.

form and the wavefront measured with wide-frequency-band and high-frequency-band capacitive voltage dividers, respectively. These voltage dividers in which a circular sensor plate insulated from the ground directly faces to the side surface of the high-voltage electrode of the test gap, are installed at the inner surface of the test tank.

The rise time of the wavefront is 20ns, the oscillation on the voltage crest is 3.5%, and droop is only 1% after the crest duration of 10µs. These characteristics coincide well with the simulated results (Fig. 4).

With increasing voltage, while the rise time tends to decrease to 18ns, the oscillation appearing on the wavefront becomes larger reaching about 8% for 1MV voltage generation. The oscillation is affected by the gas pressure of SF_6 in the gap switch. Hereafter, the effect of the properties and pressure of the switching gas on the generated waveform should be clarified.

7. Conclusion

Utilizing the advanced pulsed-power technology and elaborate circuit design, we succeed in developing the *"SPARK"*, a 1MV steep-front rectangular-impulse voltage generator. We are convinced that the flat-topped voltage crest duration is the longest, and the most ideal rectangular waveform that can be achieved in *"SPARK"*, among the currently available several-hundred-kV-class step pulse generators.

By using *"SPARK"*, the following research in the field of discharge physics for many gas insulation media, including compressed SF_6, will be possible:

(1) Measurement of discharge time lag (statisticaland formative time lags).

(2) Clarification of discharge development

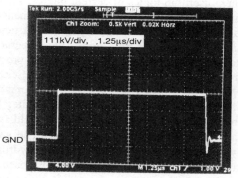

(a) whole waveform

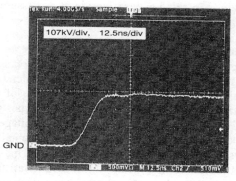

(b) wavefront

Fig. 8.   Measured output voltage waveform for positive 300kV voltage generation.

mechanism.

(3) Quantitative evaluation of voltage-time characteristics.

References

[1] K. Feser, *et al.*: "Steep Front Impulse Generators", Proc. of the Third ISH, 41.06, Milan, Italy (1979)

[2] J. R. Woodworth, *et al.*: "UV-Laser Triggering of 2.8-Megavolt Gas Switches", IEEE Trans. on Plasma Science, **Vol.PS-10**, No.4, pp.257-261 (1982)

Address of the Author:
Masafumi Yashima
CRIEPI, 2-11-1, Iwado-kita, Komae-shi, Tokyo
201-8511   JAPAN

High Speed Gas Breakdown in Plasma Closing Switches

A R Dick[1], S J MacGregor[1], M T Buttram[2], R C Pate[2], P A Patterson[2], L F Rinehart[2], K R Prestwich[3]

[1] Dept. of Electronic and Electrical Engineering, University of Strathclyde, 204, George Street, Glasgow, G1 1XW, Scotland
[2] Sandia National Laboratories, P O Box 5800, Albuquerque, New Mexico, 87185, USA
[3] Kenneth R Prestwich Consulting, 12201 Cedar Ridge NE, Albuquerque, New Mexico, 87112, USA

Abstract

High-speed, plasma closing switches are capable of exhibiting rapid (sub-nanosecond) voltage collapse over a range of voltages and pressures. This property makes them useful in topics of current interest such as laser drivers and ultra-wide band radar equipment. In order to optimise switch behaviour for a particular application, knowledge of how the operating parameters of the switch affect the breakdown process is required. The experimental investigations described in this paper were performed with the gas pressure being varied between 0.15 and 3 bar absolute. For these investigations a constant electric field was maintained, prior to breakdown occurring. The gases tested were sulphur hexafluoride, helium, hydrogen and nitrogen. The results showed that sub-nanosecond voltage collapse times were achieved in these gases within this pressure range. The fastest voltage collapse was obtained in sulphur hexafluoride at a pressure of 0.15 bar.

Introduction

Under pulse conditions, the voltage which can be applied to a spark gap can be 2-3 times greater than that achieved under DC charging. The value of the overvoltage may be a function of the gas type, gap spacing, polarity, and the rate of rise of the voltage. It is well known that the level of overvoltage obtained increases as the pulse rise time reduces [1]. This has been attributed to the delay introduced (statistical time-lag) while waiting for an initiatory electron to appear [1, 2]. Thus it should be possible, by applying a fast rising pulse to a spark gap to achieve significant overvoltage and subsequent ionisation, leading to rapid collapse of the channel.

This paper reports on experiments that have been carried out to investigate how the rate of voltage collapse of a plasma channel depends upon these parameters. An experimental system has been designed to permit the observation of high-speed gas breakdown. The experimental approach takes into account distributed effects such as transmission line behaviour and wave behaviour, which have to be

High Voltage Engineering Symposium, 22–27 August 1999
Conference Publication No. 467, © IEE, 1999

considered when breakdown is occurring on a sub-nanosecond timescale.

The experimental arrangement includes diagnostic probes that are placed in the vicinity of the channel formation region in order to provide information on the voltage collapse across the switch gap. The information from these probes is recorded and digitised on Tektronix SCD5000 transient digitisers. It can be subsequently processed to compensate for the high frequency losses which occur as a signal propagates through the cables connecting the probes to the digitisers. The compensated information can be used to study the time evolution of the voltage collapse across the electrodes and to determine the influence of the operating conditions upon high-speed gas breakdown.

Experimental Layout

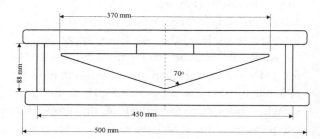

Figure 1: Sketch of switch.

The main component of the experimental apparatus is the switch assembly. This has the topology of a conical transmission line. In this topology the lower electrode consists of an aluminium disk which has a diameter of 500 mm and is 25 mm thick. The upper electrode is an aluminium cone with a half-angle of 70° and a diameter at its base of 370 mm. The upper edge of the cone has been rounded to reduce the magnitude of the electric field at its surface. Similarly the apex of the cone has been rounded to have a 2.5 mm radius of curvature. The side wall of the switch is a hollow nylon cylinder of thickness 12 mm and an outer diameter of 450 mm. With these dimensions, the switch has a safety limit on the pressure of 6 bar. A schematic of the switch assembly is shown in Figure 1.

The results presented in this paper were obtained using a 3 mm gap in the switch. Electrostatic modelling was carried out to find the field distribution along the central switch axis, between the upper and lower electrodes. For the modelling process, a potential difference of 1 volt was applied between the electrodes. The result of this analysis is shown in Figure 2. This modelling predicted that the switch configuration produced a field enhancement factor of 1.8 on the stressed (upper) electrode.

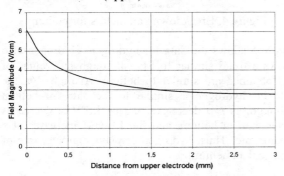

Figure 2: Field distribution along switch axis

It is essential to employ a transmission line switch topology. When the switch gap breaks down, a radial electromagnetic wave is transmitted out from the spark axis through the conical transmission line, which has a constant impedance. This means the radial wave does not encounter any discontinuities until it reaches the limit of the electrodes. The switch has been designed to ensure that a fast, sub-nanosecond voltage collapse can be completely monitored by the diagnostics, before a reflected wave returns from the outer limit of the electrodes to the spark axis. In the present switch there is a delay time of 1.4 ns before the reflected pulse returns to the axis. In such circumstances, where breakdown is occurring rapidly, the impulse generator impedance and external circuit impedance are transit time decoupled from the switch and may be neglected.

The impedance of a conical transmission line is given by [3]:

$$Z_0 = \frac{\zeta_0}{2\pi\sqrt{\varepsilon_r}} \ln\left[\cot\left(\frac{\theta}{2}\right)\right] \qquad (1)$$

where,

Z_0 is the transmission line impedance
ζ_0 is a constant ($= 120\pi$)
ε_r is the relative permittivity of the dielectric surrounding the electrodes
θ is the half-angle made by the conical electrode.

For an angle of 70° in a gaseous dielectric the line impedance $Z_0 = 21.4\ \Omega$.

A single-stage, non-inverting pulser was used to charge the switch. The energy store was an 80 nF capacitor. This capacitor was charged with a positive DC voltage. The pulser produced an output waveform having a 10-90% risetime of 40 ns and a 90-10% decay time of 140 μs. Thus, the pulser produced a flat-top waveform, with a well defined voltage maximum, for several hundred nanoseconds.

Diagnostics
To measure the voltage collapse across the switch gap, three D-dot monitors were flush mounted in the base plate. One monitor was constructed from a length of RG 402 semi-rigid microwave cable and had a bandwidth in excess of 30 GHz [4]. This monitor was placed at a distance of 10 mm from the switch axis. A second monitor was constructed from a soldered end of a length of URM 67 high voltage cable. This monitor had a greater surface area than the first monitor and so produced an output signal with a greater amplitude than the RG 402 based monitor. This allowed the URM 67 based monitor to record a signal in relatively low field regions where the signal produced by the RG402 monitor had an insufficient amplitude to be recorded by the SCD5000 digitisers. Since the area of the central conductor has been increased, the upper frequency limit of the monitor will have been reduced (> 9 GHz) [5]. This monitor was placed at a distance of 20 mm from the switch axis. A third monitor was constructed using a modified N-type bulkhead adapter with a small plate attached to the central conductor. This monitor was placed 30 mm from the switch axis and had an upper frequency limit > 6 GHz.

The high frequency signal components produced by these monitors can face significant attenuation along the cables carrying them to the recording instruments. To recover these losses, and so restore the signal to its original condition, a compensation technique can be employed [6]. The compensation procedure can also be implemented to correct for the response limitations in the SCD5000s. To enable the SCD5000 digitiser to record the breakdown pulse, and be triggered by it, its internal delay line had to be used. This restricted the instrument's analogue bandwidth to about 2 GHz [7].

Results
The gap spacing was monitored with the signal produced by the D-dot probe that was 30 mm from the switch axis. A charging voltage, insufficient to cause breakdown, was applied to the switch. The waveform

was recorded and integrated. The amplitude of the integrated pulse was measured and compared to the amplitude of the charging pulse. The calculated voltage division ratio was proportional to the gap spacing.

To ensure as constant a value of the electric field E as possible, a charging voltage of 55 kV was used in all cases. Only those results where the charging voltage reached its plateau value were recorded. The E/p values shown in Figures 3-6 were calculated using the mean field, which in this case was 183 kV/cm. This investigation is part of an ongoing study [8] which has previously examined SF_6, helium and hydrogen, and has now been extended to include low pressure breakdown in nitrogen.

Two parameters are shown on the graphs. These are the peak value (peak dV/dt) and the full-width, half-maximum (FWHM) of the recorded pulse. Since the recorded pulse is a scaled representation of the differential of the voltage collapse, the peak value gives an indication of the rate of voltage collapse and the FWHM gives an indication of the duration of the voltage collapse. The results reported here have not yet been compensated for either the signal cable attenuation or the bandwidth limitations of the SCD5000 digitisers.

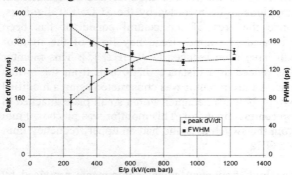

Figure 3: Breakdown parameters for SF_6

Figure 3 shows the results obtained for SF_6. The graph shows a trend for the peak dV/dt to increase with increasing E/p, and for the FWHM to decrease with increasing E/p. The implication of this is that the rate of voltage collapse is increasing with increasing E/p. The results for the highest values of E/p appear to show a maximum in the peak dV/dt data.

Figure 4 shows the data for helium. The same trends are apparent in helium as were observed in SF_6. However, the highest E/p values are almost an order of magnitude less than in SF_6. For low E/p, the peak dV/dt is higher than in SF_6. However this value changes slower than in SF_6 so that at high E/p it is

substantially less than in SF_6. The value of FWHM is greater than SF_6.

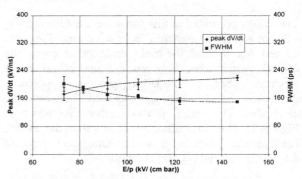

Figure 4: Breakdown parameters for helium

Figure 5 shows the data for hydrogen. Hydrogen also shows a trend similar to SF_6, although the values of E/p obtained are similar to helium. At high E/p, the data for hydrogen is similar to the data for helium although, as E/p decreases, the peak dV/dt decreases faster than in helium, and the FWHM increases at a greater rate. Thus, hydrogen displays a greater rate of change than helium.

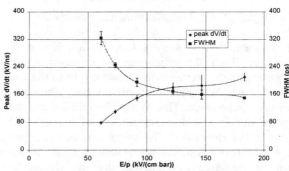

Figure 5: Breakdown parameters for hydrogen

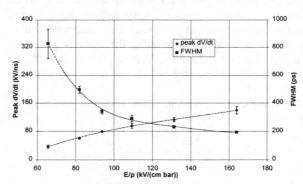

Figure 6: Breakdown parameters for nitrogen

The data for nitrogen is shown in Figure 6. The same trend is observed in nitrogen as was observed in the other gases. The measured parameters indicate that breakdown is slower in nitrogen, in the recorded range of E/p, than the other gases. However the data also indicates that, at high E/p, the voltage collapse time is decreasing at a greater rate in nitrogen than in the other gases.

It is important to note that the data does not show the energy losses incurred in the formation of the plasma channel, but these losses are dependent on the rate of voltage collapse across the channel. The data shows that as the pressure is reduced, the voltage collapse time decreases, and so the energy losses incurred will also decrease. This is true for each of the gases but it can be seen that the fastest breakdown can be achieved in SF_6. This is because of the much higher values of E/p which can be reached compared with hydrogen, helium or nitrogen. However, care has to be taken at high E/p, especially in SF_6, to maintain a good finish on the electrode surfaces as surface roughness can reduce the E/p attainable and also introduce additional effects such as multi-channelling.

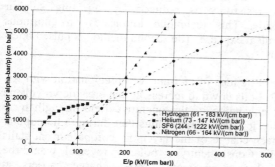

An indication of the development of the experimental results can be found from Figure 7. This figure is a plot of α/p (or $\overline{\alpha}/p$) against E/p for each of the gases. The data have been obtained from tables of published data [9]. The values quoted in the legend are the experimentally obtained values. The figure shows that in the experimental range of E/p values, helium has the slowest variation. This is reflected in the variation of its experimental parameters. At low E/p, hydrogen has a lower ionisation coefficient than helium, but at higher E/p its ionisation coefficient is comparable to helium. Nitrogen has a lower ionisation coefficient than hydrogen but its rate of change is greater. The experimental values of E/p for SF_6 are mostly off the scale of this graph, so it is more difficult to interpret in this context. However, the net ionisation coefficient is the greatest of the gases.

Conclusions

A conical transmission line switch has been used to measure the voltage collapse resulting from the electrical breakdown of several gases. These gases were SF_6, helium, hydrogen and nitrogen. For all the experimental measurements the same electric field was applied and the same gap spacing maintained. The only parameter allowed to vary was the pressure. The results showed that all of the gases exhibited rapid voltage collapse in the pressure range 0.15 to 3 bar. All of the gases showed that, as E/p increased, that is, as the pressure was reduced, the rate of voltage collapse increased. Nitrogen showed the slowest breakdown with helium showing the least variation. Hydrogen, at low E/p, had a slower breakdown than helium, but as E/p increased its rate of voltage collapse became comparable to helium. SF_6 showed the greatest range of the gases, reaching much higher values of E/p. At these high values of E/p it produced the fastest breakdown of all the tested gases. The relative behaviour of the gases is broadly reflected in the variations of the ionisation coefficients.

Acknowledgement

This work has been supported through funding from Sandia National Laboratories, USA (Contract No. AU-0932), and by the University of Strathclyde.

References

[1] Gibert A et al: "Dielectric behaviour of SF_6 in non-uniform fields", *J. Phys. D: Appl. Phys.*, 26, pp 773-781, 1993

[2] Somerville I C and Tedford D J: "Negative ion detachment and its effect on the statistical time lag in gases", *7th Int. Conf. On Gas Discharges and their Applications*, pp 325-327, 1982

[3] King R W P: "The Conical Antenna as a Sensor or Probe", *IEEE Trans. Electromagn. Compat.*, EMC-24, pp 8-13, 1983

[4] Burkhart S: "Coaxial E-Field Probe for High Power Microwave Measurement", *IEEE Trans. Microwave Theory Tech.*, MTT-33, pp 262-265, 1985

[5] Baum C E et al: "Sensors for Electromagnetic Pulse Measurements Both Inside and Away from Nuclear Source Regions", *IEEE Trans. Antennas Prop.*, AP-26, 1978

[6] Patterson P E, Aurand J F and Frost C A, "15 GHz Bandwidth Enhancement of Transient Digitisers in a Portable Data Acquisition System", *Sandia National Labs Technical Report SAND93-2280*, 1994

[7] *"Tektronix SCD5000 Transient Waveform Recorder User Manual"*, Beaverton, OR: Tektronix

[8] Dick A R et al: "An Investigation into High Speed Gas Breakdown", *23rd Int. Power Modulator Symp.*, pp 202-205, 1998

[9] Kuffel E and Zaengl W S *"High Voltage Engineering Fundamentals"*, Oxford: Pergamon Press, Ch. 5, 1988

Separation of Fine Dust with AC-Energized Electrostatic Precipitators

Hoferer, B.; Weinlein, A.; Schwab, A.J.

Institute of Electric Energy Systems and High-Voltage Technology
University of Karlsruhe, Kaiserstraße 12, D-76128 Karlsruhe, Germany

Abstract

In addition to studies with DC-energized electrostatic precipitators (ESPs) this paper presents the results of experimental investigations with a plate-type laboratory-scale AC ESP. These precipitators are relatively insensitive to the dust resistivity. The paper's theo-retical part deals with the basic physical mechanisms of the separation of particles with AC energization. This is followed by an overview of general methods for AC-precipitation. Finally, this paper describes experi-ments with a modified laboratory-scale AC ESP, excelling by the absence of back corona. Fractional efficiency measurements were conducted for various geometric and electric parameters and constant gas conditions.

Introduction

DC-energized electrostatic precipitators represent the most effective means to separate solid and fluid particles from flue gas.

The optimum functionality of the electrostatic precipitation is governed by five mechanisms which interact simultaneously. The electric charge carriers charging the dust particles in the gas flow are generated in the neighbourhood of the discharge electrodes. The char-ged particles are transported towards the collecting electrode of opposite polarity due to the Coulomb for-ces. The precipitation results in dust layers which must be removed from the electrodes by periodic rapping with hammer blows.

In order to economically cope with future emission standards, the capital expenditures must be lowered by reducing the ESP's overall size. To meet the demanded clear-gas dust concentration with a smaller collecting-area a reduction of system immanent effects causing poor precipitator efficiency is required. Those are insufficient charging of fine-dust particles, reentrainment of previously separated dust and back corona. In the DC- mode these effects can only be reduced by enormous technical and economical efforts.

High Voltage Engineering Symposium, 22–27 August 1999
Conference Publication No. 467, © IEE, 1999

Motivation and Background

The motivation for employing AC-energized electrostatic precipitators arises from the detrimental back corona in the DC- mode. Two technical options are presently available to counteract back corona while maintaining precipitator efficiency.

On the one hand, flue gas conditioning is applied, i.e dust resistivity is reduced by injecting sulphur trioxide. This technique implicates an enormous chemical expense. The second method is pulsed oper-ation.

The basic idea of an improved precipitation with AC ESPs boils down to increasing the breakdown voltage and enhancing of the operative field strength by a dielecric layer. It is obvious that during a homopolar surface charge of the layer the corona current stalls itself. Consequently a polarity reversal would be desirable to generate further charge carriers at increased breakdown voltage. An AC-power supply with a power frequency of 50 Hz offers precisely this feature.

Fundamental Principles of AC-Corona

AC energization deploys the polarity effect in a non-uniform field resulting in a higher negative breakdown voltage.

Beyond, there exists a greater negative ion mobility resulting in the rectifier effect of the AC corona. According to Krug [Kru69], corona starts in the negative halfwave entailing a negative direct-current offset.

Further, increasing voltage results in the onset of the positive corona involving a stronger positive direct-current. These currents are superimposed resulting in a positive net current. Shortly afterwards, the configuration sparks over. By means of a dielectric layer the breakdown voltages can be significantly increased.

Laboratory-Scale AC-Electrostatic Precipitator

The AC ESP is illustrated in Figure 1. Its basic design is similar to that of a DC ESP. The main difference is a dielectric layer on the collecting electrode. Air heating and air humidification equipment ensure the essential operating conditions.

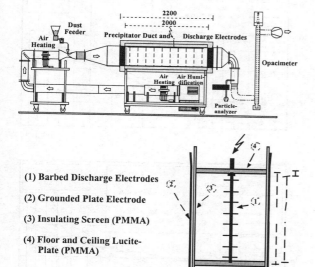

(1) Barbed Discharge Electrodes

(2) Grounded Plate Electrode

(3) Insulating Screen (PMMA)

(4) Floor and Ceiling Lucite-Plate (PMMA)

Figure 1: Laboratory-scale AC ESP (top) and insight into duct (bottom)

The dust feeder and the duct are followed by an optical particle-analyzer and an opacimeter. Thereafter, the dust-gas mixture exits the system. Acrylic glass (PMMA) allows an optical observation of dust pattern and back ionization detection. The dielectric layer permits an overvoltage-factor of about 1.5.

The laboratory-module is a horizontal, plate-type ESP with a length of 200 cm, a duct height of 40 cm and an electric duct width of 30 cm. A frame allows a steady variation of the discharge electrode distance. The studies were carried out with a highly resistive limestone fraction with an average diameter of $d_{P,50} = 6$ µm added to the gas, Figure 1.

The power supply consisted a conventionel high-voltage test transformer connected in series with a variable transformer. The voltage was measured by a capacitive voltage divider, the r.m.s. current and its waveform by a current-viewing resistor.

An optical particle-analyzer (Polytec HC 15) permits the specification of the precipitator efficiency for specific particle size. To obtain the size distributions both the partial flows of the polluted and of the cleaned gas were analyzed with and without energization.

In the outlet a probe for an isokinetical partial flow sampling is located directing it to a very small measuring volume of the optical particle-analyzer, Figure 2.

Each particle streaming through this measuring volume is illuminated by a halogen lamp. The scattered light of each particle is converted by a sensitive photomultiplier to an electrical signal and counted in one of 128 data channels.

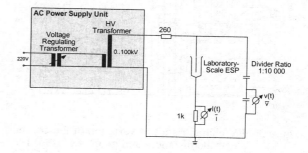

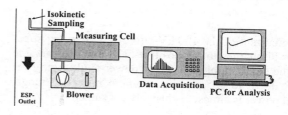

Figure 2: AC-Power supply (top) and measuring system for the detection of particle size distribution (bottom)

Results

Current and Voltage Characteristics

Below the corona onset the current and voltage characteristic show a capacitive current, rising linearly with the applied voltage, Figure 3, top. Barbed discharge electrodes show a stronger current density in the corona range than wire discharge electrodes. In the current and voltage waveforms, respectively, an increment in corona intensity in the peak regions of the voltage is apparent with increasing voltage, Figure 3, bottom.

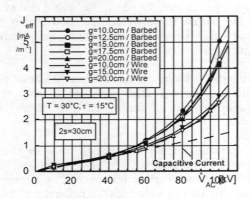

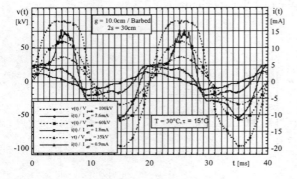

Figure 3: J/V charactristics of variation of the corona electrode distances and geometies (top) I/V waveshapes for peak voltages of 35, 60 and 100 kV, g = 10,0 cm (bottom)

The current of the 100 kV curve shows strong streamer discharges in the positive halfwave. Remarkable is the considerably decreased power-angle from over 80° down to 15°. In the optimum operating range, real power is almost exclusively consumed.

Fractional Efficiency Measurements

The results demonstrate predominantly that all geometries show the same pattern of fractional efficiency measurements, which is in contrast to the DC ESP. For a peak voltage of 100 kV the fractional efficiency curves of the five barbed geometries lie very close together, Figure 4, top. The maxima are about 45-60%, the minima about 5-22%. The conspicuous difference to measurement with DC energization is the similarity of the curves of the different geometries. In the finest dust range the 10 cm barbed geometry shows the best results (22%), however the poorest fractional efficiency for the largest particle sizes (45%). The 20 cm geometry

behaves reversely, about 5% in the finest dust range and over 60% for the largest particles sizes. Attention should be payed to the power consumption which amounts to 500 W for all barbed geometries.

For example, a wire geometry is shown in Figure 4. For a corona electrode distance of g = 15 cm at a peak voltage of 100 kV, there is almost zero precipitation.

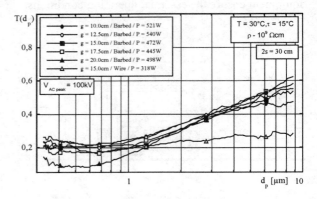

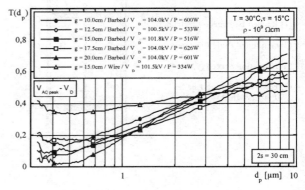

Figure 4: Fractional efficiency for $V_{AC\,peak} = 100$ kV (top) and $V_{AC\,peak} \approx V_{D\,peak}$ (bottom)

The qualitative curve patterns are reproduced with measurements directly below the spark-over voltage, Figure 4, bottom. In contrast to previous assumptions, the finest dust range is not impaired for any of the barbed geometries. The 20 cm barbed geometry attains a maximum fractional efficiency of over 70%, however a minimum efficiency of about 2% in the finest dust range. The 10 cm barbed geometry starts at 12% and increases to 60% at a power consumption in both cases of 600 W. In contrast to this, the 15 cm wire geometry has a minimum of 33% and a maximum of 48% with a power consumption of 330 W. For precipitation onset a minimum voltage of 80 kV is required. As expected, in AC ESPs the efficiency is increased for higher voltages, however decreased in the fine dust range, Figure 5, top . In the finest dust fractions the reduction of the fractional efficiencies could be explained by the contribution of the secondary gas flow, perpendicular to the primary gas flow, due to the higher voltage which causes a reentrainemt of previously separated dust.

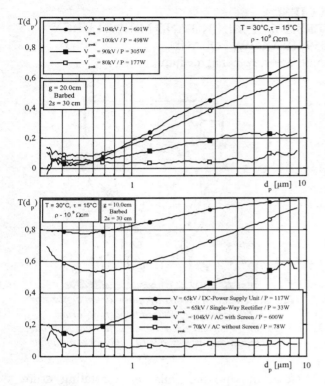

Figure 5: Fractional efficiency compared at various AC voltages for one geometry (top) and various operating modes (bottom)

The direct comparison of the fractional efficiencies in various operating modes shows better results in the DC- mode which consumes less power and operates at lower voltages than in the AC-mode, Figure 5, bottom. The minimum of the fractional efficiency measurements of a DC ESP has a value of 77%, in the one-way rectifier operating mode of 52% and in the AC-mode without screen of 7%. The maxima for the largest particle sizes are about 98% (DC-mode), 93% (rectifier-mode) and 10% (AC-mode without screen). Obviously, with a significantly inferior fractional efficiency, a 4-5 times higher power consumption and a 1.6 times higher voltage, implicating an increased insulating requirement, render the discussion of AC-electrostatic precipitation with a power frequency of 50 Hz very difficult. This result applies to all geometries.

Conclusions

These investigations payed special attention to fractional efficiency measurements of an AC-energized ESP. Summerizing the results, it can be stated that AC precipitation of fine dust is possible. However, compared with DC precipitation, the separation of fine dust with an AC ESP shows only unsatisfactory efficiency for the chosen conditions. The obtained results demonstrate that AC-electrostatic precipitation cannot be regarded as a strong alternative technology for the separation in the fine dust range. The first reason lies in the insufficient charging of the fine dust in the available time of max. 10 ms. The second reason is the very large secondary gas flow which, on the one hand, assists the separation and, on the other hand, frees already separated dust.

Moreover, the results and conclusions of these examinations show that using an alternating bipolar voltage waveform for the reduction of back corona is certainly advantageous. There remains the question how improved charging of the finest particles can be achieved. In order to prove this effect, further AC investigations must be conducted.

References

[Ehr87]   Ehrlich, R.M.; Melcher, J.R.
AC Corona Charging of Particles
IEEE, Transactions on Industry Applications Vol. IA-23 (1987), pp. 103-107

[Hei87]   Heinz, D.; Fischer, F.
Zur Elektrofiltration mit hochgespanntem Wechselstrom
Staub - Reinhaltung der Luft 47 (1987), pp. 181-185

[Kru69]   Krug, H.
Physikalische Grundlagen zur Wirkungsweise wechselspannungsbetriebener Elektrofilter
Luftverunreinigung 1969, pp. 40-46

[Mil93]   Miller, J.; Schwab, A.
Gepulste Elektrofilter zur Verbesserung der Staubabscheidung
Kfk-PEF 116 Forschungszentrum Karlsruhe, 1993

[Mil96]   Miller, J.; Schwab, A.
Improved Discharge Electrode Design Yields Favourable EHD-Field with Low Dust Layer Erosion in Electrostatic Precipitators
6th ICESP, Budapest 1996

[Mil98]   Miller, J.; Hoferer, B.; Schwab, A.
The Impact of Corona Electrode Configuration on Electrostatic Precipitator Performance
Journal of Electrostatics 44 (1998) pp. 67-75

NEW APPROACH TO HIGH VOLTAGE ELECTRODYNAMIC DRUM SEPARATOR
PART II: EXPERIMENTAL VERIFICATION

ANTONI CIEŚLA

ELECTRICAL POWER INSTITUTE, UNIVERSITY OF MINING AND METALLURGY
al. Mickiewicza 30, 30-059 KRAKÓW, POLAND
tel (0048 12) 617-37-73, e-mail: aciesla@uci.agh.edu.pl.

Abstract

One of kinds of electroseparation is a conductance method, depending on differences in charge vs time characteristic: separated components are of different electric conductances - consequently, of various time constants of charging. Prevalent construction of the device is the „corona" drum separator (Huff's invention). This paper follows the theoretical analysis of [1]. The drum separator is considered as a control object. Some experimental results are given and analysed.

1. INTRODUCTION

The high voltage electrodynamic drum separator consists of 2 functional parts: of charging and of electrophoretical separation ones (Fig.1).

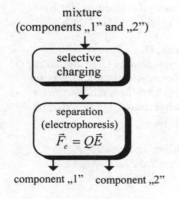

Fig. 1. Principle of electrodynamic separation based on the electrophoresis phenomenon

The key problem of the separation - selective charging of mixture components - can depend on various physical phenomena: in a practical manner: electrostatic induction and a „corona" occurrence. The prevalent construction in mineral processing (benefication) technology is the „corona" drum separator invented by Huff [2]. A scheme of this separator is presented in Fig. 2.
A mixture pour out of the hopper and passes through the corona zone (the grains assume negative charges).

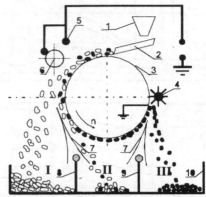

Fig. 2. *Scheme of electrodynamic drum separator: 1- feed container, 2 - feed hopper, 3 - drum, 4 - brush, 5 - corona electrode, 6 - deflecting electrode, 7 - separation barriers, 8,9,10 - bins for separation products*

Next, the mixture falls on a rotating drum. Grains of small resistivities (consequently, of small relaxation times) allow the charges to conduct away rapidly. Subsequently, they take up some charges of opposite sign and are repelled by the field forces into the bin I. However, grains of large relaxation time (of great resistance) lose little of no charge: they are „pinned" to the rotating drum by means of „image" forces. Eventually, the grains are removed with the brush into the bin III. The intermediate product falls into the bin II and may be returned to the feed hopper. A theoretical model of the process is presented by the author in [1].

2. EXPERIMENTAL INVESTIGATIONS

2.1 Research methodology

The device presented above may be analysed as a control object - carrying into effect electroseparation as a technological process (Fig.3.).
The state of the process is defined by state variables x(t). The object can be controlled by some number of input values, called control values u(t). Other input values, so called disturbances z(t), are undesirable for the control process.

High Voltage Engineering Symposium, 22–27 August 1999
Conference Publication No. 467, © IEE, 1999

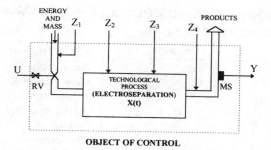

Fig. 3. *Technological process as a control object*
x,u,y - vectors of variables of state, control and output
(respectively), z_1 ... z_4 - vectors of disturbances,
RV - regulation valves, MS - measurement sensors

The most important input values for the separation process are, among others: voltage V, geometry of the electrodes (distance D), rotational speed of the drum ω, angle of inclination of the corona electrode α, concentration of the tested component in the feed β. The disturbances are as follows: temperature T, pressure p, moisture δ. The state variables are as follows: grain charge Q, electric field $\vec{E}$ as well as granulation of the feed d. The separation process described above results in a product of flow off γ and tested component concentration λ: so, quality of the separation process can be defined.

The research has concerned the effect of selected parameters (rotational speed of the drum, angle of inclination of the corona electrode) on the separation process. As a material, narrow granulation of granite 0,3 ÷ 0,5 mm has been chosen. The temperature and moisture were invariables of the tests. The series of the tests results in some histograms of grains distributions among the bins of the receiver - for various operational conditions. A sketch of the separator and definitions of variables are presented in Fig. 4.

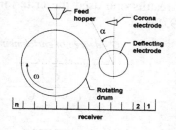

Fig.4. Sketch of the separator (rotational speed of the drum and deflecting angle of the corona electrode are marked)

2.2. Influence of the drum rotation speed

The greater drum rotational speed the larger centrifugal force effecting a grain - and it gets a father bin of the receiver (see Fig. 5a.).

a)

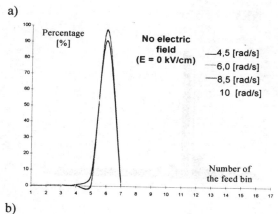

b)

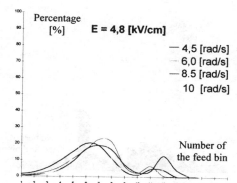

Fig. 5. *Envelopes of masses distributions in particular receiver bins for various drum angular speeds and different electric field values*

Also, the time of the contact between the grain and the drum decreases - so, the charging conditions change radically. They have no time to lose electric charge during the contact with the earthed drum: so, they stay on the drum surface for larger rotational angle, attracted with electric field forces. Next, the increasing electric field strengthens the „glue" effect (also, as a result of more intensive „corona" charging). Simultaneous action of mechanical and electrical forces results in the distribution shape - radically unlike that presented in Fig. 5b. Univocal interpretation of the results is difficult because of the mentioned above parallel action of many factors.

2.3. Influence of the corona electrode angle inclination

The greater inclination angle α (Fig.4.) the more intensive charging of grains in the corona zone. Greater initial grain charge changes - as a result - induction conditions. Some histograms of granite mass distributions for 2 different values of corona electrode deflection angles are presented in Fig. 6.

a)

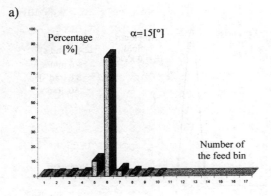

b)

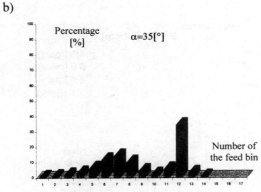

Fig.6. *Histograms of granite mass distribution for:*
E=3,1 [kV/cm] i ω=4,5 [rad/s].
a) α=15[°], b) α=35[°]

The angle increase results in radical change of the receiver mass distribution: majority of particles are conveyed on the drum surface to the farther (of higher numeration) receiver bins. It confirms the thesis of increased charging intensity in the corona zone.

3. RESEARCH OF SEPARATION OF A MIXTURE DIELECTRIC - METAL

Separation of a mixture polyethylene - cooper was tested. Such selection of components, of different electrical properties, guarantees various behaviour in the drum separator. It is a conclusion from theoretical analysis of phenomena taking place in the drum separator. Exemplary trajectories for cooper and polyethylene particles are presented in Fig. 7. A computer simulation has led to conclusion that the trajectories for cooper and polyethylene (materials of various electric resistivities) are different . It guarantees good results of the separation. So, separation of artificial mixture and the analysis have been carried out. Histograms for mass distribution of cooper and poly-

ethylene for E = 0 kV/cm and E = 5, 75 kV/cm are presented in Fig. 8.

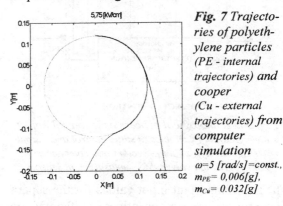

Fig. 7 *Trajectories of polyethylene particles (PE - internal trajectories) and cooper (Cu - external trajectories) from computer simulation ω=5 [rad/s]=const., m_{PE}= 0,006[g], m_{Cu}= 0.032[g]*

a)

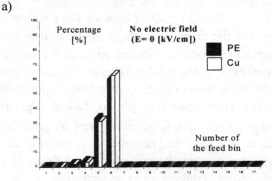

b)

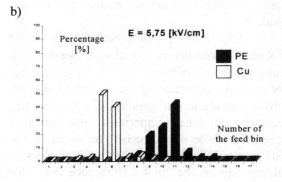

Fig. 8. Histograms of mass distribution of cooper and polyethylene for various electric fields
(other parameters are constant)

For evaluation of results one used the criterion presented in Fig. 9. [3].

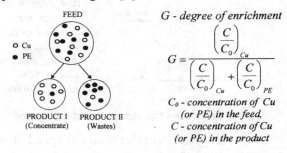

G - degree of enrichment

$$G = \frac{\left(\dfrac{C}{C_0}\right)_{Cu}}{\left(\dfrac{C}{C_0}\right)_{Cu} + \left(\dfrac{C}{C_0}\right)_{PE}}$$

C_0 - concentration of Cu (or PE) in the feed,
C - concentration of Cu (or PE) in the product

Fig. 9. *Criterion of evaluation for two-components mixture*

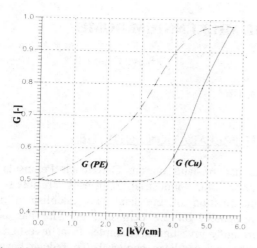

Fig. 10. *Degree of enrichment G of the concentrate (Cu) and wastes(PE) versus electric field E*

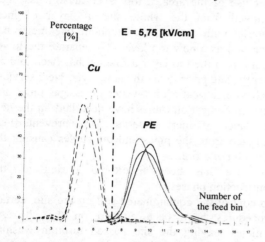

Fig. 11 *Envelopes of polyethylene and copper mass distribution reflecting repeatability of the process*

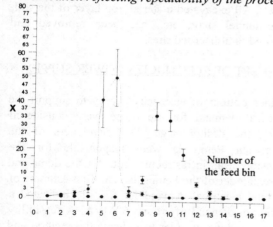

Fig. 12. *Envelopes of distribution of average values of masses in the receiver bins (deviation from average values is marked)*

Analysis of the Fig. 10 leads to conclusion that the greater electric field the better parameters of the concentrate. For constructional reasons,

usage of electric field over 6 kVcm⁻¹ was impossible but for our research it is not necessary. To test the repeability of the separation process one has carried out several series of the experiments. Envelopes of polyethylene and copper mass distribution for a series of 5 experiments are presented in Fig. 11. Envelopes of distribution of values of mass (where deviation from average value is marked) are presented in Fig. 12.

4. SUMMARY

Both theoretical analyses and experiments univocally confirm the usefulness of electrodynamic drum separators for separation of mixed granular solids. The necessary condition is that separated materials are of different resistivities. The experiments have proved out that essential influence on the behaviour of grains (treated as a control object) have: electric field, rotational speed of the drum and deflection angle between the corona electrode and the drum. Univocal interpretation of the results is difficult because a lot of mechanical and electrical factors act paralelly.

The presented method of the separation can be successfully used in electrotechnology (it is confirmed by separation of dielectric (polyethylene) and metal (copper). So, recycling of copper wastes, after production, is possible. Economical reasons confirm importance of the presented research.

REFERENCES

[1] **Cieśla A.:** *High Voltage Electrodynamic Drum Separator. Part I: Theoretical Model.* Proceedings ISH'99, London 1999

[2] **Ralston O. C.** : *Electrostatic Separation of Mixed Granular Solids.* Amsterdam - New York 1961.

[3] **Cieśla A.:** *Electrodynamic Drum Separator Used for Separation of Dielectrics with Various Resistivities* (in Polish) Proceedings of III Conference : Postępy w elektrotechnologii, Szklarska Poręba, 1998, Oficyna Wydawnicza Politechniki Wrocławskiej, pp. 39 - 43.

HIGH VOLTAGE SUPPLY TO THE MILLENNIUM DOME

Jelica Polimac   Steve Cantwell

London Electricity, UK

ABSTRACT

A dome is constructed on the Greenwich Peninsula in London as part of the celebration for the new millennium. The dome site occupies the area which had been used for industrial purposes. This resulted in land contamination and restricted redevelopment of the area. To make use of the area, the site has been cleaned and new services infrastructures, electricity, water, gas and telecommunications, have been provided in a new layer of inert material.

This paper describes high voltage electricity supply for the millennium dome including site preparation activities. Electrical equipment, incorporated in the architectural design of the dome, has been installed to provide a safe and reliable electricity supply during a millennium exhibition and later to a new residential and commercial development on the Greenwich Peninsula.

INTRODUCTION

An old power station and gas works had been in operation on the Greenwich Peninsula in the past. This resulted in contamination of the land. The presence of contamination, toxic, flammable gasses and vapours affected utilisation of the area. The emission of heat from electrical cables and service pipes mixed with flammable gasses, present in the soil, might initiate a fire. To make use of the land again , the site had to be covered with a layer of inert material of sufficient thickness to prevent ignition of the material below. Service cables and pipes had to be laid in inert material to reduce fire and explosion hazards.

The dome is constructed on the Greenwich Peninsula as part of the new millennium celebration. An exhibition will be organised in the dome during the year 2000. All services to the dome, were designed to be incorporated in the architectural design of the whole project providing safe and reliable supply.

Twelve cylinders, located around the dome, have been provided for services. The cylinders, standing on metal legs, are of metal construction. All services equipment had to be installed above ground level on one of three levels of the cylinders.

One of specific issues to be addressed at the early stage was the ecology of the proposed site.

SITE PREPARATION ACTIVITIES

The use of land on the Greenwich Peninsula in London for industrial purposes resulted in contamination of the land. To establish a new development on the peninsula the site has been thoroughly investigated. Samples from investigation boreholes have been analysed. To reduce hazards which may arise during development and during later use of the area, a top layer of soil has been removed from the whole site. A layer of inert material with a thickness of about one metre has been laid as a new top layer. To separate the lower layers from the top layer a fine net has been laid all over the site prior to lay the new layer. New services such as electrical cables, water pipes, gas pipes and telecommunication cables have been laid in the new top layer of inert material. This prevents heat emission from the pipes and the cables causing the initiation of a fire.

The site has been monitored during all the construction phases.

To control the contaminated soil on the site, certain procedures have been applied on the site. All vehicles leaving the site passed through washing facilities where contaminated soil was removed particularly from the bottom of the vehicle and the wheels. All excavated material, top layer of the soil, cable tunnel shaft, etc., had been removed and disposed of at licensed sites.

CONCEPT OF ELECTRICITY POWER SUPPLY

Various options of electricity supply to support the electrical demand for the dome were considered during the design stage. The developers of the Greenwich Peninsula were responsible for the provision of all infrastructure both for the dome and the remainder of the Peninsula. The first estimate of the total electrical demand for the site indicated a maximum demand of around 60MVA. A feasibility study was done in order to indicate the method and estimated costs of providing this demand to the developer. This included three options for primary substations located within the proposed site. Subsequently, a formal request for an estimated demand of 44MVA was made by the developer and further negotiations were held to establish possible means of supplying this demand without the need for a primary substation. The selected option proposed

High Voltage Engineering Symposium, 22–27 August 1999
Conference Publication No. 467, © IEE, 1999

an 11kV distribution network around the dome and serving the rest of the Greenwich Peninsula. This was made possible by reinforcing an existing 132/11kV substation which was being developed at the time. The 11kV cables have been installed in a new cable tunnel constructed under the river Thames to serve the Peninsula.

The 11kV distribution network is fed from the primary 132/11kV substation, as shown on figure 1. A total of twelve 11kV circuits have been provided to meet the anticipated demand. Automatic switching facilities are provided in the primary substation for a loss of infeed transformers such that any loss would not cause an interruption of supply to any of the outgoing 11kV feeders. The transformers have a cyclic rating of 130% of the rated output.

The distribution network for the Peninsula consists of nine 11kV substations. Eight of them are housed in cylinders, located adjacent to the perimeter road around the dome. They are designated as cylinder substations. The remaining substation is installed in a permanent structure and supplies a separate 11/0.4kV distribution system for the peninsula adjacent to the dome.

Cylinder substations supplying the dome are interconnected into a ring in order to increase the reliability of supply. Under normal operational conditions they are split into three groups to prevent paralleling of the 132/11/11kV transformers, with normally open interconnectors connecting each group. Substations of the same group are connected in parallel with normally closed interconnectors within the group. Each substation is supplied from the primary substation by an 11kV cable. The incoming cables are rated for fifty percent of overload. Loss of an incoming circuit of a group will therefore be covered by the remaining incoming circuits of the group. Loss of all three incoming circuits of the same group will not interrupt the power since the interconnectors between the groups can be closed.

Each substation has three busbar sections as illustrated on figure 2. A fault on a bus section could be eliminated by isolation of that bus section allowing undisturbed operation of the remaining two busbar sections.

11KV SUBSTATIONS

The 11kV cylinder substations are housed, together with other services, in cylinders (Figure 3). They are installed on second floor, eleven metres above ground level. To make their installation easier and to the required timescale, the 11kV switchboard, the DC panel and ancillary equipment were installed in a prefabricated switchgear enclosure. Space was restricted for the enclosures and for the cables connected to the switchboard. Therefore, 11kV

switch panels were designed with minimum dimensions and cable works between each cylinder structure were carefully planned.

The enclosures were installed into the completed cylinder structures. A temporary platform was built to support an enclosure weighing ten tons during installation. The same support platform was used for all eight substations. A crane was used to lift each enclosure from ground level to the support platform. Then the enclosure was slid into its final position by means of two metal rails incorporated in the cylinder structure.

All incomer and interconnector switch panels are equipped with cable unit differential relays. Some interconnectors are equipped with a numerical multifunction overcurrent and earth fault protective relay to select and trip a faulty busbar. All feeder switch panels are provided with numerical multifunction overcurrent and earth fault protective relays. In that way a fault can be cleared quickly isolating a minimum part of the network.

The cylinder substations are high risk from a fire risk aspect. All electrical equipment has been selected to reduce fire risk. Two means of escape from each substation are provided. Fire detectors are installed in all the substation enclosures and are connected to a local fire protection unit in each cylinder. All the local units are connected to a central fire protection board in the dome.

All metal parts in the substations are connected to the cylinder earth bar where the whole cylinder structure is connected. The cylinder earth bars are connected to the main earthing system of the dome.

HIGH VOLTAGE CABLES

High voltage cables for the millennium project can be classified into two groups as follows:
- cables in a cable tunnel
- cables around the dome.

Cables in a cable tunnel

A total of twelve incoming 11kV cables are installed in a dedicated cable tunnel. The cables in the tunnel have very low probability of third party damage and weather related incidents. A damaged cable is easily accessible for repairs.

The tunnel is built under the river Thames. It contains a shaft at each end and an intermediate shaft. The tunnel is 2.44 metres in diameter and is constructed of precast concrete segments. The tunnel accommodates twelve 11kV circuits supplying the millennium dome and the rest of the Greenwich Peninsula. Space has been dedicated for future cables in the lower portion of the tunnel.

All twelve 11kV circuits for the millennium project are formed of single core cables grouped in trefoil formation. Ten circuits are installed on the roof of the tunnel and two circuits are installed under pilot cables. The cables are suspended in open slings all the way through the tunnel. They can move freely under normal or fault conditions. The cross section of the tunnel is shown on figure 4.

Cables around the dome

All 11kV incoming cables from the tunnel shaft located on the Greenwich Peninsula to the cylinder substations and the cables for the interconnectors are laid direct in the ground. They are laid in a corridor formed in the top layer of uncontaminated soil. Up to six circuits run adjacent to each other in the corridor. The cables are of a bigger cross section than the cables in the cable tunnel to achieve the same current rating for both of them. All cables are single core cables installed in trefoil formation. A pilot cable is fixed to each trefoil group.

Cables rise from ground level in a service core within each cylinder to the substation, as illustrated on figure 3. Steelwork is provided to support the cables in the core and under the substation.

CONCLUSIONS

The millennium dome built on a former industrial site is an example how contaminated sites can be utilised and redeveloped. Thorough site investigation and special site preparation are necessary activities for any redevelopment on contaminated land. The special site preparation includes laying of a new top layer of inert material to minimise fire and explosion hazards associated with contaminated land. It is necessary to lay new services of electricity, water, gas and communication in inert material.

Services to supply the millennium dome are incorporated in unusual cylinder structures above ground level. To minimise resource requirements and installation time, the 11kV distribution substations were erected in a prefabricated switchgear enclosures in the manufacturer works.

Security of supply to the dome is provided by means of a distribution network of smaller switchboards adjacent to the concentration of electrical demand. The substations are interconnected into a ring to achieve a reliable supply.

A new cable tunnel has been built, accommodating twelve 11kV circuits, to provide the electricity supply to the Greenwich Peninsula for the millennium celebration and for the future development. Careful specification of the cables and use of the latest suspension technology for cable installation in the tunnel will ensure that the number of cable faults and time for cable repairs are reduced to a minimum.

REFERENCES

[1] 'London Electricity': 'Design proposal for 11kV Millennium distribution network', 1997
[2] BSEN 61330: 'High voltage/low voltage prefabricated substations', 1996
[3] D Crowhurst, PF Beever: 'Fire and explosion hazards associated with the redevelopment of contaminated land', Building Research Establishment, 1987
[4] 'British Drilling Association': 'Guidance notes for the safe drilling of landfills and contaminated land', 1992
[5] 'Limitation of fire risk in substations at 132kV and below and in enclosed cableways', Engineering recommendation

FIGURES

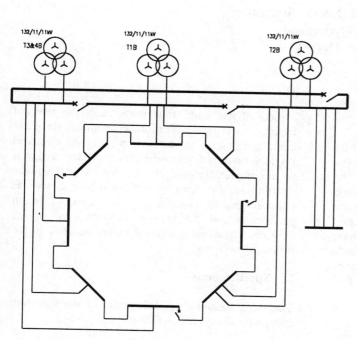

Figure 1. Single line diagram of 11kV millennium distribution network

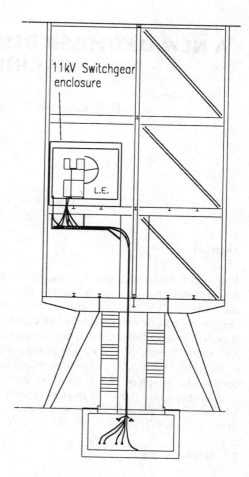

Figure 3. Cross section of a cylinder

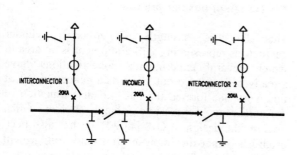

Figure 2. Diagram of an 11kV substation

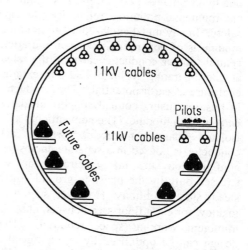

Figure 4. Cross section of the Millennium cable tunnel

A NEW OPTIMISED DESIGN OF INTEGRATED THREE PHASE GAS INSULATED BUS DUCT

R.Rajaraman, S.Satyanarayana, A.Bhoomaiah, H.S.Jain

BHEL R&D Division
Hyderabad
INDIA

Abstract

The paper highlights the techniques adopted for design optimisation of a three phase, single enclosure, 145kV class gas insulated bus duct. The design is based on extensive electrostatic field studies in three dimensions using special software. An unique feature of this design is the introduction of individual rib insulators for supporting each phase conductor, in place of the conventional single conical shaped insulator. A model bus duct has been fabricated and successfully tested for its specified dielectric duties.

1.   Introduction

In the recent past, the technology of Gas Insulated Switchgear (GIS) has gained increasing application, due to advantages like compactness, immunity from environmental pollution, reliability and operational safety. In any metal enclosed substation, the bus bar module is of coaxial cylinder configuration, with a central tubular conductor carrying the rated current at the operating voltage and an external metal enclosure at earth potential. The conductor is held in position, using conical support insulators at the ends of each module. The conventional system is of isolated phase construction, where each phase conductor is housed in a separate enclosure. In this configuration, the inter-electrode gap would be stressed only by the phase-to-ground voltage under steady state conditions. However, such a layout has to allow for extra floor area between phases, towards minimum working space required. Hence, the design can be optimised by mounting all the three phases inside a common enclosure, and maintain the dielectric integrity in a size of chamber which is economical to manufacture. The main advantages of three phase construction are the reductions in the number of chambers, welded joints and their associated gas seals, thus minimising the risks of leakage. The design is more economical since the manufacturing costs, installation time and electrical losses get reduced. However, the dielectric design is more complex because of the asymmetrical voltage distribution, governed by phase-to-phase as well as phase-to-ground voltages. The critical design parameters and design approach involved in the development of such a bus duct are outlined in the following sections.

2.   Specifications

The basic technical specifications of the bus duct chosen for design are as follows:

Nominal voltage	132 kV
Highest system voltage	145 kV
Rated current	1600 A
Impulse withstand voltage	650 kV
Power frequency withstand voltage	275 kV
Partial discharge level (max.)	1 pC
Operating pressure	450 kPa, SF_6

3.   Design of bus bar module

The dielectric strength of co-axial cylinder geometries representing GIS equipment is an area in which several experimental investigations have already been carried out. The data generated out of this work has formed the basis of determination of the maximum permissible voltage stresses which govern the design of GIS [1], [2]. It has also been established that the impulse withstand voltage will be the deciding factor in working out the enclosure and conductor diameters, as well as the inter-electrode gap. Based on the recommended values, a stress level in the range of 28 to 30 kV/mm corresponding to 650 kV BIL was chosen as the governing parameter in the design of bus bar

High Voltage Engineering Symposium, 22–27 August 1999
Conference Publication No. 467, © IEE, 1999

dimensions. Since the configuration of three phase assembly is not axi-symmetric, a special software based on boundary element method (BEM) was used, for three dimensional simulation(3-D) of the geometry and analysis of electric field pattern [3]. In the model bus duct chosen for design, provision was made for taking out connections from each phase conductor individually to the high voltage test source, in the form of three drawn out ports on the enclosure periphery. Two typical conductor layouts were analysed, one in standard three phase formation and the other in the shape of right angled isosceles triangle. The configurations were first simulated in 3-D geometry. The corresponding electric stress distributions computed with one phase energised at 650 kV impulse are presented in Fig.1 and Fig.2. The maximum stress obtained was 13.33kV/mm for 120° formation and 11.73kV/mm for 90° formation. Moreover, the latter offers more clearance between the external connection of any particular phase and the other two phases. Hence, the right angled triangle configuration was selected.

4. Design of insulator

For the design of support insulator, a maximum permissible stress of 2.0 kV/mm on the insulator surface was adopted as the criterion [4]. The conventional design would be in the form of a single conical insulator supporting all the three phases [5]. However, such a design would pose limitations like large size, high material content, heavy weight and difficulty in handling, apart from high cost associated with the mould development. Hence, an alternate viable design was worked out, using rib shaped insulators to support each phase conductor separately. Fig.3 shows the 3-D simulation of the bus bar assembly with rib insulators. The entire geometry has been meshed into a number of elements, for analysis by BEM. Various cases were studied, with different rib angles, material thicknesses and creepage lengths and the design yielding the minimum surface stress was identified . Fig.4 shows the stress distribution on the insulator of the phase energised with 650 kV impulse.

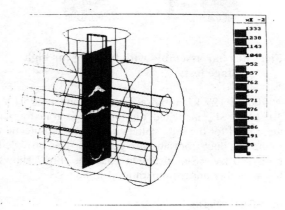

Fig.1 Electrostatic stress distribution in standard three phase formation.

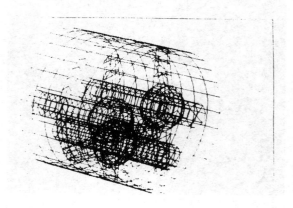

Fig.3 Representation in 3-D geometry with elements.

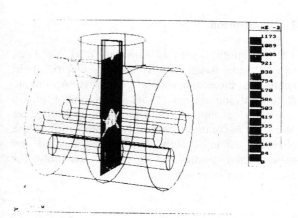

Fig.2 Electrostatic stress distribution in 90° formation.

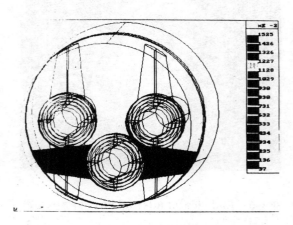

Fig.4 Stress distribution on surface of HT rib insulator.

5. Experimental set-up

A model bus bar assembly was fabricated comprising a stainless steel enclosure of 498 mm inner diameter, and aluminium conductors of 64mm outer diameter. For insulators, a suitable die for casting was first designed and developed using CNC part program. The insulators were manufactured out of alumina filled epoxy. A photographic view of the insulator is given in Fig.5. Based on electrostatic field studies, stress relief shields were designed for conductor terminations, metal insert regions of insulators and junctions of conductors with external connections. An internal view of the bus bar assembly is shown in Fig.6.

Fig.7 Bus bar assembly with bushing for high voltage tests

A 900 kV, 27 kJ impulse generator and a 325 kV, 30 kVA SF$_6$ insulated discharge free transformer were used for the high voltage tests. The connection to the bus duct from impulse generator was provided by a 170 kV, SF$_6$ insulated bushing. Fig.7 shows the assembly under preparation for the tests.

6. Test results

The model bus duct successfully withstood the impulse tests at 650 kV and power frequency tests at 275 kV. The partial discharge performance was also found to be within the specified level of 1 pC at the test voltage of 93 kV.

Fig.5 Photographic view of rib insulator

7. Conclusion

The optimised design of GIS bus duct has been carried out, based on exhaustive electric field analysis in three dimensional geometry. An unique feature of the design is the development of rib insulators to support the phases individually, in place of a common conical insulator for all the phases. The design has been verified by experimental studies on a model assembly for dielectric performance.

Fig.6   End view of internal bus bar assembly

Acknowledgement

The authors are grateful to Mr.D.Suryanarayana, General Manager for his involvement and suggestions. Thanks are due to Mr.P.P.Varada-charyulu for his contribution in the manufacture of insulators. The team members are thankful to Mrs.K.Sarojini Subhashini for her painstaking efforts in typing the manuscript and to BHEL Management for permission to publish this paper.

References

1. P.Hogg, W.Schmidt and H.Strasser, "Dimensioning of SF_6 metal clad switchgear to ensure high reliability", paper no.23-10, CIGRE 1972.

2. S.Zelingher and R.Matulic, "EHV gas insulated line: parameters determination based on system performance criteria", IEEE Transactions on Power Apparatus and Systems, Vol.100, No.11, pp.4515-4523, Nov. 1981.

3. S.Satyanarayana, R.Rajaraman, A..Bhoo-maiah and H.S.Jain, "Analytical and experimental studies on design optimisation of 145 kV gas insulated bus duct", 4th Workshop & Conference on EHV Technology, Bangalore, India, July 1998.

4. J.M.Braun, G.L.Ford, N.Fujimoto, S.Rizzelto and G.C.Stone, "Reliability of GIS EHV epoxy insulators: The need and prospects for more stringent acceptance criteria", IEEE Transactions on Power Delivery, Vol.8, No.1, Jan. 1993, pp. 121-131.

5. K.Itaka, T.Hara, T.Misaki and H.Tauboi, "Improved structure avoiding local field concentration on spacers in SF_6 gas", IEEE Transactions on Power Apparatus and Systems, Vol 102 , No.1, pp 250-255, Jan. 1983.

Address for correspondence:

R.Rajaraman
DGM-High Voltage Engg.
BHEL R&D Division
Vikasnagar
Hyderabad – 500 093
INDIA.

STABLENESS OF ELECTRICAL OPERATION CHARACTERISTICS OF ELECTROSTATIC LENTOID PRECIPITATOR*

Chen Shixiu   Sun Youlin

Dept. of Electrical Engineering
Wuhan University of Hydraulic
& Electrical Engineering
Wuhan Hubei 430072
P. R. China

Chen Xuegou

Wuhan University of Technology
Wuhan Hubei 430070
P. R. China

ABSTRACT

Electrostatic Lentoid Precipitator is a new type of ESP. It possesses many better performances. This new type of ESP is composed of three electrodes having different potentials. To get a better dust collecting performance, it is important for lens electrode to maintain a stable potential. Therefore, following questions should be answered: What is varying regularity of potential of lens electrode versus applied voltage? What mechanism is the varying regularity? How does the potential of lens electrode maintain the stable potential while the applied voltage maintains unchanged?

1.   Introduction

Electrostatic Lentoid Precipitator(ESLP) is a new type of ESP developed by Prof. Chen Xuegou in 1986[1]. This new type of ESLP possesses both good electrical performances and good electrohydrodynamics (EHD) performances. So, it can collect dust of both high and low specific

volume is smaller than that of the conventional wire-plate ESP and so is the cost. Its dust collection process and its mechanism can be found in literature [1] [2] [3] [4]. Electric field inside the ESLP is the so called electrostatic lentoid field(ESLF) energized by three electrodes: negative corona electrode, lens electrode and positive corona electrode. Its electrode structure is shown in Fig.1.

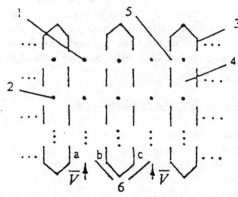

1. Negative electrode 2. Positive electrode
3. Lens electrode 4. Dust collection chamber
5.Lentoid opening   6.Gas duct

Fig.1 Electrode structure of ESLP

The two terminals of high voltage power supply

* The Project Supported by Natural Science Foundation of China

resistivities with high dust collection efficiency. Its

High Voltage Engineering Symposium, 22–27 August 1999
Conference Publication No. 467, © IEE, 1999

are connected to the positive corona electrode and negative corona electrode, respectively. The lens electrode is suspended by porcelain insulators to isolate from the other two electrodes. Fig.2 shows the ESLF.

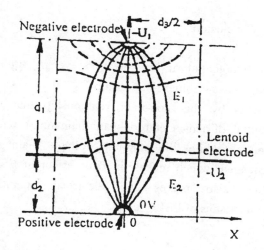

Real lines are force lines, dashed lines
are equipotential lines
Fig.2 Electrostatic lentoid field

To get focus effect, the ESLF should be a protrude lentoid field, which means that the potential lines near the lens opening should protrude towards gas duct, as shown in Fig.2. This demand can be satisfied naturally. At the same time, we have found that the potential of the lens electrode affects the focus effect of ESLF greatly. Therefore, it is important to find out the varying regularity of the potential of the lens electrode versus applied voltage. There is still another question: Because the lens electrode is suspended by porcelain insulators, whether its potential can maintain a stable potential while the applied voltage maintains unchanged?

2. Potential characteristic of lens electrode versus applied voltage and its mechanism

Fig.3 shows the potential characteristic of lens electrode versus applied voltage, also its schematic of experimental set-up.

It can be seen from Fig.3 that, the relationship of the potential of lens electrode versus applied voltage is a non-liner one. As well known, in electric field of

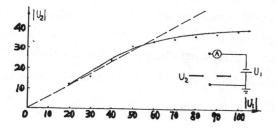

Fig.3 Relation of potential of lens electrode versus applied voltage

space charge-free, the potentials of lens electrode or on a some point should be proportional to the applied voltage, which implies that the image of electric field in region of space charge-free remains unchanged, But in the case of corona discharge, the potentials of lens electrode or on a some point are no longer proportional to the applied voltage. This implies that space charge has deformed the original electric field. Experiment shows that when applied voltage is 20 kV and the current is zero, $\left|U_2\right|$, the potential of lens electrode is 12 kV. We can draw a ray passing through the origin and the point of (20, 12), which is a dashed ray as showed in Fig.3. This dashed ray indicates that the potential of the lens electrode is proportional to the applied voltage in the case of space charge-free. It can be seen from the Fig.3 that the potential of lens electrode is above the dashed ray while the applied voltage $\left|U_1\right|$ in the range 22 to 55 kV. This implies that negative corona first occurs, some electrons and ions move to the lens electrode, making the lens electrode negative charged, so the $\left|U_2\right|$ rises. While $\left|U_1\right|$ is in the range bigger than 55kV, the rising tendency of $\left|U_2\right|$ decreases gradually. Curve $\left|U_2\right|$ crosses with dashed ray. At this cross point, the potential of the lens electrode equals to that of lens electrode in the case of space charge-free, so the net charge on lens electrode should be zero. Therefore, it can be concluded that positive corona has also occurred. At this cross point, The lens electrode accepts negative charges as well as positive charges but the net charge is zero. On the right of the cross point, the positive corona becomes stronger and stronger, thus more and more positive ions move to the lens electrode, then net charge on the lens electrode is positive. So the curve $|U_2|$ is below the dashed ray. With the $|U_1|$ increasing, the rising tendency of $|U_2|$ becomes more and more flat, which indicates that most part of the applied voltage has distributed to space between negative corona electrode and lens electrode. Therefore it is beneficial to increase electric field strength in space between the negative corona electrode and the lens electrode, also to increase the negative corona and the electric wind.

3. Stability of the potential of the lens electrode

The non-liner relation of potential of lens electrode versus applied voltage can make the ESLF work stably. If some reason makes potential of lens electrode descend but applied voltage $|U_1|$ maintains some value unchanged in the range of $|U_1|$ being bigger than 55kV, then the electric field strength in the space between the lens electrode and positive corona electrode will be weaker, also will be the positive corona. Thus the number of emitted positive ions will decrease, also dose the net positive charge on the lens electrode. Therefore the potential of lens electrode will rise. In this way, the potential of lens electrode can be adjusted by itself to maintain a stable potential. Vice versa, if potential of lens electrode rises due to some reason but the applied voltage maintains unchanged, the electric field strength in space adjacent to the positive corona electrode will be strengthened, which will make the positive corona stronger and the net positive charge on lens electrode will increase, Thus the potential of lens electrode will descend. So, we can see that the varying process of potential of lens electrode under the condition of applied voltage maintaining unchanged is a process of negative feedback, that is, the adaptive process, which ensures the $|U_2|$ maintain unchanged as long as $|U_1|$ being unchanged. Therefore the focus effect of the ESLF will be stable, which makes the electrical operation characteristics of ESLP stable.

4. Conclusion

a. The potential of lens electrode versus applied voltage is a non-liner relationship.
b. It is by this non-liner relationship that the potential of lens electrode can maintain a stable potential while the applied voltage maintains unchanged, which makes th focus effect of ESLF stable, also makes the electrical operation characteristics of ESLP stable.

References:
[1]. Chen Xuegou, et al., a research on the mechanism of dust collection in electrostatic lentoid field, Proceedings of international conference on modern electrostatics, pp41-44, 1988, Beijing, China.
[2]. Chen Shixiu, Chen Xuegou, Electrostatic lentoid field and its optimizing computation, Pro. of the 6th ISH,14-05,1989,8, New Orleans, USA.
[3]. Chen Shixiu, Chen Xuegou, Why does ESLP possesses better performance? Pro. Of the 7th international conference on electrostatic precipitation, Sept.1998, Korea.
[4]. Chen Shixiu, Chen Xuegou, Electrostatic lentoid precipitator and its performance, Pro. Of the 7th International conference on electrostatic precipitation,Sept.1998, Korea.

Chen Shixiu, he is an Associate professor, Ph.D. He majors in high voltage engineering. Because of his contributions to the science and application of electrostatic precipitation, he received the Senichi Masuda Award at the 7th International conference on Electrostatic Precipitation held in Sept.1998 in Korea.   e-mail: gysys@wuhee.edu.cn
Tel: 86 27 87892101

Zero Phase Current Measurements at Neutral Earth of Transformer by EM Propagation for Possible Earthquake Prediction

Takemitsu   Higuchi

Research center of The Kansai Electric Power Co.,Inc. in Hyogo prefecture, JAPAN

Keywords: EM propagation at earth, Ground current in Power Systems

1. Introduction

The four years since more than 5 thousand people died in the Great Hanshin−Awaji Earthquake when it hit the Kobe area in 1995 have passed swiftly. In Japan, however, we must be constantly alert to the possibility of disastrous earthquakes. To minimize the injuries and damages, caused by such natural disasters, it is imperative to establish an earthquake prediction system.

It has been reported that in Greece, tha VAN method of measuring telluric current resulted in successful earthquake prediction. When crushed, reportedly, rocks irradiate electromagnetic waves. Such experiments and other references, in Japan and abroad, have revealed electromagnetic variations in the earth depending on crust distortion stress. Building a large−scale extensive information−processing network of observation points may enable detecting abnormal telluric current and spatial electromagnetic waves as signs of earthequakes. Successful detection should improve prediction reliability.

In substations, located at both ends of a long−distance power transmissinon line, the electric power company concerned grounds the neutral earth lines of the three−phase trans− formers in order to protect the power transmission system and to prevent injury to workers. The power transmission line, the earth, and the neutral earth lines of transformers compose a large loop circuit. Such a loop reaches at least tens of kilometers with some loops reaching 200 to 300 kilometers. Measurements were made to assist the detection of the telluric current that circulates deep in the earth. An impulse generator (IG) was used as an artificial lightning generator to input pulse current to the earth. A simulated transformer earth line was tested for electro− magnetic−wave detection. The present report concerns these measurements.

2. Transformer neutral earth lines and measuring system

At 60 Hz, neutral earth current is zero because it is the vectorial sum of three currents that have varying phases of 120 degrees, from each other. At other frequencies, however, DC, harmonics, and high−frequency current constantly circulate through the earth.

A clamp CT(current transformer) was mounted on the primary neutral earth line of the main transformer. CT output was inputted to an automatic recorder, and used to trace variations with time. The tracing speed of the automatic recorder was set to 2 cm/h. The automatic recorder was calibrated to a tracing amplitude of 1 A/cm.

The CS (current sensor) had frequency characteristics of 0 to 300 kHz. The measuring system was of an automatic continuous recording type that monitored the current constantly.

Railways and gas pipes, normally, shunt DC. In everyday life, many switching devices, such as inverters and DC−to−DC converters, are in use.

Every switching operation produces harmonics, all of which are superimposed on the background DC. As noise, such electromagnetic waves are detected by the CT.

High Voltage Engineering Symposium, 22–27 August 1999
Conference Publication No. 467, © IEE, 1999

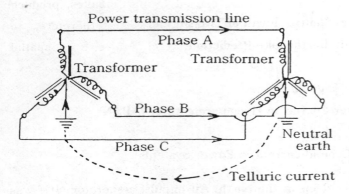

Fig.1 Three-phase Power Transmission Line
and Transformer Circuit

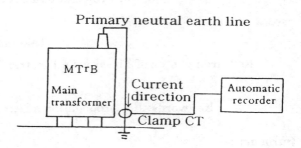

Fig.2 Method for Telluric Current Measurement

3. Telluric current measurements

From November 1996 to January 1997, telluric current was measured at the transformer neutral earth of the 500-kV system Inagawa Substation in Inagawa-cho, Hyogo, Japan. The following describes the measurements.

From the end of the year, through the New Year holidays, to the middle of January, telluric current proportional to load current, greatly varied with power consumption. Such variations in telluric are described.

At 22:36 on January 8, 1997, a small earthquake, registering a magniude of 1(M1) or so, hit the Hyogo area. This report focuses on its infulence on telluric current variations.

We geted frequently the greatly neutral current before earthwake.

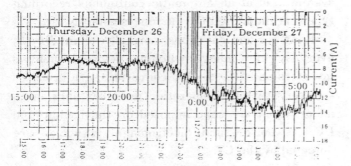

Fig.3 neutral current on load

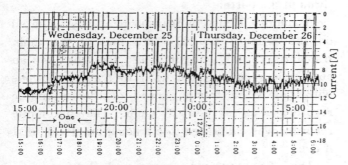

Fig.4 neutral current on thursday

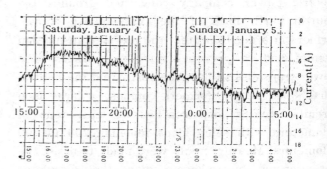

Fig.5 neutral current on saturday

Device operation surges, in particular, rise sharply to form pod-shaped traces due to the automatic recorder pens winging past the maximum. In contrast, many spatial elrctoro- magnetic waves, caused by earthquakes, produce moderate traces. It is possible, therefore, to distinguish device operation surges from spatial electromagnetic waves.

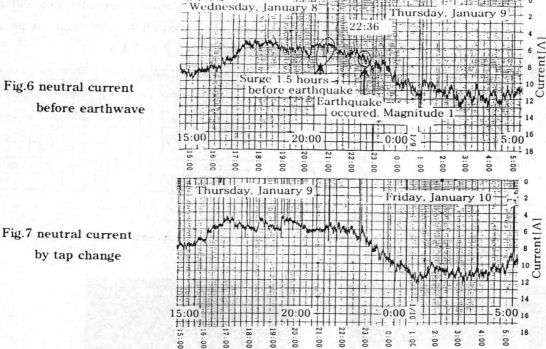

Fig.6 neutral current
before earthwave

Fig.7 neutral current
by tap change

Current within the range of DC to 300 kHz, was recorded continuously. From 20:00 to 22:00 on Saturday, December 28, the curent remained at the minimum(zero). From 16:00 to 19:00 on Monday, December 30, the current also remained at 0 A. On January 1 the current reached 2 A; on January 2.5 A. On January 3, it reached the substantial amperage of 8 A as on December 26 and 27. From the end of the year through the New Year holidays, factories did not consume power, as Fig.5 Load current was the lowest throughout the year. The neutral earth current, proportinal to load current, is shown in the graphs. In Fig.8, at 17:20 and 22:00, the current fell sharply because the tap changer was activated for transformer voltage regulation. At midnight, the lighter the load, the higher the voltage. The tap changer reduces voltage automatically. A rapid increase in current occurred and leapt from 2 A to 8 A. This was because of the tap change for the transformer to which the measuring instruments were connected. Transient current, due to the tap change, induced a surge which was traced on the record chart sheet. Device operation surges, lightning, and other spatial electromagnetic waves also induced surges.

On the other hand, neither spatial high-frequency waves induced on the long-distance power transmission line, nor high-frequency switching pulses produced in the power system, were removed by the capacitors of the measuring instrument transformers(PD) in the substation, or blocked by inductance of the transformer windings. They could not reach the neutral earth line. In contranst, low-frequency harmonics (approximately 1kHz or less) passed through capacitors and windings, and reached the neutral earth. From the chart sheet, it was not evident that the earthquake (magnitude 1) caused any remarkable change. In Fig.6, at 22:36 on January 8, the earthquake occurred. At 21:00, before the earthquake, the current reached 5.8 A. At 22:30 and 22:36, the current reached 7.4 A. At 22:45 the current reached 9.2 A; at 23:00 the current reached 9.1 A. Such a change is greater than the variation of $\triangle$ 2.4 A, as shown in Fig.5, from 6.6 A at 20:00, January 4, to 9 A at 22:00, January 4.

On January 8, the variation of $\triangle$ 3.4 A was recorded. At 23:00, a larger current of 9.2 A occurred. At 1:10, the current peaked at 12.7 A. As shown in Fig.6, on Friday, December 27, the largest daily current of 14.5 A occurred; whereas, as shown in Fig.4 of the chart sheet, around 21:00,

negative surges were frequently recorded ($\triangle$ 1. A). The surge cannot be attributed to device operations, because step−shaped rapid increase in current were not superimposed on the surges and because the automatic recorder pen did no swing past the maximum. Variations remained within the range of 1 A to 1.4 A.

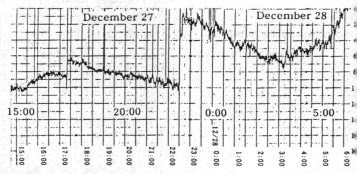

Fig.8 neutral current by tap change

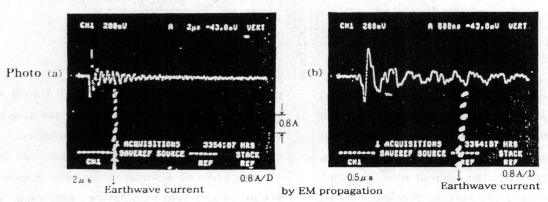

Photo (a)   (b)

0.8 A

2μ s   0.8 A/D 0.5μ s   0.8 A/D

Earthwave current   by EM propagation   Earthwave current

Fig.9 Propogation wave in earth ground from IG

5. Summary

An earthquake, with a magnitude of 1, and with an observation acceleration of 3.8 gal, hit the region close to Inagawa Substation. One and a half hour before the earthquake, frequent electro−magnetic pulses were induced on the transformer neutral earth. An increasing tendency of telluric current, in general, was observed. Such electromagnetic pulses and telluric current seemed to correlate with the earthquake and was the only data that could be used for prediction. We intend to collect a larger amount of seismic data, and

compare them with acceleration data. It has been reported that at 300 kHz or less, electromagnetic waves are transmitted through the transformer neutral earth line without damping. We estimate that seismic information can be obtained from the transformer neutral earth line.

If electromagnetic waves emit from the surface or topsoil, it is possible to locate the electro− magnetic wave source from its arrival time and the difference between the spatial electromagnetic waves and the earth current.

NEW APPROACH TO HIGH VOLTAGE ELECTRODYNAMIC DRUM SEPARATOR
PART I: THEORETICAL MODEL

ANTONI CIEŚLA

ELECTRICAL POWER INSTITUTE, UNIVERSITY OF MINING AND METALLURGY
al. Mickiewicza 30, 30-059 KRAKÓW, POLAND
tel (0048 12) 617-37-73, e-mail: aciesla@uci.agh.edu.pl.

Abstract

Electric field has a lot of applications in technology. One of them is electrodynamic separation: electric field influences selectively granular solids of different electric moments or charges.
A mathematical model of the separation process in the high voltage drum separator is presented in the paper. Particles are charged both with induction and corona phenomena: next, they are separated with field forces. The computational results are presented and analysed.

1. INTRODUCTION

Dielectric particles staying at electric field are influenced by the field, in accordance with (1):

$$\vec{F}_e = \vec{\mu} \cdot \nabla \vec{E} + Q \cdot \vec{E} \qquad (1)$$

where:

$\vec{F}_e$ [N] - electric field force, $\vec{\mu}$ [Cm] -electric moment of the particle, $\vec{E}$ [Vm^{-1}] - electric field, Q [C] - electric charge of the particle.

The first component of (1) defines a diaphoresis force (appearing in heterogeneous fields) but the second is an electrophoresis force - influencing a charged particle only. The forces of the influence and main methods of dielectric particles charging are schematically presented in Fig. 1.

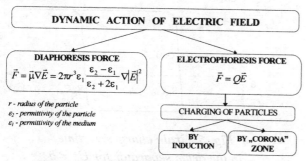

Fig. 1. *Dynamic influence of electric field (a scheme)*

A mathematical model of separation process in the high voltage electrodynamic separator (Fig 2) will be presented hereunder. The device can be used for separation of mixed granular dielectrics of different electrical properties. There are two various constructions of separator: the Blake - Morscher's one (without a „corona" electrode) and more widespread the Huff's device (containing a „corona" electrode) [1].

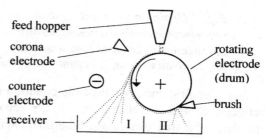

Fig. 2. *High voltage electrodynamic drum separator*

2. MATHEMATICAL MODEL OF ELECTRIC FIELD DYNAMICS IN THE DRUM SEPARATOR

2.1. Charging of grains

Mixture of grains falls out of the hopper $(\vec{E} = 0)$ into direct electric field $(\vec{E} = const)$ generated between two cylindrical electrodes (one of them rotates). The grains in contact with rotating electrode assume electric charge by induction. A scheme of the „grain - electrode" system is presented in Fig. 3. [2].

The grain charge vs time as response for „gate" function is defined as follows [2]:

$$Q(t) = U * \frac{C_p C_z}{C_p + C_z} + $$
$$+ U * \frac{C_p^2}{C_p + C_z} \left(1 - \exp\left(\frac{-t}{R_z\left(C_p + C_z\right)} \right) \right) \qquad (2)$$

where:

t [s] - time (t = 0 just as the grain comes into contact with the rotating electrode), U [V] - voltage between the electrodes, C_p [F] - the capacitance of the „grain - counter electrode"

High Voltage Engineering Symposium, 22–27 August 1999
Conference Publication No. 467, © IEE, 1999

system, C_z [F] - the capacitance of the „grain - rotating electrode" system, R_z [Ω] - the resistance of the „surface of the grain - rotating electrode" system.

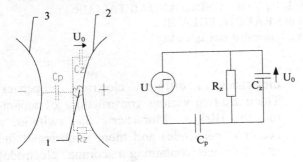

Fig. 3. *Grain on the rotating drum in an electrodynamic drum separator (the situational sketch and the scheme)*
1 - grain, 2 - rotating drum, 3 - counter electrode

The first component of the equation (2) defines a charge of polarization but the second one defines a free charge vs. time characteristic. Full analysis of the charging process can be found among others in [3]. The characteristics of „conductor" (small values of R_z, T_z) charge and „non-conductor" (great values of R_z, T_z) charge are presented in Fig. 4.

$$T_z = R_z\left(C_z + C_p\right) \approx R_z C_z , \qquad C_p << C_z \quad (3)$$

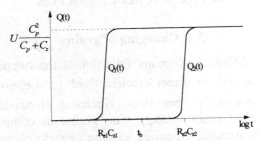

Fig. 4. *Grain charges vs. time in the drum separator for $U_0 = 0$*

The analysis leads one to regard that the separation takes place when the contact time t_b between the mixture and drum is longer than the time constant of the „conductor" grain but shorter than the time constant for the „non-conductor" grain (grain of the „conductor", i.e. the component „1" - z_1; grain of the „non-conductor", i.e. the component „2" - z_2).

In the other construction of the electrodynamic drum separator (the Huff's device) the mixture passes trough the corona discharge and assumes some charges. The time of the passing is not shorter than 1/10 s; so, the charge Q_0

assumes the maximal possible value. Next, the feed falls on the rotating drum. Grains having small relaxation times („conductors") allow the charge to conduct away rapidly. Subsequently they take up some charges of opposite sign and are repelled by the field forces into the bin I of the receiver. However, grains having large relaxation times („non-conductors") lose little or no charge: they are „pinned" to the rotating drum by means of image forces. Eventually, these grains are removed with the brush into the bin II.

The electrical scheme for the „grain - drum electrode" system is the same as for Blake - Morscher's separator but initial conditions (a charge of the capacitance C_z) must be taken into account. So, the charge vs. time characteristic is defined by the formula:

$$Q(t) = U * \frac{C_p C_z}{C_p + C_z} +$$
$$+ U * \frac{C_p^2}{C_p + C_z}\left(1 - \exp\left(\frac{-t}{R_z\left(C_p + C_z\right)}\right)\right) - \quad (4)$$
$$- Q_0 * \exp\left(\frac{-t}{R_z\left(C_p + C_z\right)}\right)$$

where:

$$Q_0 = \left(1 + 2\frac{\varepsilon_z - \varepsilon_0}{\varepsilon_z + \varepsilon_0}\right)\varepsilon_0 E \pi r^2 \quad (5)$$

Q_0 [C] - charge of the grain assumed in corona zone, [3], ε_z [Fm^{-1}] -permittivity of the grain, ε_0[Fm^{-1}] -permittivity of free space, r [m] - radius of the grain.

The courses of free charges vs. time are presented in Fig. 5.

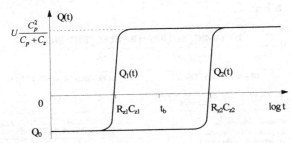

Fig. 5. *Grain charges vs. time in the drum separator for $U_0 \neq 0$*

Alike Blake - Morscher device, separation takes place if:

$$R_{z1} C_{z1} << t_b << R_{z2} C_{z2}.$$

2.2. Motion of grains in electrical field

2.2.1. Electric field distribution in the separator

A particle, charged according to the formula (2) or (4) is passing trough electric field between two cylindrical electrodes (Fig.6). Potential of the electric field for M point (out of the electrodes) is defined by the formula [4]:

$$V_M = \frac{Q}{2\pi\varepsilon_o}\ln\frac{r_2}{r_1} \qquad (6)$$

where:

r_1, r_2 [m] - the distances between the M point and the poles O_1, O_2 respectively:

$$r_1 = \sqrt{\left(a - h_1 + x\right)^2 + y^2}$$
$$r_2 = \sqrt{\left(h_1 + a - x\right)^2 + y^2} \qquad (7)$$

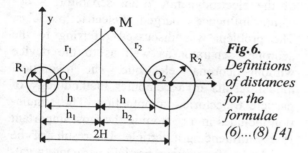

Fig.6. *Definitions of distances for the formulae (6)...(8) [4]*

Components of the electric field in the M point:

$$E_x = -\frac{\partial V_M}{\partial x} = \frac{Q}{2\pi\varepsilon_o}\left[\frac{a - h_1 + x}{\left(a - h_1 + x\right)^2 + y^2} + \frac{h_1 + a - x}{\left(h_1 + a - x\right)^2 + y^2}\right] \qquad (8)$$

$$E_y = -\frac{\partial V_M}{\partial y} = \frac{Q}{2\pi\varepsilon_o}\left[\frac{y}{\left(a - h_1 + x\right)^2 + y^2} + \frac{y}{\left(h_1 + a - x\right)^2 + y^2}\right]$$

The formula (8) is used to determine the field distribution between the electrodes of the drum separator.

2.2.2. Forces influencing the grains

The equation for the forces influencing a particle has the form [5]:

$$m\frac{d\overline{v}}{dt} = \sum \overline{F} \qquad (9)$$

where:

m [g] - mass of the particle,

v [ms⁻¹ - velocity of the particle.

A particle staying on the rotating drum surface as well as passing trough electric field is influenced by forces of the following modules:

- of electric field force	$F_{el} = QE$	(10)
- of „image" force	$F_{im} = Q^2/4\pi\varepsilon_0\,(2R_2^{\,2})$	(11)
- of centrifugal force	$F_c = m\omega^2 R_2$	(12)
- of gravitational force	$F_g = mg$	(13)
- of medium dynamic resistance force	$F_d = 6\pi\mu r v$	(14)

where : R_2 [m] - radius of rotating drum; ω [s⁻¹] - velocity of rotation; g [ms⁻²] - acceleration of gravity, μ [Pa·s] - dynamic coefficient of medium viscosity.

A scheme of forces influencing a particle during the contact with rotating drum as well as during free fall is presented in Fig. 7.

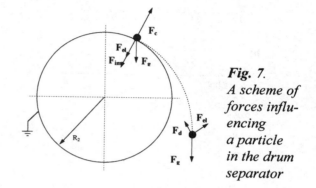

Fig. 7. *A scheme of forces influencing a particle in the drum separator*

3. NUMERICAL CALCULATION OF PARTICLE TRAJECTORIES

The trajectories of particles in the separators have been calculated on the basis of the field distribution (formula (8) as well as the forces influencing the particles (formulae (10)... (14)). For calculations, the software package „Matlab" was used.

The results of the calculations (several trajectories) for various dynamic conditions are presented in Fig. 8.

On the basis of the particle trajectories, some curves of the particle lift off angle versus the voltage between electrodes, for several particle resistances have been determined (the speed of the rotation is constant). The definition of the angle φ is shown in Fig. 9 but the results of computations are presented in Fig. 10. Dependence between grain resistances and the trajectories is evident. The resistance influences the assumed charge value and - as a result - the time of stay of the particle on the rotating drum and the lift-off angle. So, the method of the particle trajectories computations is important and it affords possibilities to define the range of application of the electrodynamic drum separator. Full analysis of the application ranges of the drum separators as well as proposition of a new method (the separation with use of alternating field) are given in [2].

a)

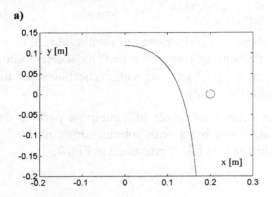

b)

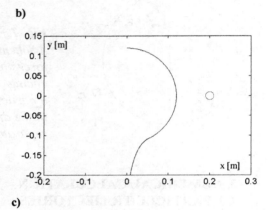

c)

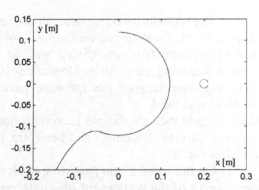

Fig. 8. *Exemplary particle trajectories (computer computation); $R_z = var$; a) $R_z = 1E4$ [Ω], b) $R_z = 4E6$ [Ω], c) $U = R_z = 7E6$ [Ω], $U = 20$ [kV] = const; $\omega = 8$ [rads^{-1}] = const*

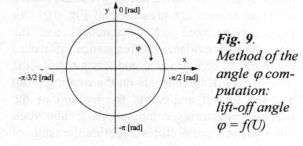

Fig. 9. *Method of the angle φ computation: lift-off angle $\varphi = f(U)$*

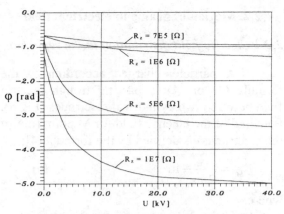

Fig. 10. *Lift-off angle φ vs. voltage U for selected particle resistances R_z*

4. SUMMARY

The analysis refers to essential problem of the electrodynamic drum separators - dynamic influence charged dielectric particles. The problem was discussed referring to the „corona" separator as well as to the device without „corona" electrode. The theoretical model results in a computer methodology of particle trajectories computation. The simulation includes in a reckoning the most important forces influencing a particle. The results of the simulation, for various particle parameters values, show effect of these parameters on grain trajectories in the drum separator. The achieved data confirm correctness of the model as well as usefulness of the computer program. Parameters of the process can be determined on the basis of electrical properties of particles.

REFERENCES

[1] **Ralston O. C.** : *Electrostatic Separation of Mixed Granular Solids*. Amsterdam - New York, 1961.

[2] **Szczerbiński M.**: *Electrodynamic Conductance Separation with use of Alternating Fields*, Journal of Electrostatics, (1983), pp. 175 - 186

[3] **Rewniwcew W. I.** : *Fiziczeskije osnowy elektriczeskoj separacji*. Izd. Nedra. Moskwa, 1983.

[4] **Zahn M.**: *Electromagnetic Field Theory: Problem Solving Approach* (Polish edition by PWN, Warszawa, 1989).

[5] **Cieśla A.**: *Computation of the Trajectories of Charged Particles in the Electric Field of Flat-Plate Separator (Reynolds Number, Re > 1)* (in Polish). Archiwum Elektrotechniki, 1979, Vol XXVIII, No. 2, pp. 407 - 420.

NANOSECOND FORMING LINE BASED ON
COMPACT POWER HIGH-VOLTAGE PULSE CAPACITORS

Georgy S. Kuchinsky, Mikhail N. Kozhevnikov, Oleg V. Shilin,
Ludmila T. Vekhoreva, Nikolay V. Korovkin, Ekaterina E. Selina, Anton A. Potienko

St.-Petersburg State Technical University, St.-Petersburg, Russia

Abstract

Specific problems regarding the high-voltage pulse capacitors for quasi-rectangular nanosecond pulse formation at a low resistive load are discussed. Such a capacitor could be presented as a ladder-circuit which functions as a strip transmission line discharging to a matched load. The capacitor consists of conventional low inductive capacitor sections. The packages of these sections are connected in parallel by means of plate conductors which form the main (longitudinal) inductance of circuit cells. The inductance of the packages (the transverse inductance) affects negatively the shape of the formed pulse. The capacitor construction is improved by making the circuit cells non-uniform. As a result the negative effect of the transverse inductance could be cancelled. Furthermore, the non-uniform capacitor consists of a half number of the cells as compared with the uniform one and provides the same front and tail steepness and flat-top stability. A numerical calculation technique is developed which could be applied for such non-uniform circuit calculation. The experimental results are in very good agreement with the theoretical behavior and demonstrate the following characteristics achieved with compact non-uniform capacitors: pulse length of 30 ns to 1 ms, front steepness of up to 10^{13} A/s, resistive load of 1 to 10 Ohm, charging voltage of up to 200 kV.

Introduction

Compact power high-voltage pulse capacitors for quasi-rectangular nano- and microsecond impulses formation are widely used when extremely short front and tail of impulse and a high stability of impulse flat-top are required. We consider that it is reasonable to name such a capacitor as pulse capacitor with transmission (forming) line properties or artificial forming line (AFL). Generated pulses may have a range from tens nanoseconds to microseconds and amplitudes up to hundreds of kV. AFL design allows to connect them in series or parallel with the low joint inductance to form current sources of up to hundreds kA and voltage sources of up to MV with the front steepness of up to 10^{13} A/s. Pulses with these characteristics are required for feeding power gas lasers, particle beam sources, high-voltage (current) flash test setups etc.

Until recently coaxial or strip transmission lines with liquid or solid dielectric media were used for these purposes. The insulation of lines were designed to withstand total charging voltage, which resulted in

low values of electric field strength and specific energy. Therefore their size and cost were rather high.

Artificial forming line

The conventional low inductive capacitor section could be considered as a strip transmission line and could be applied for quasi-rectangular impulses formation at a matched load. By connecting such sections in series, a source for required charging voltage could be designed. Because of the less thickness of the section dielectric, electric field strength and specific energy could be greatly increased as compared with the lines which are designed to carry total charging voltage. However the experience in developing of sources for the pulse voltage range over 10 kV shows that when connecting the reasonable large number of sections the active losses in the section foil affect negatively the pulse shape, viz, front and tail lengths increase [1].

Ohmic losses could be reduced by means of parallel-series connection of sections (Fig.1).

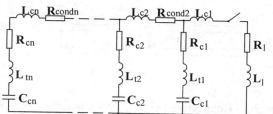

Fig.1 Circuit with parallel-series connection of sections.

This circuit could be considered as an equivalent transmission line ladder-circuit and it has been realised as a set-up shown in Fig.2 [1,2].

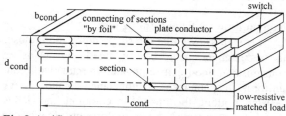

Fig.2 Artificial forming line.

This AFL consists of n_c cells, each including n_s sections joined in series to the packages. The packages are connected in parallel by means of plate conductor which forms the main (longitudinal) inductance of circuit cells L_c. It is important that the sections foil inductance is the main (pulse shape forming) inductance, in case the transmission line properties of series-joined sections are utilised. But in the case of parallel-series joining of sections, the main inductance is created by plate conductor and the sections foil inductance impairs the pulse shape. For AFL properly operating as an equivalent transmission line ladder-circuit the sections must have, as far as possible, properties of concentrated capacitance. Thus, while analysing the AFL discharge it is impossible not to take into account the inductance of the packages (the transverse inductance) L_t which affects negatively the shape of the forming pulse. The cell transverse inductance consists of the sections foil inductance combined with the sections joint inductance. As the sections and their joints design may be different, the transverse inductance value varies from fractions of the main cell inductance to the value exceeding the main cell inductance. When connecting sections "by foil" (Fig.3.a), the ratio between transverse and longitudinal inductances is about $L_t/L_c = 1/3$. The foil inductance could be reduced by making on the section butts one or several pairs of joints evenly distributed along the foil (Fig.3.b) [3].

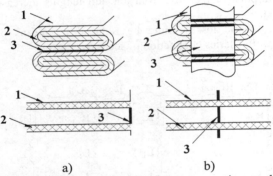

a)   b)

Fig.3 a - Connecting "by foil", b - connecting on the section butts. 1 - Foil, 2 - dielectric, 3 - joint.

But it entails the growing of the joint inductance. Calculation and experiment show that transverse inductance begins to impair the pulse shape (to increase the front and the tail length and the amplitude of fluctuations on the flat-top of the impulse) when ratio is about $L_t/L_c = 0.2$ [1].

Non-uniform artificial forming line

It is well known that the shape much more closer to rectangular may be obtained by using a transmission line (or its equivalent circuit) with non-uniform capacitances, inductances and wave resistance along the line. It is common knowledge that two conventional types of forming circuits can be used - the resonance and the antiresonance circuits. A resonance circuit consists of few non-uniform cells and forms a satisfactory rectangular pulse shape. But such a circuit is not convenient for realization because of wide variations of the cells capacitance. In [4] the ladder-circuit with non-uniform cell parametres has been shown. There is only five cells in it and it forms the pulse whose shape is closer to rectangular than that of the pulses formed by uniform ladder-circuit containing greater number of cells. But the authors do not describe the method of obtaining cell parameters, noting that to obtain non-uniform line parameters is quite a problem which could be solved only with numerical techniques. Analytical methods could be applied only for finding out the approximate solution because of great number of coupled cells, each containing several parameters. That is why this problem had been postponed for the future.

Finding of high-voltage non-uniform AFL cell parameters for quasi-rectangular pulse formation is the main task of this investigation.

It is important to note that the majority of the methods of pulse shape correction is based on one or the other method of closing the transient characteristic of a concentrated-element circuit in on the transient characteristic of a transmission line which is accepted as the best pulse-shaping device. But generally speaking, such a point of view gives neither estimation of this closing nor assurance that the highest degree of closeness of the pulse being formed to rectangular shape would be achieved (when the number of cells is fixed). The method suggested in this work is based on closing the transient characteristic of non-uniform AFL ladder-circuit directly in on the rectangular impulse.

The problem of obtaining non-uniform AFL cell parameters is solved by minimization of the special functional in the time or the spectral domain [5]. This functional could be presented as an area between the curve corresponding to the impulse formed by non-uniform AFL and the curve corresponding to the required (rectangular) impulse. When the circuit consists of n cells, the problem of functional minimization has 2n parameters. Since n=2, the problem is no longer unimodel. The main difficulty in solving this problem is finding out the global minimum of the functional. To obtain the global minimum of the functional, the following steps have been taken. First, the area of seeking of the L_c and C_c parameters was localized reasoning from common sense and from the supposed nature of distribution of parameters along the line. Second, the method of initial approximation of parameters for the procedure of minimization of the functional was developed. As the functional has a lot of local minimums, a correct localization of the area of possible parameters and an accurate choice of the initial approximation of the parameters are of vital importance in solving of the functional minimization problem. The accuracy of finding of the functional minimum is ensured by the predetermined number of steps within whose duration the functional changing is not more then predetermined value.

This procedure of obtaining of the optimal parameters of the non-uniform AFL was applied to the AFL circuits (without taking into account the transverse inductance) containing from 3 to 10 cells. As a result the optimal parameters were obtained. The nature of changing of L_c and C_c parameters along the line is shown in Fig.4.

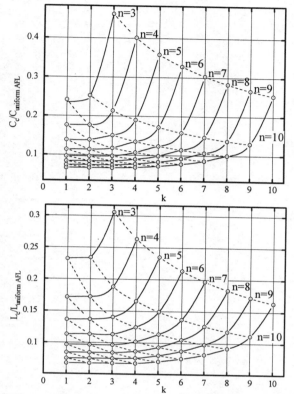

Fig.4 The nature of changing of L_c and C_c parameters along the non-uniform AFL with $n_c=3\div10$ without taking L_t into consideration.

The figure shows that capacitance C_c, main inductance L_c and wave resistance of cells gradually increase from the matched load along the line. The non-uniform AFL which contains the obtained optimal parameters consists of a half number of the cells as compared with the uniform AFL and provides the same front and tail steepness and flat-top stability (Fig.5).

Further calculation shows even greater negative influence of the transverse inductance on the pulse shape in non-uniform AFL then in uniform one. Then the transverse inductance was taken into account in the minimization procedure and, as a result, the negative effect of the transverse inductance had been cancelled at the expense of greater non-uniformity of line. In Fig.6 one of the pulses illustrates the L_t influence in the non-uniform AFL circuit whose parameters have been obtained without taking L_t into account. The second one illustrates the result of additional optimization of line parameters when taking L_t into account (AFL consists of 5 cells).

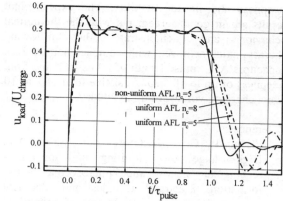

Fig.5 Pulses for uniform AFL where $n_c=5$; 8 and non-uniform AFL where $n_c=5$.

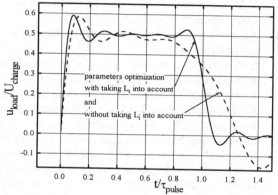

Fig.6 Pulses for non-uni AFL where $n_c=5$, $l_t/l_c=1/3$.

During the optimization process the transverse inductance was set in different ways (constant for all the cells or connected with the main inductance of cells or others) and varied widely. Essentially in all the cases of the transverse inductance value from fractions of the main cell inductance to the value even exceeding the main cell inductance its negative influence was managed to be canceled.

On the basis of this investigation the some models of non-uniform AFL were developed and produced. The experimental results are in very good agreements with the theoretical behavior and demonstrate the following characteristics achieved with compact high-voltage non-uniform AFL: pulse length of 30 ns to 1 ms, front steepness of up to 10^{13} A/s, resistive load of 1 to 10 Ohm, charging voltage of up to 200 kV.

Conclusions

It is shown that the shape of quasi-rectangular impulse could be improved by modifying the equivalent transmission line ladder-circuit. Also the negative influence of cells transverse inductance on the pulse shape is shown. A method of obtaining of non-uniform artificial line parameters has been developed. As a result the negative effect of the cells transverse inductance could be cancelled. Furthermore, the non-uniform artificial line consists of a half number of the cells as compared with the uniform one and provides the same front and tail steepness and flat-top stability. The experimental results are in good agreements with the theoretical behavior and demonstrate the following characteristics achieved with compact non-uniform capacitors-AFL: pulse length of 30 ns to 1 ms, front steepness of up to 10^{13} A/s, resistive load of 1 to 10 Ohm, charging voltage of up to 200 kV.

References
1. Kuchinsky G.S., Vekhoreva L.T., Shilin O.V. High-voltage power forming lines for quasi-rectangular nano- and microsecond pulse formation. Electrical technics and electrical power engineering. Problems of reliability. Works of SPbGTU, № 460, 1996.
2. Burtcev V.A., Vasilevsky M.I., Vodovozov V.M. etc. Capacitance sources with transmission line properties. Electricity, №7,1989.
3. Kuchinsky G.S. High-voltage pulse capacitors.-L.: Energy, 1973.
4. Evtyanov S.I., Redkin G.E. Pulse modulator with artificial line. - M.: Soviet radio, 1973.
5. Kuchinsky G.S., Shilin O.V., Vekhoreva L.T., Kozhevnikov M.N., Korovrin N.V., Selina E.E., Potienko A.A. Non uniform nano- and microsecond high-voltage power forming lines. Electrical technics and electrical power engineering. Problems of control of high-voltage power systems. Works of SPbGTU, № 471, 1998.

STEEP-FRONTED OVERVOLTAGES IN INVERTER-FED INDUCTION MOTORS: NUMERICAL IDENTIFICATION OF CRITICAL PARAMETERS

C.Petrarca, G.Lupò
Dip. di Ing. Elettrica
Università di Napoli
Via Claudio 21 Napoli
Italy

L.Egiziano, V.Tucci
Dip. di Ing. Inf. e Ing. Elettrica
Università di Salerno
Via Ponte Don Melillo Fisciano
Italy

M.Vitelli
Dip. di Ing. dell'Informazione
Seconda Università di Napoli
Via Roma Aversa
Italy

Abstract

Due to the recent trends in power-systems, steep-fronted switching surges are often encountered in many applications as, for example, in induction motors fed by pulse-width modulated (PWM) inverters. In such a case, the effect of these voltage transients can result in undesirable electrical and thermal stresses in the stator windings, which can lead to the premature failure of the insulating material between the coils and, definitely, to the failure of the whole component.

Up to now, the dimensioning of the interturn insulation of such motors is based empirically on previous service experience; a more rational design can be performed if a more exact knowledge of the voltage distribution within the stator winding is reached.

In the present paper the first results of a simulation carried out in order to predict the voltage distribution between the coils of a stator winding fed by a PWM inverter are presented. Such distributions depend on a very large number of parameters (insulation dimension, coils size, length of connecting cable, shape of the applied voltage, etc.). In the following the influence played on such distribution by the cable length and applied voltage rise-time is discussed.

Introduction

Because of the recent tendency in power-systems, high switching surges are more often encountered than in the past, as a result of the introduction of new components, among which:

1) modern vacuum interrupters;
2) low loss cables for the interconnection of machines and switch-gears;
3) adjustable-frequency inverters.

In particular, adjustable speed drives operate nowadays with PWM inverters using the insulated gate bipolar transistors (IGBT) which are able to generate steep fronted pulses (2500 V/μs) at high switching frequencies (close to 20kHz) [1].

As a consequence, the electrical insulation of induction motors fed by such devices is subjected to enhanced stresses (both thermal and electric) with respect to those deriving from a pure sinusoidal wave shape which can significantly reduce the life-time of the insulation materials.

An experimental confirmation of the above assumptions has been given by the authors in a recent paper [2]: the application of repetitive unipolar pulse voltages on stator winding samples can sensibly affect the dielectric properties of the insulating material and the partial discharge activity.

For a rational design of inverter fed motors, information is needed not only about the degradation mechanisms of the dielectrics but also about the stresses that such insulation will have to withstand in service. The voltage distribution along the winding is influenced by surge rise time and shape, coil insulation dimensions, coil size and shape, etc.

In particular, due to the "impulsive" nature of the inverter output voltages, with rise times of about 10-100ns, the geometric dimensions of the winding become comparable with the wavelength λ of the electromagnetic signal involved. This implies that the potential distribution inside the windings is not "instantaneous" but derives from the superposition of the forward wave travelling from the inverter to the load and that reflected from the motor, travelling in the opposite direction.

To obtain a general understanding of how interturn voltages are distributed along the stator winding, the winding itself can be regarded as a transmission line; a steep fronted wave travels into a winding in a similar manner to a wave being propagated along the transmission line.

Adaye and Cornick [3] proposed an equivalent network of the machine winding consisting of a chain of coils each of which described as a lossy 2-port network; however this model does not take into account the effect of electric and magnetic coupling between the turns.

Wright, Young and McLeay proposed an extension of such a model by treating each coil as a series of five multiconductor transmission lines; the junctions of these lines were analysed using scatter matrix theory [3].

It is thus evident that the study of such configurations is very complex in that it involves a large number of parameters. Wright et al. in [4] analysed the effect of varying coil and surge parameters, such as the surge front time, the dimension of the insulation, the number of coils and turns. In their work, however, they did not take into

High Voltage Engineering Symposium, 22–27 August 1999
Conference Publication No. 467, © IEE, 1999

account the effect of varying the cable length or the insulating material. The present paper is a first attempt to clarifying such aspects.

Theoretical approach

In fig. 1 the general arrangement of an inverter connected to a stator winding via a loading cable is depicted. The cable, represented as single transmission line, is connected to a multiconductor transmission line, representing the turns of one coil of the stator winding. .In a single stator coil two different regions can be evidenced: region 1, in which the coil sides are enclosed inside the slots in the magnetic core structure; region 2, in which the end turns are not confined to slots: the conductors in both regions are represented by two different multiconductor transmission lines connected at junction J1, J2, J3 and J4.

As a first approximation, losses are incorporated through resistive elements R_i at the coil junctions in region 2, while the dielectric is assumed to be perfect and the cable losses are represented by the DC resistance R_{DC}.; the coil parameters are assumed to be frequency-independent.

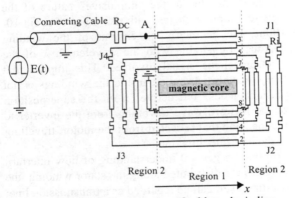

Fig.1:schematic representation of cable and winding

The stator core is assumed as an equipotential surface at zero voltage. Under typical high frequency excitation (about 10 MHz) it acts like a barrier to the magnetic flux density and mutual coupling between the slot portions of coils in adjacent slots can thus be assumed to be negligible. Furthermore, since a single slot acts like an earthed boundary, the capacitive coupling between coils and between windings lying in different slots can also be neglected.

The multiconductor transmission-line equations, written in the Laplace-domain with homogeneous initial conditions, are of the form

$$\begin{cases} \dfrac{dV^2(x,p)}{dx^2} = (R+pL)(G+pC)V(x,p) \\ \dfrac{dI^2(x,p)}{dx^2} = (R+pL)(G+pC)I(x,p) \end{cases} \quad (1)$$

where V and I are the vectors representing,

respectively, the voltages with respect to the reference conductor and the currents of all conductors, x is the longitudinal co-ordinate, R, L, C and G are, respectively, the resistance, inductance, capacitance and conductance matrices per unit length.

By the application of modal analysis [6] the above system can be subdivided in N systems of N uncoupled two conductor transmission lines. In absence of losses, the equation governing the i-th system can be written as:

$$\begin{cases} \dfrac{dV_{mi}^2(x,p)}{dx^2} = l_{mi}c_{mi}V_{mi}(x,p) \\ \dfrac{dI_{mi}^2(x,p)}{dx^2} = c_{mi}l_{mi}I_{mi}(x,p) \end{cases} \quad (2)$$

where V_{mi} and I_{mi} are the modal voltage and current at the i-th line and l_{mi} and c_{mi} are the i-th components of the modal diagonal inductance (L_m) and capacitance (C_m) matrices. The characteristic impedance Z_{mi} is real, $Z_{mi} = (1/Y_{mi}) = (\sqrt{l_{mi}}/\sqrt{c_{mi}})$; the velocity of propagation of the i-th mode is $v_{mi} = (1/\sqrt{l_{mi}c_{mi}})$ and the time delay is $T_{mi} = (v_{mi}/d)$. The solution can be written in terms of forward (+)and backward (-) travelling wave components and can be expressed as a function of the terminal quantities I_{0mi} I_{dmi} V_{0mi} V_{dmi}, that is as function of I_{mi} and V_{mi} evaluated at x=0 and x=d:

As a consequence, the problem of solving the multiconductor transmission line is replaced by the problem of solving N single transmission lines by finding the terminal quantities I_{0mi} I_{dmi} V_{0mi} V_{dmi}. We can write:

$$\begin{cases} V_{0mi}(p) - Z_{mi}I_{0mi}(p) = [V_{dmi}(p) - Z_{mi}I_{dmi}(p)] \cdot e^{-pT_{mi}} \\ V_{dmi}(p) + Z_{mi}I_{dmi}(p) = [V_{0mi}(p) + Z_{mi}I_{0mi}(p)] \cdot e^{-pT_{mi}} \end{cases} \quad (3)$$

In order to get the solution in the time domain, it is necessary to perform the inverse transformation of eq. (3). The final system in the time domain is:

$$\begin{cases} v_{0mi}(t) - Z_{mi}i_{0mi}(t) = [v_{dmi}(t-T_{mi}) - Z_{mi}i_{dmi}(t-T_{mi})] \\ v_{dmi}(t) + Z_{mi}i_{dmi}(t) = [v_{0mi}(t-T_{mi}) + Z_{mi}i_{0mi}(t-T_{mi})] \end{cases} \quad (4)$$

By adding suitable initial and boundary conditions for I_{0mi} I_{dmi} V_{0mi} V_{dmi}, the terminal currents and voltages can be evaluated by knowing their value at the time instant $(t-T_{mi})$.

In order to employ the above theory, the inductance (L) and capacitance (C) matrices must be evaluated. In the present paper the inductance matrix L has been calculated by means of the relationship: $L \cdot C_o = \varepsilon_o \mu_o \cdot 1$, once the capacitance matrix C_o in a lossless homogeneous medium (ε_o,μ_o) is known. The capacitance C has been approximated by considering each couple of conductors as an indefinite parallel-plate capacitor. Calculation by means of field analysis methods can also be employed [7] and will be the topic of future works.

Numerical results and discussion

As discussed above, the connecting cable Cc acts like a transmission line, that is it significantly contributes to wave propagation and wave distortion. As a result, it has a considerable effect on the voltage at the terminals of the motor and on the voltage stressing the insulation between adjacent conductors. In order to investigate on the influence of the cable length on the voltage distribution, a simulation program has been used with the parameters reported in Tab.1; at present only one stator coil has been considered, with five conductors per slot, although the method can be extended to a larger number of coils and windings.

Cable	Winding
Capacitance:131 nF/km	Conductor size:1.2*11mm
Inductance: 0.3 mH/km	Slots length: 0.5 m
DC resistance: 8 Ω/km	
Length: 1, 2, 4, 8, 16 m	

Tab.1: Simulation parameters for cable and winding

The shape of the applied voltage is shown in Fig.2. The set of rise-times t_s (25, 50, 100, 200, 400 ns) has been chosen in order to cover a very wide range of steep front surges. In the following all the voltages will be expressed as percentages of the maximum value of such a surge.

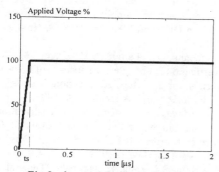

Fig.2: shape of the applied voltage

The effect of the cable length variation is clearly visible in Fig. 3 in which the voltage between the motor terminal A and ground is shown both for a 1 meter cable length (thick line) and 4 meter cable length (thin line), when applying a surge of 50ns rise time. The electrical stress is clearly different in the two cases: while with the shorter cable the insulation of the motor is subjected to few pulses of decreasing amplitude, with the longer cable the insulation has to withstand a much greater number of repetitive unipolar pulses of very high amplitude (about 150%).

However, a very strong influence on the electrical stress on the motor insulation is also played by the steepness of the applied voltage or, better, by the

duration of the rise time ts, as shown in Fig.4, where it is visible that a longer rise time does not produce overvoltages at the motor terminals.

The above considerations are summarised in Fig. 5 where the maximum overvoltage V_{max} at the motor terminal is reported as function of the cable length and of the rise time ts. V_{max} increases with lower rise times and assumes its higher values when the cable is 8 metres long: in that case the insulation is stressed not only due to very high peak values (about 170 % p.u.) but also due to the high frequency of such voltage pulses.

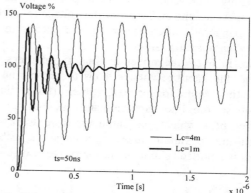

Fig.3: Voltage to ground vs. time at the terminals of the motor. Influence of cable length Lc (ts=50ns; thick line: Lc=1m; thin line: Lc=4m).

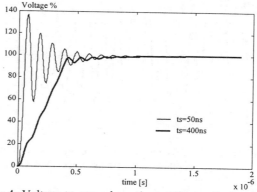

Fig.4: Voltage to ground vs. time at the terminals of the motor. Influence of rise time ts. (Lc=2m; thick line: ts=400ns; thin line: ts=50ns)

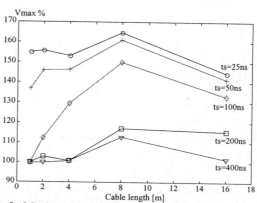

Fig.5: Maximum overvoltage at the motor terminal vs. cable length as function of the rise time ts.

The same considerations hold for the interturn voltage, that is the voltage across two different conductors in the same slot. The evaluation of such a stress is of great importance in that it is one of the main causes of inception of partial discharges in electrical insulation.

Furthermore, it is useful to remind that both the amplitude of the voltage and its shape have a great role in provoking and sustaining the partial breakdown in a defect inside the insulation [8]; in particular, the discharge frequency is much higher when the applied voltage across the insulation is oscillating. For such a reason, it is evident from Fig.6 that the time evolution of the voltage is surely more dangerous when the cable is 8 metres long than when it is much shorter.

Again, in Fig.7 the effect of longer rise times is visible: the interturn voltage tends to become non-oscillating with longer rise times (thick line: ts=400ns; thin line: ts=50ns) and assumes much smaller peak values, as summarised in Fig.8 in which the maximum interturn voltage vs. cable length is reported as function of the rise time ts.

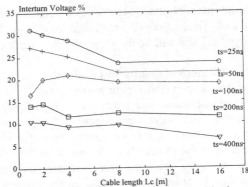

Fig.8: Maximum interturn voltage vs. cable length as function of the rise time ts.

Conclusions

In the present paper the first results of a simulation carried out on a stator winding fed by a PWM inverter are reported. It is shown how the length of the connecting cable, together with the rise time of the applied voltage, has a considerable effect on the electrical stress of the motor insulation. With longer cables the voltage stress upon the insulation assumes the form of repetitive pulses at high frequency; with shorter rise times the overvoltage sensibly increases. Some critical conditions with respect to partial discharges inceptions can thus beachieved which are worth considering

References

[1] A. Mbaye, F. Grigorescu, T. Lebey. Bui Ai, "Existence of Partial Discharges in Low-voltage Induction Machines supplied by PWM Drives, IEEE Trans. On Diel. and Elec. Ins., Vol. 3, No.4, pp.554-560, 1996

[2] L. Egiziano, C. Petrarca, V. Tucci, M. Vitelli: "Investigation on Performances of Insulation Materials for Inverter-Fed Traction Motors", 1998 CEIDP, pp.564-567, Atlanta, USA, 1998.

[3] R.E. Adaye, K.J. Cornick,: "Distribution of Switching Surges in the Line-end Coils of Cable-connected Motors", IEEE Trans. On El. Pow. Appl., Vol. 2, No. 1, pp.11-21, 1979.

[4] MT Wright, S.J. Yang, K. McLealy, "General theory of fast-fronted interturn voltage distribution in electrical machine windings", IEE proc., Vol.130, No.4, 1983, pp.245-256.

[5] MT Wright, S.J. Yang, K. McLealy, "The influence of coil and surge parameters on transient interturn voltage distribution in stator windings", IEE proc., Vol. 130, No. 4, 1983, pp. 257-264.

[6] Clayton R. Paul, *Analysis of Multiconductor Transmission Lines*, John Wiley and Sons.

[7] P.G. McLaren, H. Oraee, "Multiconductor transmission-line model for the line-end coil of large AC machines", IEE proc., Vol. 132, No. 3, 1985, pp. 149-156.

[8] Bartnikas, McMahon, "Corona Measurements and Interpretation", *Engineering Dielectrics*, Vol.I, ASTM Press, Philadelphia, 1979

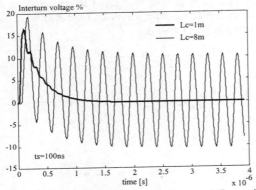

Fig.6: Interturn stress between conductors 5-7 vs. time. Effect of cable length Lc. (ts=100ns; thick line: Lc=1m; thin line: Lc=8m)

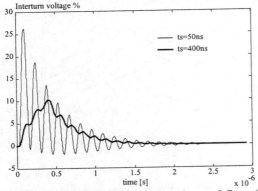

Fig.7: Interturn voltage between conductor 5-7 vs.time. Effect of rise time ts. (Lc=2m; thick line: ts=400ns; thin line: ts=50ns)

AN OPTIMIZATION MODEL FOR REACTIVE POWER CONTROL OF SUPER HIGH VOLTAGE GRID SYSTEMS USING PSEUDO-INVERSE METHOD

Najim Abbod Hamudi

University of Baghdad –IRAQ

ABSTRACT

This paper develops an optimization model to control the excessive MVAR generation by super high voltage grid systems for maintaining the nodal voltages within the required acceptable margin. The model enhances an approximate solution on the condition of balanced real power.

The Pseudo-Inverse Method of optimization is implemented as an optimal means of solving non-square systems of equations based on Lagrange's Theory of optimization. It recommends the locations and ratings of the minimum required reactors from the preinstalled ones that to be in service for optimum nodal reactive power and voltage controls.

The model had been tested on the Iraqi Super High Voltage Grid System (400kV), and it proved to be efficient in the obtained results and reliable in convergence.

INTRODUCTION

Present power flow optimization methods can be classified into two categories, one is providing exact solutions, and the other, approximate solutions. Exact methods (1,2,3) take into account both real and reactive flows in obtaining the solution. While approximate methods (4,5) achieve simplified representations and possibly computational efficiencies by ignoring either the real or reactive equations. Approximate models are normally tailored for particular applications and do not have the generality inherent in the exact models.

The reduced gradient method of Dommel and Tinny (2), the Fletcher Powell method as developed by Sasson for power flow applications (3), and Carpentier's method based on satisfying the Khun-Tucker conditions (1), are all extremely accurate and widely applicable.

While some have been developed to become computationally very efficient, a number of approximate models dealing exclusively with either the real or reactive power equations, have also been developed. Hano, et Al (4) and Kumai, et Al (5), used primarily the reactive equations to develop a method for real time control of voltages and reactive powers.The need to adjust voltage magnitudes and reactive powers at times so that the overall solution is operationally implementable, led to the consideration of reactive equations.

MATHEMATICAL FORMULATION

Figure (1) is simulating a super high voltage grid system by representing the nodes into four categories from the point of view of their voltages

High Voltage Engineering Symposium, 22–27 August 1999
Conference Publication No. 467, © IEE, 1999

[either they are within (m) or out (o) of the acceptable margin] and VAR absorption installations [either there are (r) or none (w)]

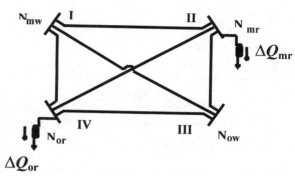

Figure (1) Schematic Single Line Diagram of the Optimization Model .

Considering the system in steady-state condition and the real power mismatch ΔP is zero, then the following linearized equations describe the effect of Q on V for the four categories of the system:-

$$\frac{dQ_{mw}}{dV_{mw}}\Delta V_{mw} + \frac{dQ_{mw}}{dV_{mr}}\Delta V_{mr} + \frac{dQ_{mw}}{dV_{ow}}\Delta V_{ow} + \frac{dQ_{mw}}{dV_{or}}\Delta V_{or} = 0 .....(1.a)$$

$$\frac{dQ_{mr}}{dV_{mw}}\Delta V_{mw} + \frac{dQ_{mr}}{dV_{mr}}\Delta V_{mr} + \frac{dQ_{mr}}{dV_{ow}}\Delta V_{ow} + \frac{dQ_{mr}}{dV_{or}}\Delta V_{or} = \Delta Q_{mr} .....(1.b)$$

$$\frac{dQ_{ow}}{dV_{mw}}\Delta V_{mw} + \frac{dQ_{ow}}{dV_{mr}}\Delta V_{mr} + \frac{dQ_{ow}}{dV_{ow}}\Delta V_{ow} + \frac{dQ_{ow}}{dV_{or}}\Delta V_{or} = 0 .....(1.c)$$

$$\frac{dQ_{or}}{dV_{mw}}\Delta V_{mw} + \frac{dQ_{or}}{dV_{mr}}\Delta V_{mr} + \frac{dQ_{or}}{dV_{ow}}\Delta V_{ow} + \frac{dQ_{or}}{dV_{or}}\Delta V_{or} = \Delta Q_{or} .....(1.d)$$

System (1) can be rewritten as below :-

$$\begin{bmatrix} dQ_{mw}/dV_{mw} & dQ_{mw}/dV_{mr} & 0 & 0 \\ dQ_{mr}/dV_{mw} & dQ_{mr}/dV_{mr} & -E & 0 \\ dQ_{ow}/dV_{mw} & dQ_{ow}/dV_{mr} & 0 & 0 \\ dQ_{or}/dV_{mw} & dQ_{or}/dV_{mr} & 0 & -F \end{bmatrix} \begin{bmatrix} \Delta V_{mw} \\ \Delta V_{mr} \\ \Delta Q_{mr} \\ \Delta Q_{or} \end{bmatrix} = \begin{bmatrix} b1 \\ b2 \\ b3 \\ b4 \end{bmatrix} ...(2)$$

Where E and F are unit diagonal matrices. Reducing System (2) using Newton-Gauss technique, it will be :-

$$\begin{bmatrix} A & 0 \\ B & -F \end{bmatrix} \begin{bmatrix} \Delta Q_{mr} \\ \Delta Q_{or} \end{bmatrix} = \begin{bmatrix} C1 \\ C2 \end{bmatrix} .....(3)$$

Or $\quad D\ \Delta Q_R = C\ .....(4)$

Where $D = \begin{bmatrix} A & 0 \\ B & -F \end{bmatrix}, \Delta Q_R = \begin{bmatrix} \Delta Q_{mr} \\ \Delta Q_{or} \end{bmatrix}, and, C = \begin{bmatrix} C1 \\ C2 \end{bmatrix}$

The ranks of the main and submatrices are :-

Main Matrix	
D	$N_o \times N_r$
ΔQ_R	$N_r \times 1$
C	$N_o \times 1$

Where $\quad N_r = N_{mr} + N_{or}, N_o = N_{or} + N_{ow}$

Sub Matrices			
A	$N_{ow} \times N_{mr}$	F	$N_{or} \times N_{or}$
B	$N_{or} \times N_{mr}$	$C1$	$N_{ow} \times 1$
E	$N_{mr} \times N_{mr}$	$C2$	$N_{or} \times 1$
ΔQ_{or}	$N_{or} \times 1$	ΔQ_{mr}	$N_{mr} \times 1$

The objective function of this model is :-

$$[\Delta Q_R]^T [\Delta Q_R] = Minimum .....(5)$$

To approach ΔQ_R optimum, the following algorithms will be followed during the iterative computational process :

Algorithm I :- if $N_o = N_r$, then:-

$$\Delta Q_R = D^{-1}C .....(6)$$

Algorithm II :- if $N_o < N_r$, then:-

$$\Delta Q_R = D_1^+ C .....(7)$$

Where $D_1^+ = D^T (DD^T)^{-1} .....(8)$

Algorithm III :- if $N_o > N_r$, then:-

$$\Delta Q_R = D_2^+ C .....(9)$$

Where $D_2^+ = (D^T D)^{-1} D^T .....(10)$

D_1^+ and D_2^+ are the Pseudo-Inverses of the rectangular matrix $D(6,7)$, and their mathematical derivations are presented in Appendix-I.

MODEL'S FLOW CHART

The flowchart of the optimization model is shown in figure (2). The dotted block represents the part where ΔQ_R is computed .

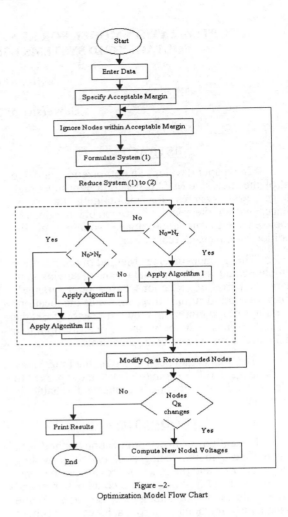

Figure –2-
Optimization Model Flow Chart

MODEL'S PERFORMANCE

This optimization model had been tested successfully on the Iraqi Super High Voltage Grid System shown in Figure (3) at Minimum Load Condition of the year 1996 .

To recognize the significance of the optimization model , three modes for MVAR control were implemented on the system at the same load condition as follows :-

Mode I :- In this mode , No MVAR absorption assistance was made to the system ($\Delta Q_R = 0$) .

Mode II :- All MVAR absorption resources were linked to the system (ΔQ_R =Maximum) .

Mode III :- In this , the optimization model was tested on the system to achieve the optimum performance of the MVAR absorption resources by recommending the locations and ratings of the minimum required reactors from the preinstalled ones.

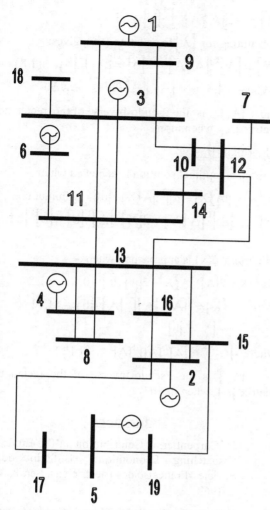

Figure-3- Iraq Super High Voltage Grid System

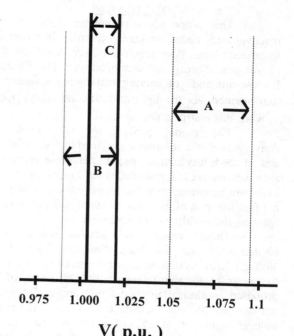

V(p.u.)

Margin A :Nodal Voltages Margin at (Mode I).
Margin B :Nodal Voltages Margin at (Mode II).
Margin C :Nodal Voltages Margin at (Mode III).

Figure –4- Nodal Voltages Margins for the three modes of MVAR control at Minimum Load of the Iraqi Super High voltage Grid System

Figure (4) translates nodal voltages in p.u. of the system into margins, while figure (5) translates reactive power generation / absorption required by the power stations for the three modes at minimum load condition of the system.

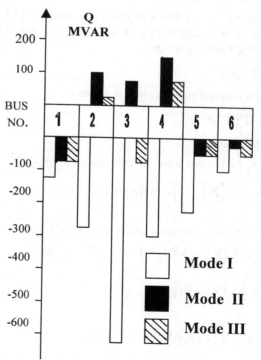

Figure –5-Reactive Power Generation/Absorption by the power stations for the three modes of MVAR control.

CONCLUSION

The proposed optimization model for the reactive and nodal voltage controls, is based on linearized load flow reactive power equations and Lagrange's Theory of optimization. The Pseudo-Inverse method of solving rectangular matrices is implemented via the algorithms of computing $[\Delta Q_R]$, when this matrix is not square.

The model proved to be reliable in convergence, the solution is obtained in few iterations and it needs less memory requirements due to that the Jacobian matrix is formulated with ½ of the linearized load flow equations (reactive power equations only) in the first iteration of the solution and it is reduced when ignoring the healthy nodes iteratively.

Three modes of MVAR control are implemented on the Iraqi National Super High Voltage Grid System at its worst operating condition, to recognize the significance and the efficiency of the proposed optimization model concerning the nodal voltage and MVAR controls. It has given better nodal voltage regulation at 25% of savings of total reactors ratings and has improved the performance of the generating stations from the point of view of reactive power generation/absorption, then it improves the stability of the system.

APPENDIX- I

Having the system of equations $[A][X]=[b]$, Where

$[A]=\{a_{ij}\}$, $i=1....m$ and $j=1....n$

Its method of solution is dependant on whether $[A]$ is square or not, and it can be obtained by one the following three algorithms :-

Algorithm I :-

if $[A]$ is a square matrix, ($m=n$), then :-

$[X]=[A]^{-1}[b]$.....(I.1)

Algorithm II :-

If m<n, an optimal solution is achieved when:-

$$\sum_{i=1}^{n} X_i^2 = [X]^T[X]= \text{Minimum}.....(I.2)$$

Optimum $[X]$, is approached by Lagrange's theory :-

$$L = [X]^T[X]+[\lambda]^T[[A][X]-[b]]....(I.3)$$

$$\frac{\partial L}{\partial \lambda} = [A][X]-[b]=0.....(I.4)$$

$$\frac{\partial L}{\partial X} = 2[X]+[A]^T[\lambda]=0.....(I.5)$$

Equation (I.5) gives :-

$$[X]=-1/2[A]^T[\lambda]....(I.6)$$

Subsituting for $[X]$ in equation (I.4), results:-

$$[b]=-1/2[A][A]^T[\lambda]$$
$$[\lambda]=-2[[A][A]^T]^{-1}[b]....(I.7)$$

Subsituting for $[\lambda]$ in equation (I.6), gives: -

$$[X]=[A]^T[[A][A]^T]^{-1}[b]=[A_1]^+[b]....(I.8)$$

Where: - $[A_1]^+=[A]^T[[A][A]^T]^{-1}.....(I.9)$

and $[A_1]^+$ is the Pseudo-Inverse of the rectangular matrix $[A]$, when m<n .

Algorithm III :-

If m>n, an optimal solution is achieved when :-

$$\phi(X)=[[A][X]-[b]]^T[[A][X]-[b]]=\text{Minimum}$$
$$=[X]^T[A]^T[A][X]-2[X]^T[A]^T[b]+[b]^T[b]....(I.10)$$

Minimum $\phi(x)$ is approached when :-

$$\partial\phi/\partial X = 2[A]^T[A][X]-2[A]^T[b]=0$$

then $[X]=[[A]^T[A]]^{-1}[A]^T[b]....(I.11)$

$=[A_2]^+[b]$

where:- $[A_2]^+=[[A]^T[A]]^{-1}[A]^T....(I.12)$

and $[A_2]^+$ is the Pseudo-Inverse of the rectangular matrix $[A]$, when m>n.

References

1. J. Carpentier, "Contribution a'l' Etude du Dispatching Economique "Bulletin de la Societe Francaise des Electriciens, Ser. 8, Vol. 3, 1962.

2. H.W.Dommel, W.F. Tinney,"Optimal Power Flow Solutions" IEEE; PAS, Vol. 87, 1968, PP. 1866-1876.

3. A.M. Sasson"Combined Use of the Powell and Fletcher-Powell Nonlinear Programming Methods for Optimal Load Flows", IEEE, PAS, Vol. 88, 1969, PP. 1530-1537.

4. I. Hano, Y. Tamura, S. Narita, "Real Time Control of System Voltage and Reactive Power", IEEE,PAS, 88, 1969, PP. 1544-1559.

5. K. Kumari and K. Ode, "Power System Voltage Control by Using a Process Control Computer", IEEE, PAS, 87, 1968, PP. 1985-1990.

6. V.A. Stroev, I.S. Rokotian "The Algorithms for Alleviating Overloads Using Pseudo-Inverse Method" Paper from Moscow Power Engineering Institute, 1991.

7. Brian D.Burctay, "Basic Optimization Methods", PP(47-90), Edward Arnold (publishers), 1985.

LV POWER LINE CARRIER NETWORK SYSTEM FOR AMR

M Saso ,   H Ouchida ,   M Iso   ,   T Naka   , Y Shimizu

Tokyo Electric Power Co. Inc. , Toko Electric Co. Ltd. , Osaki Electric Co. Ltd. , Fuji Electric Co. Ltd. , Japan

INTRODUCTION

TEPCO (The Tokyo Electric Power Co., Inc.) has been researching several AMR (Automatic Meter Reading) systems and introduced LV AMR system using telephone lines as a trunk line and LV power lines as a branch line among approximately 3,000 customers at mountainous areas, where the electric power demand density is low and the meter reading efficiency is poor on account of no access other than on foot.

During several years of commercial operation, we sometimes encountered an unavailability of reading the metering data due to the disconnection of telephone line, change in customer's telephone number, etc.

To solve these communication problems at a low cost and improve the automatic metering efficiency, we are developing a new LV AMR system with the regular communication route backed up by another telephone line and low voltage power line.

This paper describes the features and performance of conventional and newly developed AMR system.

CONVENTIONAL SYSTEM

System Configuration

Fig. 1 shows the configuration of the conventional LV AMR system. The host computer installed at TEPCO branch office reads the watt-hour meter every month automatically via modem, telephone exchange, telephone line, NCU, transponder and terminal unit.

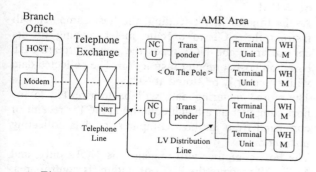

Fig.1 - Conventional system configuration

Communication Method

Public telephone lines are used as communication route between TEPCO branch office and field NCU supported by no-ringing service from NTT (Nippon Telegraph and Telephone Corporation), and from Transponder to each terminal unit, distribution line carrier is applied in order to reduce the charge of telephone line and from the fact that the telephone line is not distributed on account of electric power demand, such as public lighting facilities, etc.

No-ringing service is such that, the Telephone Exchange sends a special signal to call NCU without ringing the telephone bell of the line subscriber. Moreover, in case someone will be using the telephone line while the host computer is communicating with NCU, one's conversation will take precedence and the host computer will discontinue its communication.

As for distribution line carrier method, we adopted a ground return method where the signal is conveyed by the neutral wire, since Japanese houses mainly use single-phase 3-wire 100/200V with insulated wires, and all LV distribution lines of TEPCO contain a neutral wire notwithstanding whether it is transformer area or not, for example. Accordingly, we were able to secure transmission route over the transformer supply area for a long distance.

The interface related to communication is as follows.

TABLE 1. - Communication specifications

	Telephone line	LV distribution line carrier	WHM communication
Transmission method	No-ringing communication	Ground return system	2-wire current loop
Data rate	1200bps	100bps	1200bps
Frequency	1,700400Hz	9,500200Hz	
Communication method	Half duplex, polling type		
Standard	ITU-T V.23	FSK modulation	

Problems of Conventional System

Conventional AMR system were suffered from

High Voltage Engineering Symposium, 22–27 August 1999
Conference Publication No. 467, © IEE, 1999

unavailability of meter reading due to the stoppage of telephone line related to the nonpayment of telephone charge by the subscriber, disconnection of telephone line by the customer's change in telephone number, communication failure cause by fluctuation of ground resistance, change of communication level caused by the LV distribution line construction work, etc. Such events were experienced more than anticipated.

TABLE 2 - AMR Performance of this two years

	Total applied points	Total cases of difficulty in AMR	AMR ratio
Apr. 1997Mar. 1998	12,472	243	98.1%
Apr. 1998Sep. 1998	6,528	77	98.8%

Since, the areas applying AMR system were far away from the branch office and has no other access for meter reading than on foot, developing of the new system with the efficiency of automatic meter reading improved became necessary, indeed.

NEW SYSTEM

Conventional AMR system's unavailability for meter reading was from the disconnection or disturbance of communication line. To overcome these problems, utilization of TEPCO's LV distribution line network, conventional device configuration and performance were restudied. Now, a new device requiring a simple maintenance on the host computer for managing the latest network information were developed. But Communication specifications are same as conventional system shown in Table.1.

System Configuration

Fig. 2 shows the configuration of the new system.
New device, NCT(Network Control Terminal unit) and PLT(Power Line carrier Terminal unit), are developed for replacing NCU, transponder and terminal unit of the conventional system, but retaining the conventional communication interface method.
Host computer will communicate with NCT via public telephone line, and NCT will communicate with PLT, installed with WHM (Watt Hour Meter) via distribution line carrier using the LV distribution line neutral wire.

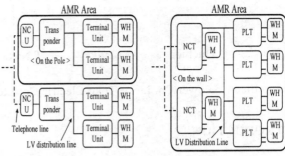

Conventional System New System (Improved)

Fig. 2 - System configuration in AMR area

Technique of Utilizing the Distribution Line as Communication Network

TEPCO's LV distribution lines are meshed, linked as a network. To utilize this network for communication, we developed a technique of monitoring the signal transferred through the LV distribution line, since the signal may be changed by the modification of distribution line due to construction work, communication range availability may change by the variation of ground resistance which comes from weather condition, etc.
By using this new monitoring method, the precise information on the communication network with its circumstances and conditions can be obtained. In other words, variation of LV distribution network can easily be recognized and incorporated based on such information.
The details are described hereunder.

Basic AMR Method of New System. NCTs will collect the data from PLTs, and transfer the data to the host computer at TEPCO branch office.
Each NCT has its own control area, where it shall be responsible for communicating with PLTs, in which the communication through the LV distribution line carrier is ensured as a regular route. The said area is designated and registered with the ID numbers of WHMs.
At the time of meter reading, the host computer will instruct the NCT to execute meter reading for its control area, and NCT reads the meter through PLT based on host computer instruction. The collected data will temporarily be stored at NCT.
The stored metering data will be transmitted to the host computer as a batch data, upon NCT's receipt of command from the host computer for data collecting instruction.
The host computer will manage the NCTs only, and NCT will manage the PLTs of within its control area, and thus, the role of effective devices control are defined for the communication of the AMR area.

Network Condition Monitoring and Utilizing for Communicating Purpose. Each NCT has a function of supervising the communication between host computer and other NCTs. If ID number of a WHM which are not within its control area or ID of another NCT has been recognized, NCT will store the latest monitored information, up to 10 for other NCTs and up to 40 for PLTs not within its control area. The collection of metering data will take place when the host computer gives a instruction to other NCTs or PLTs for meter reading.

This information allows to grasp and manage the latest information of the devices on which communicating of NCT with other NCTs including their control areas are available. This arrangement configures a linking communication route by distribution line carrying method in addition to a communication route via public telephone line from the host computer.

With these methods, the new system can easily keep and manage a communication network resorting to public telephone line and LV distribution line in spite of host computer managing NCTs only.

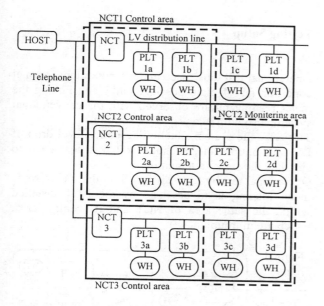

Fig. 3 - Keeping and managing the communication network

Using a LV Distribution Line Network as a Communication Route. If meter reading becomes impossible by the regular communication route due to some kind of failure on communication line, the host computer will select another NCTs or PLTs, which are able to communicate with disconnected NCT, based on the monitored network information. The command for re-transmitting the data of disconnected NCT will be sent from host computer to the selected NCT or PLT, and the selected NCT or PLT deliver the re-transmission instruction to disconnected NCT for collecting the data within disconnected NCT control area.

In case a communication with disconnected NCT is unavailable even by re-transmission through the distribution line carrier by the selected NCT or PLT, then the selected NCT will directly read the meter of PLT within the control area of disconnected NCT.

This arrangement allows AMR even at a failure of regular telephone line via another telephone line or LV distribution line.

As a supplementary effect, the use of the re-transmitting function allows a long-distance transmission of LV distribution carrier.

Instruction method by telegram

The following shows the telegram format of this system.

ID No. 1 is for ID of NCT itself, and ID Nos. 2-6 are for IDs of NCTs or PLTs to be re-transmit through the LV distribution line. The Pointer is consisting of 2 digits, one for ID position and the other for number of IDs. In case ID position $\neq$ number of IDs, the telegram is judged as a re-transmitting task, and in case ID position = number of IDs, the telegram is judged as own data and a communication will be made within the WHM of its control area.

On account of telegram configuration, the re-transmission is available up to 5 stages.

STX	Mode	Pointer	ID No. 1	ID No. 2	...	ID No. 6	Data part	ETX	BCC

Fig. 4 - Telegram format

Example of Actual Re-route Meter Reading by the Model

Fig. 5 shows an example of re-routing when the host computer can not directly communicate with NCT 2 for reading the NCT 2 collected metering data of its control area.

At first, the host computer will instruct all the NCTs for meter reading. However, only NCT 1 and NCT 3 will read the meter against the host computer's instruction, and NCT 2 will fail due to the disconnection of telephone line with host computer.

Since, the host computer can not communicate with NCT 2 for meter reading, the host computer will then select the NCT or PLT which can communicate with NCT2 based on network information, and deliver a meter reading instruction to NCT 2.

The following shows a concrete procedure according to the status of communication line.

In case NCT 1 can communicate directly with NCT 2, the host computer will collect NCT 2 meter reading data via NCT 1.

In case NCT 1 cannot communicate directly with NCT 2, but can communicate via PLT 2, the host computer will will collect NCT 2 meter reading data via NCT 1 and PLT2.

In case NCT 2 can not communicate with any other NCTs and PLTs, the NCT 1 will collect the metering data of PLT2 and NCT 3 will collect the metering data of PLT 3, directly.

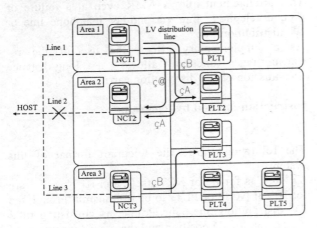

Fig. 5 - Re-route meter reading

Function And External View Of NCT And PLT

Functions. Table 3 shows the functions of NCT and PLT.

TABLE 3 - Functions of NCT and PLT

	NCT	PLT
Interface	Telephone : 1 LV distribution line carrier : 1 WHM : 3	LV distribution line carrier : 1 WHM : 3
Own ID	NCT-ID : 1 unit WHM-ID : 3 units	WHM-ID : 3 units
Re-transmitting function	Yes	Yes
Monitoring function	Yes Other NCT : 10units Other PLT : 40 units	No
Area control	Yes 50 PLTs	No

External View. Photo 1 shows the front view of WHM and NCT. NCT and PLT have the same size so that the actual installation work can be easily done. In addition, they are so compact that they can be piled at the rear of WHM and installed on the wall together with WHM.

Photo.1 - WHM and NCT(behind WHM)

NEW AMR SYSTEM FIELD TEST

After development of new device and trial product completion of host computer, NCTs and PLTs, trial run as a functional test to verify the whole system was done.

In detail, approximately 30 NCTs and their controlling PLTs were installed in the field.

Testing Setup

Operational test took place under 5 conditions, normal condition and 4 main possible faults considering the meter reading failure observed with the conventional system, also with two way of re-routing.

To verify the system functioning under level drop of LV distribution line, a choke coil was connected to simulate the signal attenuation.

Operational test by Fig. 6 is based on that the re-route meter reading is to be done by the NCT installed within the same area of NCT and its control area, which are in trouble.

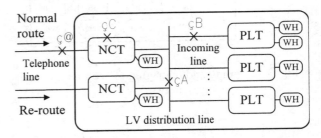

I-1:   When usual
I-2:   At telephone line fault
I-3:   At level drop of LV distribution line
I-4:   At incoming line fault
I-5:   At NCT fault

Fig. 6 - AMR Test with backup from NCT installed in the same area

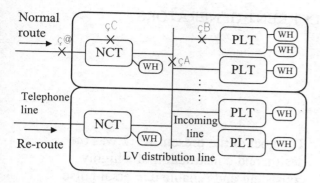

II-1:   When usual
II-2:   At telephone line fault
II-3:   At level drop of LV distribution line
II-4:   At incoming line fault
II-5:   At NCT fault

Fig. 7 - AMR Test from Backup by NCT in
Another Area

Test Results

Table 4 shows the results of meter reading at each testing patterns.

TABLE 4 - Results of meter reading at each
testing pattern

	Tested cases	Read meters		Non read	Comment
		Normal route	Re-route		
I-1	32	32		0	
I-2	32	0	32	0	
I-3	32	15	17	0	
I-4	32	31	0	1	Incoming line's faulty PLT
I-5	32	0	31	1	Faulty NCT
II-1	32	32		0	
II-2	32	17	15	0	
II-3	32	24	8	0	
II-4	32	31	0	1	Incoming line's faulty PLT
II-5	32	0	31	1	Faulty NCT

Observations

We could collect all metering data via normal route and re-route except connected to the faulty devices.
But, the test also revealed that, in case of repeat metering via re-route, it took for a while to receive the metering data. Accordingly, it is expected, for example, to build a routine method of judging the device fault in order to minimize the number of retries, etc., to minimize the required the metering data reading time .

FUTURE SUBSET

At present, we are further in verifying the functions of the system. We were able to simplify the overall system and reduce the cost by reviewing the functions of each device while improving their performance compared to the conventional system.
The new system utilizing LV distribution network suits to mountainous areas where the wireless communications are difficult, etc., and after the complete verification, the new system is scheduled to take over the conventional system considering its advantages in all respects.

HIGH VOLTAGE INDUSTRIAL PULSED CURRENT GENERATORS

Yuri Livshitz and Oren Gafri

Pulsar Ltd., Israel

In recent years, there has been intensive development of uses for pulsed electrical energy. The potential applications that could utilize such a process requires suitable equipment to handle the high pulsed currents and/or high pulsed magnetic fields generated.

Such generators need to provide currents in the range of hundreds of kA, magnetic fields around 50 T, and a pulse width (first half of the period of oscillation) from 5 to 100 μsec. For industrial uses, the generators need a pulse repetition of approximately 0.01 Hz, and a design life of over 1 million pulses.

The principles of such a generator are described in the works of Frungel, Zaienz, Schneerson, and Knopfel, among others.

The goal of the present work was the design and construction of a highly consistent and reliable universal pulse current generator for industrial use. Pulsar Ltd. of Israel has successfully constructed such generators of high pulsed current and high pulsed magnetic fields, in stationary and mobile units. The various designs and applications have been patented worldwide.

The base model was a 20 kJ system with two options of own frequency oscillation: 35 kHz and 100 kHz, with the two respective voltages being 9 kV and 25 kV.

The main challenge was to achieve the lowest own inductance for the discharge circuit using a low inductance load. Four switches were

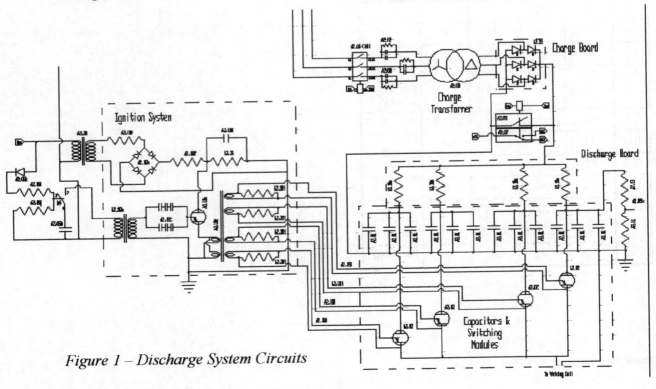

Figure 1 – Discharge System Circuits

High Voltage Engineering Symposium, 22–27 August 1999
Conference Publication No. 467, © IEE, 1999

used in parallel.

The ignition system is a four-canal transformer on the coaxial cables with 4 outputs.

The current from the batteries flow through the switches, which are connected by flat conductors to a section of the capacitor's battery as well as with a mutual collector. An illustration of the principal circuits of the discharge system and the ignition system are shown in Fig. 1.

The ignition pulse is illustrated in Fig. 2.

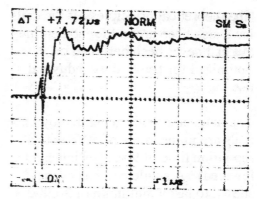

Figure 3 – Discharge of 4 switches

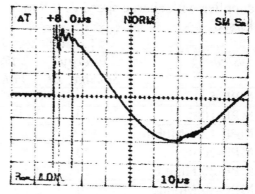

Figure 4 – Discharge of 4 switches

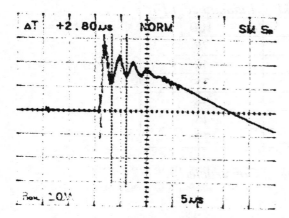

Figure 2 – Ignition Pulse

The activation characteristic for each of the four switches was registered by measuring the discharge current and voltage on the load, as shown in Figures 3 and 4. The intervals between the activation of the switches is dependent on the working voltage and ranges from 0.5 μsec to 1.0 μsec for a 8.5 kV charge voltage and from 1.0 μsec to 1.5 μsec for a 5.5 kV charge voltage.

The above was compared to the discharge of a single switch. From the measured results it appears that as the charge voltage is decreased, the duration of the current front is increased (see Fig. 5).

General view of the system: Fig. 6.

Data for the various systems constructed so far are shown in Table 1.

These industrial-level systems can be used effectively and economically for a wide variety of applications. The high pulse magnetic generators can be used for forming, crimping, cutting, joining and even welding of certain materials. Various patented coils have been created to properly channel the force to give the desired effect. The types of coils include:

- Multi-turn coil,
- One-turn coil,
- Multi-turn coil with a field-shaper,
- One-turn coil with pulse transformers.

The theoretical principles for the coils have also been around for some time, but had never been designed or optimized for heavy-duty industrial use.

The magnetic fields generated by these coils are in the range of 50 T with a repetition time of 9 sec (which can be lowered). A single coil has been shown to last a minimum of 30,000 pulses.

The systems carry high voltage power supplies with charge speed of 0.5 to 5 kJ/sec.

The systems can also be used for high voltage electrical discharge. The range of applications for the electric discharge generators is much broader than the magnetic pulse generators. However to even begin listing them is beyond the scope of this paper.

For more information contact the authors at pulsar@pulsar.co.il, or visit the website at www.pulsar.co.il.

System	Energy Storage, kJ	Working Voltage, kV	Own Inductance, nHn
MPF-5	5	10	100
MPF-10	10	10	78
MPF-12.5 Mobile	12.5	25	40
MPF-20	20	8.5	35
MPF-20 High F	20	25	38
MPF-40 Mobile	40	8.5	45
MPF-100	100	25	30

Table 1 – Pulsar High Voltage Pulse Generators Constructed

Figure 5 - Parameter Drift

Figure 6 – Pulsar High Voltage Pulse Generator

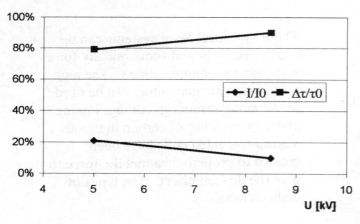